# METAL-ORGANIC FRAMEWORKS

# METAL-ORGANIC FRAMEWORKS

## Design and Application

Edited by

LEONARD R. MacGILLIVRAY
University of Iowa

WILEY
A JOHN WILEY & SONS, INC., PUBLICATION

Published by John Wiley & Sons, Inc., Hoboken, New Jersey
Published simultaneously in Canada



***Library of Congress Cataloging-in-Publication Data:***

Metal-organic frameworks: design and application / edited by Leonard R. MacGillivray.
p. cm.
Includes index.
ISBN 978-0-470-19556-7 (cloth)
1. Supramolecular organometallic chemistry. 2. Organometallic polymers. 3. Porous materials. I. MacGillivray, Leonard R.
QD882.M48 2010
547′.0504426–dc22 2009049259

10 9 8 7 6 5 4 3 2 1

# CONTENTS

# PREFACE

The field of metal-organic frameworks, or MOFs, is undergoing accelerated and sustained growth. I personally became acquainted with MOFs, or more generally coordination polymers, as an undergraduate research student while at Saint Mary's University, Halifax, Nova Scotia, Canada, from 1991 to 1994. The process of mixing readily available metal precursors with organic linkers—many of which fell under the heading of being commercially available—to produce a wide array of extended frameworks clearly then, and now, captured the imagination of chemists and materials scientists worldwide.

From a fundamental standpoint, there is an important link between MOF chemistry and the field of inorganic chemistry. In many ways, MOF chemistry enables chemists to connect previously existing coordination complexes so as to make a conceptual link into the field of materials chemistry. This link has now evolved to afford applications ranging from catalysis to energy storage. Organic chemists are also able to contribute to the mix by crafting ligands with properties that one ultimately plans to express within the walls of MOFs. Solid-state chemists and X-ray crystallographers provide insights into the structures of MOFs so that the process of designing and synthesizing MOFs can be refined so as to ultimately control a targeted property and give rise to function.

My personal draw to MOFs was, in retrospect, also inspired by the field of supramolecular chemistry, particularly as it relates to the rational design of solids, or crystal engineering. The early 1990s witnessed supramolecular chemistry envelop the process of self-assembly, with a crystal being regarded as a supermolecule par excellence.[1] Metal–ligand bonding is reversible and, thus, fits within the realm of supramolecular chemistry. Self-assembly involves subunits of a larger superstructure being repeated in zero-dimensional (0D), 1D, 2D, or 3D space, with the solid state being a perfect resting place for intermolecular forces to dominate. Today, many of the boundaries between these areas have become increasingly more difficult to distinguish, which can be expected as more is being uncovered and as more emphasis is placed on properties and function.

It is, thus, with great pleasure that I am able to assemble a multi-author monograph that includes authoritative contributions from leading research laboratories in the field of MOF chemistry. My goal is to provide insights into where the field of MOFs began to take root and provide an account of the fundamentals that define where the field has come and is able to go. Indeed, MOFs provide chemists a means to think about how to utilize coordination space to mimic the chemistry of zeolites with an added degree of organic function. These possibilities have become apparent in key developments and important advances that are outlined in the chapters that follow.

Fujita (Chapter 1) and Eddaoudi (Chapter 2), for example, document the first reports of MOFs, or coordination networks, particularly those that exhibit catalysis, the emergence of heteroaromatic ligands, and how carboxylates provided an important entry to increasingly robust solids. Batten (Chapter 3) demonstrates a role of symmetry in defining and understanding the simple and complex frameworks that result from the solid-state assembly process that affords a MOF. Next, Schroder (Chapter 4) addresses the design and synthesis of extended frameworks of increasingly structural complexity in the form of highly connected MOFs based on lanthanide ions. Kitagawa (Chapter 5) then shows how the internal structures of coordination networks can be rationally modified and tailored with organic groups while Lin (Chapter 6) documents some of the first systematic applications of MOFs as they relate to the generation of nonlinear optic materials. A great challenge facing mankind is making efficient use of energy. MOFs have emerged as potentially useful platforms for facing this challenge in the form of gas storage, separation, and conversion. Thus, Kim (Chapter 7) and Zhou (Chapter 8) address how MOFs interact with small gas molecules (e.g., $H_2$) and how these materials may be integrated into schemes for energy utilization. In a related topic, Friscic (Chapter 9) tackles the emerging issue of mechanochemical, or solvent-free, "green" preparation of MOFs while work by our group demonstrates how the walls of extended frameworks can be designed to serve as platforms for light-induced chemical reactions (Chapter 10). Finally, Snurr (Chapter 11) addresses how the field of computational chemistry can be used to understand, and ultimately, aide the design of MOFs, with targeted applications in separations, gas uptake, and materials characterization. Carefully chosen references serve to guide the reader through the extensive literature, which makes the field accessible to a wide and varied audience.

My initial interests in the chemistry of MOFs, and supramolecular chemistry and solid-state chemistry in general, stemmed from an experience as an undergraduate researcher. It is for this reason that I dedicate this monograph to the

undergraduate research experience and to all of those that support undergraduate research.

LEONARD R. MACGILLIVRAY

*Iowa City, IA*
*March 2010*

## REFERENCE

1 Dunitz, J. D. *Pure Appl. Chem.* **1991**, *63*, 177.

# CONTRIBUTORS


**Stuart R. Batten**, School of Chemistry, Clayton Campus, Bldg. 19, Monash University, 3800 Australia

**Neil R. Champness**, School of Chemistry, The University of Nottingham, University Park, Nottingham, NG7 2RD, UK

**Hyungphil Chun**, Department of Applied Chemistry, College of Science and Technology, Hanyang University, 1271 Sadong, Ansan 426-791, Republic of Korea

**David J. Collins**, Department of Chemistry and Biochemistry, Miami University, Oxford, OH 45056, USA; Department of Chemistry, Texas A&M University, College Station, TX 77843, USA

**David Dubbeldam**, Department of Chemical and Biological Engineering, Northwestern University, 2145 Sheridan Road E136, Evanston, IL 60208, USA

**Saikat Dutta**, Department of Chemistry, University of Iowa, Iowa City, IA 52242, USA

**Mohamed Eddaoudi**, Department of Chemistry, University of South Florida, 4202 East Fowler Avenue, CHE204, Tampa, FL 33620, USA

**Jarrod F. Eubank**, Department of Chemistry, University of South Florida, 4202 East Fowler Avenue, CHE204, Tampa, FL 33620, USA

**Tomislav Friščič**, Department of Chemistry, University of Cambridge, Lensfield Road, Cambridge CB2 1EW, UK

**Houston Frost**, Department of Chemical and Biological Engineering, Northwestern University, 2145 Sheridan Road E136, Evanston, IL 60208, USA

**Makoto Fujita**, Department of Applied Chemistry, School of Engineering, University of Tokyo, 7-3-1 Hongo, Bunkyo-ku, Tokyo, 113-8656, Japan

**Ivan G. Georgiev**, Department of Chemistry, University of Iowa, Iowa City, IA 52242, USA

**Sujit K. Ghosh**, Department of Synthetic Chemistry and Biological Chemistry, Graduate School of Engineering, Kyoto University, Nishikyo-ku, Kyoto 615-8510, Japan; Department of Chemistry, Indian Institute of Science Education and Research (IISER), Pune, India

**Peter Hubberstey**, School of Chemistry, The University of Nottingham, University Park, Nottingham, NG7 2RD, UK

**Hyunuk Kim**, National Creative Research Initiative Center for Smart Supramolecules, Department of Chemistry and Division of Advanced Materials Science, Pohang University of Science and Technology, Pohang, 790-784, Republic of Korea

**Kimoon Kim**, National Creative Research Initiative Center for Smart Supramolecules, Department of Chemistry and Division of Advanced Materials Science, Pohang University of Science and Technology, Pohang, 790-784, Republic of Korea

**Susumu Kitagawa**, Department of Synthetic Chemistry and Biological Chemistry, Graduate School of Engineering, Kyoto University, Nishikyo-ku, Kyoto 615–851 Japan; Kitagawa Integrated Pore Project, Exploratory Research for Advanced Technology (ERATO), Japan Science and Technology Agency (JST), Shimogyo-ku, Kyoto 600-8815, Japan; Institute for Cell Materials Sciences (iCeMS), Kyoto University, Sokyo-ku, Kyoto, Japan

**Wenbin Lin**, Department of Chemistry, CB3290, University of North Carolina at Chapel Hill, Chapel Hill, NC 27599, USA

**Xiang Lin**, School of Chemistry, The University of Nottingham, University Park, Nottingham, NG7 2RD, UK

**Shengqian Ma**, Department of Chemistry and Biochemistry, Miami University, Oxford, OH 45056, USA; Department of Chemistry, Texas A&M University, College Station, TX 77843, USA

**Leonard R. MacGillivray**, Department of Chemistry, University of Iowa, Iowa City, IA 52242, USA

**Martin Schröder**, School of Chemistry, The University of Nottingham, University Park, Nottingham, NG7 2RD, UK

**Randall Q. Snurr**, Department of Chemical and Biological Engineering, Northwestern University, 2145 Sheridan Road E136, Evanston, IL 60208, USA

**A. Őzgűr Yazaydin**, Department of Chemical and Biological Engineering, Northwestern University, 2145 Sheridan Road E136, Evanston, IL 60208, USA

**Shuting Wu**, Department of Chemistry, CB3290, University of North Carolina at Chapel Hill, Chapel Hill, NC 27599, USA

**Hong-Cai Zhou**, Department of Chemistry and Biochemistry, Miami University, Oxford, OH 45056, USA; Department of Chemistry, Texas A&M University, College Station, TX 77843, USA

# 1

# FROM HOFMANN COMPLEXES TO ORGANIC COORDINATION NETWORKS

Makoto Fujita
*Department of Applied Chemistry, School of Engineering, University of Tokyo, 7-3-1 Hongo, Bunkyo-ku, Tokyo 113-8656, Japan*

## 1.1 INTRODUCTION

Recently, there has been a considerable upsurge in the study of porous hybrid organic–inorganic materials referred as *organic coordination networks*. Porous organic coordination networks can be prepared via self-assembly of a connector (a metal) and a link (a ligand) where connectors and ligands are bound together through a metal–ligand bond to form porous crystalline structures. It is in the pores that a wide range of processes can be carried out. One salient feature of coordination networks is the possibility to *design* the pores. Now it is possible to create pores with determined sizes just by selecting ligands with determined shapes, and particular environments through functionalization of the ligands. Therefore, certain reactions that are not possible to be carried out in solution are now being carried out within the porous space.

Another feature of organic coordination networks is their flexibility, which differentiates them from the robust frameworks of zeolites. This flexibility enables a dynamic behavior in porous coordination networks, which facilitate structural modifications (guest exchange or chemical reactions within the pores) without loss of its structural integrity. Hence, chemical reactions that occur in the pore can be monitored *in situ* in great detail by X-ray crystallography and other spectroscopic techniques.

*Metal-Organic Frameworks: Design and Application*, Edited by Leonard R. MacGillivray

The growing interest in such materials both scientifically and economically is not surprising due to the remarkable physicochemical properties that have been reported during the last two decades. Organic coordination networks are useful in a wide range of applications. For instance, these porous networks can be used in processes such as selective separation, catalytic reactions, guest exchange, and gas storage. Thus, due to its importance, in this review we have summarized the history of the coordination networks from the first documented examples until the latest advances in this field.

## 1.2 DISCOVERY OF A COORDINATION NETWORK

Initially porous and open-framework coordination networks attracted considerable attention as post-zeolite materials. Recent progresses of coordination networks are remarkable in that many intriguing properties and functions, for example robust and flexible framework, framework transformation, pore post-modification, selective molecular recognition, gas adsorption, and catalysis, have been reported. This review follows the history of coordination networks from the beginning, namely, Hofmann complex.

### 1.2.1 Hofmann Complex

The first coordination network having a chemical formula of $Ni(CN)_2(NH_3){\cdot}C_6H_6$ was discovered by Hofmann and Küspert in 1897.[1] They obtained the complex as a crystal by slow layering diffusion of $C_6H_6$ into an $NH_3$ solution of $Ni(CN)_2$. Pfeiffer in 1927[2] and Feigl in 1944[3] speculated that the structure of Hofmann complex was a Ni monomer coordinated by benzene as a side-on form, $Ni(CN)_2(NH_3)(\eta^6\text{-}C_6H_6)$ (Scheme 1.1).

Finally in 1954 Powell and coworkers clarified the structure of the Hofmann complex by X-ray analysis (unit cell dimensions $a = b = 7.242$ Å; $c = 8.277$ Å$^3$; $\alpha = \beta = \gamma = 90$). The crystal structure was a square network bridged by CN groups encapsulating benzene in a channel.[4] A partial structure of the Hofmann complex is shown in Figure 1.1. Arrays of Ni covalently linked through CN groups form two-dimensional layers that are parallel to each other. From each layer two amine groups protrude toward the adjacent layer, creating a series of voids where benzene molecules are included.

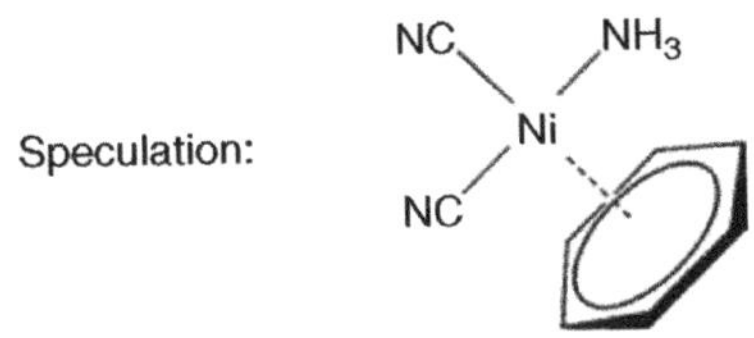


**SCHEME 1.1**

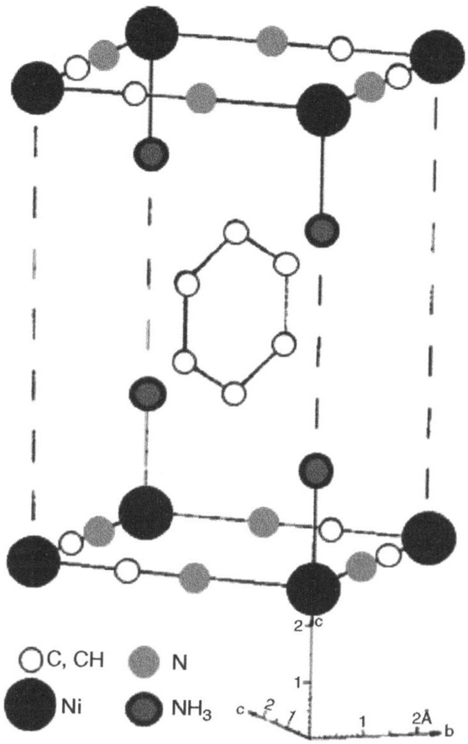


**FIGURE 1.1** A partial crystal structure of the Hoffman complex showing benzene encapsulation within the cavity.

### 1.2.2 Variation of the Hofmann Complex

Thereafter various Hofmann type complexes have been reported. Iwamoto and coworkers reported $M^1M^2(NH_3)_2(CN)_4 \times \mathbf{G}$ ($M^1$: Ni, Zn, Cd, Cu, Mn, Fe, Co; $M^2$: Ni, Pd, Pt; **G**: benzene, aniline, pyrrole, thiophene) in 1967–1968.[5,6] They paid attention to the components of Hofmann complex that can be divided into three parts: $[Ni(NH_3)_2]^{2+} + [Ni(CN)_4]^{2-} + 2C_6H_6$. Therefore, they prepared Hofmann type complexes according to the following Scheme 1.2. The structures were identified by powder X-ray diffraction analysis.

Walker and Hawthorne proposed expanded *n*-alkylamine Hofmann complexes in 1966 (Figure 1.2).[7] The complexes were synthesized by addition of *n*-alkylamines to a suspension to anhydrous nickel cyanide (Scheme 1.3). The crystalline samples were studied using the powder X-ray diffraction technique.

$$[M^1(NH_3)_2]^{2+} + [M^2(CN)_4]^{2-} \xrightarrow{\mathbf{G}} M^1M^2(NH_3)_2(CN)_4{\cdot}2\mathbf{G}$$

**SCHEME 1.2**

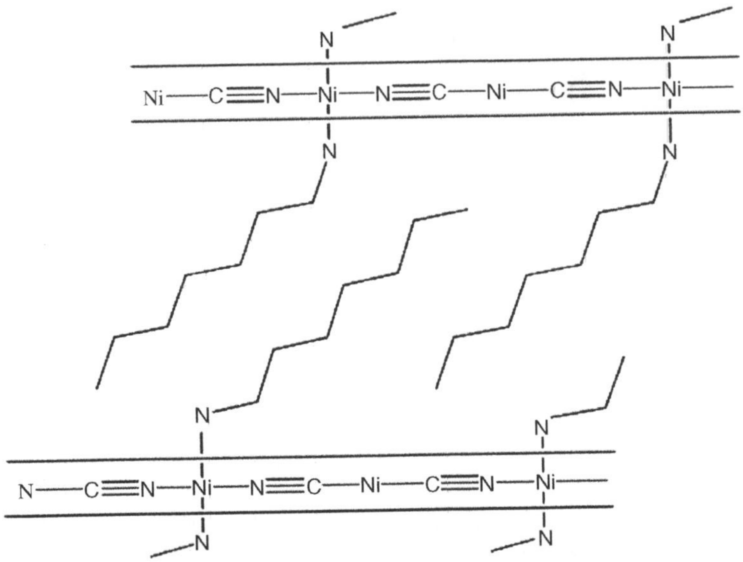

**FIGURE 1.2** Proposed structure of the expanded Hofmann complexes using *n*-alkylamines.

In 1968–1975, Iwamoto and coworkers also reported expanded Hofmann complexes bridged by ethylenediamine (en)[8] (Scheme 1.4). The structure was determined by X-ray analysis (Figure 1.3). The coordination network can encapsulate aromatic guests such as aniline, benzene, thiophene, and pyrrole.

In 1977, Mathey prepared aromatic diamine complexes as shown in Scheme 1.5.[9] Depending on the length of diamines, the length of the *c*-axis also varies (Figure 1.4). The diamine complexes show selective encapsulation for aromatic guests and solvents. For example, [Ni(4,4′-bipyridyl)Ni(CN)$_4$] encapsulate 0.8G (G: benzene, naphthalene, anthracene, $CHCl_3$, $CH_2Cl_2$, $CH_3OH$, but not phenanthrene, $CCl_4$).

$$Ni(CN)_2 + n\text{-alkylamine} \longrightarrow Ni(CN)_2 \cdot n\text{-alkylamine}$$

**SCHEME 1.3**

$$CdCl_2 + K_2M(CN)_4 + en \longrightarrow [Cd(en)M(CN)_4] \cdot 2C_6H_6$$

**SCHEME 1.4**

$$Ni(CN)_2 + \text{diamine} \xrightarrow[24\text{ h}]{110°C} Ni(CN)_2(\text{diamine})$$

**SCHEME 1.5**

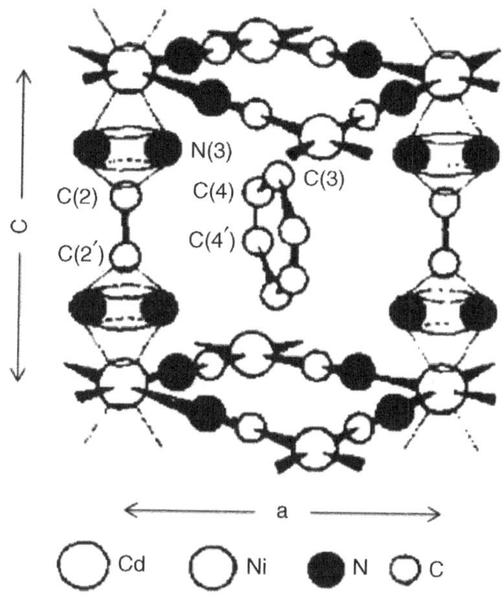


**FIGURE 1.3** Representation of $[Cd(en)Ni(CN)_4]\cdot 2C_6H_6$.

Likewise Iwamoto and coworkers expanded from Hofmann complex into 1,2-diaminopropane (pn) complex in 1980,[10] dimethylamine complex in 1982–1984,[11] and 1,ω-diaminoalkane complex (1,ω-diaminoalkane: $H_2N(CH_2)_nNH_2$, $n = 4$–8) in 1984–1985.[12] Each complex encapsulated specific aromatic compounds. In addition, they prepared many related complexes by combination of metal ions with bridging ligands.[13]

As a whole, Iwamoto and colleagues developed unique chemistry of intriguing series of Hofmann complexes. He is one of pioneers to show promising future visions of designable coordination networks as a new class of materials.

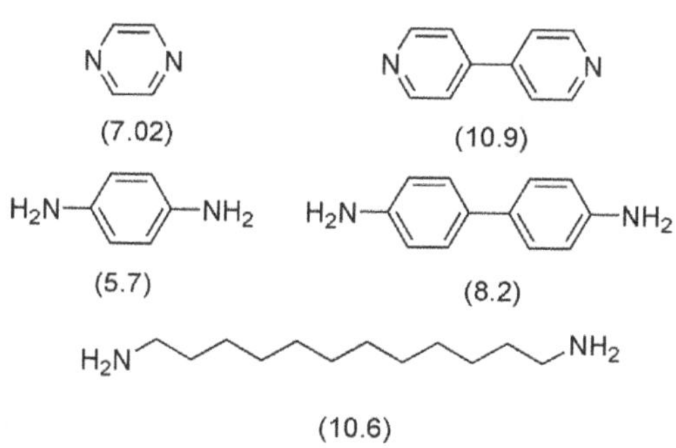


**FIGURE 1.4** Diamines and *c*-axis values (in parentheses, Å).

## 1.3 ORGANIC COORDINATION NETWORK: ORGANIC MODIFICATION OF THE HOFMANN COMPLEX

Hofmann complex inspired scientists to develop strategies aiming to find a new class of materials. Worldwide the effort of many researchers paved the way for finding a new route to synthesize materials having larger cavities. For instance, if a –CN group could be replaced with organic linkers, a variety of coordination networks having finely tuned cavities could be prepared.

### 1.3.1 Organic Coordination Network: The First Example

In 1989, Robson reported the first organic coordination network by complexation of anions with tetrahedral bridging ligands.[14] The aim of their work was to prepare three-dimensional solid polymeric materials with cavities by linking centers together with either a tetrahedral or an octahedral array of valencies. They prepared an infinite framework $\{Cu^{I}[C(C_6H_4{\cdot}CN_4)]\}^+$ of tetrahedral centers linked together by rod-like units. The rod-like units were obtained by substitution of the acetonitrile ligands in $[Cu^{I}(CH_3CN)_4]^+$ by 4,4′,4″,4-tetracyanotetraphenylmethane.

X-ray crystallographic analysis revealed a diamond-like structure containing disordered nitrobenzene and $BF_4^-$ ions in the cavities (Figure 1.5). The framework has adamantane-like cavities with a volume of approximately 700 Å$^3$. It was estimated

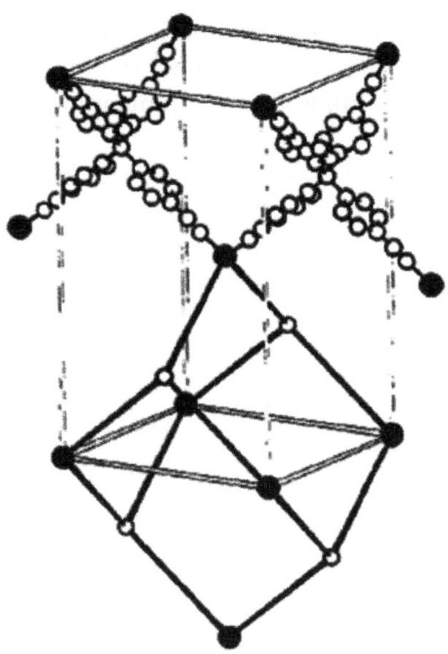

**FIGURE 1.5** Tetragonal unit cell with parts of adjacent unit cells of $\{Cu^{I}[C(C_6H_4{\cdot}CN_4)]\}^+$. Gray circles denote the Cu atoms. The adamantane-like cavity is highlighted (black sticks). Nitrobenzene molecules and $BF_4^-$ are omitted for clarity.

that the framework represents one third of the volume of the crystal while the remaining two thirds correspond to the nitrobenzene and $BF_4^-$ ions.

With this work Robson and coworkers established a new strategy capable of designing new solids containing large cavities by linking organic molecules with determined size and shape.

## 1.4 M-BIPYRIDINE SQUARE GRIDS: TWO-WAY LINK. TOWARD NEW FUNCTIONS AND APPLICATIONS OF ORGANIC COORDINATION NETWORKS

The preparation of a macrocyclic polynuclear complex $[(en)Pd(4,4'\text{-bpy})]_4(NO_3)_8$ (bpy = bipyridine) with the ability to recognize organic molecules in aqueous environment was an important step toward the applicability of such organic coordination networks.[15] Such applicability was demonstrated in 1994 with the synthesis of a two-dimensional square network solid $[Cd(4,4'\text{-bpy})_2(NO_3)_2]$ containing large cavities with the possibility of guest encapsulation (Scheme 1.6a).[16] Crucially, the first catalytic process within a porous coordination network was demonstrated by treating benzaldehyde and cyanotrimethylsilane with a $CH_2Cl_2$ suspension of powdered $[Cd(4,4'\text{-bpy})_2(NO_3)_2]$ (40°C, 24 h), which gave 2-(trimethylsiloxy)phenylacetonitrile) in 77% yield (Scheme 1.6b). Later on, we reported the cyanosilylation of imines catalyzed by $[Cd(4,4'\text{-bpy})_2(H_2O)_2](NO_3)_2{\cdot}4H_2O$ (Scheme 1.6c).[17]

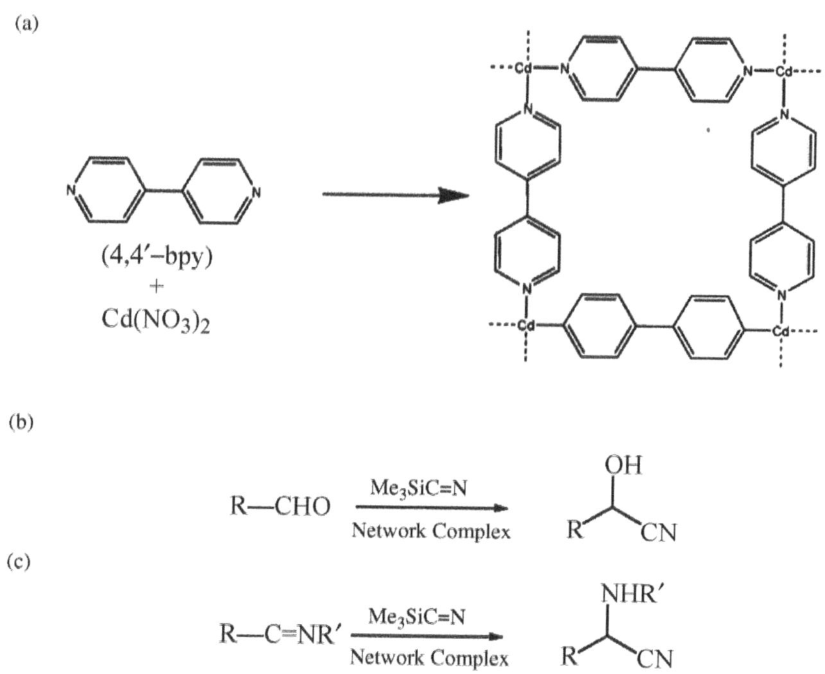


SCHEME 1.6

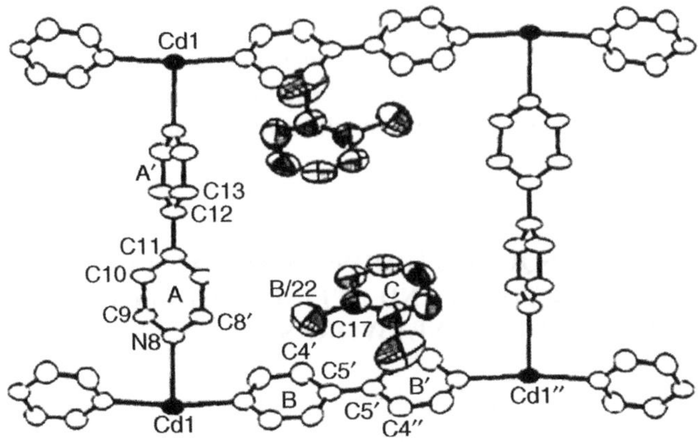


**FIGURE 1.6** View of the complex $[Cd(bpy)_2](NO_3)_2(C_6H_4Br_2)_2$ showing 50% probability ellipsoids. Nitrate ions have been omitted for clarity.

The ability to include guest molecules within the cavities was observed by preparation of a clathrate with *o*-dibromobenzene. The inclusion of the aromatic guest was confirmed by single-crystal X-ray diffraction. The structure was described as a graphite-like stacking of two-dimensional layers on top of each other (i.e., interplanar distance *ca.* 6.30 Å). One layer consists of an edge-sharing, perfectly planar square with a Cd(II) ion and 4,4′-bpy at each corner and side, respectively. Two *o*-dibromobenzene are included in each square void (Figure 1.6).

At the almost same time, Zaworotko and Yaghi reported organic coordination networks in 1995.[18,19] Zaworotko and coworkers reported the formation of a coordination network $[Zn(4,4'\text{-bpy})_2]SiF_6$ with large non-interpenetrated channels (Figure 1.7). The effective pore size ($8 \times 8$ Å$^2$) is comparable to the pore sizes of

**FIGURE 1.7** ORTEP representation of a square channel viewed along the *c* crystallographic axis. The dimensions of the channels are the same as the dimensions of the unit cell *ca.* $11.396 \times 11.396$ Å$^2$.

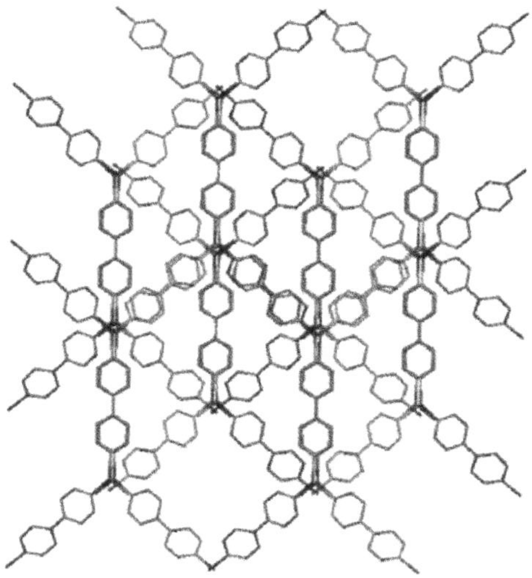

**FIGURE 1.8** Crystal structure of $[Cu(4,4'\text{-bpy})_{1.5}](NO_3)\cdot(H_2O)_{1.25}$ viewed along the [100] crystallographic direction. Nitrate anions and water molecules have been omitted for clarity.

large zeolites. The volume corresponding to the pores is about the 50% of the total volume. Interestingly, the pores are hydrophobic, which, in principle, should be able to include hydrophobic molecules with dimensions in the order of the pore size.

In the same year, Yaghi and coworkers succeeded on the hydrothermal synthesis preparation of another coordination network containing three identical interpenetrated channels. Single-crystal X-ray diffraction revealed the extended network of $[Cu(4,4'\text{-bpy})_{1.5}]NO_3\,(H_2O)_{1.25}$ in which trigonal planar $Cu^I$ centers are linked by rod-like 4,4′-bpy ligands forming an extended 3D porous network (Figure 1.8). The interpenetrated networks did not fill all the available space as observed in many other solids, therefore leaving an important volume of the solid as rectangular channels. Those channels are filled with nitrate anions and water molecules and are linked by hydrogen bond interactions. Ion exchange properties were investigated. Nitrate ions were replaced in aqueous solutions with other ions such as $SO_4^{2-}$ and $BF_4^-$. Under inert atmosphere, the coordination network was observed to be stable up to 180°C and for hours in water at 70°C.

The synthesis and crystal structures of new porous coordination networks with gas adsorption properties were first reported by Kitagawa and coworkers in the late 1990s.[20] Three structures $[M_2(4,4'\text{-bpy})_3(NO_3)_4](H_2O)_x$ (where M = Co, Ni, and Zn) were characterized by X-ray crystallography. The coordination network with Co metal was determined by single-crystal X-ray diffraction, and the isomorphous porous coordination networks with Ni and Zn metals were demonstrated by powder X-ray diffraction (Figure 1.9). These solids have two channels running parallel to the *b* and *c* crystallographic axes whose dimensions are $3 \times 6$ and $3 \times 3\,\text{Å}^2$, respectively. The channels are filled with water molecules that are not bonded to the crystal framework.

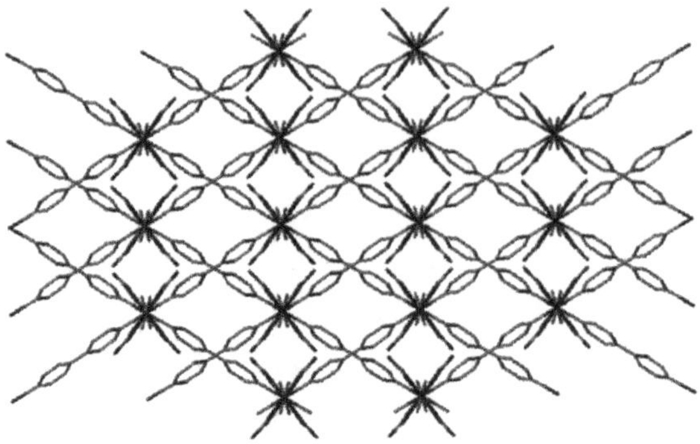

**FIGURE 1.9** Crystal structure corresponding to $[Co_2(4,4'\text{-bpy})_3](NO_3)_4{\cdot}(H_2O)_4$ viewed along the *c*-axis. Hydrogen atoms have been omitted for clarity.

Gas-adsorption properties were studied due to the 3D porous nature of such coordination network. Under dried vacuum, the adsorption activity for $CH_4$, $N_2$, and $O_2$ was investigated. Clearly, diffusion of the different gases was observed (Figure 1.10). For 1 g of the anhydrous sample at 30 atm, about 2.3 mmol of $CH_4$ and 0.80 mmol of $N_2$ or $O_2$ were adsorbed. Importantly, powder X-ray diffraction confirmed that during the gas adsorption process there was no loss of crystallinity.

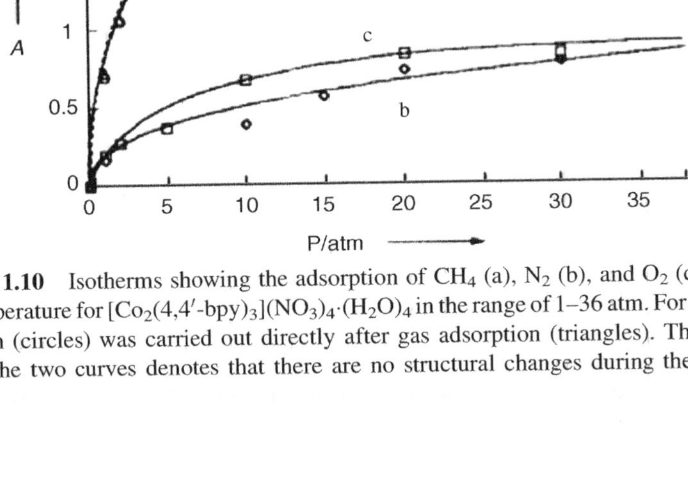


**FIGURE 1.10** Isotherms showing the adsorption of $CH_4$ (a), $N_2$ (b), and $O_2$ (c) carried at room temperature for $[Co_2(4,4'\text{-bpy})_3](NO_3)_4{\cdot}(H_2O)_4$ in the range of 1–36 atm. For the $CH_4$ gas desorption (circles) was carried out directly after gas adsorption (triangles). The similarity between the two curves denotes that there are no structural changes during the adsorption process.

## 1.5 SINGLE-CRYSTAL-TO-SINGLE-CRYSTAL PHENOMENA IN POROUS COORDINATION NETWORKS

After application to catalysis, our interests moved on to single-crystal-to-single-crystal phenomena (SCSC), because organic coordination networks often crystallize very well. Crystallographic studies are useful as they provide valuable information about the relative atomic and molecular positions within a crystalline solid. In this regard, detailed information can be obtained from SCSC transformations. For instance, such solid-state processes enable one to analyze framework transformations and molecular interactions between guest–guest and host–guest within the pores in detail.[21]

### 1.5.1 SCSC Guest Removal

The first successful SCSC guest removal was reported by Kepert in 2000.[22] They prepared $[Ni_2(4,4'\text{-bpy})_3](NO_3)_4 \cdot 6EtOH$. The interlocked bilayer framework has $6 \times 3$ Å$^2$ channels, which can reversibly uptake $H_2O$ at room temperature with other guests such as EtOH, MeOH, and *i*-PrOH (Figure 1.11). The desolvation process at 100°C causes just a 2.4% decrease in cell volume and the mosaicity of the crystal remains intact.

### 1.5.2 SCSC Framework Dynamics

Unlike zeolite, one of salient features of organic coordination networks is their flexible frameworks capable of SCSC drastic framework transformation. We reported in 2000 the synthesis of an open $20 \times 20$ Å$^2$ square-grid coordination network. $[Ni(4,4'\text{-bis(4-pyridyl)biphenyl})_2](NO_3)_2 \cdot 4(o\text{-xylene})$ was prepared via self-assembly of a rod-like ligand, 4,4′-bis(4-pyridyl)biphenyl, with $Ni(NO_3)_2$ in *o*-xylene.[23] The framework has remarkable thermal stability and maintains its integrity up to 300°C. SCSC sliding of 2D-coordination layers of $[Ni(4,4'\text{-bis(4-pyridyl)biphenyl})_2](NO_3)_2 \cdot 4(o\text{-xylene})$[24] triggered by guest exchange was successfully observed by X-ray analysis (Figure 1.12).[25] The crystals of $[Ni(4,4'\text{-bis(4-pyridyl)biphenyl})_2](NO_3)_2 \cdot 4(o\text{-xylene})$ showed guest specificity and exchanges of *o*-xylene for mesitylene but not for *m*-xylene, or 1,3- or 1,2-dimethoxybenzene. The crystals of a mesitylene-inclusion network, $[Ni(4,4'\text{-bis(4-pyridyl)biphenyl})_2](NO_3)_2 \cdot 1.7(\text{mesitylene})$, were prepared by immersion of $[Ni(4,4'\text{-bis(4-pyridyl)biphenyl})_2](NO_3)_2 \cdot 4(o\text{-xylene})$ in mesitylene for 6 h. The single-crystal analysis of $[Ni(4,4'\text{-bis(4-pyridyl)biphenyl})_2](NO_3)_2 \cdot 1.7(\text{mesitylene})$ revealed considerable sliding of the layers relative to the 2D-framework of $[Ni(4,4'\text{-bis(4-pyridyl)biphenyl})_2](NO_3)_2 \cdot 4(o\text{-xylene})$ (Figure 1.13). Edge-to-face aromatic interactions exist at the corner of the channels in both $[Ni(4,4'\text{-bis(4-pyridyl)biphenyl})_2](NO_3)_2 \cdot 4(o\text{-xylene})$ and $[Ni(4,4'\text{-bis(4-pyridyl)biphenyl})_2](NO_3)_2 \cdot 1.7(\text{mesitylene})$ but not along the walls. The dimensions of the channels in $[Ni(4,4'\text{-bis(4-pyridyl)biphenyl})_2](NO_3)_2 \cdot 1.7(\text{mesitylene})$ are much bigger than those of $[Ni(4,4'\text{-bis(4-pyridyl)biphenyl})_2](NO_3)_2 \cdot 4(o\text{-xylene})$. The sliding motion of the 2D-layers could be monitored by X-ray powder and single-crystal diffraction. The crystallographic results and GC analysis clearly indicate that the framework

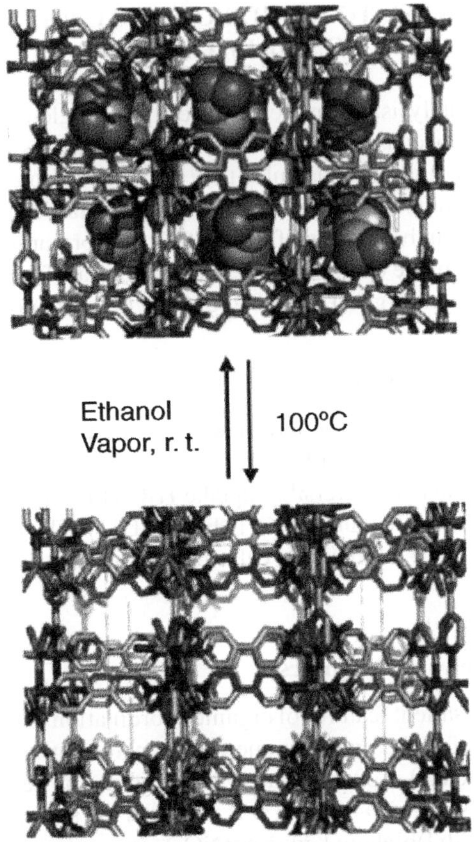


**FIGURE 1.11** Guest inclusion/removal in $[Ni_2(4,4'\text{-bpy})_3](NO_3)_4 \cdot 6EtOH$ viewed down the *a*-axis. Ethanol molecules (space-filling model) are disordered.

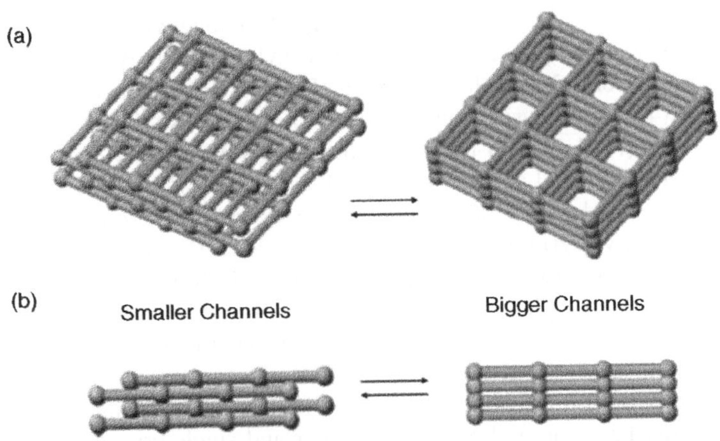


**FIGURE 1.12** Sliding motion of 2D-layers: (a) top and (b) side views.

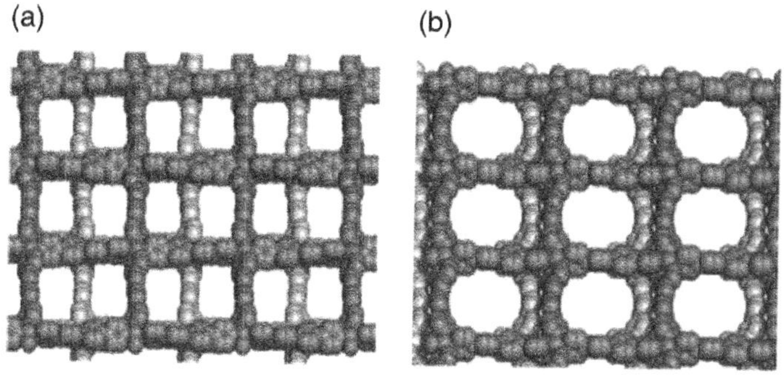


**FIGURE 1.13** Packing of 2D layers viewed along the *c*-axis in (a) [Ni(4,4′-bis(4-pyridyl)-biphenyl)$_2$](NO$_3$)$_2$·4(*o*-xylene) 4 and (b) Ni(4,4′-bis(4-pyridyl)biphenyl)$_2$](NO$_3$)$_2$·1.7(mesitylene). Solvent molecules are omitted for clarity.

transformation proceeds in a crystal-to-crystal fashion. The first step is guest exchange, and the second step is the sliding of the layers.

Two years later, in 2002, Suh and coworkers also reported SCSC dynamic framework transformation.[26] They prepared a coordination porous bilayer-open-framework (BOF-1), $[Ni_2(C_{26}H_{52}N_{10})]_3[BTC]_4{\cdot}6C_5H_5N{\cdot}36H_2O$, which was synthesized from the reaction of dinickel(II) bismacrocyclic complex $[Ni_2(C_{26}H_{52}N_{10})(Cl)_4]{\cdot}H_2O$ and sodium 1,3,5-benzenetricarboxylate ($Na_3BTC$) in water in the presence of DMSO and pyridine.

X-ray diffraction revealed two-dimensional layers formed between the Ni(II) macrocyclic unit and the $BTC^{3-}$ ions are separated by the *p*-xylyl groups that act as pillars (Figure 1.14).

The coordination network has 3D channels with an effective channel width of *ca.* 11.12 Å (Figure 1.14 top) that are filled with water and pyridine molecules. The void volume corresponding to the channels was estimated to be 61% of the total volume. Upon exposure of $[Ni_2(C_{26}H_{52}N_{10})]_3[BTC]_4{\cdot}6C_5H_5N{\cdot}36H_2O$ at 75°C for 1.5 h, all the aromatic and nearly all water molecules were removed to yield $[Ni_2(C_{26}H_{52}N_{10})]_3[BTC]_4{\cdot}4H_2O$. Despite the significant structural reorganization (i.e., the cell volume changed from 5974.3 to 3877.9 Å$^3$) the crystallinity was maintained. It was observed by X-ray diffraction that the thickness of the bilayer was reduced considerably due to the tilting of the *p*-xylyl molecules (Figure 1.14 bottom). Interestingly, powder X-ray diffraction revealed that the process is reversible when $[Ni_2(C_{26}H_{52}N_{10})]_3[BTC]_4{\cdot}4H_2O$ was immersed in a solution containing water and pyridine during only 5 min.

### 1.5.3 Robust Hydrogen-Bonded 2D-Grid Network

In 2005, we reported a robust 2D-grid network considerably stabilized by hydrogen bonding.[27] The bidentate bridging ligand was simply modified by introducing ethylene glycol chains as hydrogen bonding sites. The modification of the link enabled us to perform the *in situ* crystallographic observation of reversible apical ligand

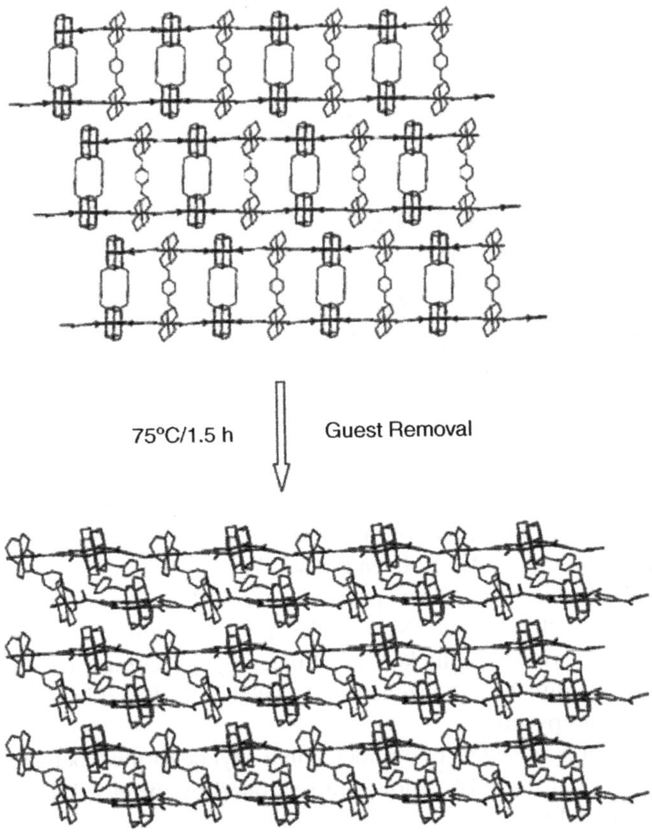


**FIGURE 1.14** X-ray structures of BOF-1. View of $[Ni_2(C_{26}H_{52}N_{10})]_3[BTC]_4{\cdot}6C_5H_5N{\cdot}36H_2O$ along the *a*-axis (top). View along the *b*-axis of $[Ni_2(C_{26}H_{52}N_{10})]_3[BTC]_4{\cdot}4H_2O$ showing the tilting of the *p*-xylyl groups as a consequence of the water loss (bottom). Water and pyridine molecules have been omitted for clarity.

exchange at a cobalt center at 150°C. It was shown how the aqua and nitrate ligands can be reversibly coordinated to the cobalt center in a SCSC fashion.

A rod-like pyridyl ligand containing ethylene glycol side chains (**1**) was synthesized (Scheme 1.7). Slow complexation of $Co(NO_3)_2{\cdot}6H_2O$ in MeOH over a solution

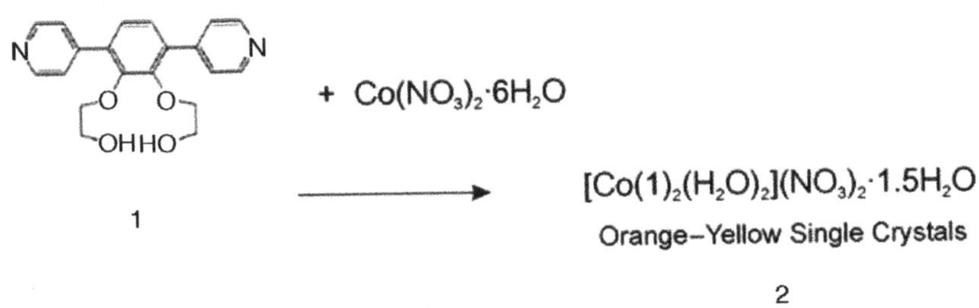


**SCHEME 1.7**

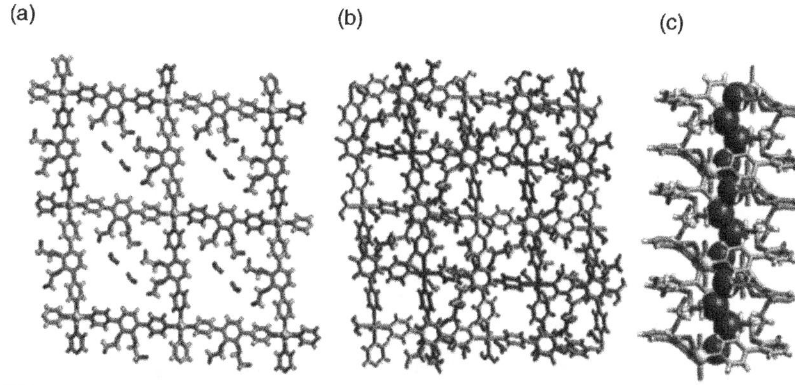


**FIGURE 1.15** (a) Square-rigid network of $[Co(\mathbf{1})_2(H_2O)_2](NO_3)_2{\cdot}1.5H_2O$. Guest molecules ($H_2O$) have been omitted for clarity. (b) The two-layer stack of **2**. (c) The infinite 1D hydrogen bonding interaction (O–H ··· O) between layers in $[Co(\mathbf{1})_2(H_2O)_2](NO_3)_2{\cdot}1.5H_2O$. The oxygen atoms of the hydroxyl groups are represented by a space-filling model.

of **1** in toluene produced orange–yellow single crystals of $[Co(\mathbf{1})_2(H_2O)_2]$-$(NO_3)_2{\cdot}1.5H_2O$ (**2**).

Crystallographic analysis revealed the formation of square-grid network structure where the grid sheet layers are stacked on top of each other (Figure 1.15a). The two layers are in an offset fashion on both edges (Figure 1.15b). Interestingly, the ethylene glycol chains play two important roles. First, they add considerable stabilization to the crystal due to 1D hydrogen-bond (O–H ··· O) arrays that interpenetrate the layers (Figure 1.15c). Second, they also introduce moderate flexibility within the cavity by allowing guest molecules to behave as in a solution-like environment. In fact, a similar square-grid coordination network without functionalized bridging ligands gradually deteriorated above room temperature due to the loss of guest molecules. Importantly, the increased stability and flexibility of $[Co(\mathbf{1})_2(H_2O)_2](NO_3)_2{\cdot}1.5H_2O$ allowed us to directly observe a reversible apical-ligand-exchange reaction at the hinge metals by single-crystal X-ray crystallography.

Two water molecules are coordinated to the cobalt(II) center at the apical positions (aqua form) with four pyridyl groups at the equatorial positions (Figure 1.16a). Upon heating at 150°C a single crystal of $[Co(\mathbf{1})_2(H_2O)_2]$-$(NO_3)_2{\cdot}1.5H_2O$ during 1 day, the yellow crystal turned red without loss of crystallinity. Crystallographic analysis revealed that the apical water molecules were substituted by two nitrate ions and extruded from the crystal to form the nitrate complex $[Co^{II}(\mathbf{1})_2(NO_3)_2]$ (Figure 1.16b). It was observed that after leaving $[Co^{II}(\mathbf{1})_2(NO_3)_2]$ at room temperature, it returned to the initial aqua form. The crystal changed from red to yellow in color and the crystallinity was maintained. Crystallographic and elemental analysis showed that the new obtained aqua form $[Co(\mathbf{1})_2(H_2O)_2](NO_3)_2{\cdot}1.5H_2O$ is almost identical to the original structure of $[Co(\mathbf{1})_2(H_2O)_2](NO_3)_2{\cdot}1.5H_2O$ (Figure 1.16c).

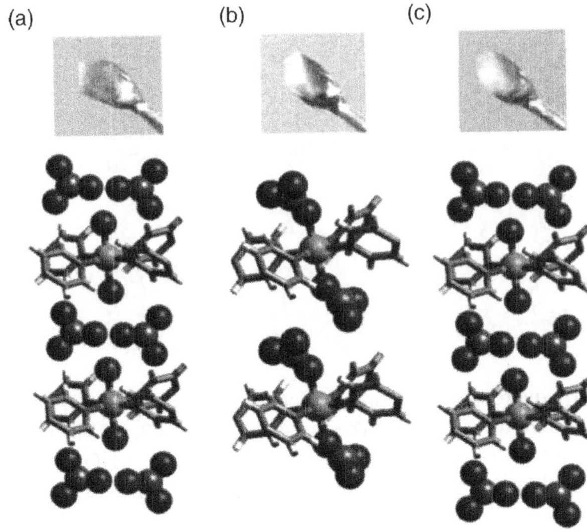


**FIGURE 1.16** Photographs corresponding to the actual crystals and molecular structures around the cobalt ion: (a) original crystal $[Co(\mathbf{1})_2(H_2O)_2](NO_3)_2{\cdot}1.5H_2O$, (b) the crystal after heating at 150°C for 24 h $[Co^{II}(\mathbf{1})_2(NO_3)_2]$, and (c) the crystal after exposure of $[Co^{II}(\mathbf{1})_2(NO_3)_2]$ to air $[Co(\mathbf{1})_2(H_2O)]_2(NO_3)_2{\cdot}1.5H_2O$. Two other disordered nitrate ions have been omitted for clarity.

## 1.6 EXPANSION FROM TWO- TO THREE-WAY LINK: CONSTRUCTION OF TPT COORDINATION NETWORKS

Self-assembly of a two-way link with metal ion connectors provides predictable architecture. What about a three-way link? The triazine ligand 2,4,6-tris(4-pyridyl)-triazine (TPT) was chosen as a three-way link. In fact, a three-way link of TPT gives more variety of discrete coordination cages having a large void capable of encapsulating hydrophobic guests. We reported molecular $M_6L_4$ cages (M: *cis*-protected Pd or Pt; L: TPT) in 1995.[28] Meanwhile we attempted to prepare coordination networks consisting of metal ions and TPT, but had no success.

In 1996, Robson and coworkers succeeded in the preparation of a TPT coordination network.[29] A porous coordination network in which two independent cubic nets are interpenetrated was synthesized by diffusion of a solution of $[Cu(CH_3CN_4)]ClO_4$ in acetonitrile into a solution of TPT in $CHCl_3$/1,1,2,2-tetrachloroethane. Dark red crystals of $[Cu_3(TPT)_4](ClO_4)_3$ were analyzed by single-crystal X-ray diffraction.

The repeating structural unit in the network is formed by six Cu centers at the corners of a regular octahedron. Large octahedral chambers are formed in which each chamber is connected through six Cu metal vertices to six others (Figure 1.17). An infinite arrangement of cubic chambers is then produced. At the center of each collection of eight chambers, there is a cavity that can accommodate a chamber of an independent but identical infinite network (Figure 1.18). Importantly, the octahedral chambers are left unoccupied, which generates a solid with large empty voids that can include other molecules.

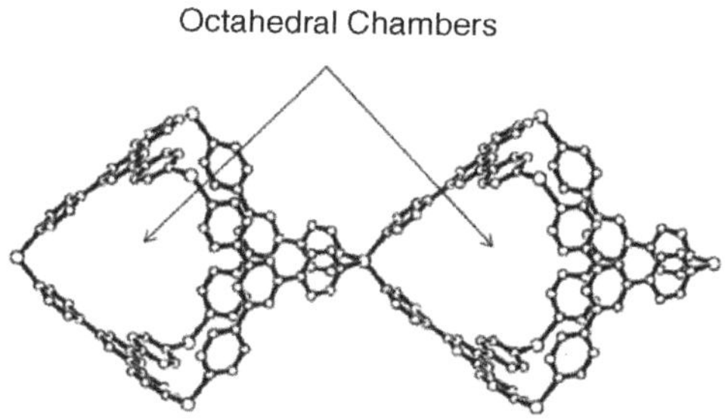


**FIGURE 1.17** Representation of two octahedral chambers in the structure of $[Cu_3(TPT)_4]$-$(ClO_4)_3$·solvate. $Cu^I$ centers (large circles). Smaller circles represent C and N.

### 1.6.1 SCSC Guest Removal in TPT Coordination Networks

In 2002, we reported SCSC dynamic framework transformation on guest removal.[30] $[(ZnI_2)_3(TPT)_2]\cdot 6C_6H_5NO_2$ was prepared by slow complexation of $ZnI_2$ with TPT in a mixture of methanol and nitrobenzene. The network consists of two doubly interpenetrated 3D frameworks. The network shrinks on guest removal and swells on guest encapsulation without loss of crystallinity. For example, crystals of $[(ZnI_2)_3(TPT)_2]\cdot 6C_6H_5NO_2$ show a remarkable ability to exchange nitrobenzene for various organic molecules such as benzene, mesitylene, *cis*-stilbene, and $CHCl_3$ without deterioration of crystallinity.

Desolvated crystals, $[(ZnI_2)_3(TPT)_2]$, were prepared by leaving single crystals of $[(ZnI_2)_3(TPT)_2]\cdot 6C_6H_5NO_2$ (Figure 1.19a) at room temperature to equilibrate with atmosphere for a day. Crystallographic analysis of the guest-free structure revealed

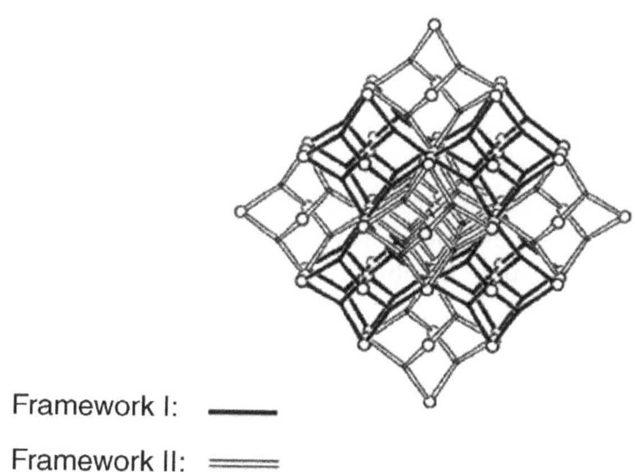


**FIGURE 1.18** Plot of two interpenetrating identical frameworks. Black connectors correspond to eight chambers of one framework, while white connectors correspond to the second framework. The central cavity corresponding to the white framework is connected octahedrally to six other white chambers.

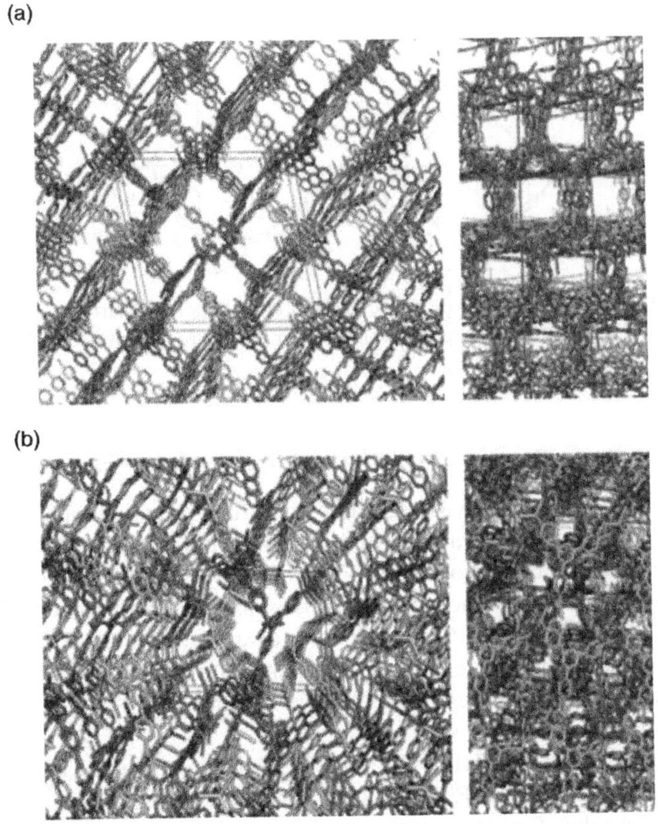


**FIGURE 1.19** Crystal structures of $[(ZnI_2)_3(TPT)_2]\cdot 6C_6H_5NO_2$ and $[(ZnI_2)_3(TPT)_2]$. (a) View of $[(ZnI_2)_3(TPT)_2]\cdot 6C_6H_5NO_2$: left, in the (010) direction; right, in the (101) direction. (b) View of the desolvated framework $[(ZnI_2)_3(TPT)_2]$: left, in the (010) direction; right, in the (101) direction.

that the monoclinic crystal system of the solvate structure changes into triclinic in $[(ZnI_2)_3(TPT)_2]$ and the framework considerably compresses (Figure 1.19b). The void volume decreases from 61 to 23% in the desolvation process. When crystals of $[(ZnI_2)_3(TPT)_2]$ were placed in nitrobenzene for a day, the network re-expanded to the initial structure with nitrobenzene while retaining crystallinity. These results demonstrate the robustness and flexibility of the interpenetrated framework of $[(ZnI_2)_3(TPT)_2]\cdot 6C_6H_5NO_2$.

### 1.6.2 SCSC Guest Exchange in TPT Coordination Networks

Recently, we reported the guest exchange of large organic molecules within the 3D framework of a coordination network, $[(ZnI_2)_3(TPT)_2]\cdot 5.5$(nitrobenzene), which proceed in a SCSC fashion. As a result of the guest exchange, a charge–transfer (CT) interaction between the host and the guest was observed.[31]

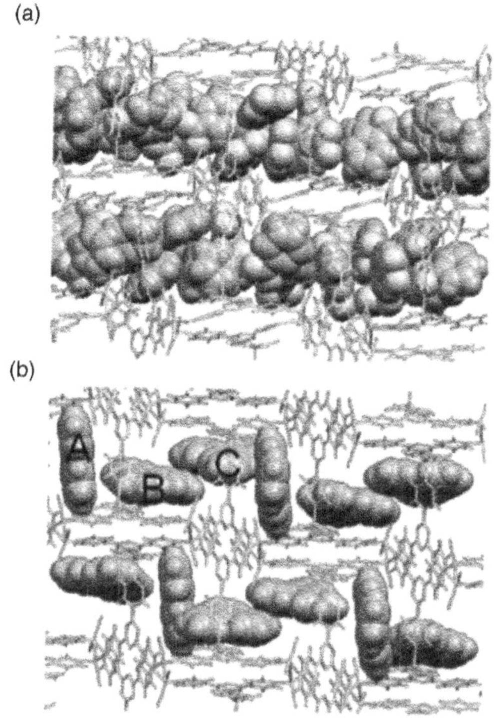


**FIGURE 1.20** Crystal structures of $[(ZnI_2)_3(TPT)_2]\cdot 5.5$(nitrobenzene) viewed down the *b* axis: (a) Before guest exchange (nitrobenzene are encapsulated in the channels), (b) After guest exchange for triphenylene. Cyclohexane molecules are omitted for clarity.

Previous guest exchange processes that occur in a SCSC manner have been focused on small molecules, like solvent molecules, mostly because of the small pore sizes.

However, the guest inclusion process of large molecules that can be monitored via X-ray crystallography is crucial in order to rationalize the design of new solid materials with interesting functional properties.

Nitrobenzene molecules are densely packed along large pores of $[(ZnI_2)_3(TPT)_2]\cdot 5.5$ (nitrobenzene) (Figure 1.20a). Upon immersion of $[(ZnI_2)_3(TPT)_2]\cdot 5.5$(nitrobenzene) in a saturated cyclohexane solution of triphenylene at room temperature during 1 day, the crystals turned from colorless to pale yellow while maintaining its shape and crystallinity.

Single-crystal X-ray diffraction revealed the structure of the inclusion complex $[(ZnI_2)_3(TPT)_2]\cdot 1.5$(triphenylene)$\cdot 2.5$(cyclohexane) in which the guest triphenylene molecules were found in three different positions (Figure 1.20b).

Interestingly, the unit cell parameters are almost the same as those of $[(ZnI_2)_3(TPT)_2]\cdot$ 5.5(nitrobenzene), indicating that the 3D network did not change. Likewise, anthracene and perylene also replace nitrobenzene in $[(ZnI_2)_3(TPT)_2]\cdot 5.5$-(nitrobenzene) to give the inclusion complexes $[(ZnI_2)_3(TPT)_2]\cdot 1.4$(anthracene)$\cdot 2.2$-(cyclohexane) and $[(ZnI_2)_3(TPT)_2]\cdot$(perylene)$\cdot$(cyclohexane)$\cdot 1.5$(nitrobenzene), respectively, without loss of crystallinity. Crystallographic analysis showed efficient $\pi$–$\pi$ stacking of the

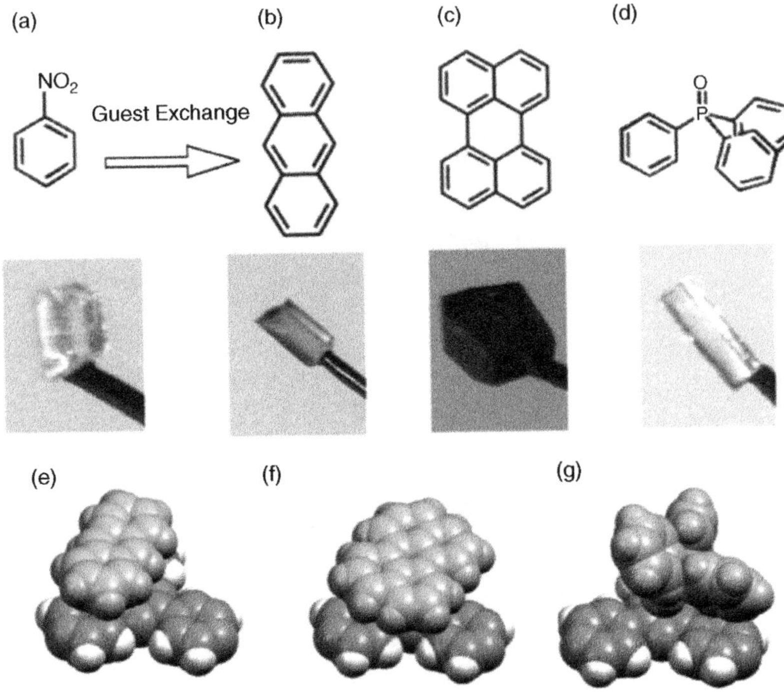


**FIGURE 1.21** Crystal color changes of $[(ZnI_2)_3(TPT)_2]\cdot 5.5$(nitrobenzene) on guest exchange: (a) nitrobenzene (colorless), (b) anthracene (black), (c) perylene (gray), (d) triphenylphosphine oxide (colorless). The partial structures of the inclusion compounds: (e) anthracene, (f) perylene, and (g) triphenylphosphine oxide.

guest and ligand. Close ligand–guest contact suggests donor– acceptor interaction between electron-rich guests and the electron-deficient TPT ligands. This was supported by UV–Vis studies. DFT calculations also show that the symmetry of the guest HOMOs effectively matches the LUMO of the simplified model framework of $[(ZnI_2)_3(TPT)_2]\cdot$ 5.5(nitrobenzene). The inclusion complexes of planar aromatic guest with CT interactions are distinctly colored whereas the inclusion complex of triphenylphosphine oxide is colorless due to the lack of CT interaction (Figure 1.21).

### 1.6.3 SCSC Photochromism in Coordination Networks

A crystalline organic coordination network can be used to control a certain molecular property by encapsulation in a pore. We demonstrated that a non-chromic compound can become chromic by just encapsulation via guest exchange (Figure 1.22).[32] Salicylideneanilines have chromic properties depending on the substitution patterns of the aromatic rings. In general, the derivatives with non-planar conformation ($\phi > 25°$) are photochromic whereas those with planar conformation ($\phi < 25°$) are thermochromic.[33] To control the chromic properties of salicylideneanilines, numerous derivatives have been synthesized by introducing various substituents on the chromophore. We studied the inclusion complex of ($o$-HO-$C_6H_4$)CH=N($m$-Cl-$C_6H_4$),

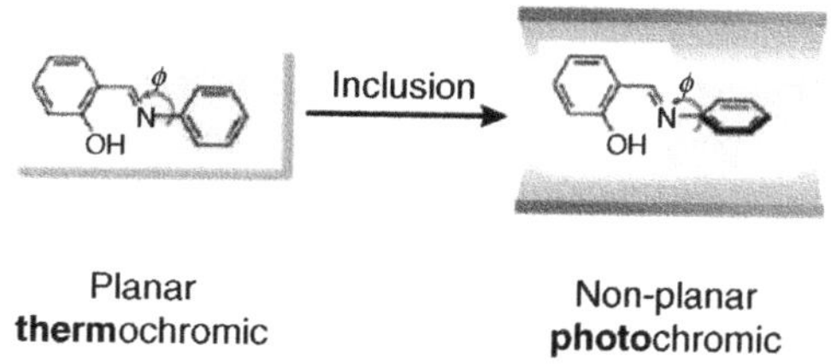


**FIGURE 1.22** Switching from thermochromic to photochromic behavior of salicylideneanilines by guest inclusion.

which is a thermochromic compound in crystalline state. This inclusion compound was prepared by SCSC guest exchange from nitrobenzene clathrate $[(ZnI_2)_3(TPT)_2]\cdot5.5(C_6H_5NO_2)$. The crystallographic analysis of the clathrate complex $[(ZnI_2)_3(TPT)_2]\cdot1.3$(*o*-HO-$C_6H_4$)CH=N(*m*-Cl-$C_6H_4$)1.0 (*t*-BuOH) revealed the exchange of nitrobenzene in $[(ZnI_2)_3(TPT)_2]\cdot5.5(C_6H_5NO_2)$ for a mixture of (*o*-HO-$C_6H_4$)CH=N(*m*-Cl-$C_6H_4$) and *t*-BuOH (Figure 1.23). The guest molecule (*o*-HO-$C_6H_4$)CH=N(*m*-Cl-$C_6H_4$)

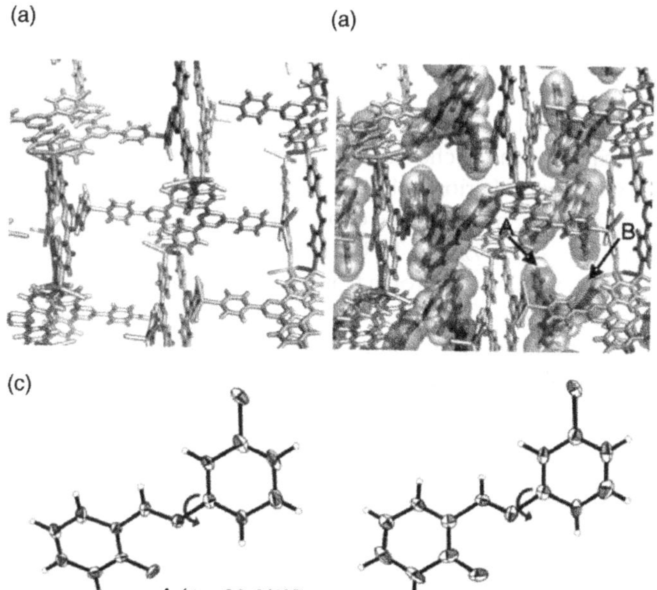


**FIGURE 1.23** Crystal structure of the porous network $[(ZnI_2)_3(TPT)_2]$ enclathrating (*o*-HO-$C_6H_4$)CH=N(*m*-Cl-$C_6H_4$) in the pore: (a) view along the (101) direction. Included (*o*-HO-$C_6H_4$)CH=N(*m*-Cl-$C_6H_4$) molecules and solvents in the pores are omitted. (b) View along the *b*-axis. The porous network of $[(ZnI_2)_3(TPT)_2]$ is shown in stick model. Two crystallographic independent molecules of (*o*-HO-$C_6H_4$)CH=N(*m*-Cl-$C_6H_4$) (A and B), included in the pore of the network, are shown in stick and space-filling model. *t*-BuOH molecules co-existing in the pore are omitted for clarity. (c) Thermal ellipsoid plots of (*o*-HO-$C_6H_4$)CH=N(*m*-Cl-$C_6H_4$). The ellipsoids are drawn at the 30% probability level. The arrows represent the dihedral angle ($\phi$) defined by the two planes of C=N and *m*-chlorophenyl groups.

exists in two crystallographic independent positions, A and B, within a column parallel to the $a + c$ direction. The three-dimensional network remains unchanged after the guest exchange and the lattice parameters are almost the same as those of $[(ZnI_2)_3(TPT)_2]\cdot 5.5(C_6H_5NO_2)$. It is important to note that the dihedral angles of (*o*-HO-$C_6H_4$)CH=N(*m*-Cl-$C_6H_4$) in the coordination network is considerably distorted [$\phi = 23.3(3)$ and $29.3(4)^\circ$ at positions A and B, respectively] from that found at the single-crystal X-ray analysis of (*o*-HO-$C_6H_4$)CH=N(*m*-Cl-$C_6H_4$) alone ($\phi = 5^\circ$). Thus, the limited space in the channel significantly twisted the conformation of (*o*-HO-$C_6H_4$)CH=N(*m*-Cl-$C_6H_4$) (Figure 1.23c). As expected, photoirradiation ($\lambda = 366$ nm) of the crystal of the inclusion complex at $-153^\circ$C induced chromism from pale yellow to red. The photoisomerization of (*o*-HO-$C_6H_4$)CH=N(*m*-Cl-$C_6H_4$) (*cis*-enol form) into (*o*-O=$C_6H_4$)CH=CH-NH(*m*-Cl-$C_6H_4$) (*trans*-keto form), which causes the photochromism, was confirmed by single-crystal microscopic IR spectroscopy.

## 1.7 BIPOROUS COORDINATION NETWORKS

Selective guest inclusion of two different guest molecules can be achieved if two different channels exist in a given crystal. For instance, with such biporous materials selective guest inclusion can be readily obtained depending upon the different nature of the two channels.

In 2004, we prepared a biporous coordination network by complexation of $ZnI_2$ with TPT and triphenylene (Scheme 1.8).[34] Importantly, most porous coordination network materials reported prior to this study contain only one type of channel. Those materials are referred as monoporous coordination network.

Single crystals of $[(ZnI_2)_3(TPT)_2(\mathbf{2a})]\cdot x$(nitrobenzene)$\cdot y$(methanol) (**2a** = triphenylene, $x \approx 4$, $y \approx 2$) were grown from a triple-layered solution with a methanol solution (0.5 mL) of $ZnI_2$ (0.03 mmol) as the top layer, methanol (0.5 mL) as the middle layer, and a nitrobenzene/methanol solution (4:1, 5 mL) of TPT (0.02 mmol) and triphenylene (**2a**, 0.1 mmol) as the bottom layer. In this biporous

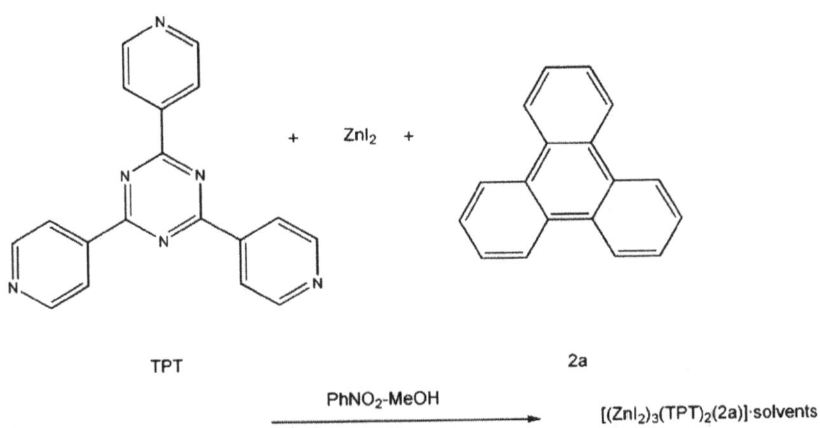


**SCHEME 1.8**

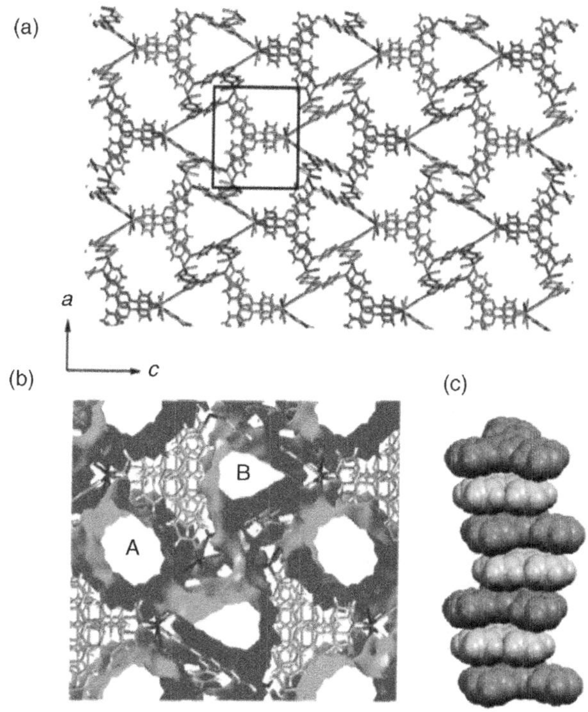


**FIGURE 1.24** Crystal structure of $[(ZnI_2)_3(TPT)_2(\mathbf{2a})]\cdot x$(nitrobenzene)$\cdot y$(methanol): (a) view of the interpenetrated networks along the $b$-axis, (b) two distinct channels A and B, (c) the infinite aromatic stacking between TPT and triphenylene at the inserted square in (a).

material, triphenylene is considered as forming part of the host substructure rather than being a guest, since the intercalated triphenylene is not replaced by common aromatic compounds under guest exchange process due to the strong $\pi$–$\pi$ stacking.

The structure of complex $[(ZnI_2)_3(TPT)_2(\mathbf{2a})]\cdot x$(nitrobenzene)$\cdot y$(methanol) contains two distinct channels (A and B) within the framework. Channel A is roughly cylindrical and is lined with the hydrogen atoms of infinitely stacked TPT and **2a**. Channel B, however, is roughly trigonal prismatic, where two of the three walls are the $\pi$ faces of TPT and the third is the edges of TPT and **2a** (Figure 1.24a–c). The approximate dimensions of the channels are 7.4 Å × 5.5 Å (channel A) and 7.3 Å × 5.5 Å (channel B).

### 1.7.1 Biporous Network: Guest Uptake

Interestingly, the two channels adsorb their own preferred guest molecules from a mixture: channel A of as-synthesized $[(ZnI_2)_3(TPT)_2(\mathbf{2a})]\cdot x$(nitrobenzene)$\cdot y$(methanol) is filled with nitrobenzene and methanol, which are considerably disordered and cannot be fully located by crystallographic analysis, whereas channel B is filled with nitrobenzene (Figure 1.25). Dipping crystals of $[(ZnI_2)_3(TPT)_2(\mathbf{2a})]\cdot x$(nitrobenzene)$\cdot y$(methanol) into a saturated cyclohexane solution of naphthalene at room temperature

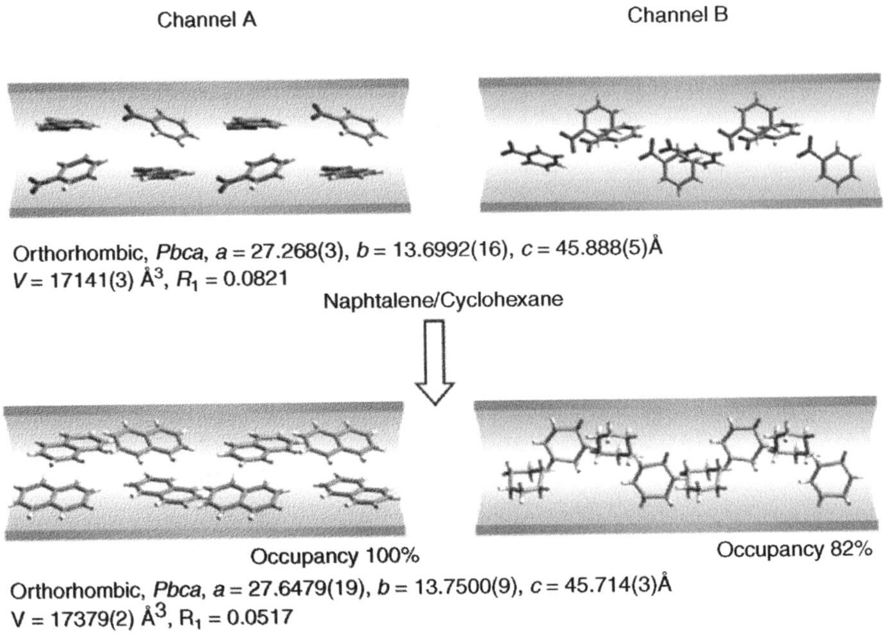


**FIGURE 1.25** Selective guest exchange of nitrobenzene for naphthtalene and cyclohexane in $[(ZnI_2)_3(TPT)_2(\mathbf{2a})]\cdot x$(nitrobenzene)·$y$(methanol). Partially disordered molecules are omitted for clarity.

for 2 days caused channels A and B to be selectively filled with naphthalene and cyclohexane, respectively, to give guest exchanged crystals of $[(ZnI_2)_3(TPT)_2(\mathbf{2a})]\cdot$ $x$(cyclohexane)·$y$(naphthalene) ($x \approx 1.3$ and $y \approx 2.3$).

After the guest exchange process, the crystallinity was maintained, which allowed high-quality X-ray diffraction data to be collected (i.e., $R_1 = 0.052$). The crystallographic analysis revealed the formation of columnar arrays of naphthalene (100% occupancy) and cyclohexane (82% occupancy) in channels A and B, respectively (Figure 1.25). Similarly, the absorption of two different guests by the two channels was also observed for an azulene/cyclohexane mixture: channels A and B selectively adsorbed azulene (100% occupancy) and cyclohexane (100% occupancy), respectively. A wide range of scientific and technologic applications are likely to be possible for such biporous coordination network.

### 1.7.2 Biporous Network: Cartridge Synthesis

Although numerous porous coordination networks have been synthesized to date,[15,16,35] the functionalization of the pores at will without changing the network frameworks is still a difficult task in spite of the high designability of organic ligand components. This is because the porous network formation is very sensitive to the ligand structure.

For instance, the formation of a completely different and unpredictable structure is often observed when a small functional group is introduced to the ligand. We have

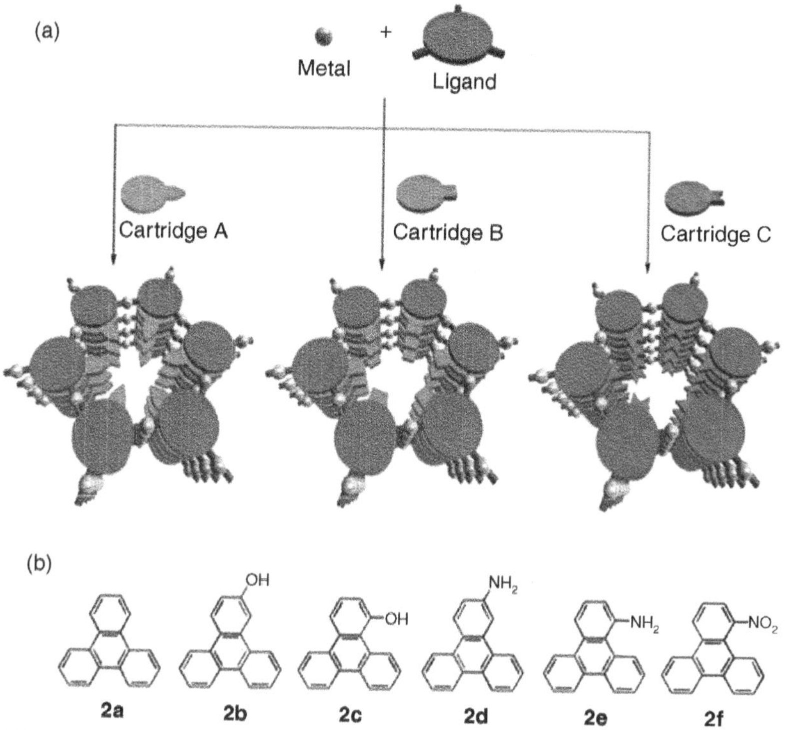


**FIGURE 1.26** (a) Cartoon representation of the cartridge synthesis. (b) The library of the cartridges.

previously synthesized a unique porous coordination network in which the pores are surrounded by aromatic bricks. The bricks consist of alternatively layered TPT and triphenylene. TPT forms infinite 3D network via coordination to $ZnI_2$, whereas triphenylene is involved in the 3D framework forming charge–transfer $\pi$–$\pi$ interactions with TPT. We reported that the non-covalently intercalated triphenylene (**2a**) in this porous complex can be replaced with functionalized triphenylenes (**2b–f**) to form biporous networks with different internal pore environment (**3a–f**) without causing any change in the porous network structure (Figure 1.26). In these porous complexes, intercalated triphenylenes **2b–f** are regarded as the cartridges of functional groups. We note that the internal environment of the channels can be easily modified simply by replacing the cartridges. Thus, this strategy is a new approach toward the synthesis of a family of porous networks, which we regard as *cartridge (modular) synthesis.*[36]

### 1.7.3 Biporous Network: Pore Modification and Specific Molecular Recognition

Guest exchange properties differ considerably from the clathrate containing OH functionalized triphenylene (**3b**) (Figure 1.27a, b) and the clathrate without functionalized triphenylene (**3a**). For instance, the pore A in **3b** and **3c** has an acidic

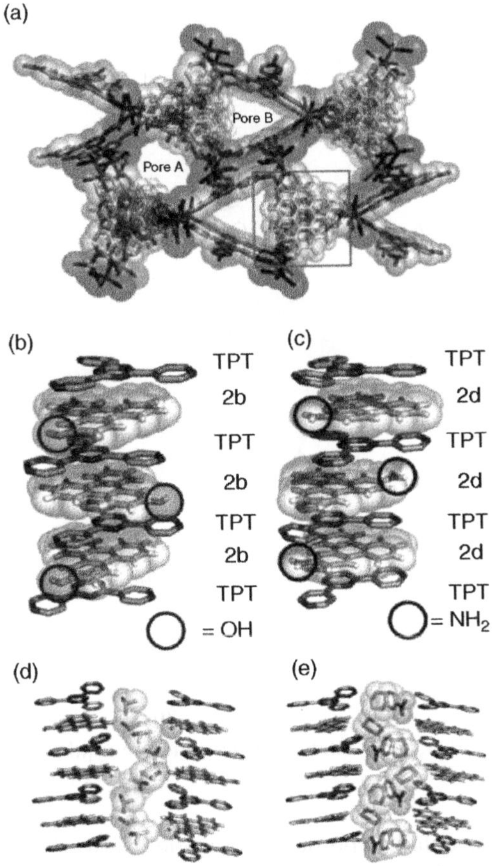


**FIGURE 1.27** (a) Porous structure of **3b**. Guests in the pores are omitted for clarity. (b) The view of the aromatic layers of **3b** (square region in (a)) from pore A. Acidic OH groups are pointed toward the pore A. (c) The same view of **3d**. Basic $NH_2$ groups are pointed toward the pore A. (d) Channel of the guest (propan-2-ol) formed in the pore A of **3b** via hydrogen bonding with phenolic OH groups. (e) Channel of the guest (cyclohexane) formed in the neutral pore A of **3a**.

environment due to the presence of the acidic OH groups, whereas in **3a** the same pore A has a less polar environment. On the other hand, the complex **3d** and **3e** has a basic pore (Figure 1.27c) due to the presence of the $NH_2$ group while the biporous network **3f** has a polar pore arising from the $NO_2$ group.

When single crystals of **3b** were immersed in a mixed solvent of propan-2-ol and cyclohexane (1:39), propan-2-ol was selectively adsorbed by pore A, thus replacing the nitrobenzene molecules filling the pore A of the as-synthesized **3b** (Figure 1.27d). Interestingly, even in such a low concentration of propan-2-ol, it was selectively adsorbed within the pore A via hydrogen bond formation of propan-2-ol and the acidic phenolic hydroxyl group of cartridge **2b** through a water molecule as discussed later.

Such a selective uptake of propan-2-ol by pore A was not observed for **3a** because its pore A is hydrophobic; instead, cyclohexane was taken up by the pore A of **3a**

(Figure 1.27e). The pore B of **3a** and **3b** are both hydrophobic. Thus, nitrobenzene in the pore B of as-synthesized **3a** and **3b** was fully exchanged with cyclohexane.

No significant loss of crystallinity was observed during the guest exchange process, which allowed good crystallographic analysis. In the pore A of **3b**, propan-2-ol is hydrogen bonded via one water molecule with the OH group of the cartridge **2b**.

### 1.7.4 Crystalline Molecular Flasks

A pore of a crystalline organic coordination network is so fluid that it can play a role as a molecular flask, because reagents and solvents can be introduced inside of pores by guest exchange like flasks on a bench. The difference between molecular flask and laboratory flask is the size of space. The crystalline molecular flask is a nanometer-size container where a kinetic effect can be expected for chemical reactions. Moreover, a minimum amount of solvents and reagents is required for reactions. Another fascinating feature is that the SCSC process enables direct observation of chemical reactions by X-ray analysis.[37] Therefore, a porous organic coordination network can be called *crystalline molecular flasks*.

***1.7.4.1 SCSC Chemical Transformation: Direct Observation of Unstable Imine*** The direct observation of unstable molecules that are difficult to isolate as crystals can be carried out by *in situ* preparation of labile compounds in a single crystalline state. This method, however, is not straightforward because of the following reasons: (i) the reactivity of the substrates is considerably reduced in the solid state; (ii) bulky reagents cannot enter into the pores; and (iii) often, crystallinity dramatically decreases as the solid-state reaction proceeds. Thus, successful examples are extremely limited.

We demonstrated the concept of single crystalline molecular flasks using a cartridge coordination network. Thanks to the pseudo-solution state in the pores, bulky substrates can easily diffuse into the crystals without reducing the crystallinity of the network. We studied the reaction of acetaldehyde with amines to form acetaldehyde imines that are, normally, easily hydrolyzed or isomerized into enamines.[38] The aromatic amines were incorporated in the columnar array of aromatic ligands of an as-synthesized porous network complex. The amino group, which is oriented toward the pore, reacted with the acetaldehyde that diffused into the crystal, and unstable imines were formed in the pore in a SCSC manner. Our *in situ* SCSC crystallography studies revealed not only the formation of labile imine structures but also the dynamic rotor-like motion of the amine substrates during the reaction. We also emphasize that the crystal-to-crystal reaction in the pore provides a key technique for the post-functionalization of porous coordination networks.

To carry out such *in situ* experiments, a porous coordination network was prepared by treating TPT and 1-aminotriphenylene (**2e**) with $ZnI_2$ in a nitrobenzene–methanol gradient solution. Red single crystals with a composition of $[(ZnI_2)_3(TPT)_2$(**2e**)] were isolated in a good yield (Figure 1.28). In this network complex, ligand TPT and amine **2e** were alternatively and infinitively stacked along the crystallographic *b*-axis (Figure 1.29). There are two types of 1D-pores (A and B) with different chemical

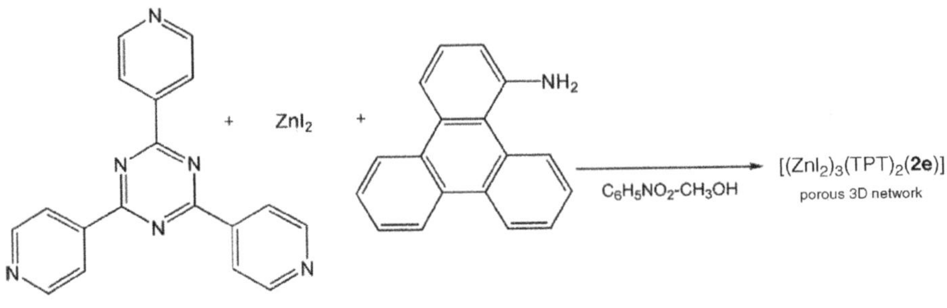


TPT 2e

**FIGURE 1.28** Preparation of porous network complex [$(ZnI_2)_3(TPT)_2$(**2e**)].

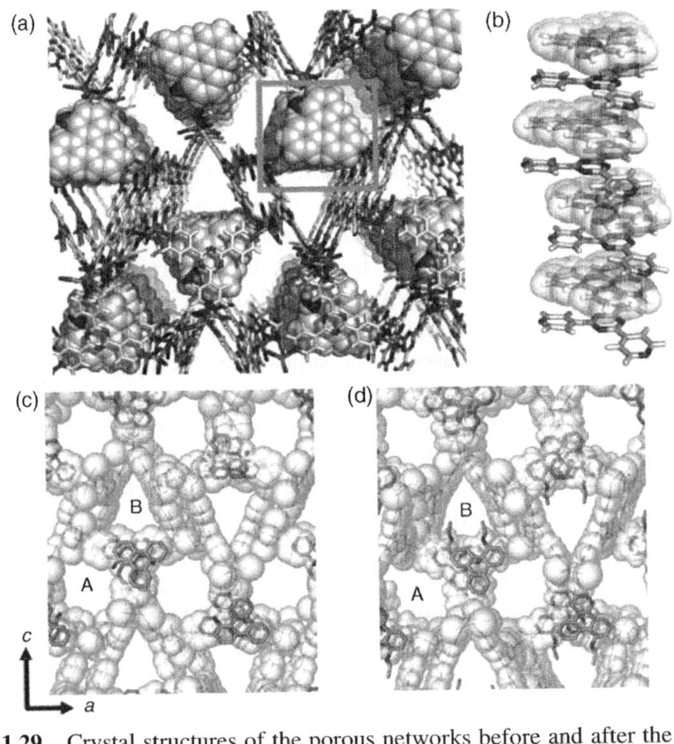


**FIGURE 1.29** Crystal structures of the porous networks before and after the SCSC imine formation. (a) The porous network of [$(ZnI_2)_3(TPT)_2$(**2e**)] viewed along the *b*-axis. The porous network of [$(ZnI_2)_3(TPT)_2$] is shown as sticks. Imbedded 1-aminotriphenylene molecules (**2e**) are shown as space-filling models. Nitrobenzene molecules in the pore are omitted for clarity. (b) The infinite aromatic stacking of ligand TPT and amine **2e** at the highlighted square region in (a). (c) Crystal structure of the porous network of [$(ZnI_2)_3(TPT)_2$(**2e**)] viewed along the *b*-axis. The porous network of [$(ZnI_2)_3(TPT)_2$] is shown in gray as lines and translucent space-filling models. Imbedded 1-aminotriphenylene molecules (**2e**) are shown as sticks. Nitrobenzene molecules in the pore are omitted for clarity. (d) Crystal structure of the porous network of [$(ZnI_2)_3(TPT)_2$(**3a**)]. Nitrobenzene molecules are omitted for clarity.

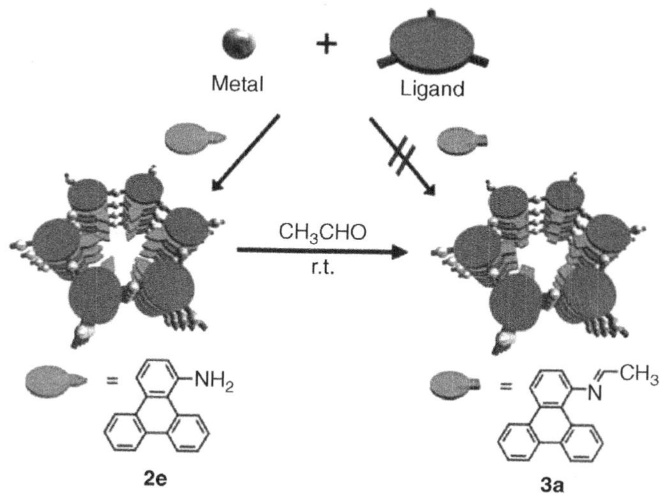


**FIGURE 1.30** Formation of labile imines by SCSC reactions in the pores of a porous coordination network.

properties and shapes. The amino groups of **2e** are located in pore A but not in pore B (Figure 1.29c).

Condensation of the amino group of **2e** with acetaldehyde produced the unstable imine **3a**. Needle-shaped crystals of $[(ZnI_2)_3(TPT)_2(\mathbf{2e})]$ were immersed in a nitrobenzene solution of acetaldehyde (20 wt%) at room temperature (Figure 1.30). Interestingly, the diffusion of the aldehyde into the pores induced a color change in the original crystal. The crystal turned from red to yellow in 4 h. The change of crystal color arises from the decrease in the donor ability of aromatic amine **2e** by the imine formation. Interestingly, we observed that the color changes from both ends of the crystal and this change progresses toward the center of the crystal. This occurs along the long axis of the crystal, which corresponds to the *b* crystallographic axis (i.e., the channels run parallel to the *b*-axis) (Figure 1.31). After isolating the crystals, the complete conversion to imine **3a** in the network was confirmed by elemental analysis,

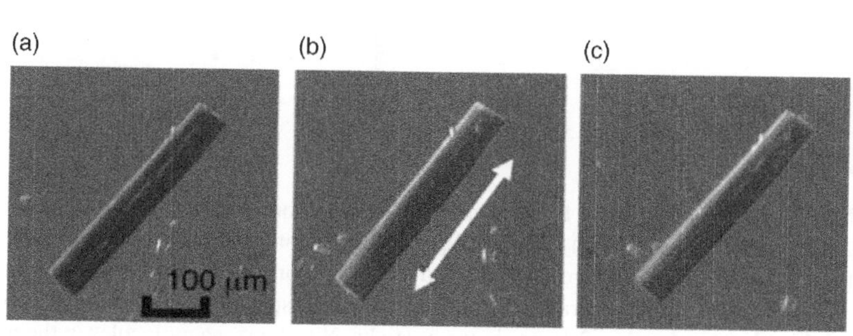


**FIGURE 1.31** Photographs of a single crystal of the clathrate $[(ZnI_2)_3(TPT)_2(\mathbf{2e})]$ (a) before reaction. (b) 15 min after immersing into acetaldehyde–nitrobenzene (1:4) solution at room temperature. The arrow represents pore direction in $[(ZnI_2)_3(TPT)_2(\mathbf{2e})]$. (c) After 30 min.

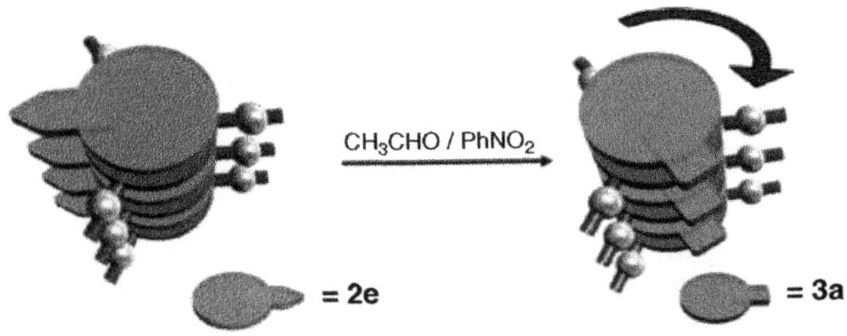


**FIGURE 1.32** Rotation of amine **2e** via SCSC transformation into **3a**.

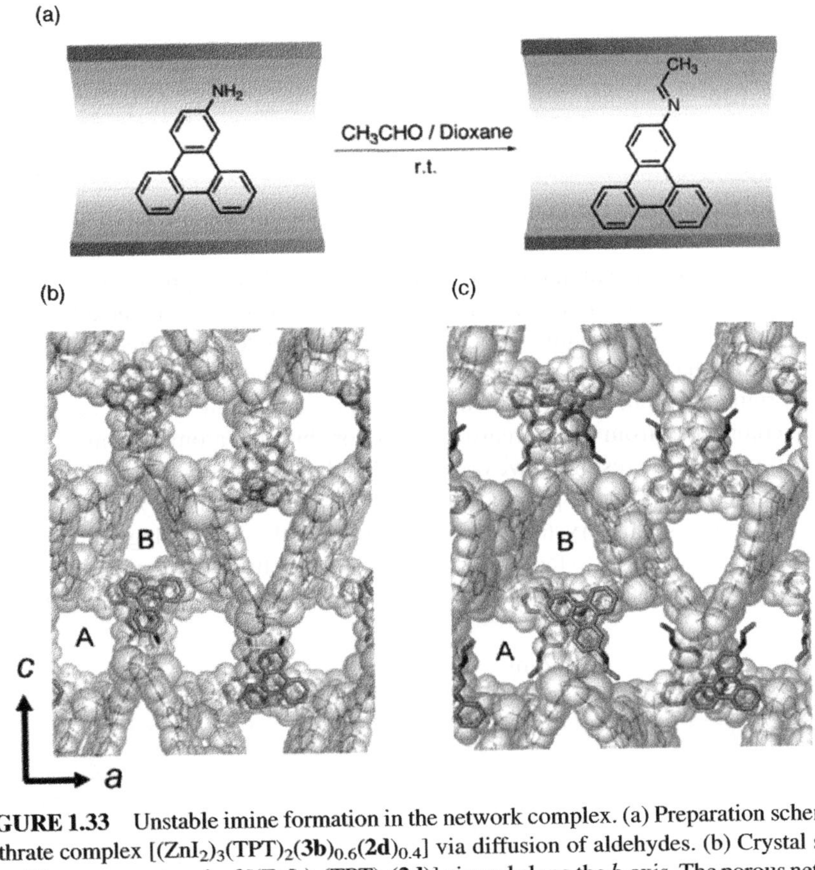


**FIGURE 1.33** Unstable imine formation in the network complex. (a) Preparation scheme of clathrate complex $[(ZnI_2)_3(TPT)_2(\mathbf{3b})_{0.6}(\mathbf{2d})_{0.4}]$ via diffusion of aldehydes. (b) Crystal structure of the porous network of $[(ZnI_2)_3(TPT)_2(\mathbf{2d})]$ viewed along the *b*-axis. The porous network of $[(ZnI_2)_3(TPT)_2]$ is shown in gray as lines and translucent space-filling models. Imbedded 2-aminotriphenylene molecules (**2d**) are shown as sticks. Nitrobenzene molecules in the pore are omitted for clarity. (c) Crystal structure of the porous network of $[(ZnI_2)_3(TPT)_2(\mathbf{3b})_{0.6}(\mathbf{2d})_{0.4}]$. Unreacted amine **2d** (40% in pore A) and dioxane in the pore as a solvent are omitted for clarity.

microscopic IR analysis, and extraction of the product **3a**. After the imine formation within the channels, the single crystal did not show any appreciable change in size and morphology. Therefore, the single crystal was suitable for X-ray diffraction.

Crystallographic analysis revealed the formation of unstable imine **3a** within the porous network. The solid-state conversion of **2e** into **3a** was quantitative in agreement with microscopic IR analysis. The amino group of **2e** before the reaction exists only in pore **A**, but, surprisingly, the imino group after the reaction is observed both in pore **A** (44%) and in pore **B** (56%). Thus, rotation of the imbedded **2e** must occur during the reaction (Figure 1.32).

The SCSC formation of a more labile imine in the pore was also examined with 2-aminotriphenylene (**2d**). The single crystalline network complex $[(ZnI_2)_3(TPT)_2(\mathbf{2d})]$ was prepared and subjected to the condensation with acetaldehyde by immersing the crystals in a dioxane solution of acetaldehyde (Figure 1.33a). The expected imine **3b** itself is too unstable to be isolated. However, the formation of **3b** within the porous cavity was successfully observed. The conversion was approximately 60%. In the case of imine **3a** formation, the triphenylene core did not rotate during the reaction of amine **2d** with acetaldehyde (Figure 1.33b, c).

Further, in order to understand the rotation of the aromatic amines during the imine formation, we studied the reaction of **2e** using two additional aldehydes. *In situ* crystallography showed that the imino groups from hexanal exist in pore **A**, whereas those from *p*-anisaldehyde are found only in pore **B** (Figure 1.34). While the aldehydes

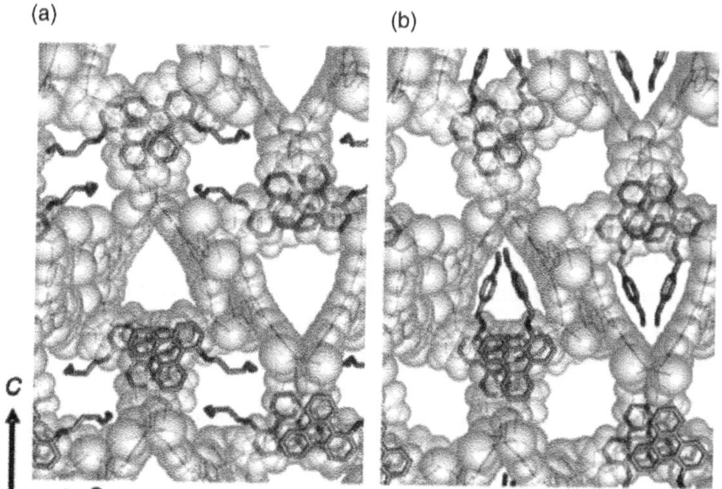


**FIGURE 1.34** Crystal structures of clathrate complexes $[(ZnI_2)_3(TPT)_2(\text{imine})_x(\text{amine})_y]$ prepared by the crystalline state reaction of **2e** with aldehydes via diffusion of aldehydes. (a) Reaction with hexanal ($x = 1$, $y = 0$). Substituted triphenylene molecules are shown as sticks. Unreacted hexanal molecules in both pores are omitted for clarity. (b) Reaction with *p*-anisaldehyde ($x = 0.5$, $y = 0.5$). Substituted triphenylene molecules are shown as sticks. Unreacted amine **2e** (50% in pore A) and *p*-anisaldehyde (in pore A and B) are omitted for clarity.

can diffuse into both pores, the reaction selectively proceeds in one specific pore where the imine is more stabilized. In the case of *p*-anisaldehyde, the amine must first rotate before the condensation can occur. Once the reaction takes place, the new bulky imino substituent prevents the rotation.

## 1.8 CONCLUDING REMARKS

This review described historical development of organic coordination networks from the first example (Hofmann complex) to the latest developments in this interesting research area. The precise structural knowledge obtained from X-ray crystallography has been a key point to develop a strategy to design organic coordination networks. It has been shown how the selection of organic building ligands linked by metal connectors can form a large variety of porous structures that can accommodate a wide range of guest molecules. Interesting processes such as catalytic reactions, gas adsorption, and the formation of unstable species within the pore space of some coordination networks have been reviewed. Furthermore, cartridge synthesis of networks, which enables the design of pores with different environments, will make significant difference in molecular recognition, separation, and catalytic activity. Finally, we introduced the concept of crystalline molecular flasks using porous coordination networks, which will provide more opportunity to achieve new stereoselective reactions in pores and clarify reaction mechanisms. Visualizing reactions by X-ray will make considerable difference in designing tailor-made reactions. Seeing is believing.

## REFERENCES


1. Hofmann, K. A.; Küspert, F. Verbindungen von Kohlenwasserstoffen mit Metallsalzen. *Anorg. Chem.* **1897**, *15*, 204.
2. Pfeiffer, P. Organische Molekülverbindungen. *Enke, Stuttgart* **1927**, 213.
3. Feigl, F. *Anais Assoc. Quím. Brasil* **1944**, *3*, 72. As cited in reference 4.
4. Rayner, J. H.; Powell, H. M. Structure of molecular compounds. Part X. Crystal structure of the compound of benzene with an ammonia-nickel cyanide complex. *J. Chem. Soc.* **1952**, 319.
5. Iwamoto, T.; Miyoshi, T.; Miyamoto, T.; Sasaki, Y.; Fujiwara, S. The metal ammine cyanide aromatics clathrates. I. The preparation and stoichiometry of the diamminemetal(II) tetracyanoniccolate(II) dibenzene and dianiline. *Bull. Chem. Soc. Jpn* **1967**, *40*, 1174.
6. Iwamoto, T.; Nakano, T.; Morita, M.; Miyoshi, T.; Miyamoto, T.; Sasaki, Y. The Hofman-type Clathrate: $M(NH_3)_2M'(CN)_4 \cdot 2G$. *Inorg. Chim. Acta* **1968**, *2*, 313.
7. Walker, G. F.; Hawthorne, D. G. Complexes between *n*-alkylamines and nickel cyanide. *Trans. Faraday Soc.* **1967**, *63*, 166.
8. Miyoshi, T.; Iwamoto, T.; Sasaki, Y. The structure of catena-μ-ethylenediaminecadmium(II) tetracyanoniccolate(II) dibenzene clathrate: $Cd(en)Ni(CN)_4 \cdot 2C_6H_6$. *Inorg. Chim. Acta* **1972**, *6*, 59.

9. Mathey, Y.; Mazières, C.; Setton, R. Formation of 3-dimensional structures using bifunctional ligand bridges between nickel cyanide planes. *Inorg. Nucl. Chem. Lett.* **1977**, *13*, 1.

10. Nishikiori, S.; Iwamoto, T.; Yoshino, Y. Three-dimensional metal complex structures with ambident propylenediamine ligands serving as the hosts of the aromatic guest molecules. Hofmann-pn and pn-Td type clathrates. *Bull. Chem. Soc. Jpn* **1980**, *53*, 2236.

11. (a) Nishikiori, S.; Iwamoto, T. Inclusion of aromatic guest molecules with bulky substituents in layered metal complex host *trans*-bis(dimethylamine)cadmium(II) tetra-*catena*-μ-cyanonickelate(II). *Chem. Lett.* **1982**, 1035. (b) Nishikiori, S.; Iwamoto, T. Crystal structures of the *o*-, the *m*-, and the *p*-toluidine clathrates in bisdimethylamine-cadmium(II) tetracyanonickelate(II) host. *Chem. Lett.* **1983**, 1129. (c) Nishikiori, S.; Iwamoto, T. Crystal structure of Hofmann-Dma-type benzene clathrate bis(dimethylamine)cadmium(II) tetracyanonickelate(II)-benzene (2/1). *Chem. Lett.* **1984**, 319.

12. (a) Hasegawa, T.; Nishikiori, S.; Iwamoto, T. Inclusion compounds formed between α,ω-(long-carbon-chain)-diaminecadmium(II) tetra-cyanonickelate(II) host and aromatic guest molecules. *J. Inclusion Phenomena* **1984**, *1*, 365. (b) Hasegawa, T.; Nishikiori, S.; Iwamoto, T. Isomer selection of 1,6-diaminohexanecadmium(II) tetracyanonickelate(II) for *m*- and *p*-toluidine. Formation of 1,6-diaminohexanecadmium(II) tetracyanonickelate(II)-*m*-tolouidine (1/1) inclusion compound and bis(*p*-toluidine)-1,6-diaminohexanecadmium(II) tetracyanonickelate(II) complex. *Chem. Lett.* **1985**, 1659.

13. (a) Nishikiori, S.; Iwamoto, T. Novel three-dimensional host structure revealed in the crystal structure of *N*-methyl-1,3-diaminopropanecadmium(II) tetracyanonickelate(II)-cyclohexane (2/1). *Chem. Lett.* **1987**, 1127. (b) Kitazawa, T.; Nishikiori, S.; Kuroda, R.; Iwamoto, T. Two novel metal-complex host structures consisting of cyanocadmate coordination polyhedra. Clay-like and Zeolite-like structures. *Chem. Lett.* **1988**, 459. (c) Kitazawa, T.; Nishikiori, S.; Kuroda, R.; Iwamoto, T. Novel clathrate compound of cadmium cyanide host with an adamantane-like cavity. Cadmium cyanide-carbon tetrachloride (1/1). *Chem. Lett.* **1988**, 1729.

14. Hoskins, B. F.; Robson, R. Infinite polymeric frameworks consisting of three dimensionally linked rod-like segments. *J. Am. Chem. Soc.* **1989**, *111*, 5962.

15. Fujita, M.; Yazaki, J.; Ogura, K. Preparation of a macrocyclic polynuclear complex, $[(en)Pd(4,4'\text{-bpy})]_4(NO_3)_8$, which recognizes an organic molecule in aqueous media. *J. Am. Chem. Soc.* **1990**, *112*, 5645.

16. Fujita, M.; Kwon, Y. J.; Washizu, S.; Ogura, K. Preparation, clathration ability, and catalysis of a two-dimensional square network material composed of cadmium(II) and 4,4′-bipyridine. *J. Am. Chem. Soc.* **1994**, *116*, 1151.

17. Ohmori, O.; Fujita, M. Heterogeneous catalysis of a coordination network: cyanosilylation of imines catalyzed by a Cd(II)-(4,4′-bipyridine) square grid complex. *Chem. Commun.* **2004**, 1586.

18. Subramanian, S.; Zaworotko, M. J. Porous solids by design: $[Zn(4,4'\text{-bpy})_2(SiF_6)]_n \cdot x$DMF, a single framework octahedral coordination polymer with large square channels. *Angew. Chem. Int. Ed. Engl.* **1995**, *34*, 2127.

19. Yaghi, O. M.; Li, H. Hydrothermal synthesis of a metal-organic framework containing large rectangular channels. *J. Am. Chem. Soc.* **1995**, *117*, 10401.

20. Kondo, M.; Yoshitomi, T.; Seki, K.; Matsuzaka, H.; Kitagawa, S. Three-dimensional framework with channeling cavities for small molecules: $\{[M_2(4,4'\text{-bpy})_3(NO_3)_4] \cdot xH_2O\}_n$ (M = Co, Ni, Zn). *Angew. Chem. Int. Ed. Engl.* **1997**, *36*, 1725.

21. Kawano, M.; Fujita, M. Direct observation of crystalline-state guest exchange in coordination networks. *Coord. Chem. Rev.* **2007**, *251*, 2592.
22. Kepert, C. J.; Rosseinsky, M. J. Zeolite-like crystal structure of an empty microporous molecular framework. *Chem. Commun.* **1999**, 375.
23. Biradha, K.; Hongo, Y.; Fujita, M. Open square-grid coordination polymers of the dimensions 20 × 20 Å: remarkably stable and crystalline solids even after guest removal. *Angew. Chem. Int. Ed.* **2000**, *39*, 3843.
24. Later on we reported a 24 × 24 Å square ladder. Ohmori, O.; Kawano, M.; Fujita, M. A 24 × 24 Å square ladder. *CrystEngComm* **2004**, *6*, 51.
25. Biradha, K.; Hongo, Y.; Fujita, M. Crystal-to-crystal sliding of 2D coordination layers triggered by guest exchange. *Angew. Chem. Int. Ed.* **2002**, *41*, 3395.
26. Suh, M. P.; Ko, J. W.; Choi, H. J. A metal-organic bilayer open framework with a dynamic component: single-crystal-to-single-crystal transformations. *J. Am. Chem. Soc.* **2002**, *124*, 10976.
27. Takaoka, K.; Kawano, M.; Tominaga, M.; Fujita, M. In situ observation of a reversible single-crystal-to-single-crystal apical-ligand-exchange reaction in a hydrogen-bonded 2D coordination network. *Angew. Chem. Int. Ed.* **2005**, *44*, 2151.
28. Fujita, M.; Oguro, D.; Miyazawa, M.; Oka, H.; Yamaguchi, K.; Ogura, K. Self-assembly of ten molecules into nanometre-sized organic host frameworks. *Nature* **1995**, *378*, 469.
29. Abrahams, B. F.; Batten, S. R.; Hamit, H.; Hoskins, B. F.; Robson, R. A cubic (3,4)-connected net with large cavities in solvated $[Cu_3(tpt)_4](ClO_4)_3$ (tpt = 2,4,6,-Tri(4-pyridyl)-1,3,5-triazine). *Angew. Chem. Int. Ed. Engl.* **1996**, *35*, 1690.
30. Biradha, K.; Fujita, M. A springlike 3D-coordination network that shrinks or swells in a crystal-to-crystal manner upon guest removal or readsorption. *Angew. Chem. Int. Ed.* **2002**, *41*, 3392.
31. Ohmori, O.; Kawano, M.; Fujita, M. Crystal-to-crystal guest exchange of large organic molecules within a 3D coordination network. *J. Am. Chem. Soc.* **2004**, *126*, 16292.
32. Haneda, T.; Kawano, M.; Kojima, T.; Fujita, M. Thermo-to-photo-switching of the chromic behavior of salicylideneanilines by inclusion in a porous coordination network. *Angew. Chem. Int. Ed.* **2007**, *46*, 6643.
33. (a) Cohen, M. D.; Schmidt, G. M. J. Photochromy and thermochromy of anils. *J. Phys. Chem.* **1962**, *66*, 2442. (b) Cohen, M. D.; Schmidt, G. M. J.; Flavian, S. Topochemistry. Part VI. Experiments on photochromy and thermochromy of crystalline anils of salicylaldehydes. *J. Chem. Soc.* **1964**, 2041. (c) Hadjoudis, E. Photochromic and thermochromic anils. *Mol. Eng.* **1995**, *5*, 301; (d) Harada, J.; Uekusa, H.; Ohashi, Y. X-ray analysis of structural changes in photochromic salicylideneaniline crystals. Solid-state reaction induced by two-photon excitation. *J. Am. Chem. Soc.* **1999**, *121*, 5809. (e) Fujiwara, T.; Harada, J.; Ogawa, K. Solid-state thermochromism studied by variable-temperature diffuse reflectance spectroscopy. A new perspective on the chromism of salicylideneanilines. *J. Phys. Chem. B* **2004**, *108*, 4035.
34. (a) Ohmori, O.; Kawano, M.; Fujita, M. A two-in-one crystal: uptake of two different guests into two distinct channels of a biporous coordination network. *Angew. Chem. Int. Ed.* **2005**, *44*, 1962. (b) Ohmori, O.; Kawano, M.; Fujita, M. Construction of biporous coordination networks via π-π interaction. *CrystEngComm* **2005**, *7*, 255.
35. (a) Constable, E. C. Higher oligopyridines as a structural motif in metallosupramolecular chemistry. *Prog. Inorg. Chem.* **1994**, *42*, 67. (b) Dunbar, K. R.; Heintz, R. A. Chemistry of

transition metal cyanide compounds: modern perspectives. *Prog. Inorg. Chem.* **1996**, *45*, 283. (c) Whiteford, J. A.; Rachlin, E. M.; Stang, P. J. Fast atom bombardment mass spectrometry for characterizing cationic chelated species. *Angew. Chem. Int. Ed. Engl.* **1996**, *35*, 2524. (d) Batten, S. R.; Robson, R. Interpenetrating nets: ordered, periodic entanglement. *Angew. Chem. Int. Ed.* **1998**, *37*, 1460. (e) Hagrman, P. J.; Hagrman, D.; Zubieta, J. Organic-inorganic hybrid materials: from "simple" coordination polymers to organodiamine-templated molybdenum oxides. *Angew. Chem. Int. Ed.* **1999**, *38*, 2638. (f) Moulton, B.; Zaworotko, M. J. From molecules to crystal engineering: supramolecular isomerism and polymorphism in network solids. *Chem. Rev.* **2001**, *101*, 1629. (g) Eddaoudi, M.; Moler, D. B.; Li, H.; Chen, B.; Reineke, T. M.; O'Keeffe, M.; Yaghi, O. M. Modular chemistry: secondary building units as a basis for the design of highly porous and robust metal-organic carboxylate frameworks. *Acc. Chem. Res.* **2001**, *34*, 319. (h) Kitagawa, S.; Kitaura, R.; Noro;S. -I. Functional porous coordination polymers. *Angew. Chem. Int. Ed.* **2004**, *43*, 2334. (i) Kitagawa, S.; Uemura, K. Dynamic porous properties of coordination polymers inspired by hydrogen bonds. *Chem. Soc. Rev.* **2005**, *34*, 109. (j) Hoskins, B. F.; Robson, R. Design and construction of a new class of scaffolding-like materials comprising infinite polymeric frameworks of 3D-linked molecular rods. A reappraisal of the $Zn(CN)_2$ and $Cd(CN)_2$ structures and the synthesis and structure of the diamond-related frameworks $[N(CH_3)_4][Cu^{I}Zn^{II}(CN)_4]$ and $Cu^{I}$[4,4′,4″,4‴-tetracyanotetraphenylmethane]$BF_4 \cdot xC_6H_5NO_2$. *J. Am. Chem. Soc.* **1990**, *112*, 1546.

36. Kawano, M.; Kawamichi, T.; Haneda, T.; Kojima, T.; Fujita, M. The modular synthesis of functional porous coordination networks. *J. Am. Chem. Soc.* **2007**, *129*, 15418.

37. In the porous hydrogen-bonded network, modification of a network pore by single-crystal-to-single-crystal reaction has been reported: Brunet, P.; Demers, E.; Maris, T.; Enright, G. D.; Wuest, J. D. Designing permeable molecular crystals that react with external agents to give crystalline products. *Angew. Chem. Int. Ed.* **2003**, *42*, 5303.

38. Haneda, T.; Kawano, M.; Kawamichi, T.; Fujita, M. Direct observation of the labile imine formation through single-crystal-to-single-crystal reactions in the pores of a porous coordination network. *J. Am. Chem. Soc.* **2008**, *130*, 1578.

# 2

# INSIGHT INTO THE DEVELOPMENT OF METAL-ORGANIC MATERIALS (MOMs): AT ZEOLITE-LIKE METAL-ORGANIC FRAMEWORKS (ZMOFs)

MOHAMED EDDAOUDI AND JARROD F. EUBANK
*Department of Chemistry, University of South Florida, Tampa, FL 33620, USA*

## 2.1 INTRODUCTION

Porous functional materials are an important class of solid-state materials, and their prevalence and significance in our world today are undeniable. Such materials have become an integral part of our society due to their potential applications, and are as expansive as the people are diverse. Zeolites,[1] purely inorganic frameworks (i.e., aluminosilicates), are probably the most prominent and widely utilized group of porous functional materials due to their applications in petroleum refinement,[2–4] catalysis,[5–8] ion exchange[9–12] (e.g., water softening and purification) and molecular adsorption[13–15] stemming from their anionic nature and open structures (i.e., accessible periodic channels and cavities that do not interpenetrate). Nevertheless, zeolite applications have generally been limited to smaller molecule applications, with few exceptions, due to their restricted pore and cavity size ($\leq$1 nm),[16] in addition to their limited functionality[17,18] and atomic composition.[19] Over the past two decades, much progress has been made in the development of organic–inorganic hybrid materials, namely metal-organic materials (MOMs), for example, metal-organic polyhedra (MOPs) and metal-organic frameworks (MOFs) or coordination polymers. These readily modular materials (with reversible coordination bonds that aid in crystallization, mild synthesis conditions, effective design methods, diverse metal compositions,

*Metal-Organic Frameworks: Design and Application,* Edited by Leonard R. MacGillivray

sundry organic functional groups directly incorporated into the framework, large pores and cavities, and high surface areas), including the recent development of zeolite-like metal-organic frameworks (ZMOFs), have introduced new possibilities in applications that have traditionally utilized the aforementioned inorganic zeolites, such as catalysis and separations,[20] size- and shape-selective uptake,[21] and gas storage,[22,23] as well as novel applications.[24]

## 2.2 METAL-ORGANIC MATERIALS (MOMs)

MOMs have drawn considerable awareness because of the potential of exploiting properties of both organic and inorganic components within a single material, in addition to their modular nature and mild synthesis conditions.[25] Organic molecules offer tunable properties based on shape (i.e., potential ligand coordination angles), size (i.e., expansion of a ligand), and functionality (i.e., decoration of a ligand). Inorganic elements provide the potential for similar properties to traditional zeolites, including thermal and mechanical stability,[20] as well as many more like optical[26] or electronic properties.[27,28]

Another attractive feature of MOMs is the ability to target structures of particular topologies that may be desirable for certain applications by using the molecular building block (MBB) approach.[29–32] This is a design strategy for the construction of solid-state materials, where the metal ions, coordination clusters, and organic ligands are pre-designed to have specific geometry and directionality upon coordination *in situ*.[33] These so-called MBBs then can be used as just that, "building blocks," to target structures.

Essentially, there exist previously enumerated nets[34–39] (here the term "net" is used loosely to refer to discrete polygons and polyhedra, or finite systems of linked points, for example, a square or a cube, and layers and networks or infinite periodic systems of linked points, for example, 2D square lattice or 3D cubic diamond lattice) that can be dismantled into their specific primary vertices (nodes), for example, a point connected to three other points is a 3-connected or trigonal node, and edges (the linear connection between each node). Each net is given a three-letter code by O'Keeffe as a means of simplification and identification;[34] for example, a discrete cube = *cub*, 2D square lattice = *sql*, and 3D cubic diamond lattice = *dia*. Based on its geometry, each MBB can serve as a node or spacer (though much larger than the simple vertex or edge) in a given net so that the arrangement of coordinated ligands and metals and/or clusters into a framework is analogous to the net (i.e., they have the same topology); it should be noted that if the MBB is of linear geometry, it may only serve as an extended connection (linker or spacer) between two nodes.

One difficulty with this method is that for any given simple node (e.g., trigonal, tetrahedral), there exist numerous potential nets, making design in the solid state seemingly unachievable. However, experimental progress in MOMs over the years has revealed that there are networks that tend to predominate in these crystalline solids,[40,41] commonly classified as default structures, and thus these prevalent nets are the most obvious as logical targets for the would-be designer.

This method of targeting enumerated nets for construction from MBBs is known as "top-down" design, "bottom-up" synthesis.[41–43] The MBB approach has proven quite successful, as evidenced by the explosion of literature and citations related to MOMs due to burgeoning academic and industrial[44] interest. There is no doubt that this strategy to obtain solid-state materials is making a mark on the scientific field, and that the eventual design of made-to-order functional materials for targeted applications is foreseeable.

### 2.2.1 Background

MOMs have a rich history, and there have been numerous efforts to design and synthesize functional porous structures. Prior to the late 1980s, a variety of metal-organic coordination compounds were discovered (e.g., Werner complexes,[45] Hofmann clathrates,[46] Prussian blue[47]) and studied/developed for their interesting properties, including molecular inclusion and magnetism. However, no systematic approach to the construction of this class of solid-state materials had been introduced until 1989, when Hoskins and Robson proposed the design of open framework MOMs[48] based on a node-and-spacer approach, that is, tetrahedral nodes can be linked by linear molecular spacers to construct an open structure based on the extension of cubic diamond (considered the default structure for the assembly of simple tetrahedral building blocks). Hoskins' and Robson's molecular node-and-spacer concept of design was realized in 1990 upon their synthesis of a series of analogous MOFs with varied linkers and nodes (i.e., $Zn(CN)_2$, $Cd(CN)_2$, $[N(CH_3)_4]$-$[CuZn(CN)_4]$, $Cu^I$[4,4′,4″,4‴-tetracyanotetraphenylmethane]$BF_4 \cdot xC_6H_5NO_2$) having skeletons that can be simplified as a net analogous to an extended cubic diamond lattice (*dia*).[49] In the latter, both the single-metal ion and central carbon of the ligand serve as the tetrahedral nodes or tetrahedral building units (TBUs), and the cyanophenyl moieties serve as the spacers (Figure 2.1). In addition, they allude to the tunability of MOMs, suggesting the ability to target functional open materials via variations in (e.g., divergent atomic composition) and modifications to (e.g., functionalization) the nodes and spacers required to construct structures based on the same networks, as well as other networks.

### 2.2.2 Polytopic Pyridine-Like Ligands

A primary thrust in the further development of MOMs, especially throughout the 1990s, was to utilize Hoskins' and Robson's design principles, with a working knowledge of simple *N*-donor coordination chemistry, to target a variety of network architectures from single-metal ions and cyanides and/or polytopic monodentate *N*-donor ligands, such as 4,4′-bipyridine (4,4′-Bipy), with the prospect of constructing an open functional solid-state material for applications purposes. 4,4′-Bipy (and its derivatives), for example, possesses monodentate donor nitrogen atoms at opposite ends of the molecule, which coordinate in a similar linear ditopic manner to the shorter cyanide anion.[49–57] Essentially, 4,4′-Bipy acts as a longer linear spacer, and therefore extends the length of connections between nodes, ideally resulting in enlarged cavities

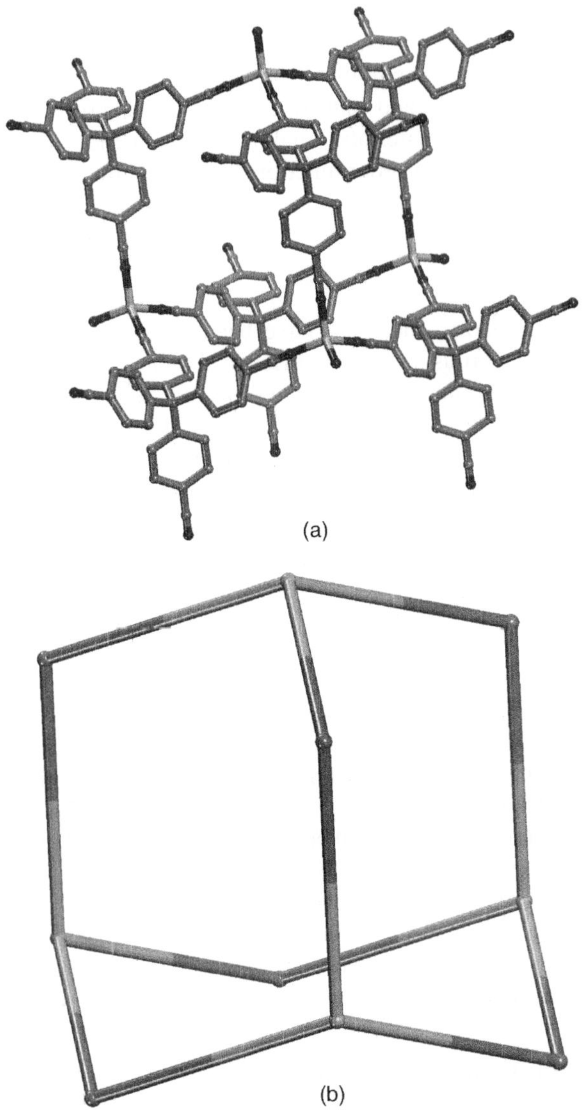


**FIGURE 2.1** (a) Hoskins' and Robson's Cu-tetracyanotetraphenylmethane *dia*-MOF (hydrogen atoms have been omitted for clarity; Cu = black, C = gray, N = dark gray) with (b) cubic diamond topology.

and more open structures compared to zeolites,[49] in addition to analogous metal cyanides.

MOMs based on a variety of nets (of various dimensions) can be synthesized from 4,4′-Bipy and single-metal ions depending on several factors, including the metal: ligand ratio, counter ions, guests, and the coordination environment and geometry of

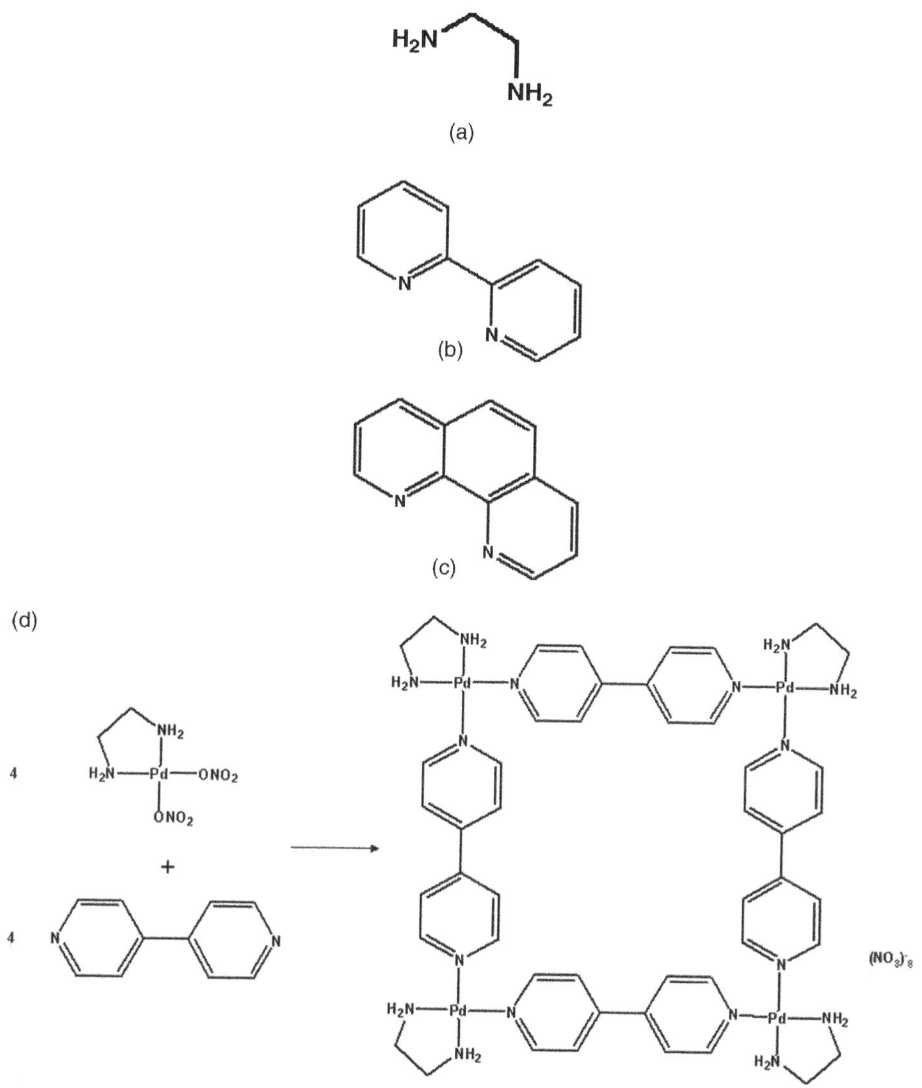


**FIGURE 2.2** Capping agents: (a) ethylenediamine, (b) 2,2′-Bipy, and (c) 1,10-phenanthroline. (d) The construction of molecular squares can be achieved by using capping agents.

the metal.[58–67] Since 4,4′-Bipy is a simple linear linker, the determination of each structure's network topology is directly governed by the coordination environment and geometry of the metal ion. One method that has proven successful at controlling the metal coordination environment is the use of a terminal chelating ligand (ethylenediamine, 2,2′-Bipy, 1,10-phenanthroline, etc., as shown in Figure 2.2a–c) to cap or terminate sites on the metal to give a specific geometry. A classic example of this level of directed assembly is the molecular square synthesized by Fujita and coworkers,[65] where square planar Pd(II) ions are *cis*-capped by ethylenediamine (en), $[enPd(II)]^{2+}$, to provide the 90° angle necessary when coordinating two 4,4′-Bipy molecules in the remaining *cis* positions (Figure 2.2d). One drawback to this capping method is that it generally results in low-dimensional structures, as evidenced by the vast repertoire of corresponding discrete structures.

The use of nonlinear polytopic pyridine-based molecules introduces ligand geometry into the construction mix, where *N*-donor ligands have the potential to act as three-, four-, or six-connected linkers (Figure 2.3).[68–80] These nonlinear ligands can then serve as additional nodes when coordinated to the metal ions, allowing for the synthesis of analogous and novel topologies when compared to 4,4′-Bipy, depending on the size and shape of the ligand (limited only by the synthesis capabilities of the organic chemist) and the coordination environment and geometry of the metal ion.[81] As with 4,4′-Bipy-based MOMs, the metal-ligand assembly can be directed by numerous factors, resulting in myriad architectures, as evidenced by the plethora

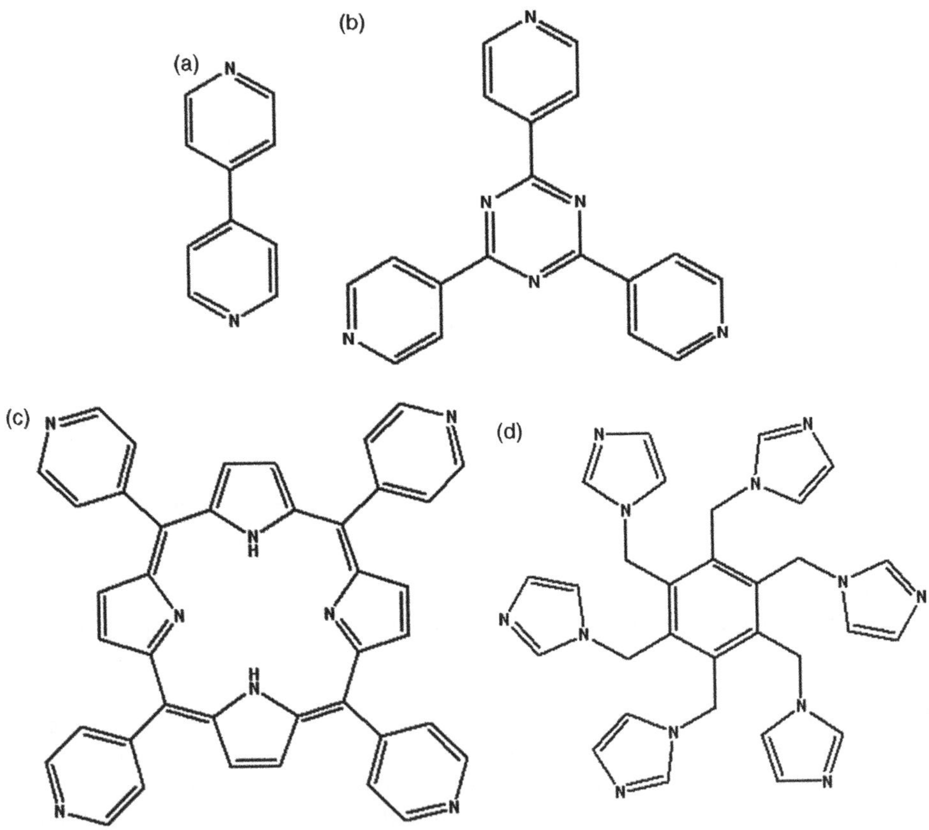


**FIGURE 2.3** Nitrogen-donor organic ligands: (a) linear 4,4′-Bipy; (b) 3-connector 2,4,6-tris(4-pyridyl)-1,3,5-triazine; (c) 4-connector 5,10,15,20-tetrakis(4-pyridyl)porphyrin; and (d) 6-connector 1,2,3,4,5,6-hexakis(imidazol-1-ylmethyl)benzene.

of discrete MOMs (cages, bowls, boxes, tubes, catenanes, and spheres) that can be targeted from the simple *cis*-capped square planar $[enPd(II)]^{2+}$ and a range of *N*-donor ligands.[65]

Although the first porous 4,4′-Bipy-based MOM was reported by Kitagawa and coworkers in 1997 (though the surface area and pore volume were not determined due to high-pressure conditions), permanently porous monodentate (4,4′Bipy-like)-based MOMs are still scarce,[82,83] as these materials typically are unstable and irreversibly lose crystallinity, undergo a phase change, or alter their morphology upon exchange or removal of guests[64] and have been complicated by interpenetration.[33] While many of the 4,4′-Bipy-like ligands are quite rigid, the lack of permanent porosity has traditionally been correlated to the flexible nature of the M–N coordination bonds/angles, which usually results in more flexible frameworks and has limited their utility as robust porous materials.[84]

### 2.2.3 Carboxylate-Based Ligands

In the mid-1990s, the potential of polytopic carboxylate-based bridging ligands began to be explored, since carboxylic acids can be deprotonated for charge balance, precluding the need for extra-framework counterions,[84] and can bind metals in a variety of ways, including monodentate fashion like the *N*-donor ligands (Figure 2.4). As a result, carboxylate-based ligands were utilized to target similar architectures analogous in size and shape to the previously utilized *N*-donor ligands (Figure 2.5). In addition, the ability of carboxylates to form more than one coordination bond added to their allure, since the possibility of multidenticity[85] (e.g., chelation or bis-monodentate coordination), well established in discrete carboxylate-based coordination complexes,[86–90] increased the chances of generating a robust framework upon coordination with a single-metal ion or multiple metals to form a polynuclear metal-carboxylate cluster.

A milestone was achieved in 1998, when Yaghi and coworkers reported the first permanently microporous (evidenced by reversible type I gas (typically $N_2$ or Ar) adsorption isotherms, which are characteristic of microporous materials[91]) MOF, $Zn(BDC)\cdot(DMF)(H_2O)$ or MOF-2, based on linear ditopic 1,4-benzenedicarboxylate (BDC) ligands where the carboxylates each coordinate in a bis-monodentate fashion to two square pyramidal $Zn^{2+}$ ions to form, *in situ*, a known 4-connected dinuclear metal-carboxylate cluster (square paddlewheel, $Zn_2(O_2CR)_4L_2$ (L = axial terminal ligand)). The network structure was only 2D, but the layers were held together by hydrogen bonds between the terminal axial aqua ligands of each cluster of one layer and the carboxylate O atoms of the neighboring layer. The structural integrity was maintained upon dehydration and evacuation, likely due to new bonds between the would-be open-metal sites of one layer and the same carboxylate O atoms of the adjacent layer.[92]

This breakthrough launched a new era in the progression of MOMs, especially carboxylate-based MOFs, leading to an upsurge in the exploitation of square paddlewheel and other known polynuclear metal-carboxylate complexes,[86–90] as well as the quest for novel clusters, through either their generation *in situ* or, to a lesser extent,

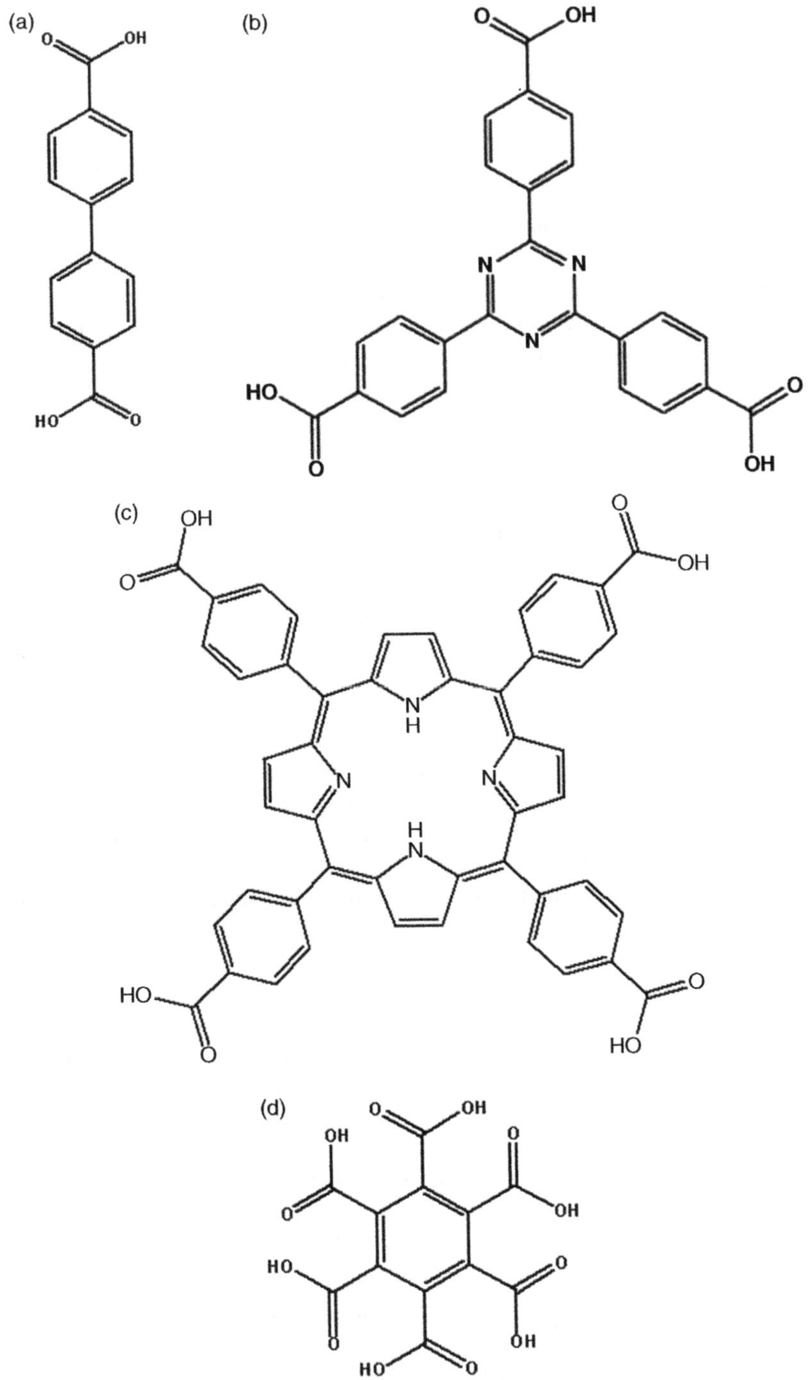


**FIGURE 2.4** Carboxylate-based organic ligands: (a) linear 4,4′-biphenyldicarboxylic acid; (b) 3-connector 2,4,6-tris(4-carboxyphenyl)triazine; (c) 4-connector 5,10,15,20-tetrakis(4-carboxyphenyl)porphyrin; and (d) 6-connector mellitic acid.

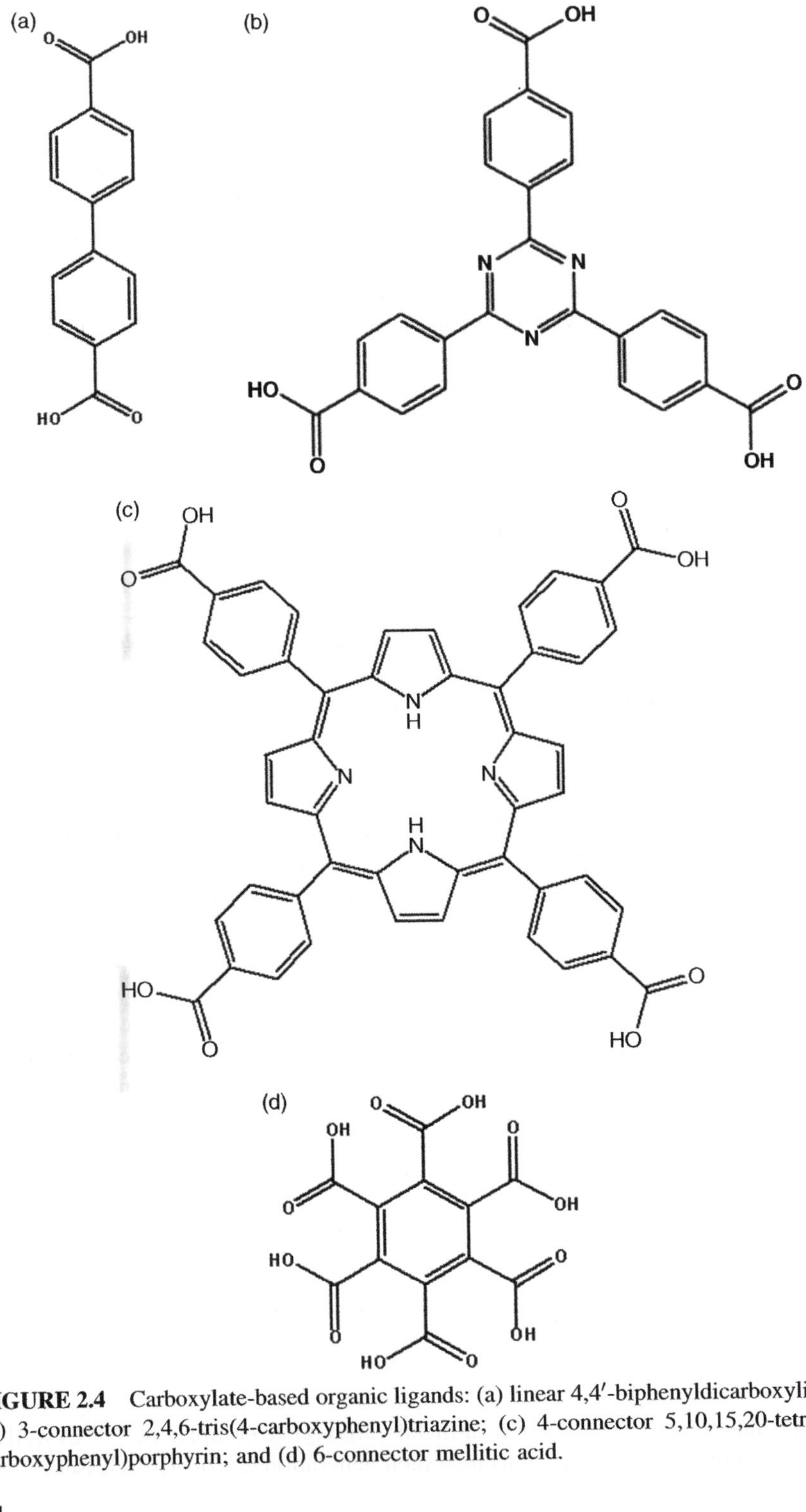

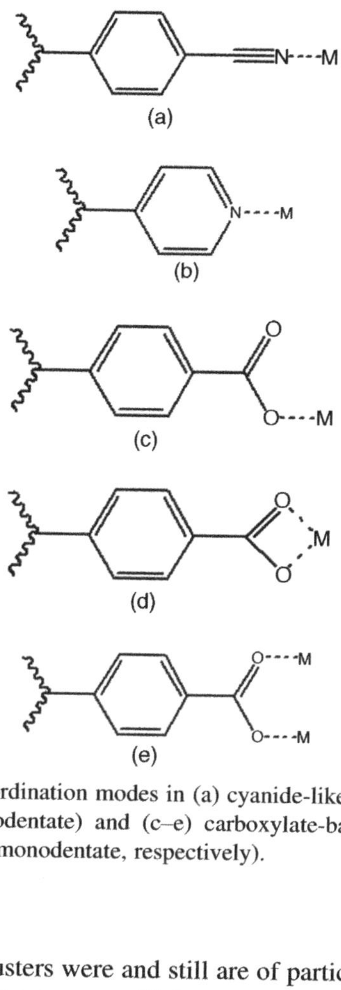


**FIGURE 2.5** Common coordination modes in (a) cyanide-like ligands (monodentate), (b) 4,4′-Bipy-like ligands (monodentate) and (c–e) carboxylate-based ligands (monodentate, bidentate or chelate, and bis-monodentate, respectively).

ligand exchange. These clusters were and still are of particular interest in the MOF community, since they often possess multiple metal-oxygen coordination bonds that result in the generation of rigid nodes with fixed geometry that are ideal as MBBs[84] and facilitate the formation of robust (and likely, permanently porous) 3D frameworks, as well as hold potential for open (or coordinatively unsaturated) metal sites that are of interest for various applications (i.e., gas storage and catalysis).

In 1999, two key microporous 3D MOFs were reported based on the assembly of polytopic carboxylates and metal-carboxylate clusters. The first of these MOFs, $[Cu_3(TMA)_2(H_2O)_3]_n$ or HKUST-1,[93] was constructed from $Cu_2(O_2CR)_4L_2$ square paddlewheel clusters (generated *in situ*) and a tritopic ligand, trimesic acid (TMA, 1,3,5-benzenetricarboxylic acid or BTC), that possesses three-fold symmetry (trigonal). The overall framework is neutral and possesses large openings and cavities, with a Brunauer–Emmett–Teller (BET) surface area of 692.2 $m^2/g$, a Langmuir surface area of 917.6 $m^2/g$, and a single-point total pore volume (typically determined using the Dubinin–Radushkevich equation[91]) of 0.333 $cm^3/g$ ($N_2$),[93] which was comparable with most zeolites.[94] As in MOF-2, the aqua ligands bound to the axial sites on the paddlewheel cluster can be removed upon heating to provide open-metal sites or replaced by other terminal ligands (namely pyridine), while maintaining framework stability. The second of these MOFs, $Zn_4O(BDC)_3 \cdot (DMF)_8(C_6H_5Cl)$ or MOF-5,[95] was constructed from $Zn_4O(O_2CR)_6$ (octahedral symmetry) basic zinc acetate clusters (generated *in situ*) and linear BDC to give a neutral 3D MOF that has now become the prototypical MOF with an open cubic-like network topology. The apparent Langmuir surface area was estimated at 2900 $m^2/g$, the pore volume was calculated as 0.61 $cm^3/cm^3$ ($N_2$), which was higher than that of most zeolites (ranging from 0.18 $cm^3/cm^3$ for analcime to 0.47 $cm^3/cm^3$ for zeolite A),[94] and the density (0.59 $g/cm^3$) was among the lowest recorded for any crystalline material.[95]

Along the lines of Hoskins' and Robson's original conjectures, with insight from pioneering works[35–38,96,97] and a rapidly growing number of MOFs, several simplifying geometric principles and design strategies were developed toward targeting porous, robust (metal-carboxylate)-based MOFs,[25,30,42,43] and the concept of modular chemistry was introduced, where both inorganic clusters and organic ligands can be utilized as molecular building blocks to construct decorated and expanded frameworks of the same base network. However, one limitation to the construction of MOFs from metal-carboxylate clusters (and their modularity) was and is the ability to maintain the cluster via ligand exchange or consistently generate it *in situ*, which is necessary for their utilization as MBBs. If these criteria can be met, under the appropriate conditions, expanded or functionalized ligands can be used in conjunction with the targeted inorganic moiety (in this case, metal-carboxylate cluster) to construct analogous frameworks with varied pore size, shape, and functionality,[31,41] The ability to reliably target an inorganic cluster *in situ* with a variety of analogous linkers (i.e., modularity) was exemplified in 2002 through the construction of a series of isoreticular MOFs or IRMOFs[31] (all having the same primitive cubic (*pcu*) net topology, which resembles the boron net of $CaB_6$), based on the prototypical MOF-5 and numerous other extended/functionalized analogues assembled from the 6-connected octahedral basic zinc acetate MBB, generated *in situ*, and linear dicarboxylates of various lengths/functionalities (Figure 2.6).

In addition to the tetranuclear basic zinc acetate cluster[95,98,99] and the previously mentioned dinuclear paddlewheel cluster,[100–104] there are a variety of metal-carboxylate clusters that have been and can be targeted successfully as MBBs in MOFs, including trinuclear clusters (e.g., $M_3O(O_2CR)_6L_3$, basic chromium acetate, where L = ancillary terminal ligands).[105–115] These MBBs offer building blocks of various geometries (shapes) for the would-be designer to utilize in the construction of MOFs based on nets comprised of nodes of a particular geometry corresponding to the MBB (Figure 2.7). This strategy has permitted the construction of robust, very open MOFs: some of the materials in the IRMOF series have densities (0.41-0.21 $g/cm^3$) lower than that reported for any crystalline material; MOF-177 (reported in 2004 from the assembly of the tetranuclear basic zinc acetate cluster and 1,3,5-benzenetribenzoate (BTB)) had the highest observed surface area (i.e., 4500 $m^2/g$ using the Langmuir model, as opposed to the BET model) of any porous material at the time, up to ~5 times higher than the most open inorganic zeolite.[116] MIL-100 and MIL-101, with unusually

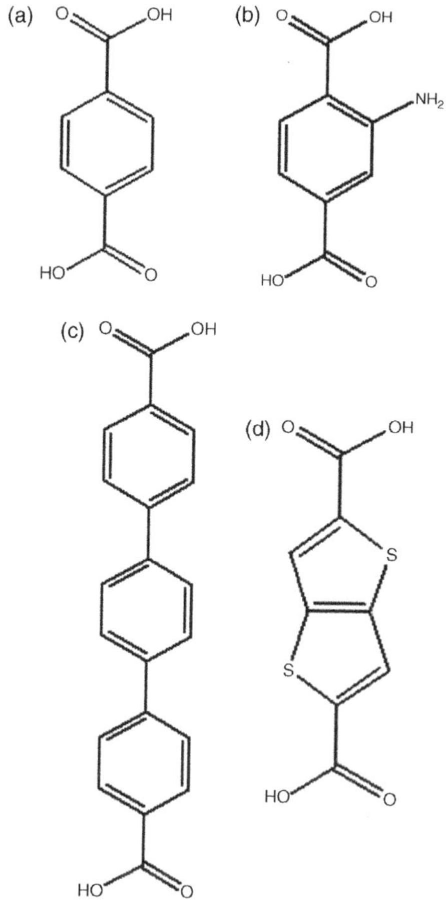


**FIGURE 2.6** Linear dicarboxylates (a–d) can be combined with the basic zinc acetate MBB *in situ* to give a series of isoreticular MOFs ((e) IRMOF-1, (f) IRMOF-3, (g) IRMOF-16,[31] and (h) IRMOF-20[123]) based upon linked octahedra. Hydrogen atoms and solvent molecules have been omitted for clarity; M = black, C = gray, N = dark gray, S = white, and O = light gray. The spheres represent the largest sphere that would fit in the cavities without touching the van der Waals atoms of the framework.

large pore volumes (~380,000 and 702,000 Å$^3$), were reported by Férey et al. in 2004 and 2005 from the assembly of trinuclear $Cr_3O(O_2CR)_6L_3$ clusters and BDC or BTC, respectively, with the latter having a Langmuir surface area of $5900 \pm 300\ m^2/g$;[109,112] nevertheless, modifications to the analysis apparatus of MOF-177 in 2006 led to an improved surface area of $5640\ m^2/g$.[117] In 2007, Long et al. also showed the importance of sample preparation and handling on the surface area of MOFs, when they measured the highest surface area ($4400\ m^2/g$ Langmuir) yet observed

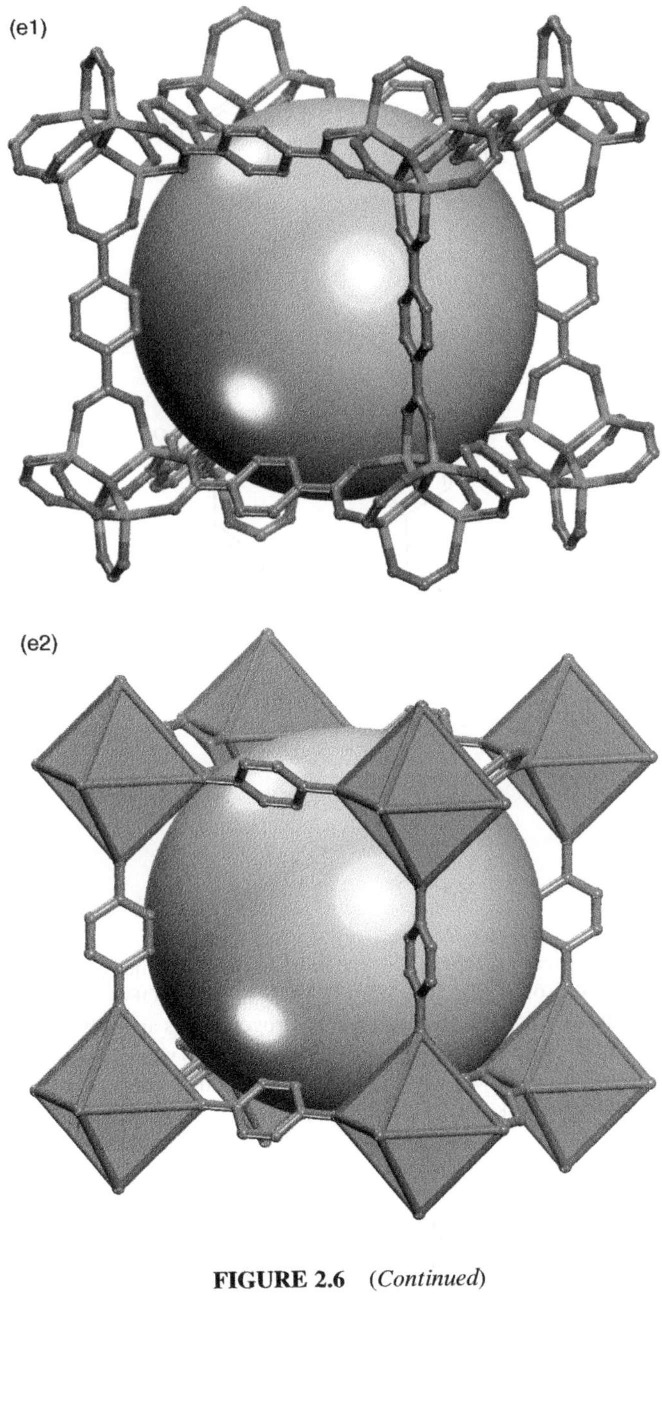


**FIGURE 2.6** (*Continued*)

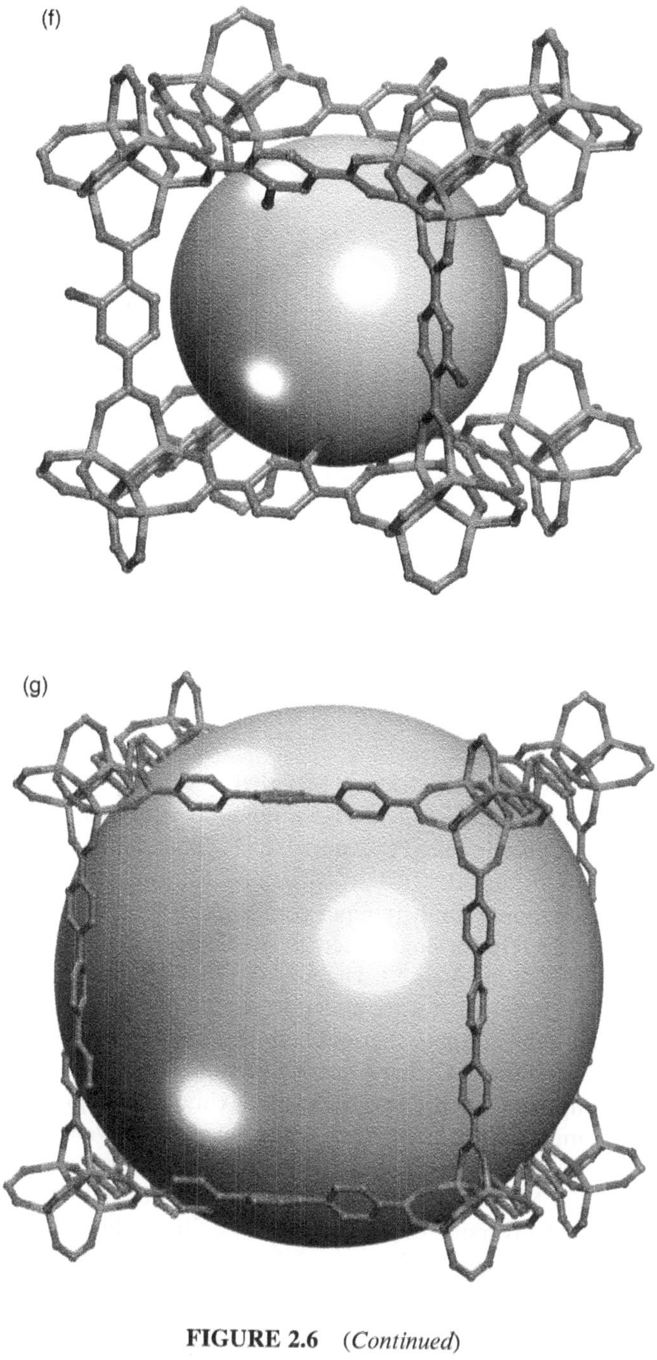

**FIGURE 2.6** (*Continued*)

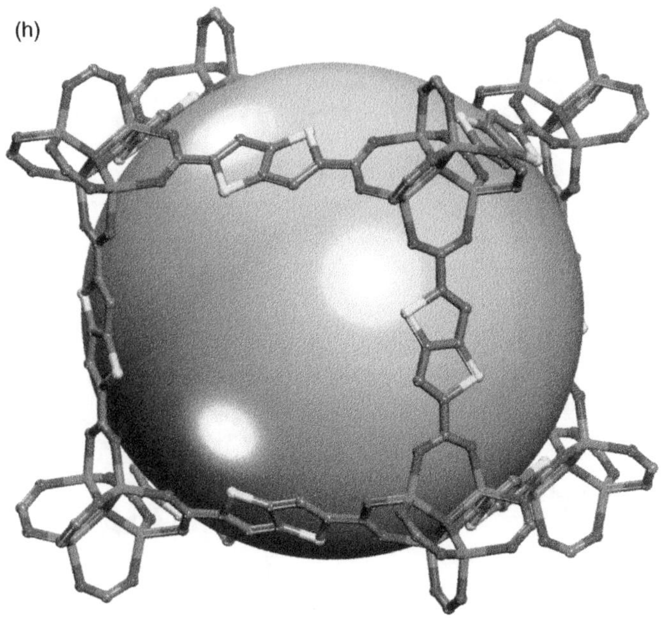


**FIGURE 2.6** (*Continued*)

for MOF-5.[118] In addition to MIL-100 and MIL-101, several MOFs constructed from carboxylate-based MBBs have recently been reported with extra-large cavities or channels in the mesoporous range (>2 nm).[119,120] Zhou and coworkers utilized a non-planar trigonal ligand, 4,4′,4″-*s*-triazine-1,3,5-triyltri-*p*-aminobenzoate (TATAB), in combination with the copper paddlewheel to generate a mesoporous MOF, mesoMOF-1, having a augmented twisted boracite network with 38.5 Å cages and 22.5 × 26.1 Å channels.[120] The mesoMOF-1 was stable in various acids (i.e., HX, X = F, Cl, Br), which were added to react with the amino groups in the TATAB ligand and induce higher thermal stability, and the acid-stabilized MOFs exhibit type IV $N_2$ isotherms, characteristic of mesoporous materials, with the HF-stabilized compound having the highest maximum $N_2$ uptake (729 $cm^3$/g). Using a unique mixed-ligand approach, Matzger and coworkers combined a specific ratio of BDC and BTB ligands under the same reaction conditions as in the respective parent MOFs, MOF-5, and MOF-177, to generate the mixed-ligand that is, UMCM-1 (University of Michigan Crystalline Material-1), which has extra-large hexagonal channels (2.7 nm × 3.2 nm) and a high apparent Langmuir surface area of 6500 $m^2$/g (higher than any value reported to date).[119] It should be noted that now some carboxylate-based mesoporous MOFs also have been obtained serendipitously from the *in situ* formation of unexpected metal clusters or rods.[121,122]

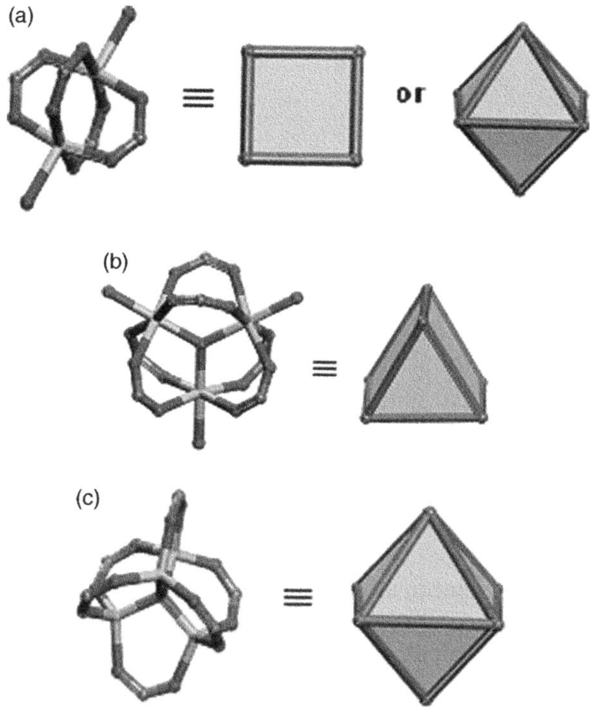


**FIGURE 2.7** Common metal-carboxylate clusters and the resulting molecular building block(s): (a) paddlewheel can act as a linear (not shown), square, or octahedral building block; (b) basic chromium acetate forms a trigonal prism building block; and (c) basic zinc acetate serves an octahedral building block. Hydrogen atoms have been omitted for clarity; M = black, C = gray, and O = light gray.

### 2.2.4 Hetero-functional Ligands

The vast repertoire of organic chemistry also allows the synthesis of organic molecules that include more than one type of potential coordination functional group (e.g., *O*-donor and *N*-donor) into their structure, hetero-functional ligands. This combination of donor types into a singular entity allows the formation of a greater diversity of MBBs, clusters, and structures. In addition, the utilization of hetero-functional ligands may permit the saturation of the metal ions with bridging ligands (i.e., no terminal ligands), and possibly help limit interpenetration, which is also possible through the use of multiple ligands with more than one type of coordination functional group (e.g., one carboxylate-based ligand and one pyridine-based ligand).[124]

Early on, researchers such as Kitagawa,[125,126] and Lin,[127] among others, realized the potential of hetero-functional ligands (e.g., pyridine-carboxylates) in the construction of stabile, porous MOMs. Lin's group utilized nicotinic acid, isonicotinic acid (Figure 2.8), and similar linear or bent ditopic derivatives to synthesize a plethora

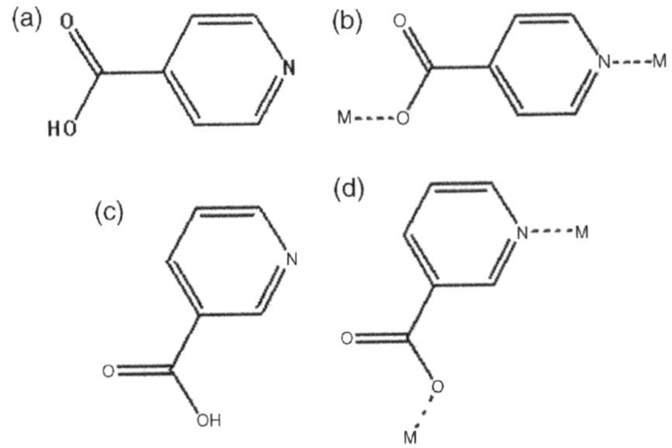


**FIGURE 2.8** (a) Isonicotinic acid and (b) one potential coordination mode of isonicotinate. (c) Nicotinic acid and (d) one potential coordination mode of nicotinate.

of MOMs with a broad range of topologies based on the assembly of these hetero-functional ligands with various single-metal ions or multi-nuclear clusters. In addition to the ability of 2,3-pyrazinedicarboxylate (pzdc) to serve as a ditopic ligand in the formation of 2D layered MOFs with single-metal ions, Kitagawa's group took advantage of additional metal sites (i.e., occupied by terminal ligands) to employ a pillared-layer strategy in the construction of 3D porous materials, where 2D sheets of $[\{Cu(pzdc)\}_n]$ are pillared by a series of linear ditopic *N*-donor ligands (i.e., pyrazine, 4,4′-Bipy, *N*-(4-pyridyl)isonicotinamide, and 1,2-dipyridylglycol). Similar pillaring strategies have been used to construct a variety of porous 3D MOFs based on several 2D sheet types from a plethora of ditopic organic ligand combinations (though linear ditopic *N*-donor ligands remain the most prevalent pillars) and single-metal ions or multi-nuclear clusters. Recently, Kim and coworkers[128] constructed a series of pillared *sql*-MOFs ($4^4$ or square lattice topology) based on $Zn_2(O_2CR)_4L_2$ MBBs, where R = numerous BDC-like linkers and L = 1,4-diazabicyclo(2.2.2)octane or 4,4′-Bipy pillars (essentially converting square paddlewheels to octahedral building units), that have high surface areas (up to 1450 $m^2/g$). Hupp and coworkers extended this strategy to include longer linkers and pillars with a high degree of functionality,[129] including ligands that possess additional metal centers (i.e., metalloligands like (*R,R*)-(−)-1,2-cyclohexanediamino-*N,N′*-bis(3-*tert*-butyl-5-(4-pyridyl)salicylidene) $Mn^{III}Cl$), but the expanded frameworks are doubly interpenetrated. In 2007, Chun and Moon used the same strategy to derive a pillared *kgm*-MOF (Kagome lattice topology) with ~1.5 nm channels and a surface area of ~2400 $m^2/g$.[130]

Our group, among others, was interested in the potential of polytopic hetero-functional ligands to construct MOFs with topologies of interest (e.g., chiral), and, in 2005, we reported one of the first neutral, non-interpenetrated *srs*-MOFs (i.e., having a chiral network analogous to the Si net in $SrSi_2$, (10,3)-*a*) from the assembly of a tritopic hetero-functional ligand, 3,5-pyridinedicarboxylate (3,5-PDC), and $Cu^{2+}$ single-metal ions.[131] In 2007, Schröder and coworkers used a ligand related to 3,5-PDC, pyridine-3,5-bis(phenyl-4-carboxylic acid) ($H_2PBPC$, Figure 2.9a), to construct

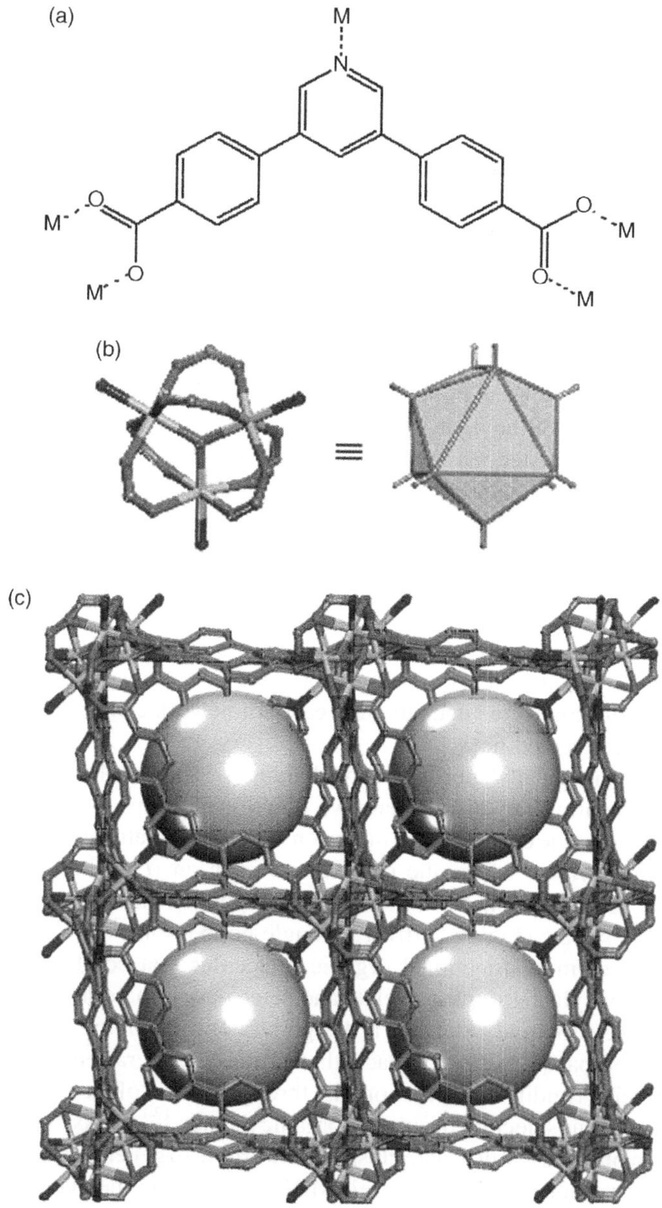


**FIGURE 2.9** (a) The hetero-functional ligand pyridine-3,5-bis(phenyl-4-carboxylate) coordinates to (b) novel 9-connected metal-(pyridine-carboxylate) clusters, generated *in situ*, that can be viewed as triaugmented trigonal prismatic building blocks in the formation of (c) (3,9)-connected MOFs. Hydrogen atoms and solvent molecules have been omitted for clarity; M = black, C = gray, N = dark gray, and O = light gray. The spheres represent the largest sphere that would fit in the cavities without touching the van der Waals atoms of the framework.

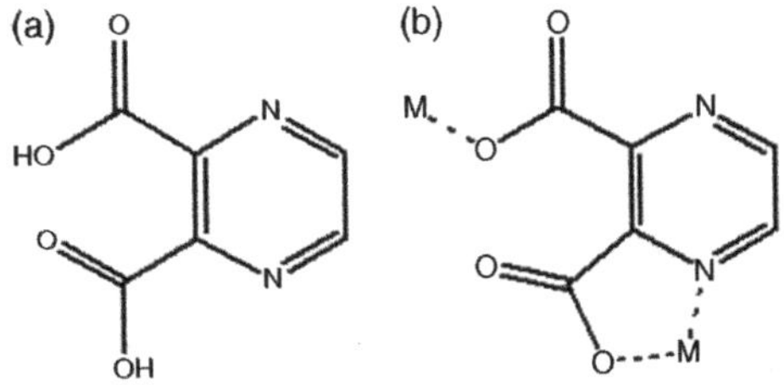


**FIGURE 2.10** (a) 2,3-pyrazinedicarboxylic acid and (b) one potential coordination mode of 2,3-pyrazinedicarboxylate.

two highly porous MOFs with surface areas up to 1553 $m^2/g$.[132] In this case, the carboxylate moieties of the ligand permit the generation of the already-known basic chromium acetate cluster, an oxo-centered trinuclear metal-carboxylate cluster, typically considered a 6-connected trigonal prismatic building block. However, the pyridine moiety of the ligand serves to coordinate the exterior axial position of the metal ions, typically occupied by a terminal ligand like water, to give a novel 9-connected triaugmented trigonal prismatic building block (Figure 2.9b). Thus, the tritopic ligand serves as a 3-connected trigonal building block and combines with the metal-(pyridine-carboxylate) cluster to give MOFs (Figure 2.9c), $[Ni_3(OH)(PBPC)_3]_n \cdot n$(solvent) and $[Fe_3(O)(PBPC)_3]_n \cdot n$(solvent), based on an unprecedented (3,9)-connected net, although each framework can also be interpreted as based on a 12-connected net.

Another advantage of hetero-functional ligands is the ability to readily target ligands with chelating moieties, where two functional groups are within a proximal distance that allows the chelation of a metal ion. This coordination mode is prevalent in (*N*-donor)-carboxylates where the carboxylate is in an $\alpha$ position relative to the aromatic nitrogen, as seen by the Kitagawa's group in the coordination of 2,3-pyrazinedicarboxylate (Figure 2.10) with single-copper ions.[125,133]

This ability of multi-functional organic molecules that combine both nitrogen donors and carboxylate groups to chelate and hetero-coordinate single-metal ions led our group to develop and establish a design strategy involving an unprecedented single-metal-ion-based MBB approach that promotes the rational construction of MOMs by forcing rigidity and directionality through control of the metal coordination sphere and judicious selection of suitable organic ligands.[134,135] As mentioned earlier, the utilization of metal-carboxylate clusters is one route to generate stability and target structures, but methods that generate rigid MBBs from single-metal ions have not been realized. In addition, the utilization of rigid single-metal-ion-based MBBs allows for the targeting of MOMs based on non-default nets, whereas the use of single-metal ions and simple organic linkers as building blocks has consistently led to the construction of default structures, depending on the shape(s) of the building blocks,[40] typically due to the flexibility and lability of the (single-metal)-ligand coordination.[95]

This single-metal-ion-based MBB approach involves prior judicious selection of these would-be ligands based on the quantity and relative position of the functional groups on the molecule depending on the desired angle and target structure. The single-metal ion geometry and coordination sphere are also key in targeting structures, and there are two basic methods: saturation of the coordination sphere with the bridging ligand, which prevents coordination of solvent or template molecules, or the use of solvent or template molecules to "cap" positions of the coordination sphere (as previously discussed and shown in Figure 2.2). In either case, only the bridging ligand serves as a linker in the formation of the net.[134,135]

In order for the hetero-functional organic ligand to saturate the coordination sphere of the single-metal ion and generate a rigid and directional MBB, certain characteristics are favored. First, the nitrogen atoms must be part of an aromatic ring, and at least one nitrogen atom must have a carboxylic acid located one carbon away. This α-position allows for *N*-, *O*- hetero-chelation to the metal ion. As part of the aromatic ring, the nitrogen atoms direct the framework topology, while the carboxylate groups secure the geometry of the metal by locking it into its position through the formation of rigid five-membered rings.[134,135]

In addition to having one hetero-chelating moiety, the organic molecule must also be polytopic, possessing at least one ancillary coordinating moiety, that is, an additional carboxylate, nitrogen atom, or hetero-chelate (Figure 2.11). The ability of a ligand to saturate the coordination sphere of the selected metal ion is also ideal, as this precludes coordination of any solvent, template, or other guest molecules,[134,135] which could lead to undesirable architectures like the metal-organic chains,[133] and allows directionality to be induced entirely dependent on the organic linker.

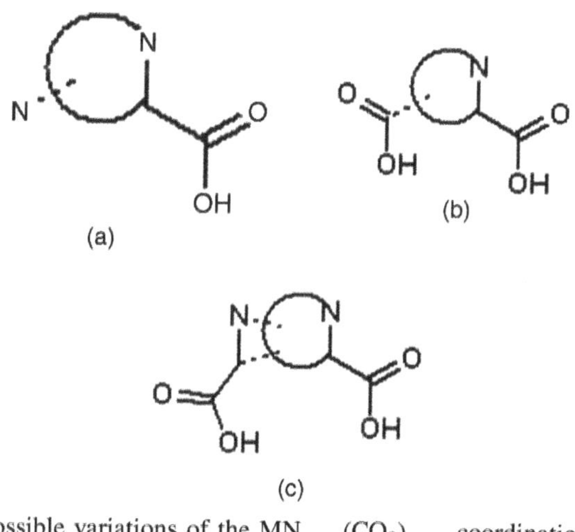


**FIGURE 2.11** Possible variations of the $MN_{x+y}(CO_2)_{x+z}$ coordination, where M is any metal with coordination of 6–8, $x$ = number of *N*-, *O*- hetero-chelating moieties, $y$ = number of ancillary nitrogen-donors, and $z$ = number of ancillary carboxylic acids: (a) $x=1, y=1, z=0$; (b) $x=1, y=0, z=1$; (c) $x=2, y=z=0$. The ancillary group position depends on the desired angle and target structure.

The use of potentially 6–8-coordinate metal ions allows for the targeting of numerous structures, depending on the ligand shape and multiplicity of functionalities; this gives $MN_{x+y}(CO_2)_{x+z}$, where M is any metal with coordination of 6–8, $x=$ number of *N*-, *O*- hetero-chelating moieties, $y=$ number of ancillary nitrogen-donors, and $z=$ number of ancillary carboxylic acids. Each hetero-coordinated single-metal ion can be rendered rigid and directional while simultaneously its coordination sphere is saturated *via* the coordination of the hetero-chelate and/or secondary functionalities of the ligands. The relative position of the ancillary functional group on the organic molecule dictates the directionality and plays a vital role in providing the angles necessary to target particular structures.[134,135]

This single-metal-ion-based MBB design strategy has proven effective in synthesizing a variety of MOMs, including discrete MOPs. Our first example was published in 2004, which was an $M_8L_{12}$ metal-organic cube (MOC or *cub*-MOP), $[Ni_8(HImDC)_{12}]^{8-}\cdot[(H_2TMDP)_4]^{8+}(DMF)_4(EtOH)_4(H_2O)_6$, synthesized from the solvothermal reaction of $Ni^{2+}$ and 4,5-imidazoledicarboxylic acid ($H_3ImDC$) in *N,N′*-dimethylformamide (DMF) and ethanol (EtOH).[136] The angular ditopic hetero-functional ligand possesses concurrently the initial *N*-,*O*-hetero-chelating moiety and an additional *N*-,*O*-hetero-chelate. Three $HImDC^{2'''}$ anions coordinate each $Ni^{2+}$ cation in a facial (*fac*) manner to saturate the Ni(II) octahedral coordination sphere (Figure 2.12). Each ligand chelates two metal ions at a large angle, which allows the ligand to function as the edges of the cube while the *fac*-$NiN_3(CO_2)_3$ MBB occupies the vertices ($NiN_3$ vertex).

The validity and versatility of this approach was supported by our rational synthesis of targeted MOMs in 2005 using the angular ditopic hetero-functional ligand 2,5-pyridinedicarboxylic acid (2,5-$H_2PDC$),[137] which possesses concurrently the initial *N*-,*O*- hetero-chelating moiety and an additional carboxylic acid in the five position. The angularity of this ligand can then be used to target two 4-connected supramolecular isomers, discrete $M_6L_{12}$ metal-organic octahedra (MOOs or *oct*-MOPs) or 2D MOFs with Kagomé lattice topology (*kgm*), depending on the orientation of the ligands upon their coordination to the single-metal ion, in this case $In^{3+}$ (Figure 2.13). The saturation of the coordination sphere by four 2,5-PDC anions generates the necessary 4-connected nodes ($MN_2(CO_2)_4$ MBB, $MN_2(CO_2)_2$ vertex), and the overall anionic charge allows the use of cationic SDAs to direct the formation of one supramolecular isomer over the other. Reaction between $In^{3+}$ and the hetero-

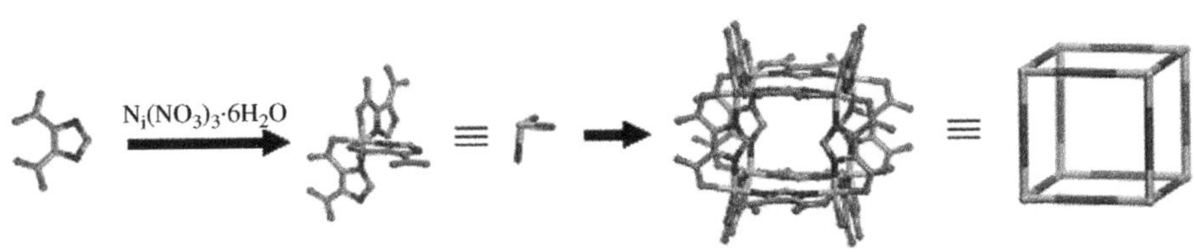


**FIGURE 2.12** The $Ni_8(HImDC)_{12}$ MOC is constructed from 12 HImDC ligands and eight octahedral single-nickel ion MBBs. Hydrogen atoms and solvent molecules have been omitted for clarity; M = black, C = gray, N = dark gray, and O = light gray.

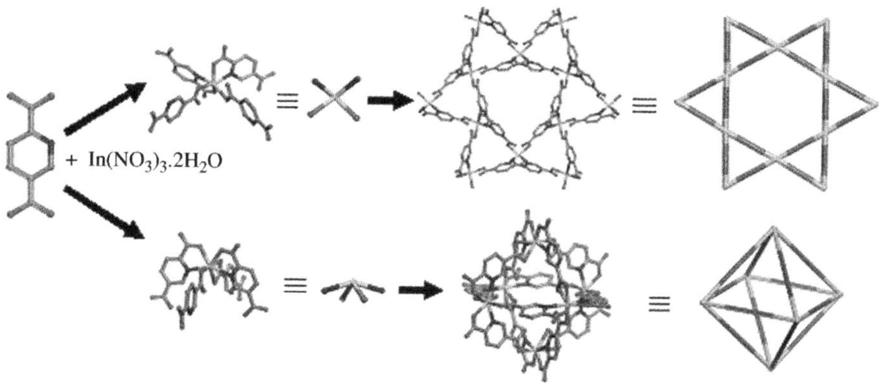


**FIGURE 2.13** 2,5-$H_2$PDC and $In(NO_3)_3{\cdot}2H_2O$ were reacted in the presence of different SDAs to give a *kgm*-MOF or *oct*-MOPs. Hydrogen atoms and solvent molecules have been omitted for clarity; M = black, C = gray, N = dark gray, and O = light gray.

functional ligand in the presence of 4,4′-trimethylenedipiperidine (TMDP) results in the *in situ* formation of the $[In(2,5\text{-}PDC)_2]_n^{-}{\cdot}n[(HTMDP)^{+}(EtOH)(H_2O)]$ *kgm*-MOF, whereas the presence of 1,2-diaminocyclohexane (DACH) produces the $M_6L_{12}[In_6(2,5\text{-}PDC)_{12}]^{6+}{\cdot}[(H_2DACH)_2]^{4+}[(H_3O)_2]^{2+}(DMF)_5(EtOH)_5$*oct*-MOP.

### 2.2.5 Recent Trends

***2.2.5.1 Polytopic Multidentate Nitrogen-Donor Ligands*** Researchers are continually searching for new materials and strategies to design and synthesize functional MOFs. One class of ligands that only recently has been utilized in MOFs is based on a particular subset of azole functional groups.[85] The azoles consist of a five-membered heterocyclic ring of mixed carbon (1–3) and nitrogen composition (2–4) with at least one nitrogen protonated, and can be extended at any position. Of particular interest are the azoles that contains a dinitrogen moiety (i.e., directly linked nitrogen atoms) available for coordination, the $N_2$-azole group ($N_2$-azR), which contains pyrazoles ($N_2C_3R$), triazoles ($N_3C_2R$), and tetrazoles ($N_4CR$), respectively (Figure 2.14a–c); imidazoles are another azole group, but the nitrogen atoms are separate. The dinitrogen moiety is desirable because it is capable of bis-monodentate coordination to metal ions, $M_2(N_2\text{-azR})$, akin to carboxylates (Figure 2.14d). As in the case of carboxylates, azolates can coordinate multiple metals to form clusters,[85] which, as discussed in the previous section, are of great importance as potential geometric building blocks in the construction of MOFs. MOFs assembled from polytopic azole ligands have thus far been serendipitous, although several metal-azolate clusters of varying geometry have been observed, generated *in situ*.[85]

There are several examples of MOMs containing trigonal building blocks. The use of pyrazoles typically has resulted in trinuclear clusters where three pyrazolates coordinate three single-metal ions in a planar fashion (Figure 2.15a), $M_3(N_2C_3R)_3$.[138,139] Tetrazoles can also produce trigonal clusters, but the only example

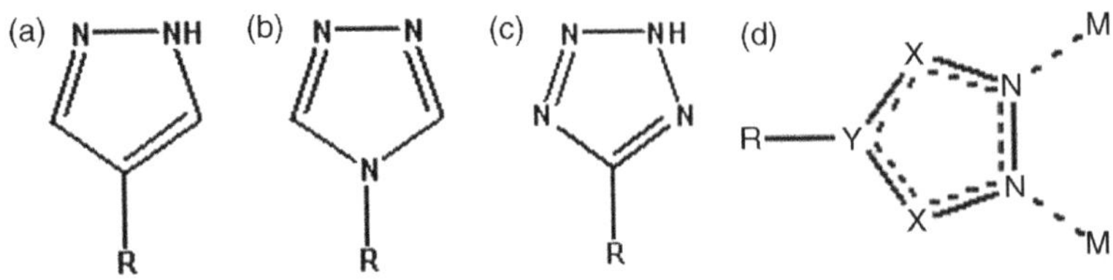


**FIGURE 2.14** $N_2$-azoles: (a) 1,2-pyrazol-4-yl, (b) 1,2,4-triazol-4-yl, (c) tetrazol-4-yl. (d) One potential coordination mode of $N_2$-azolates; X and Y = N and/or C, respectively.

in the literature consists of a dinuclear cluster where three tetrazolates coordinate two single-metals in a paddlewheel-like fashion (Figure 2.15a), $M_2(N_4CR)_3L_6$; the additional coordination sites on metals are occupied by terminal ligands.[140] Probably the most prominent examples of metal-azolate MOFs are based on the discovery by Long and coworkers of a tritopic tetrazolate ligand, 1,3,5-benzenetristetrazolate (BTT, Figure 2.15c), to form 8-connected tetranuclear cube-like clusters, $M_4Cl(N_4CR)_8L_4$ (Figure 2.15b), at each tetrazolate moiety in the presence of $MnCl_2$ or $CuCl_2$, resulting in very open MOFs, $[Mn(DMF)_6]_3[(Mn_4Cl)_3(BTT)_8(H_2O)_{12}]_2 \cdot 42DMF \cdot 11H_2O \cdot 20CH_3OH$ (Mn-BTT) or $H[Cu(DMF)_6][(Cu_4Cl)_3(BTT)_8(H_2O)_{12}] \cdot 3.5HCl \cdot 12\,H_2O \cdot 16CH_3OH$(Cu-BTT) (Figure 2.15d), each with a (3,8)-connected net resembling the structure of sodalite and BET surface areas of 2100 and 1710 $m^2/g$, respectively.[141,142] In 2008, they utilized related extended tetrazolate linkers, 1,3,5-tri-*p*-(tetrazol-5-yl) phenylbenzene ($H_3$TPB-3tz) or 2,4,6-tri-*p*-(tetrazol-5-yl)phenyl-*s*-triazine ($H_3$TPT-3tz), to give analogous expanded open MOFs with, $Cu_3[(Cu_4Cl)_3(TPB\text{-}3tz)_8]_2 \cdot 11CuCl_2 \cdot 8H_2O \cdot 120$ DMF, or without, $Mn_3[(Mn_4Cl)_3(TPT\text{-}3tz)_8]_2 \cdot 25H_2O \cdot 15CH_3OH \cdot 95DMF$ and $Cu_3[(Cu_4Cl)_3(TPT\text{-}3tz)_8]_2x \cdot$ solvent, interpenetration and Langmuir surface areas up to 1700 $m^2/g$.[143]

In addition to purely azolate-based ligands, organic chemistry can provide a range of hetero-functional azolate-based ligands, for example, pyridine-tetrazolates,[144–147] carboxylate-tetrazolates,[148–150] or imidazole-tetrazolates[151] (Figure 2.16). Many of these hetero-functional ligands have been utilized to construct MOMs. However, the majority of structures are still synthesized serendipitously (i.e., limited design strategies).

*2.2.5.2 Supermolecular Building Blocks (SBBs)* The MBB approach has proven to be a successful strategy in the design and construction of solid-state materials. As explained previously, building blocks of specific geometry (e.g., triangles or squares) can be used to target default structures based on these individual shapes[43] (e.g., unitary nets, like $SrSi_2$ or NbO) or combinations thereof (e.g., binary nets, like PtS based on triangles and quadrangles; or ternary nets,[152] based on three types of building blocks). However, it is still proving difficult to design MBBs with high coordination.[40,153] One route toward achieving this goal is the utilization of supermolecular building blocks (SBBs),[154,155] a term given to describe discrete MOPs based on Platonic, Archimedean,[156] or related (e.g., Johnson[157]) solids that can be extended into 3D.

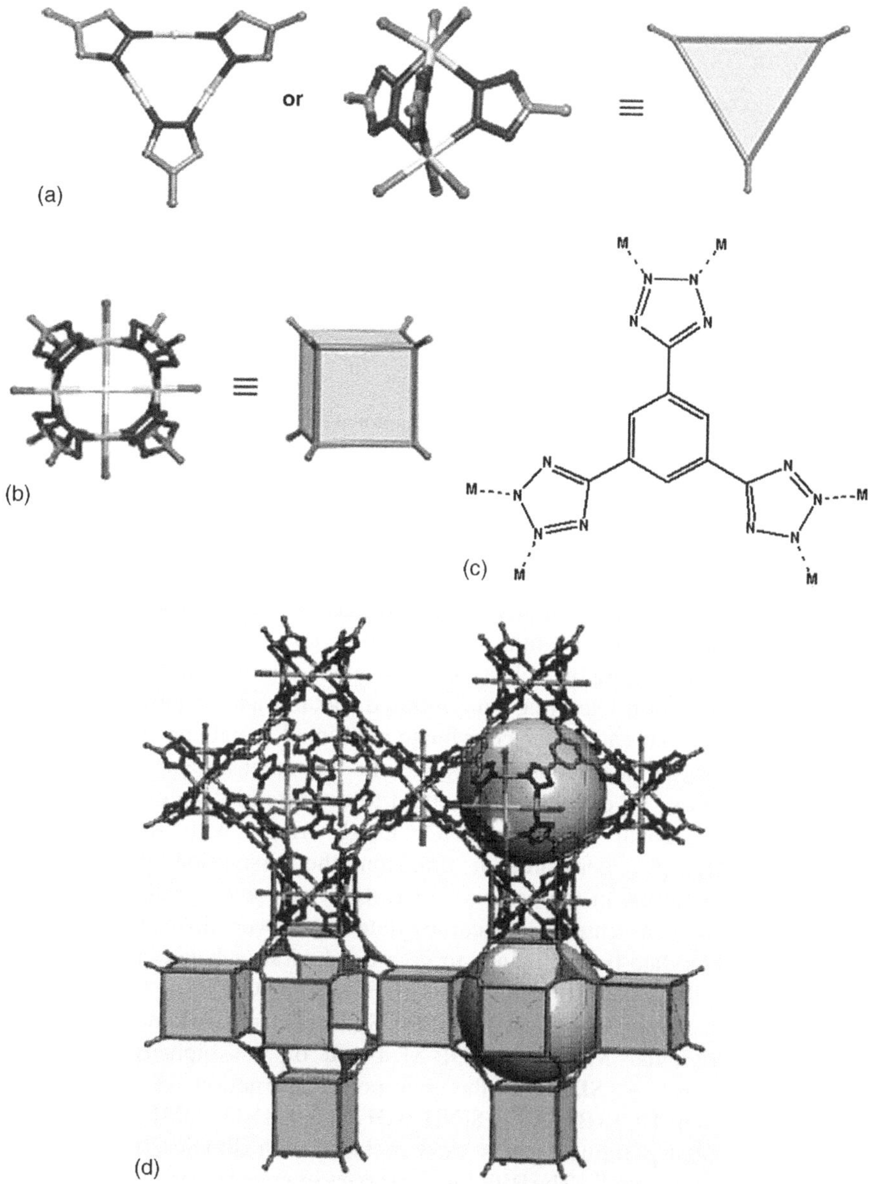


**FIGURE 2.15** (a) A $M_2(N_4CR)_3L_6$ dinuclear paddlewheel-like cluster, (b) a $M_4Cl(N_4CR)_8L_4$ tetranuclear cube-like cluster, which forms during the reaction between $MnCl_2$ or $CuCl_2$ and (c) BTT to give (d) a novel 3D MOF. Hydrogen atoms and solvent molecules have been omitted for clarity; M = black, C = gray, N = dark gray, Cl = white, and O = light gray. The spheres represent the largest sphere that would fit in the cavities without touching the van der Waals atoms of the framework.

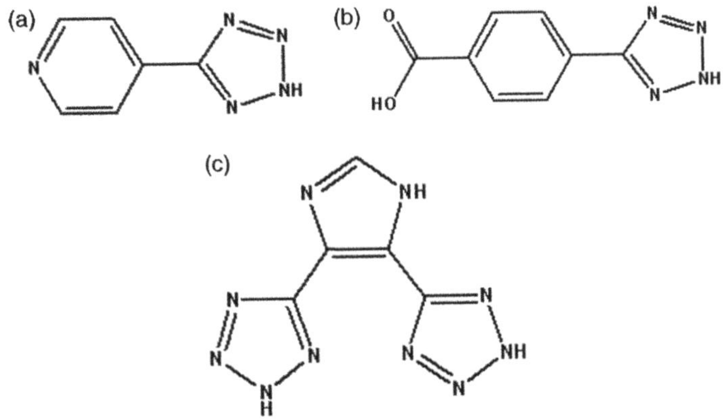


**FIGURE 2.16** Examples of various hetero-functional organic molecules that have been utilized as ligands in MOMs: (a) 4(4-tetrazolyl)pyridine, (b) 4-(4-tetrazolyl)benzoic acid, and (c) 4,5-bis(4-tetrazolyl)imidazole.

Essentially, a broad range of discrete MOPs have been synthesized over the years,[136,137,158–160] and these MOPs can be targeted as SBBs through functionalization of their exterior with potential bridging moieties,[161,162] which also has been achieved with some inorganic clusters.[163] In other words, the MOP must have peripheral coordination sites (e.g., unsaturated metals) and/or peripheral organic functional groups that can either coordinate additional metals not involved in the assembly of the MOP or serve as hydrogen bond donors or acceptors.

In 2005, Sudik et al. developed a method to consistently cap one trigonal face of the typical trigonal prismatic $M_3O(CO_2)_6L$ trimer with sulfates to give $Fe_3O(CO_2)_3(SO_4)_3(py)$, generated *in situ* from the dissolution of $Fe_2(SO_4)_3$ in DMF and triethylamine in the presence of carboxylates and pyridine (py).[160] The sulfate-capped oxygen-centered iron-carboxylate trimers were linked by simple linear or trigonal carboxylate-based ligands to give tetrahedral or heterocuboidal MOPs, respectively, of the general formula $[NH_2(CH_3)_2]_8[Fe_{12}O_4(SO_4)_{12}(link)_x(py)_{12}]\cdot G$ ($x = 4$ or 6, respectively; G = guests). In 2006, Sudik et al. utilized one of the series of tetrahedral isoreticular MOPs, IRMOP-51 based on 4,4′-biphenyldicarboxylate (BPDC) link, as an SBB in the one-pot construction of a 3D MOF, $[(Fe_3O)_4(SO_4)_{12}(BPDC)_6(BPE)_6]^{8-}\cdot 8[NH_2(CH_3)_2]^{+}\cdot 13H_2O\cdot 8DMF$ (MOF-500), where the terminal pyridine ligands were replaced with ditopic *cis*-1,2-bis-4-pyridylethane (BPE) linkers.[161] The BPE linkers serve to connect the tetrahedral MOPs in an arrangement analogous to the β-cristobalite topology, resulting in four types of pores (up to 18 Å in diameter) and nearly quadrupling the specific surface area of the parent MOP from 544 to 2274 $m^2/g$ in MOF-500.

As mentioned earlier, there have been a variety of discrete MOPs based on polytopic monodentate *N*-donor ligands, but examples of such MOPs used in the design and synthesis of extended structures is still rare. In 2006, Moon et al. reported the preparation of $M_6L_8$ octahedral MOPs from $C_3$-symmetric triangular facial ligands (i.e., *N,N′,N″*-tris(3-pyridinyl)-1,3,5-benzenetricarboxamide) and $C_4$-symmetric square tetratopic Pd(II) ions at truncated octahedron corners.[164] In 2007, the same group substituted Pd(II) with octahedral hexatopic Cu(II) ions, which allowed for extension at the exterior-oriented axial positions of the Cu(II) ions.[165] The use of copper perchlorate led to discrete MOPs analogous to the Pd(II)-based octahedron, but using copper nitrate transformed the MOPs into octahedral hexatopic SBBs by linking the exterior Cu(II) axial sites of neighboring MOPs through ditopic nitrate anions resulting in the formation of a twofold interpenetrated 3D MOF based on the augmented primitive cubic network.

The prototypical MOP is the small rhombihexahedra (24 vertices) or nanoball (Figure 2.17b) based on the assembly of square paddlewheel $[Cu_2(O_2CR)_4]$ MBBs and the angular (120°) 1,3-BDC ligand (Figure 2.17a).[158,159] Zaworotko and coworkers, among others, have attempted to decorate the exterior of the nanoball by utilizing various five-position functionalized 1,3-BDC derivatives (e.g., 5-methoxy- and 5-sulfo-1,3-BDC), which led to the assembly of 3D frameworks and 1D chains, respectively, through cross-linking of nanoballs as SBBs (Figure 2.17c).[162]

More recently, our group functionalized the 1,3-BDC ligand with a tetrazole moiety, which led to the design and synthesis of a (3,24)-connected MOF with an unprecedented *rht* topology.[155] The hetero-functional ligand simultaneously takes advantage of the ability of the 1,3-BDC moiety to form the nanoball and the tetrazole moiety to form trinuclear trigonal clusters (Figure 2.15a). As a result, the solvothermal reaction of 5-(4-tetrazolyl)isophthalic acid ($H_3TZI$) with $Cu^{2+}$ produces a highly porous *rht*-MOF (Langmuir surface area of 3223 $m^2/g$), based on the assembly of 24-connected nanoball SBBs and 3-connected Cu-tetrazolate MBBs (Figure 2.18), with cavities in the mesoporous range (i.e., the largest cage has a diameter of 23.287 Å).

Another route to utilizing MOPs as SBBs is the use of covalent linkages.[154,166] A MOF having an underlying (3,24)-connected net analogous to our *rht*-MOF has been constructed from a $C_3$ symmetric ligand with three 1,3-BDC moieties, 5,5′,5″-[1,3,5-benzenetriyltris(carbonylimino)]tris-1,3-benzenedicarboxylic acid, and Zn paddle-wheel MBBs, essentially quadruply cross-linking nanoballs; however, this MOF was not stable to loss of solvent.[167] Nanoballs also have been covalently cross-linked into a 3D MOF comparable to the *bcu* (bcc lattice) net by targeting ligands with multiple 1,3-BDC moieties covalently linked through some organic core, for example, 1,3-bis(5-methoxy-1,3-benzenedicarboxylic acid)benzene (Figure 2.19a).[166] Similar tetracarboxylic acids (Figure 2.19b, c) have been utilized to covalently cross-link metal-organic cubohemioctahedra (12 vertices) into 3D MOFs based on the12-connected *fcu* net.[154]

***2.2.5.3 Zeolite-Like Metal-Organic Frameworks (ZMOFs)*** A continual effort in metal-organic chemistry has been the design and synthesis of metal-organic frameworks having networks with topologies and properties akin to those of zeolites, purely inorganic microporous crystalline solids constructed from TBUs sharing corners. An attractive feature of zeolite-like nets is their inability to interpenetrate, a phenomenon prevalent in MOFs having larger apertures that inherently precludes open, accessible

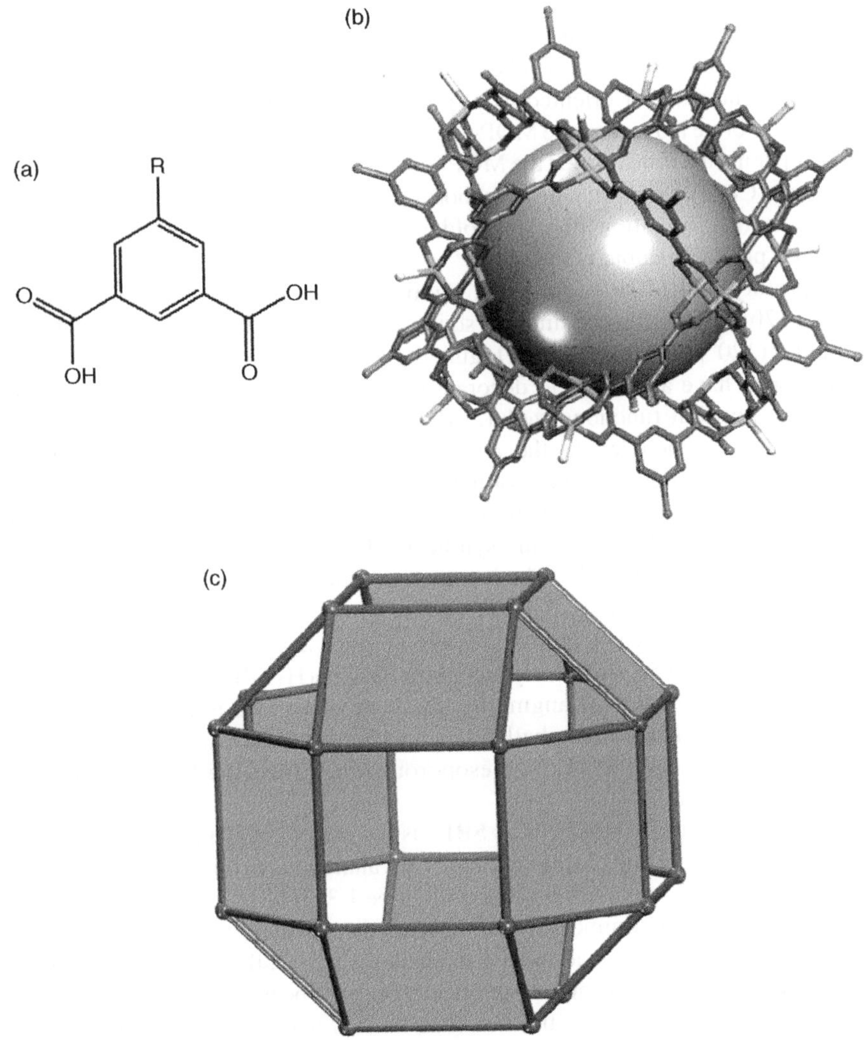


**FIGURE 2.17** (a) 5-R-1,3-$H_2$BDC, where the R group can be a variety of functional groups, can form (b) nanoballs from metal paddlewheel clusters, which can be extended through the exterior open metal sites (yellow) or ligand five-position functional groups (orange) to serve as (c) small rhombihexahedra SBBs. Hydrogen atoms have been omitted for clarity; M = black, C = gray, N = dark gray, and O = light gray. The sphere represents the largest sphere in the interior of the nanoball.

frameworks. The zeolite-like nets can be decorated and/or expanded using several strategies: replacing the single-metal ion tetrahedral vertices with larger tetrahedral clusters (i.e., supertetrahedra), which is well known in inorganic chemistry[168–170] decorating the net, and/or expanding the edges with a longer linker.

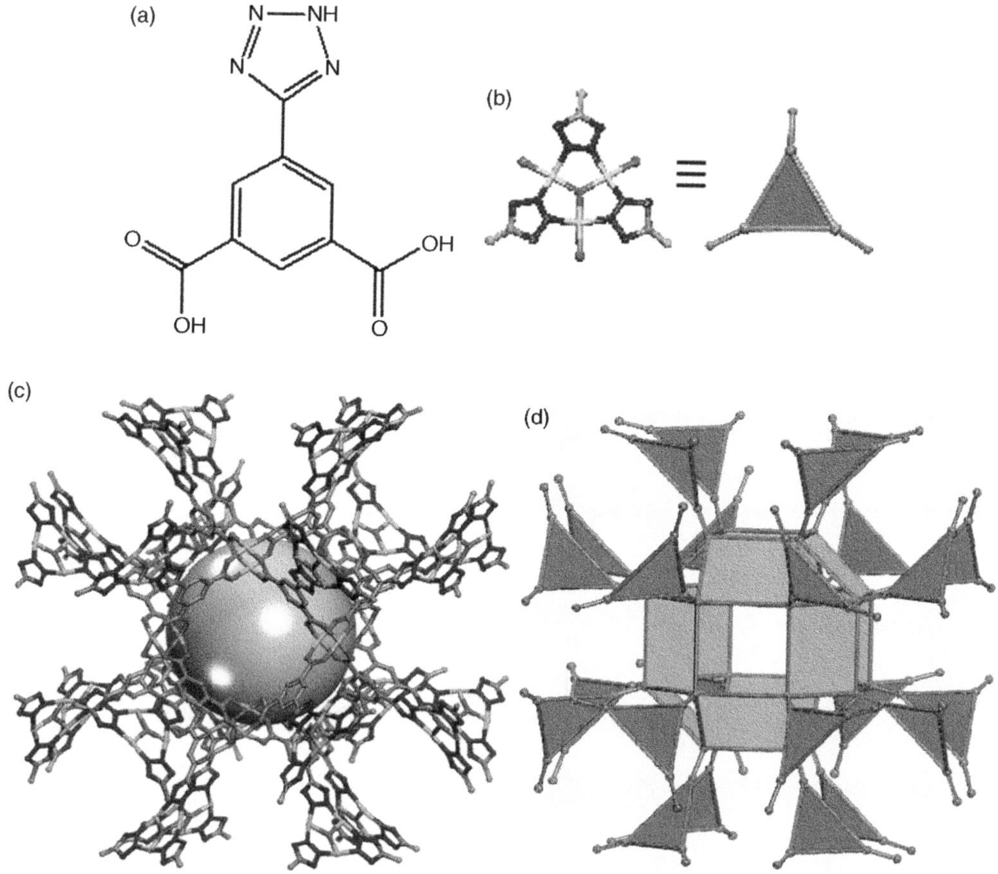


**FIGURE 2.18** (a) $H_3TZI$ can form (b) trigonal trinuclear copper MBBs at the tetrazolate moieties and (c) nanoballs at the carboxylate moieties (hydrogen atoms have been omitted for clarity; M = black, C = gray, N = dark gray, O = light gray, extension point = white) to give (d) a 24-connected SBB that connects 24 trigonal MBBs in a novel MOF based on (e) the (3,24)-connected *rht* net.

The first strategy is exemplified through the previously described MIL-100 and MIL-101, where four trinuclear $Cr_3O(O_2CR)_6L_3$ clusters, arranged in a tetrahedral geometry by BDC or BTC ligands, respectively, serve as supertetrahedral vertices that share corners to give the corresponding zeolite MTN-like structures.[109,112] Nevertheless, MOFs based on the decoration and/or expansion of the cubic diamond topology, which is known to readily interpenetrate, are extensively predominant for the assembly of TBUs. Several other MOFs have been described as zeolitic or having zeolite-like topologies, but lack true zeolite-like nets. The Long group's Mn- and Cu-BTT MOFs resemble the network of zeolite SOD (sodalite), where BTT ligands serve as panels of alternate six-membered ring faces of β cages;[141,142] however, the

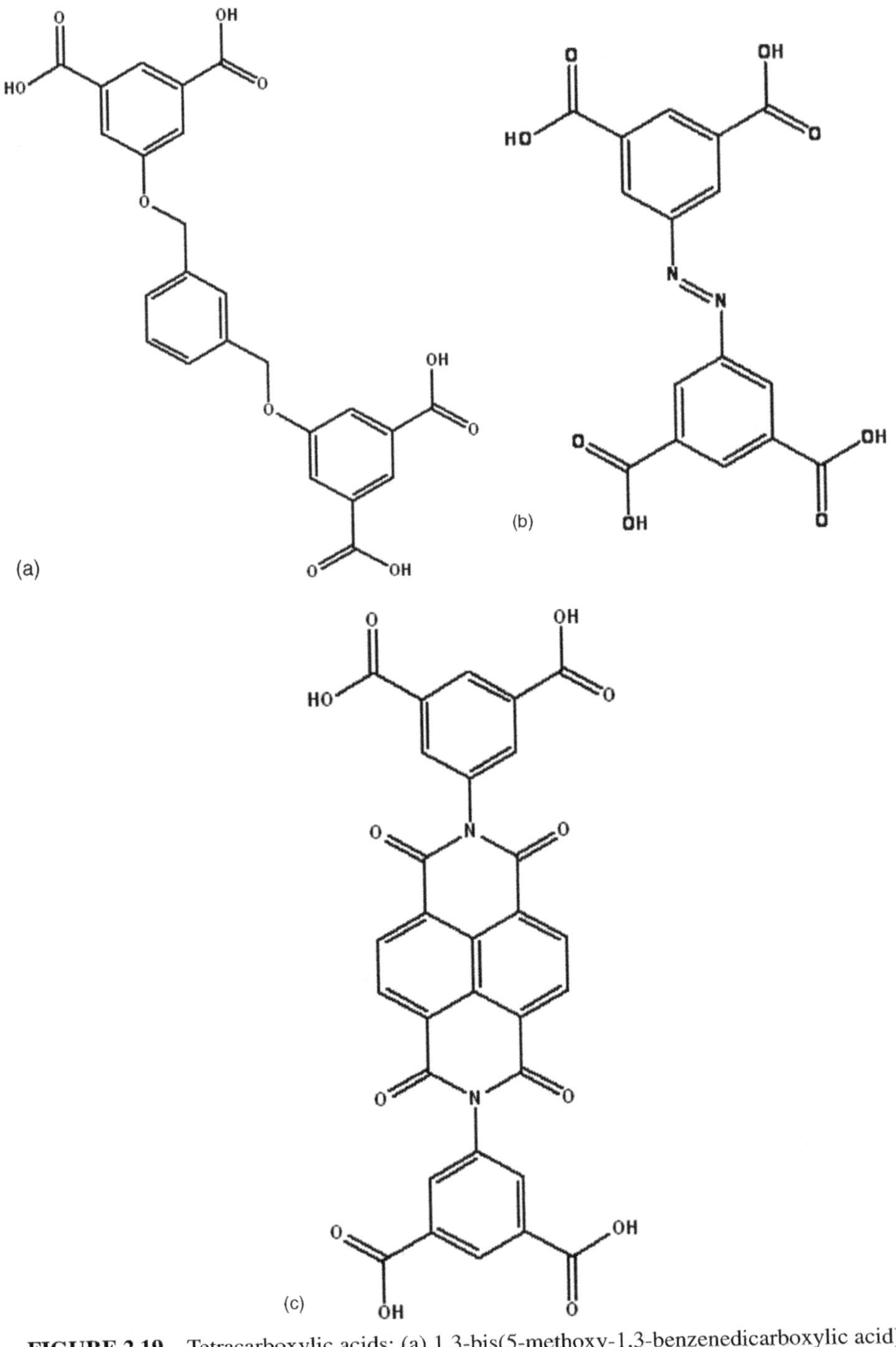


**FIGURE 2.19** Tetracarboxylic acids: (a) 1,3-bis(5-methoxy-1,3-benzenedicarboxylic acid) benzene, (b) 5,5′-azobis-1,3-benzenedicarboxylic acid ($H_4ABBDC$), and (c) $H_4BIPA$-TC.

structure does not consist of tetrahedral vertices, and their recent report of the catenation of analogous expanded sodalite-type MOFs suggests that the underlying networks are not, in fact, zeolite-like.[143]

In the edge expansion strategy, essentially, the tetrahedral-based zeolite-like nets[1] are expanded by lengthening the edges of the net with the angular organic linker,[171,172] similar to the previously mentioned enlargement of metal-cyanide compounds by using an extended organic linker like 4,4′-Bipy. In effect, the organic ligand serves as a functionalized multi-atom anionic substitute for the oxide ions ($O^{2-}$) that bridge the tetrahedral (T) silicon and/or aluminum cations in inorganic zeolites.[1] However, the organic ligand (L) also must provide an M-L-M angle comparable to the average T-O-T angle (~145°) found in typical zeolites.[1] Nevertheless, unless rendered rigid, the ligand-induced angularity can still permit the formation of the cubic diamond topology, as in the case of labile M–N coordination bonds,[84] and most attempts to construct MOFs with zeolite-like topologies based on the assembly of simple tetrahedral building blocks or units (TBUs) have dominantly led to the formation of structures with the default topology.[159,173]

Our research group applied the single-metal-ion-based MBB approach to the expansion and/or decoration of tetrahedral-based zeolite-like nets,[174,175] where the MBBs result in rigid and directional TBUs in order to preclude the construction of the default network. In this case, however, the hetero-functional ligand must be judiciously selected to contain two functionalities (at least one *N*-, *O*- hetero-chelate) that can provide the desired M-L-M angle, such as imidazole-[174] or pyrimidine-carboxylates.[176] In addition, higher coordination (6–8) metal ions are utilized instead of tetrahedral ions, so the ligands must coordinate to the metal in a manner that gives an MBB that can act as a TBU. Thus, for six-coordinate metals there must be an $MN_4(CO_2)_2$ or $MN_2(CO_2)_4$ coordination environment and $MN_4(CO_2)_4$ or $MN_2(CO_2)_6$ MBBs for eight-coordinate metals that all result in $MN_4$ or $MN_2(CO_2)_2$ TBUs, respectively.

According to these criteria, the aforementioned $H_3ImDC$ molecule was well suited to target MOFs with zeolite-like topologies, since it concurrently possesses two *N*-,*O*-hetero-chelating moieties with a potential M-L-M angle of 145° (directed by the M–N coordination) (Figure 2.20a).[174] In addition, if four $HImDC^{2'''}$ ligands saturate each single-metal ion coordination sphere (divalent or trivalent), an anionic zeolite-like MOF (ZMOF) will be produced. As in the 2,5-PDC-based supramolecular isomers mentioned previously, the anionic nature allows for the utilization of cationic SDAs, as well as exploration of applications akin to traditional zeolites.

Reaction between $In^{3+}$ and $H_3ImDC$ in the presence of different SDAs does, in fact, yield different ZMOFs.[174] Specifically, imidazole (HIm) leads to a *sod*-ZMOF, $\{[In(HImDC)_2]^{1-}\}_n{\cdot}n\{[(H'HIm)]^{1+}(DMF)_4(CH_3CN)(H_2O)_4\}$ (Figure 2.20c), and 1,3,4,6,7,8-hexahydro-2*H*-pyrimido[1,2-*a*]pyrimidine (HPP) yields a *rho*-ZMOF, $\{[In_{48}(HImDC)_{96}]^{48-}\}_n{\cdot}n\{[(H_2'HPP)_{24}]^{48+}(DMF)_{36}(H_2O)_{192}\}$ (Figure 2.21b), which are up to eight times larger than their inorganic analogues, known as SOD (sodalite) and RHO, respectively, in the Database of Zeolite Structures[1] (*sod* and *rho* by O'Keeffe[34]).

In the In-ImDC *sod*-ZMOF, each $In^{3+}$ ion is hetero-chelated by two $HImDC^{2-}$ ligands and coordinated by the ancillary nitrogen-donor from two other $HImDC^{2-}$

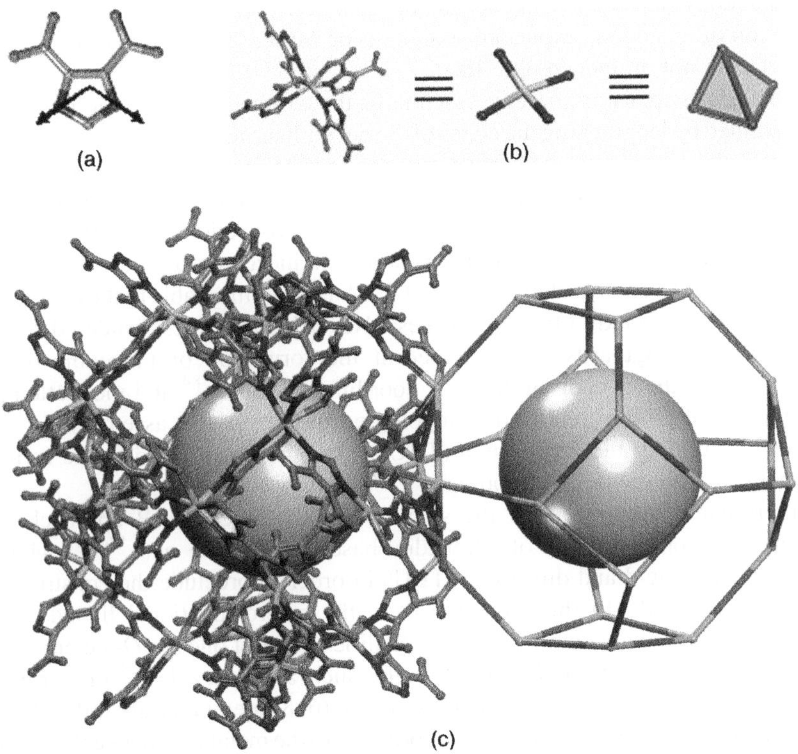


**FIGURE 2.20** (a) The angle in the ImDC ligand. (b) In-ImDC *sod*-ZMOF is composed of six-coordinate MBBs, which also can be viewed as 4-connected TBUs. (c) A fragment of the *sod*-ZMOF single crystal structure, where the spheres represent the largest sphere that would fit in the $\beta$-cavities without touching the van der Waals atoms of the framework, and the $\beta$-cages of the sodalite network connect through common four- (shown) and six-member rings. Hydrogen atoms and solvent molecules have been omitted for clarity; In = black, C = gray, O = light gray, N = dark gray.

ligands, resulting in six-coordinate In(III) ions that form the desired $InN_4(CO_2)_2$ MBBs ($InN_4$ TBUs) (Figure 2.20b). In the In-ImDC *rho*-ZMOF, each single-indium ion is eight-coordinate, saturated by the hetero-chelation of four $HImDC^{2-}$ ligands to give the desired $InN_4(CO_2)_4$ MBBs ($InN_4$ TBUs) (Figure 2.21a).[174] This anionic *rho*-ZMOF was the first material ever to contain an organic component and have a zeolite RHO-like topology,[1,34] although a few neutral examples have been synthesized since.[171,172] In addition, the *sod*-ZMOF represents the first example of a MOF with an anionic framework based on the *sod* topology, although a few examples of neutral or cationic sodalite-like MOFs have been synthesized previously.[40]

More recently, the single-metal-ion-based MBB approach to design and synthesize ZMOFs has been supported by the success with hetero-functional pyrimidine-carboxylate ligands.[176] Reaction between $In^{3+}$ and 4,6-pyrimidinedicarboxylate (PmDC,

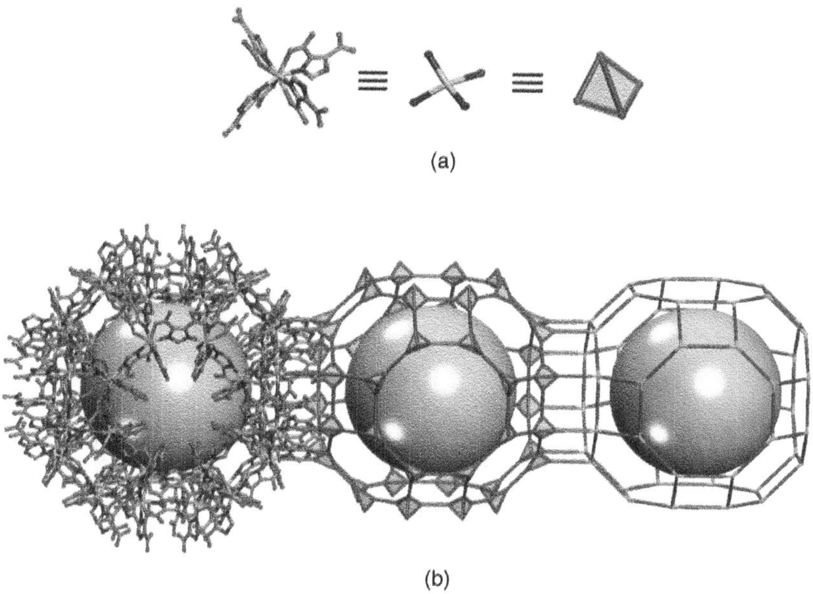

FIGURE 2.21 (a) In-ImDC *rho*-ZMOF is composed of eight-coordinate MBBs, which can be viewed as 4-connected TBUs. (b) A fragment of the *rho*-ZMOF, where the spheres represent the largest sphere that would fit in the $\alpha$-cavities without touching the van der Waals atoms of the framework and the $\alpha$-cages of the *rho* network connect through double eight-member (D8R) rings. Hydrogen atoms and solvent molecules have been omitted for clarity; In = black, C = gray, O = light gray, N = dark gray.

Figure 2.22a) under hydro-solvothermal conditions yields another anionic *sod*-ZMOF, $\{[In(PmDC)_2]^-\}_n \cdot n\{(Na^+)_{0.36}(K^+)_{1.28}[(NO_3)^-]_{0.64}(H_2O)_{2.1}\}$ (Figure 2.22c), from $InN_4(CO_2)_4$ MBBs ($InN_4$ TBUs) (Figure 2.22b). Reaction between 2-cyanopyrimidine (2-pyrimidinecarboxylate (PmC, Figure 2.22d) generated *in situ*) and $Cd^{2+}$ in the presence of piperazine (Pip) under hydro-solvothermal conditions produces another *rho*-ZMOF, $[Cd(PmC)_2]_n \cdot n[(Pip)_{0.35}(H_2O)_{5.36}]$ (Figure 2.22f), from $CdN_4(CO_2)_4$ MBBs ($CdN_4$ TBUs) (Figure 2.22e).

Since we first reported the design and synthesis of ZMOFs, there has been an upsurge in zeolitic MOFs in the literature. Primarily, nonlinear ditopic nitrogen-donor ligands (e.g., imidazole and benzimidazole derivatives, Figure 2.23) have been targeted with specific built-in angular nature to orient the metal ions at specific angles and target novel structures.[81,177–179] However, with the exception of a few early discoveries by Navarro et al., only very recently have other MOFs based on edge expansion of zeolite-like topologies been obtained.[171,172,179] Regardless, these zeolitic MOFs, typically possess neutral frameworks that negate zeolite-like (anionic) applications and the use of cationic (as in zeolites) or, less common, anionic structure-directing agents (SDAs, potentially key to accessing structures with close energetic stability), and thus require the aid of trial and error for synthesis.[171,172] Nevertheless, this method has produced MOFs with a variety

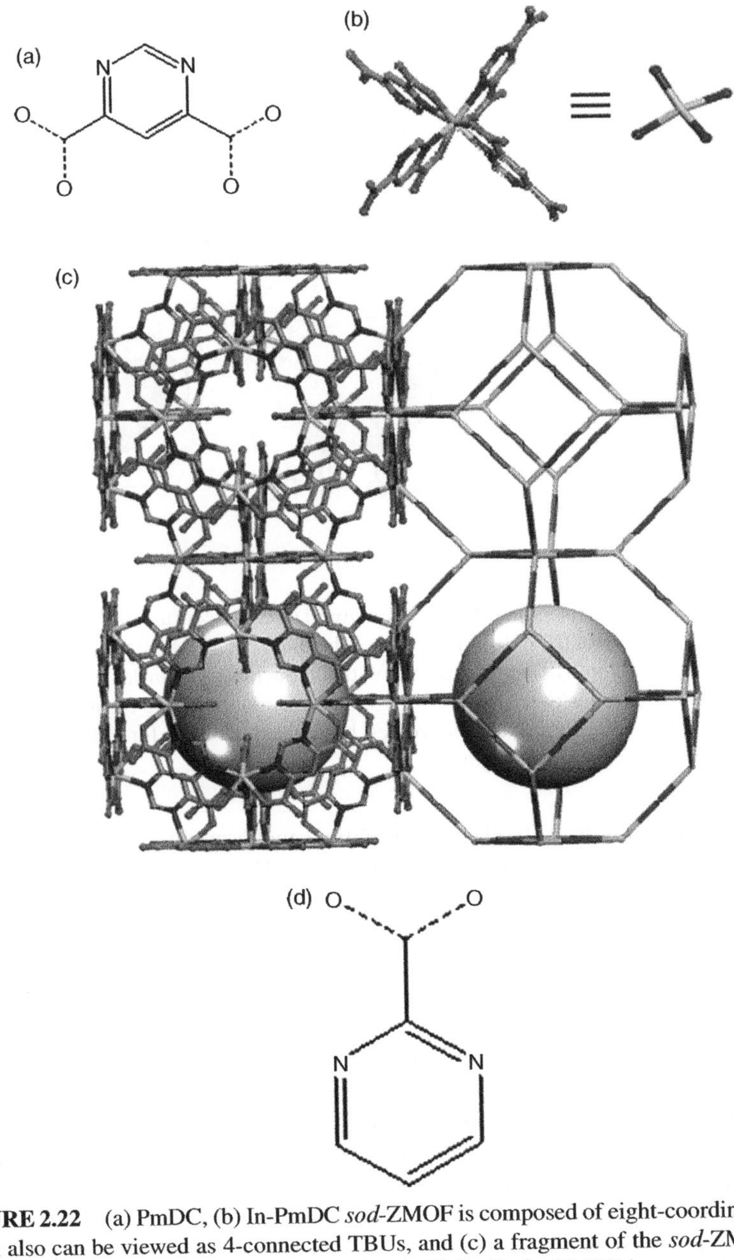


**FIGURE 2.22** (a) PmDC, (b) In-PmDC *sod*-ZMOF is composed of eight-coordinate MBBs, which also can be viewed as 4-connected TBUs, and (c) a fragment of the *sod*-ZMOF single crystal structure and the *sod* net. (d) 2-PmC, (e) Cd-PmC *rho*-ZMOF is composed of eight-coordinate MBBs, which can be viewed as 4-connected TBUs, and (f) a fragment of the *rho*-ZMOF and the *rho* net. The spheres represent the van der Waals spheres; hydrogen atoms have been omitted for clarity; M = black, C = gray, O = light gray, N = dark gray.

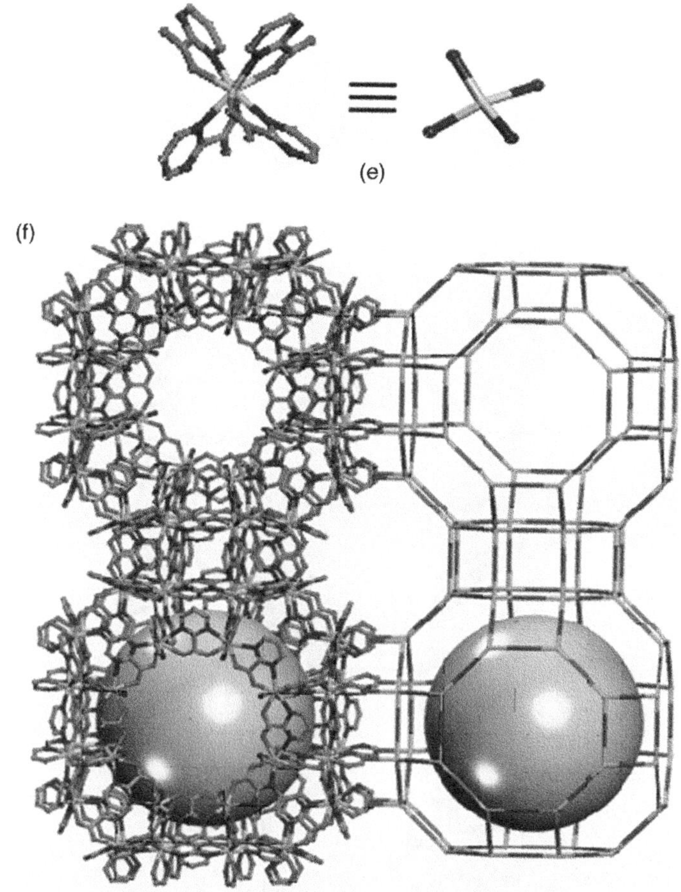


**FIGURE 2.22** (*Continued*)

of zeolite-like topologies, which are exemplified through the vast array of reported zeolitic imidazolate frameworks (ZIFs, where $M^{II}N_4$ replaces the $SiO_4$ tetrahedra of traditional zeolites) with nets related to zeolites ANA, MER, BCT, DFT, GIS, GME, and LTA,[172,180] in addition to RHO and SOD and two novel topologies, *poz* and *moz*, with extra-large cavities (A cage = 25.1 × 14.3 Å and B cage = 30.1 × 20.0 Å, and 35.6 Å, respectively) with Langmuir surface areas up to 1240 $m^2/g$.[181]

### 2.2.6 Applications

The utility of functional materials like zeolites in a variety of applications, from ion exchange to gas storage to separations to catalysis, is highly dependent on the

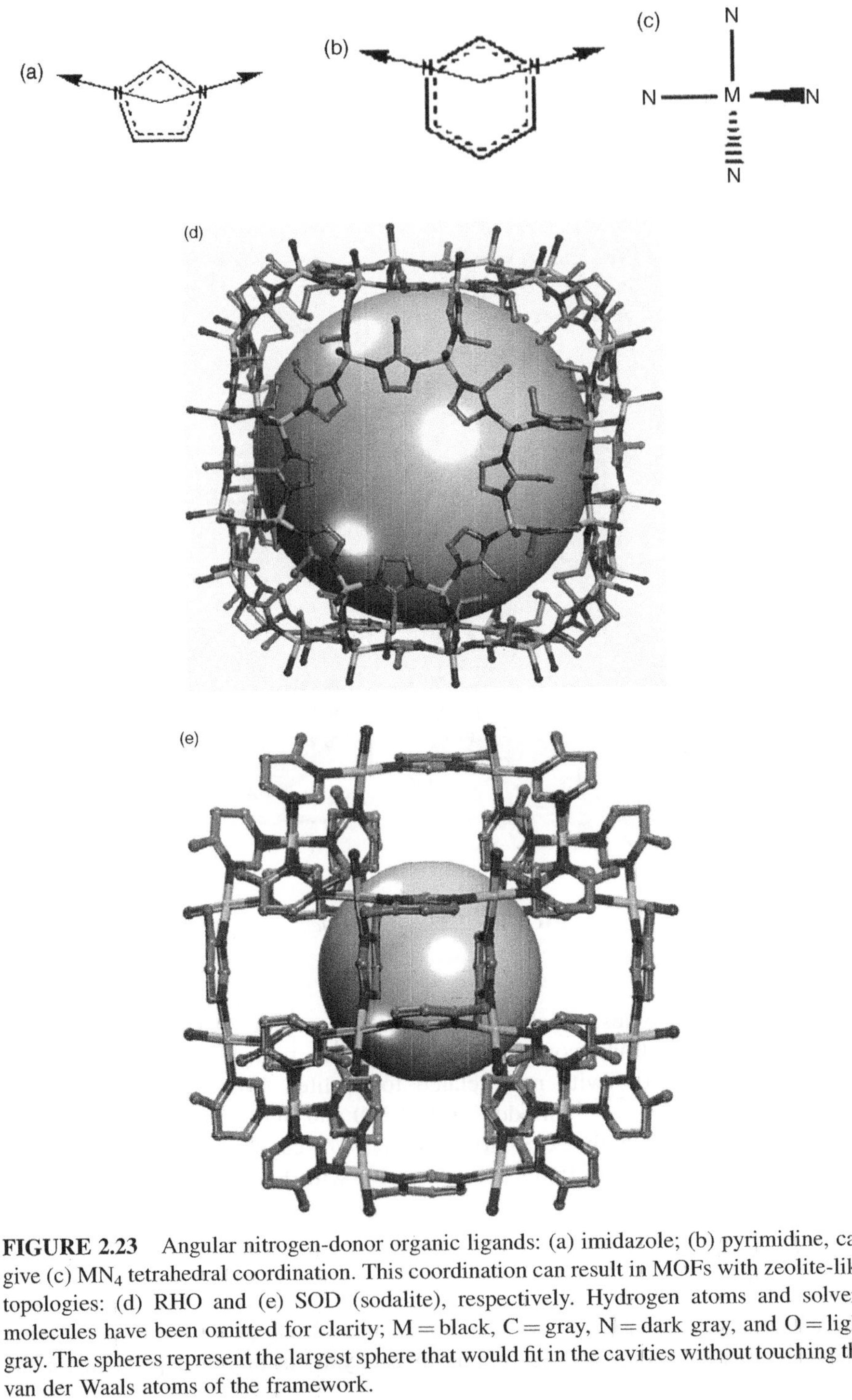


**FIGURE 2.23** Angular nitrogen-donor organic ligands: (a) imidazole; (b) pyrimidine, can give (c) $MN_4$ tetrahedral coordination. This coordination can result in MOFs with zeolite-like topologies: (d) RHO and (e) SOD (sodalite), respectively. Hydrogen atoms and solvent molecules have been omitted for clarity; M = black, C = gray, N = dark gray, and O = light gray. The spheres represent the largest sphere that would fit in the cavities without touching the van der Waals atoms of the framework.

framework's structural features, such as the framework functionality, dimensions of the pore openings, the size of the cavities, and their relative accessibility.[182,183] The restricted pore and cavity size ($\leq$1 nm) of porous zeolites has limited their potential to smaller molecule applications, with few exceptions;[16] in addition, they are limited in functionality[17,18] and atomic composition.[19] The development of MOMs, likely assisted by this limited scope of zeolites and *de facto* the need to produce more functional materials for targeted applications, and the achievement of permanently porous MOFs have opened new avenues to numerous applications that have traditionally utilized the aforementioned inorganic zeolites,[184] from gas storage[82,118] to catalysis[185–188] to drug delivery[189] to sensing,[174] as well as a novel applications.

Some of the earliest metal-cyanide compounds were shown to readily and selectively trap various molecules, although many are not stable upon loss of the guest molecules or heating.[47,190] In 1990, Hoskins and Robson showed, for the first time, the ability of a MOM to undergo ion exchange, specifically the $BF_4^-$ ion that counters the cationic diamondoid framework of $Cu^I$[4,4′,4″,4‴-tetracyanotetraphenylmethane]$BF_4 \cdot xC_6H_5NO_2$ is completely replaced upon exposure to a solution of excess $NBu_4PF_6$ ($PF_6^-$ ions).[49] Kitagawa's group reported that $\{[M_2(4,4'\text{-Bipy})_3(NO_3)_4](H_2O)x\}n$ (M = Co, Ni, Zn; $x = 4, 4, 2$, respectively) adsorbed small gaseous molecules such as $CH_4$, $N_2$, and $O_2$, and gas adsorption has been one of the primary applications of MOFs.[82] In fact, porous MOFs have some of the highest capacities of any porous material for the storage of carbon dioxide[191] and hydrogen.[117]

With the quest for new energy sources for mobile applications, hydrogen is at the forefront, and its safe and efficient storage, with initiatives and targets set by the U.S. Department of Energy,[192] are key concerns.[193–196] MOFs have been studied as potential hydrogen storage candidates,[124] with early studies on MOF-5 (IRMOF-1), IRMOF-6, and IRMOF-8 indicating high hydrogen uptake, from 4.5wt% to double and quadruple the specific $H_2$ uptake, respectively, at room temperature and 10 bar.[197] Typically, the $H_2$ sorption isotherms are determined at 77 and 87 K at 1 atm to determine the potential storage capacity, as well as estimations of the energy of interaction of $H_2$ with MOF structures by calculation of isosteric heats of $H_2$ adsorption using either the Clausius–Clapeyron equation[198] or by a virial-type expression[199] to fit $H_2$ isotherm data at the two temperatures. Increasingly more common are $H_2$ adsorption studies at higher pressures (up to 100 bar)[124] and higher temperatures (up to 298 K, room temperature).[200–204] Some of the top $H_2$ uptake values for MOFs, some higher than reported for any other porous material, are listed in Table 2.1, as well as several of the MOFs discussed earlier in this chapter, for comparison. There are several reviews that discuss this topic more in depth,[20,22,124,200–205] and, according to Férey,[205] there are only three MOFs that, to date, have been satisfactorily verified[20] for hydrogen storage at 77 K: MOF-5, HKUST-1, and MIL-53(Al).

The modular nature of MOFs should permit the substitution of intra-framework metals with lighter metals that may allow for higher hydrogen uptakes.[115,205] However, to date, there are few examples, although Férey has seen some success with the synthesis of an Al versus Cr version of MIL-53.[205] In addition, many open MOFs are ionic, which can allow for exchange of extra-framework ions, and Long has

**TABLE 2.1 Hydrogen Adsorption and Storage in Select MOFs**

| Material | $H_2$ Uptake 77 K, 1 atm (wt%) | $H_2$ Uptake 77 K, >10 atm (wt%) | $H_2$ Uptake 298 K (wt%) | Isosteric Heats of Adsorption, $Q_{st}$, (kJ/mol) |
|---|---|---|---|---|
| IRMOF-1 (MOF-5) | 1.3 | 7.1 (excess, 50 bar) | 0.4 (100 atm) | 3.8 |
| Mn-BTT | n.r. | 2.23 (900 torr)<br>6.9 (90 bar) | 1.5 (90 bar) | 10.1 |
| $Fe_3(O)(PBPC)_3$ | 1.6 | 3.05 (20 bar) | n.r. | n.r. |
| $Ni_3(OH)(PBPC)_3$ | 1.99 | 4.15 (20 bar) | n.r. | n.r. |
| In-PmDC sod-ZMOF | 0.9 | n.r. | n.r. | 8.4 |
| Cd-PmC rho-ZMOF | 1.16 | n.r. | n.r. | 8.7 |
| Cu-TZI *rht*-MOF | 2.4 | n.r. | n.r. | 9.5 |
| COF-1 | 1.28 | n.r. | 0.26 (100 atm) | n.r. |
| MOF-177 | 1.5 | 7.0 (excess, 50 bar)<br>7.5 (>70 atm) | 0.62 (100 atm) | 11.3 |
| MIL-53(Al) | n.r. | 3.8(15 bar)<br>4.5(saturation) | n.r. | n.r. |
| MIL-53(Cr) | n.r. | 3.2(16 bar) | n.r. | n.r. |
| MIL-100 | n.r. | 3.28(~26 bar) | 0.15 (~73 atm) | 6.3 |
| MIL-101 | 1.91 | 6.1(~80 bar) | 0.43 (80 atm)<br>0.51 (100 atm) | 10.0 |
| HKUST-1 | 2.28 | 3.6 (saturation) | 0.35 (100 atm) | 4.5<br>6.6 |
| *soc*-MOF | 2.61 | n.r. | n.r. | 6.5 |
| Cu-TPB | n.r. | 2.8 (excess, 30 bar) | 0.5 (excess, 68 bar) | 8.2 |
| Mn-TPT | n.r. | 3.7 (excess, 25 bar)<br>4.5 (excess, 80 bar) | 0.5 (excess, 68 bar) | 7.6 |

n.r. = not reported.

shown the utility of this approach to increasing hydrogen storage by exchanging several extra-framework metal cations in Mn-BTT and studying their hydrogen uptake and interactions.[142] A more recent route to increase the storage capacities in MOFs has been devised by Yang,[200,206–208] where a so-called spillover catalyst (containing 5 wt% Pt supported on active carbon, used for dissociation of $H_2$) is mechanically combined with the MOF material (i.e., a bridged material). Hydrogen adsorption in select MOFs via spillover at room temperature over a wide range of pressure is significantly enhanced (Table 2.2), but the mechanism is not yet fully understood.

Other key potential applications for porous MOFs are separations and catalysis. The ability to generate a wide range of structures with periodic pore sizes and shapes definitely suggests the possibility for size- and/or shape-specific separation and

**TABLE 2.2 Hydrogen Adsorption and Storage via Spillover in Select MOFs[200,206–208]**

| Bridged Material | $H_2$ Uptake 298 K, 100 atm (wt%) | Isosteric Heats of Adsorption, $Q_{st}$, (kJ/mol) |
|---|---|---|
| IRMOF-1 (MOF-5) | 3 | n.r. |
| COF-1 | 0.68 | ~16 |
| MOF-177 | 1.5 | n.r. |
| MIL-101 | 1.43 | ~21 |
| HKUST-1 | 1.12 | ~17.5 |
| IRMOF-8 | 4 | ~24.8 |

n.r. = not reported.

catalysis, similar to what has been seen for widely used zeolites. In 2002, both Fujita's and Kitagawa's groups reported the selective adsorption of MOFs for various molecules, indicating the potential of MOFs in molecular separations. Fujita's MOF, Ni(4,4′-bis(4-pyridyl)biphenyl)$_2$($NO_3$)$_2$·4(*o*-xylene), showed the ability to exchange *o*-xylene for mesitylene, but not for *m*-xylene.[209] Kitagawa's compound, [$Cu_2$(pzdc)$_2$(dpyg)]$_n$, readily adsorbs methanol and $H_2O$, but not $CH_4$.[126]

The first catalysis achieved in MOFs was reported in 1994 by Fujita et al., where their 2D MOF, {[Cd(4,4′-Bipy)$_2$]($NO_3$)$_2$}$_\infty$, showed catalysis for the cyanosilylation of aldehydes.[51] One important feature of MOFs that may help contribute to catalysis is the presence of accessible open metal sites within the material, which can induce selectivity toward reaction intermediates or guest molecules.[205] Several of the rigid MBBs used to construct MOFs, such as the previously described paddlewheel cluster [$M_2(RCO_2)_4L_2$] or basic chromium acetate cluster [$M_3O(RCO_2)_6L_3$] (Figure 2.7),[85] have the potential to provide open metal sites upon removal (via evacuation and/or heating) of the ancillary terminal ligands (L, Figure 2.24a).

In addition, the modularity of MOFs allows for the introduction of extra open metal sites via the use of metalloligands, for example, the utilization of pyridine- or carboxylate-functionalized porphyrins[83,210–212] or salen-type ligands[213,214] that can complex metal ions in a tetradentate fashion at the core (Figure 2.24b, c) and form the ancillary metal-pyridine or -carboxylate MBBs necessary for the generation of MOFs, especially those that are known to aid in the catalysis of certain reactions like olefin epoxidation.[214] As mentioned previously, the large open cavities of MOFs also allow for the exchange of extra-framework metal cations through the pores, which may add open metal sites.[143,174] It should be noted that open metal sites also have been implicated in enhanced dihydrogen uptake and interactions with MOFs,[124] and Hupp and coworkers found that doping the framework with $Li^+$ metal cations significantly increases $N_2$ and $H_2$ uptake and $H_2$ interactions (i.e., higher isosteric heats of adsorption).[215] Additionally, the large cavities permit the encapsulation of large molecules,[116,174] suggesting that metal complexes like metallated porphyrins, salen-type complexes, or others[216] may permit catalytic activity upon encapsulation in the MOF.

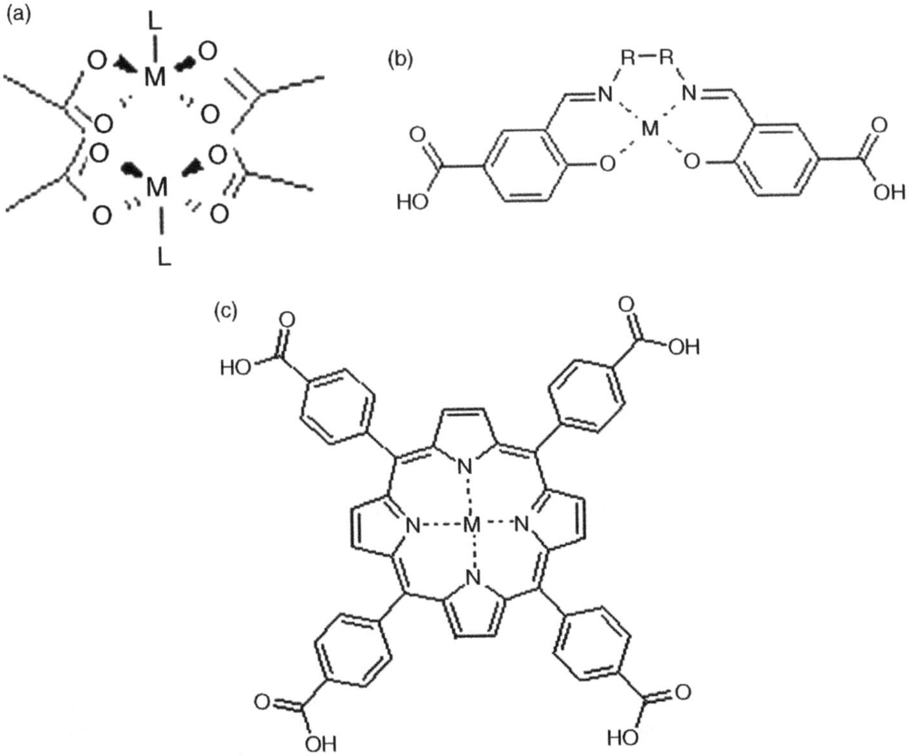


**FIGURE 2.24** Open metal sites can be introduced into MOMs in a variety of ways: (a) removal of ancillary terminal ligands on metal-carboxylate cluster MBBs, or carboxylate-functionalized metalloligands like (b) salen-type complexes (R=C in an alkyl group or aryl group) and (c) metallated porphyrins.

In addition, the tunable nature of MOMs permits to synthesize structures with built-in chirality, which may be utilized for enantioselective separation and catalysis,[107,214] an area where zeolites have seen limited success.[217,218] In fact, a few homochiral MOFs have been synthesized with some success in asymmetric catalytic activity.[107,219–221] Chiral catalysis with MOFs was first reported by Kimoon Kim's group in 2000; they successfully synthesized homochiral MOFs, {[$Zn_3O$(D-tartarate-$H)_6$]·$2H_3O$·$12H_2O$ (D-POST-1) and L-POST-1 (obtained from the ligand enantiomer)}, by utilizing enantiopure ligands, D- or L-tartaric acid, respectively.[107] In the crystal structure, $Zn_3O(RCO_2)_6L_3$ clusters are bridged and saturated by the tartarate ligands, resulting in a 3D MOF with channels of ~13.4 Å. POST-1 showed catalytic activity toward transesterification of the ester, 2,4-dinitrophenylacetate, with EtOH and several other alcohols, depending on the size. Accordingly, racemic 1-phenyl-2-propanol in enantiomorphic D- or L-POST-1 gave the corresponding chiral ester with

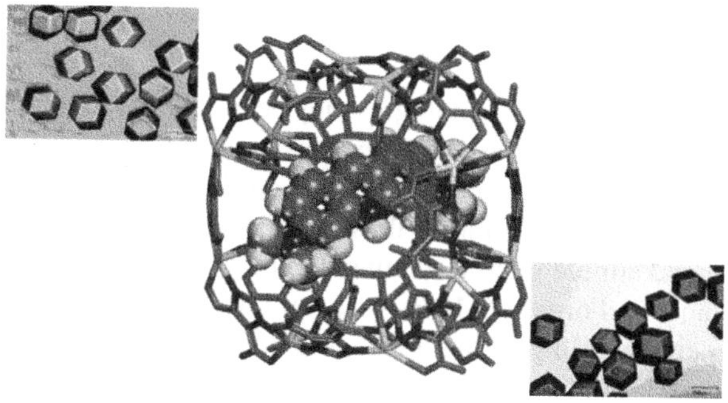

**FIGURE 2.25** Relatively large molecules like AO can freely diffuse through the D8R windows of In-ImDC *rho*-ZMOF, as supported by the color change of the single crystals from colorless to gray.

~8% ee of *S* or *R* enantiomer, respectively. Chiral MOFs can also be synthesized from achiral components, but most homochiral MOFs have been synthesized from enantiopure chiral ligands, with a majority constructed from binaphthyl-based ligands.[219]

The large apertures of some MOFs also allow for potential diffusion of large molecules[116,174,189,222], which may be used in controlled-release applications[222] like drug delivery[189] or sensing[174,222] applications, into the cavities, another area where zeolites have been limited.[174] The large apertures of MOF-177 (~11 Å) have allowed the uptake of several large molecules, including $C_{60}$ and several molecular dyes of varying size and shape (i.e., Astrazon Orange R, Nile Red, and Reichardt's dye) even though the overall framework is neutral, indicating the potential for size (and possibly shape) selectivity on a scale not previously seen.[116] Férey and coworkers have shown the potential for MOFs in drug delivery applications by the uptake and controlled release of Ibuprofen from the extra-large pores of MIL-100 (apertures of 4.8 and 8.6 Å) and MIL-101 (apertures of 12 and 16 Å),[189] and, more recently, flexible porous MOFs, MIL-53(Cr) and MIL-53(Fe).[205]

Our group has utilized the large pores of ZMOFs to adsorb large molecules for sensor applications.[174] The D8R cages of In-ImDC *rho*-ZMOF represent ~9 Å windows that allow access to the extra-large cavities, α-cages with an internal diameter of 18.2 Å. Unlike the previous examples, In-ImDC *rho*-ZMOF is anionic, making it ideal for applications involving cationic exchange. The cationic fluorophore, acridine orange (AO), is diffused into the α-cage cavities (Figure 2.25), and the electrostatic interactions with the framework preclude further diffusion of AO out of the cavities/pores, essentially anchoring the fluorophore. The extra-large dimensions allow for the diffusion of additional neutral guest molecules, and the anchored AO is utilized to sense a variety of neutral molecules, such as methyl xanthines or DNA nucleoside bases. These results demonstrate the ability of anionic MOFs to serve as (host–guest)–guest sensors, where the MOF serves as a periodic porous platform for fluorescent cations that act as the sensors.

## 2.3 CONCLUSION

Although MOMs having default topologies may be of interest for certain applications and have proven to exhibit interesting properties, structures with non-default topologies also may be of interest for applications (e.g., zeolite-like frameworks). Therefore, it makes sense that in order to target these atypical structures, a higher degree of information, beyond simple geometry, must be integrated into the MBBs. Although numerous inorganic and organic MBBs with varied connectivity are available to the would-be designer, it is still proving difficult to design MOMs. Therefore, strategies where topological control can be mandated by integrating structural information in the MBB (i.e., utilizing SBBs) are essential for the eventual design of materials for specific applications. In addition, as new MOMs are synthesized and a higher level of control obtained, more and more applications may be realized.

## REFERENCES


1. Baerlocher, Ch; McCusker, L. B. Database of Zeolite Structures, **2008**. Available at: //www.iza-structure.org/databases/.
2. Yang, X. Latest advances in application of zeolites in oil refining and petrochemical industry. *Gongye Cuihua.* **2003**, *11*, 19–24.
3. Maxwell, I. E.; Stork, W. H. J. Hydrocarbon processing with zeolites. *Stud. Surf. Sci. Catal.* **2001**, *137*, 747–819.
4. Corma, A. Zeolites in oil refining and petrochemistry. *NATO ASI Ser. C* **1992**, *352*, 373–436.
5. Glaeser, R.; Weitkamp, J. The application of zeolites in catalysis. *Springer Ser. Chem. Phys.* **2004**, *75*, 161–212.
6. Hagiwara, K. Effect of vanadium on USY zeolite destruction in the presence of sodium ions and steam—studies by solid-state NMR. *Appl. Catal. A* **2003**, *245*, N4–N5.
7. Claridge, R. P.; Lancaster, N. L.; Millar, R. W.; Moodie, R. B.; Sandall, J. P. B. Faujasite catalysis of aromatic nitrations with dinitrogen pentoxide. The effect of aluminum content on catalytic activity and regioselectivity. The nitration of pyrazole. *J. Chem. Soc. Perkin Trans.* **2001**, *2*, 197–200.
8. Hourdin, G.; Germain, A.; Moreau, C.; Fajula, F. The catalysis of the Ruff oxidative degradation of aldonic acids by titanium-containing zeolites. *Catal. Lett.* **2000**, *69*, 241–244.
9. Kuznicki, S. M.; Langner, T. W.; Curran, J. S.; Bell, V. A. Macroscopic aggregates of microcrystalline zeolites for static water softening applications. *US Pat. Appl.* US 20020077245, **2002**.
10. Kuznicki, S. M.; Bell, V. A.; Langner, T. W.; Curran, J. S. Porous static water softener containing hybrid zeolite-silicate composition. *US Pat. Appl.* US 20020074293, **2002**.
11. Le Van Mao, R.Zeolite materials with enhanced ion exchange capacity. *Can. Pat. Appl.* CA 2125314, **1995**.
12. Le Van Mao, R.; Vu, N. T.; Xiao, S.; Ramsaran, A. Modified zeolites for the removal of calcium and magnesium from hard water. *J. Mater. Chem.* **1994**, *4*, 1143–1147.

13. Breck, D. W.; *Zeolite Molecular Sieves: Structure, Chemistry and Use*; Wiley-Interscience: New York, 1974.
14. Gould, R. F. *Molecular Sieve Zeolites-I*; A.C.S: Washington DC, 1971.
15. Gould, R. F. *Molecular Sieve Zeolites-II*; A.C.S: Washington DC, 1971.
16. Paillaud, J. L.; Harbuzaru, B.; Patarin, J.; Bats, N. Extra-large-pore zeolites with two-dimensional channels formed by 14 and 12 rings. *Science* **2004**, *304*, 990–992.
17. Yamamoto, K.; Sakata, Y.; Nohara, Y.; Takahashi, Y.; Tatsumi, T. Organic-inorganic hybrid zeolites containing organic frameworks. *Science* **2003**, *300*, 470–472.
18. Jones, C. W.; Katsuyuki, T.; Davis, M. E. Organic-functionalized molecular sieves as shape-selective catalysts. *Nature* **1998**, *393*, 52–54.
19. Cheetham, A. K.; Férey, G.; Loiseau, T. Open-framework inorganic materials. *Angew. Chem. Int. Ed.* **1999**, *38*, 3268–3292.
20. Mueller, U.; Schubert, M.; Teich, F.; Puetter, H.; Schierle-Arndt, K.; Pastre, J. Metal-organic frameworks—prospective industrial applications. *J. Mater. Chem.* **2006**, *16*, 626–636.
21. Kosal, M. E.; Chou, J.; Wilson, S. R.; Suslick, K. S. A functional zeolite analogue assembled from metalloporphyrins. *Nat. Mater.* **2002**, *1*, 118–121.
22. Rowsell, J. L. C.; Yaghi, O. M. Strategies for hydrogen storage in metal-organic frameworks. *Angew. Chem. Int. Ed.* **2005**, *44*, 4670–4679.
23. Kaskel, S. Pores per modules. *Nachrichten aus der Chemie* **2005**, *53*, 394–399.
24. Lin, W. Homochiral porous metal-organic frameworks: why and how? *J. Solid State Chem.* **2005**, *178*, 2486–2490.
25. Eddaoudi, M.; Moler, D. B.; Li, H.; Chen, B.; Reineke, T. M.; O'Keeffe, M.; Yaghi, O. M. Modular chemistry: secondary building units as a basis for the design of highly porous and robust metal-organic carboxylate frameworks. *Acc. Chem. Res.* **2001**, *34*, 319–330.
26. Huang, Y.; Ding, B.; Song, H.; Zhao, B.; Ren, P.; Cheng, P.; Wang, H.; Liao, D.; Yan, S. A novel 3D porous metal-organic framework based on trinuclear cadmium clusters as a promising luminescent material exhibiting tunable emissions between UV and visible wavelengths. *Chem. Commun.* **2006**, 4906–4908.
27. Bordiga, S.; Lamberti, C.; Ricchiardi, G.; Regli, L.; Bonino, F.; Damin, A.; Lillerud, K.; Bjorgen, M.; Zecchina, A. Electronic and vibrational properties of a MOF-5 metal-organic framework: ZnO quantum dot behaviour. *Chem. Commun.* **2004**, 2300–2301.
28. Halder, G. J.; Kepert, C. J.; Moubaraki, B.; Murray, K. S.; Cashion, J. D. Guest-dependent spin crossover in a nanoporous molecular framework material. *Science* **2002**, *298*, 1762–1765.
29. Stein, A.; Keller, S. W.; Mallouk, T. E. Turning down the heat: design and mechanism in solid-state synthesis. *Science* **1993**, *259*, 1558–1564.
30. Férey, G. Building units design and scale chemistry. *J. Solid State Chem.* **2000**, *152*, 37–48.
31. Eddaoudi, M.; Kim, J.; Rosi, N.; Vodak, D.; Wachter, J.; O'Keeffe, M.; Yaghi, O. M. Systematic design of pore size and functionality in isoreticular MOFs and their application in methane storage. *Science* **2002**, *295*, 469–472.
32. Kitagawa, S.; Kitaura, R.; Noro, S. -I. Functional porous coordination polymers. *Angew. Chem. Int. Ed.* **2004**, *43*, 2334–2375.
33. Yaghi, O. M.; Li, H.; Davis, C.; Richardson, D.; Groy, T. L. Synthetic strategies, structure patterns, and emerging properties in the chemistry of modular porous solids. *Acc. Chem. Res.* **1998**, *31*, 474–484.

34. O'Keeffe, M. Reticular Chemistry Structure Resource, **2008**. Available at: //rcsr.anu.edu.au/.
35. Han, S.; Smith, J. V. Enumeration of four-connected three-dimensional nets. I. Conversion of all edges of simple three-connected two-dimensional nets into crankshaft chains. *Acta Crystallogr.* **1999**, *A55*, 332–341.
36. Han, S.; Smith, J. V. Enumeration of four-connected three-dimensional nets. II. Conversion of edges of three-connected 2D nets into zigzag chains. *Acta Crystallogr.* **1999**, *A55*, 342–359.
37. Han, S.; Smith, J. V. Enumeration of four-connected three-dimensional nets. III. Conversion of edges of three-connected two-dimensional nets into saw chains. *Acta Crystallogr.* **1999**, *A55*, 360–382.
38. Wells, A. F. *Further Studies of Three-Dimensional Nets*; American Crystallography Association: Washington, 1979 (Monograph 8, Mineralogical Society of America).
39. Wells, A. F. *Three-dimensional Nets and Polyhedra*; John Wiley & Sons: Hoboken, 1977.
40. Ockwig, N. W.; Delgado-Friedrichs, O.; O'Keeffe, M.; Yaghi, O. M. Reticular chemistry: occurrence and taxonomy of nets and grammar for the design of frameworks. *Acc. Chem. Res.* **2005**, *38*, 176–182.
41. Yaghi, O. M.; O'Keeffe, M.; Ockwig, N.; Chae, H.; Eddaoudi, M.; Kim, J. Reticular synthesis and the design of new materials. *Nature* **2003**, *423*, 705–714.
42. Kim, J.; Chen, B.; Reineke, T. M.; Li, H.; Eddaoudi, M.; Moler, D. B.; O'Keeffe, M.; Yaghi, O. M. Assembly of metal-organic frameworks from large organic and inorganic secondary building units: new examples and simplifying principles for complex structures. *J. Am. Chem. Soc.* **2001**, *123*, 8239–8247.
43. O'Keeffe, M.; Eddaoudi, M.; Li, H.; Reineke, T.; Yaghi, O. M. Frameworks for extended solids: geometrical design principles. *J. Solid State Chem.* **2000**, *152*, 3–20.
44. Mueller, U.; Lobree, L.; Hesse, M.; Yaghi, O. M.; Eddaoudi, M. Process for epoxidation of organic compounds with oxygen or oxygen-delivering compounds using catalysts containing metal-organic framework (MOF) materials. *US Pat.* 6624318, **2003**.
45. Lipowski, J. *Inclusion Compounds: Structural Aspects of Inclusion Compounds Formed by Inorganic and Organometallic Host Lattices*; In Atwood, J. L.; Davies, J. E. D.; MacNicol, D. D., Eds., Academic Press: London, **1984**; pp. 59–103 and references therein.
46. Iwamoto, T.; Nakano, T.; Morita, M.; Miyoshi, T.; Miyamoto, T.; Sasaki, Y. The metal ammine cyanide aromatics clathrates. V. The Hofmann-type clathrate: $M(NH_3)_2$-$M'(CN)_4.2G$. *Inorg. Chim. Acta* **1968**, 2, 313–316.
47. Dunbar, K. R.; Heintz, R. A. Chemistry of transition metal cyanide compounds: modern perspectives. *Prog. Inorg. Chem.* **1997**, *45*, 283–391.
48. Hoskins, B. F.; Robson, R. Infinite polymeric frameworks consisting of three dimensionally linked rod-like segments. *J. Am. Chem. Soc.* **1989**, *111*, 5962–5964.
49. Hoskins, B. F.; Robson, R. Design and construction of a new class of scaffolding-like materials comprising infinite polymeric frameworks of 3D-linked molecular rods. A reappraisal of the $Zn(CN)_2$ and $Cd(CN)_2$ structures and the synthesis and structure of the diamond-related frameworks $[N(CH_3)_4][Cu^IZn^{II}(CN)_4]$ and $Cu^I[4,4',4'',4'''$-tetracyanotetraphenylmethane$]BF_4 \cdot xC_6H_5NO_2$. *J. Am. Chem. Soc.* **1990**, *112*, 1546–1554.
50. Gable, R. W.; Hoskins, B. F.; Robson, R. A new type of interpenetration involving enmeshed independent square grid sheets. The structure of diaquabis(4,4′-bipyridine) zinc hexafluorosilicate. *J. Chem. Soc. Chem. Commun.* **1990**, 1677–1678.

51. Fujita, M.; Kwon, Y. J.; Washizu, S.; Ogura, K. Preparation, clathration ability, and catalysis of a two-dimensional square network material composed of cadmium(II) and 4,4′-bipyridine. *J. Am. Chem. Soc.* **1994**, *116*, 1151–1152.

52. Losier, P.; Zaworotko, M. J. A noninterpenetrated molecular ladder with hydrophobic cavities. *Angew. Chem. Int. Ed. Engl.* **1997**, *35*, 2779–2782.

53. Lu, J.; Paliwala, T.; Lim, S. C.; Yu, C.; Niu, T.; Jacobson, A. J. Coordination polymers of $Co(NCS)_2$ with pyrazine and 4,4′-bipyridine: syntheses and structures. *Inorg. Chem.* **1997**, *36*, 923–929.

54. Subramanian, S.; Zaworotko, M. J. Porous solids by design: $[Zn(4,4\text{-bpy})_2(SiF_6)]_n \cdot x$DMF, a single framework octahedral coordination polymer with large square channels. *Angew. Chem. Int. Ed. Engl.* **1995**, *34*, 2127–2129.

55. MacGillivray, L. R.; Subramanian, S.; Zaworotko, M. J. Interwoven two- and three-dimensional coordination polymers through self-assembly of $Cu^I$ cations with linear bidentate ligands. *J. Chem. Soc. Chem. Commun.* **1994**, 1325–1326.

56. Carlucci, L.; Ciani, G.; Prosperio, D. M.; Sironi, A. Interpenetrating diamondoid frameworks of silver(I) cations linked by N,N′-bidentate molecular rods. *J. Chem. Soc. Chem. Commun.* **1994**, 2755–2766.

57. Yaghi, O. M.; Richardson, D. A.; Li, G.; Davis, C. E.; Groy, T. L. Open-framework solids with diamond-like structures prepared from clusters and metal-organic building blocks. *Mater. Res. Soc. Symp. Proc.* **1995**, *371*, 15–19.

58. Yaghi, O. M.; Li, H. Hydrothermal synthesis of a metal-organic framework containing large rectangular channels. *J. Am. Chem. Soc.* **1995**, *117*, 10401–10402.

59. Yaghi, O. M.; Li, H. T-Shaped molecular building units in the porous structure of Ag(4,4′-bpy)·$NO_3$. *J. Am. Chem. Soc.* **1996**, *118*, 295–296.

60. Yaghi, O. M.; Li, H. Mutually interpenetrating sheets and channels in the extended structure of [Cu(4,4′-bpy)Cl]. *Angew. Chem. Int. Ed. Engl.* **1995**, *34*, 207–209.

61. Woodward, J. D.; Backov, R. V.; Abboud, K. A.; Talham, D. R. Monomers, chains, ladders, and two-dimensional sheets: structural diversity in six new compounds of Zn(II) with 4,4′-bipyridine. *Polyhedron* **2006**, *25*, 2605–2615.

62. Yaghi, O. M.; Li, H.; Groy, T. L. A molecular railroad with large pores: synthesis and structure of $Ni(4,4'\text{-bpy})_{2.5}(H_2O)_2(ClO_4)_2 \cdot 1.5(4,4'\text{-bpy}) \cdot 2H_2O$. *Inorg. Chem.* **1997**, *36*, 4292–4293.

63. Fujita, M.; Yazaki, J.; Ogura, K. Preparation of a macrocyclic polynuclear complex, $[(en)Pd(4,4'\text{-bpy})]_4(NO_3)_8$ (en = ethylenediamine, bpy = bipyridine), which recognizes an organic molecule in aqueous media. *J. Am. Chem. Soc.* **1990**, *112*, 5645–5647.

64. Kitagawa, S.; Kondo, M. Functional micropore chemistry of crystalline metal complex-assembled compounds. *Bull. Chem. Soc. Jpn* **1998**, *71*, 1739–1753.

65. Fujita, M.; Tominaga, M.; Hori, A.; Therrien, B. Coordination assemblies from a Pd(II)-cornered square complex. *Acc. Chem. Res.* **2005**, *38*, 369–378.

66. Power, K. N.; Hennigar, T. L.; Zaworotko, M. J. X-Ray crystal structure of $\{Cu[1,2\text{-bis(4-pyridyl)ethane}]_2(NO_3)_2\}_n$: the first example of a coordination polymer that exhibits the NbO 3D network architecture. *Chem. Commun.* **1998**, 595–596.

67. Carlucci, L.; Ciani, G.; Proserpio, D. M.; Sironi, A. Novel networks of unusually coordinated silver(I) cations: the wafer-like structure of $[Ag(pyz)_2][Ag_2(pyz)_5]$-$(PF_6)_3 \cdot 2G$ and the simple cubic frame of $[Ag(pyz)_3](SbF_6)$. *Angew. Chem. Int. Ed. Engl.* **1995**, *34*, 1895–1898.

68. Fujita, M.; Oguro, D.; Miyazawa, M.; Oka, H.; Yamaguchi, K.; Batten, S. R.; Ogura, K. Self-assembly of ten molecules into a nanometer-sized organic host frameworks. *Nature* **1995**, *378*, 469–471.

69. Hoskins, B. F.; Robson, R. 2,4,6-Tri(4-pyridyl)-1,3,5-triazine as a 3-connecting building block for infinite nets. *Angew. Chem. Int. Ed. Engl.* **1995**, *34*, 820–822.

70. Batten, S. R.; Hoskins, B. F.; Robson, R. Two interpenetrating 3D networks which generate spacious sealed-off compartments enclosing of the order of 20 solvent molecules in the structures of $Zn(CN)(NO_3)(tpt)_{2/3}$·solv (tpt = 2,4,6-tri(4-pyridyl)-1,3,5-triazine, solv = ∼$3/4C_2H_2Cl_4$·$3/4CH_3OH$ or ∼$3/2CHCl_3$·$1/3CH_3OH$). *J. Am. Chem. Soc.* **1995**, *117*, 5385–5386.

71. Abrahams, B. F.; Batten, S. R.; Hamit, H.; Hoskins, B. F.; Robson, R. A Wellsian 'three-dimensional' racemate: eight interpenetrating, enantiomorphic (10,3)-a nets, four right- and four left-handed. *Chem. Commun.* **1996**, 1313–1314.

72. Abrahams, B. F.; Batten, S. R.; Hamit, H.; Hoskins, B. F.; Robson, R. A cubic (3,4)-connected net with large cavities in solvated $[Cu_3(tpt)_4](ClO_4)_3$ (tpt = 2,4,6-tri(4-pyridyl)-1,3,5-triazine). *Angew. Chem. Int. Ed. Engl.* **1996**, *35*, 1690–1692.

73. Ibukuro, F.; Kusukawa, T.; Fujita, M. A thermally switchable molecular lock. Guest-templated synthesis of a kinetically stable nanosized cage. *J. Am. Chem. Soc.* **1998**, *120*, 8561–8562.

74. Abrahams, B. F.; Batten, S. R.; Grannas, M. J.; Hamit, H.; Hoskins, B. F.; Robson, R. $Ni(tpt)(NO_3)_2$—a three-dimensional network with the exceptional (12,3) topology: a self-entangled single net. *Angew. Chem. Int. Ed.* **1999**, *38*, 1475–1477.

75. Abrahams, B. F.; Hoskins, B. F.; Robson, R. A new type of infinite 3D polymeric network containing 4-connected, peripherally-linked metalloporphyrin building blocks. *J. Am. Chem. Soc.* **1991**, *113*, 3606–3607.

76. Abrahams, B. F.; Hoskins, B. F.; Michail, D. M.; Robson, R. Assembly of porphyrin building blocks into network structures with large channels. *Nature* **1994**, *369*, 727–729.

77. Sharma, C. V. K.; Broker, G. A.; Huddleston, J. G.; Baldwin, J. W.; Metzger, R. M.; Rogers, R. D. Design strategies for solid-state supramolecular arrays containing both mixed-metalated and freebase porphyrins. *J. Am. Chem. Soc.* **1999**, *121*, 1137–1144.

78. Hagrman, D.; Hagrman, P. J.; Zubieta, J. Solid-state coordination chemistry: the self-assembly of microporous organic-inorganic hybrid frameworks constructed from tetra-pyridylporphyrin and bimetallic oxide chains or oxide clusters. *Angew. Chem. Int. Ed.* **1999**, *38*, 3165–3168.

79. Kondo, M.; Kimura, Y.; Wada, K.; Mizutani, T.; Ito, Y.; Kitagawa, S. A new self-assembled porphyrin-silver(I) network. *Chem. Lett.* **2000**, 818–819.

80. Hoskins, B. F.; Robson, R.; Slizys, D. A. A hexaimidazole ligand binding six octahedral metal ions to give an infinite 3D α-Po-like network through which two independent 2D hydrogen-bonded networks interweave. *Angew. Chem. Int. Ed.* **1998**, *36*, 2752–2755.

81. Leininger, S.; Olenyuk, B.; Stang, P. J. Self-assembly of discrete cyclic nanostructures mediated by transition metals. *Chem. Rev.* **2000**, *100*, 853–908.

82. Kondo, M.; Yoshitomi, T.; Seki, K.; Matsuzaka, H.; Kitagawa, S. Three-dimensional framework with channeling cavities for small molecules: $\{[M_2(4,4'\text{-bpy})_3(NO_3)_4]\cdot xH_2O\}_n$ (M = Co, Ni, Zn). *Angew. Chem. Int. Ed. Engl.* **1997**, *36*, 1725–1727.

83. Ohmura, T.; Usuki, A.; Fukumori, K.; Ohta, T.; Ito, M.; Tatsumi, K. New porphyrin-based metal-organic framework with high porosity: 2-D infinite 22.2-Å square-grid coordination network. *Inorg Chem* **2006**, *45*, 7988–7990.

84. Eddaoudi, M.; Li, H.; Reineke, T. M.; Fehr, M.; Kelley, D. G.; Groy, T. L.; Yaghi, O. M. Design and synthesis of metal-carboxylate frameworks with permanent microporosity. *Top Catal.* **1999**, *9*, 105–111.

85. Allen, F. H. The Cambridge Structural Database: a quarter of a million crystal structures and rising. *Acta Crystallogr. B58*, **2002**, 380–388.

86. Watanabe, S. Polymorphism of beryllium oxyacetate. *Nature 163*, **1949**, 225–226.

87. van Niekerk, J. N.; Schoening, F. R. L. X-Ray evidence for metal-to-metal bonds in cupric and chromous acetate. *Nature* **1953**, *171*, 36–37.

88. van Niekerk, J. N.; Schoening, F. R. L. A new type of copper complex as found in the crystal structure of cupric acetate, $Cu_2(CH_3COO)_4{\cdot}2H_2O$. *Acta Crystallogr.* **1953**, *6*, 227–232.

89. van Niekerk, J. N.; Schoening, F. R. L.; de Wet, J. F. The structure of crystalline chromous acetate revealing paired chromium atoms. *Acta Crystallogr.* **1953**, *6*, 501–504.

90. Koyama, H.; Saito, Y. The crystal structure of zinc oxyacetate, $Zn_4O(CH_3COO)_6$. *Bull. Chem. Soc. Jpn* **1954**, *27*, 112–114.

91. Lowell, S.; Shields, J. E.; Thomas, M. A.; Thommes, M.; *Characterization of Porous Solids and Powders: Surface Area, Pore Size and Density*; Kluwer Academic Publishers: Dordrecht,The Netherlands, 2004.

92. Li, H.; Eddaoudi, M.; Groy, T. L.; Yaghi, O. M. Establishing microporosity in open metal-organic frameworks: gas sorption isotherms for Zn(BDC) (BDC = 1,4-benzenedicarboxylate). *J. Am. Chem. Soc.* **1998**, *120*, 8571–8572.

93. Chui, S. S. -Y.; Lo, S. M. -F.; Charmant, J. P. H.; Orpen, A. G.; Williams, I. D. A chemically functionalizable nanoporous material $[Cu_3(TMA)_2(H_2O)_3]_n$. *Science* **1999**, *283*, 1148–1150.

94. Breck, D. W. *Zeolite Molecular Sieves*; Kreiger: Malabar, FL, 1974.

95. Li, H.; Eddaoudi, M.; O'Keeffe, M.; Yaghi, O. M. Design and synthesis of an exceptionally stable and highly porous metal-organic framework. *Nature* **1999**, *402*, 276–279.

96. Hyde, B. G.; Anderson, S. *Inorganic Crystal Structures*; Wiley-Interscience: New York, 1988.

97. O'Keeffe, M.; Hyde, B. G. *Crystal Structures I: Patterns and Symmetry*; American Mineral Association: Washington, DC, 1996.

98. Kesanli, B.; Cui, Y.; Smith, M. R.; Bittner, E. W.; Bockrath, B. C.; Lin, W. Highly interpenetrated metal-organic frameworks for hydrogen storage. *Angew. Chem. Int. Ed.* **2005**, *44*, 72–75.

99. Sun, D.; Collins, D. J.; Ke, Y.; Zuo, J.; Zhou, H. Construction of open metal-organic frameworks based on predesigned carboxylate isomers: from achiral to chiral nets. *Chem. Eur. J.* **2006**, *12*, 3768–3776.

100. Moulton, B.; Lu, J.; Hajndl, R.; Hariharan, S.; Zaworotko, M. J. Crystal engineering of a nanoscale Kagomé lattice. *Angew. Chem. Int. Ed.* **2002**, *41*, 2821–2824.

101. Eddaoudi, M.; Kim, J.; Vodak, D.; Sudik, A.; Wachter, J.; O'Keeffe, M.; Yaghi, O. M. Geometric requirements and examples of important structures in the assembly of square building blocks. *Proc. Natl Acad. Sci. USA* **2002**, *99*, 4900–4904.

102. Abourahma, H.; Bodwell, G. J.; Lu, J.; Moulton, B.; Pottie, I. R.; Walsh, R. B.; Zaworotko, M. J. Coordination polymers from calixarene-like $[Cu_2(dicarboxylate)_2]_4$ building blocks: structural diversity via atropisomerism. *Cryst. Growth Des.* **2003**, *3*, 513–519.

103. Cotton, F. A.; Lin, C.; Murillo, C. A. Supramolecular arrays based on dimetal building units. *Acc. Chem. Res.* **2001**, *34*, 759–771.

104. Yaghi, O. M.; Davis, C. E.; Li, G.; Li, H. Selective guest binding by tailored channels in a 3-D porous zinc(II)-benzenetricarboxylate network. *J. Am. Chem. Soc.* **1997**, *119*, 2861–2868.

105. Yaghi, O. M.; Jernigan, R.; Li, H.; Davis, C. E.; Groy, T. L. Construction of a new open-framework solid from 1,3,5-cyclohexanetricarboxylate and zinc(II) building blocks. *J. Chem. Soc. Dalton Trans.* **1997**, 2383–2384.

106. Li, H.; Davis, C. E.; Groy, T. L.; Kelley, D. G.; Yaghi, O. M. Coordinatively unsaturated metal centers in the extended porous framework of $Zn_3(BDC)_3 \cdot 6CH_3OH$ (BDC = 1,4-benzenedicarboxylate). *J. Am. Chem. Soc.* **1998**, *120*, 2186–2187.

107. Seo, J. S.; Whang, D.; Lee, H.; Jun, S. I.; Oh, J.; Jeon, Y. J.; Kim, K. A homochiral metal-organic porous material for enantioselective separation and catalysis. *Nature* **2000**, *404*, 982–986.

108. Barthelet, K.; Riou, D.; Férey, G. $[V^{III}(H_2O)]_3O(O_2CC_6H_4CO_2)_3 \cdot (Cl, 9H_2O)$ (MIL-59): a rare example of vanadocarboxylate with a magnetically frustrated three-dimensional hybrid framework. *Chem. Commun.* **2002**, 1492–1493.

109. Férey, G.; Serre, C.; Mellot-Draznieks, C.; Millange, F.; Surble, S.; Dutour, J.; Margiolaki, I. Molecular modeling: a hybrid solid with giant pores prepared by a combination of targeted chemistry, simulation, and powder diffraction. *Angew. Chem. Int. Ed.* **2004**, *43*, 6296–6301.

110. Serre, C.; Millange, F.; Surblé, S.; Férey, G. A route to the synthesis of trivalent transition-metal porous carboxylates with trimeric secondary building units. *Angew. Chem. Int. Ed.* **2004**, *43*, 6285–6289.

111. Sudik, A. C.; Cote, A. P.; Yaghi, O. M. Metal-organic frameworks based on trigonal prismatic building blocks and the new "acs" topology. *Inorg. Chem.* **2005**, *44*, 2998–3000.

112. Férey, G.; Mellot-Draznieks, C.; Serre, C.; Millange, F.; Dutour, J.; Surble, S.; Margiolaki, I. A chromium terephthalate-based solid with unusually large pore volumes and surface area. *Science* **2005**, *309*, 2040–2042.

113. Volkringer, C.; Loiseau, T. A new indium metal-organic 3D framework with 1,3,5-benzenetricarboxylate, MIL-96 (In), containing $\mu_3$-oxo-centered trinuclear units and a hexagonal 18-ring network. *Mater. Res. Bull.* **2006**, *41*, 948–954.

114. Ma, S.; Wang, X.; Manis, E. S.; Collier, C. D.; Zhou, H. Metal-organic framework based on a trinickel secondary building unit exhibiting gas-sorption hysteresis. *Inorg. Chem.* **2007**, *46*, 3432–3434.

115. Liu, Y.; Eubank, J. F.; Cairns, A. J.; Eckert, J.; Kravtsov V.Ch.; Luebke R.; Eddaoudi, M. Assembly of metal-organic frameworks (MOFs) based on indium-trimer building blocks: a porous MOF with *soc* topology and high hydrogen storage. *Angew. Chem. Int. Ed.* **2007**, *46*, 3278–3283.

116. Chae, H. K.; Siberio-Perez, D. Y.; Kim, J.; Go, Y. B.; Eddaoudi, M.; Matzger, A. J.; O'Keeffe, M.; Yaghi, O. M. A route to high surface area, porosity and inclusion of large molecules in crystals. *Nature* **2004**, *427*, 523–527.

117. Wong-Foy, A. G.; Matzger, A. J.; Yaghi, O. M. Exceptional $H_2$ saturation uptake in microporous metal-organic frameworks. *J. Am. Chem. Soc.* **2006**, *128*, 3494–3495.

118. Kaye, S. S.; Dailly, A.; Yaghi, O. M.; Long, J. R. Impact of preparation and handling on the hydrogen storage properties of $Zn_4O$(1,4-benzenedicarboxylate)$_3$ (MOF-5). *J. Am. Chem. Soc.* **2007**, *129*, 14176–14177.

119. Koh, K.; Wong-Foy, A. G.; Matzger, A. J. A crystalline mesoporous coordination copolymer with high microporosity. *Angew. Chem. Int. Ed.* **2008**, *47*, 677–680.

120. Wang, X. -S.; Ma, S.; Sun, D.; Parkin, S.; Zhou, H. -C. A mesoporous metal-organic framework with permanent porosity. *J. Am. Chem. Soc.* **2006**, *128*, 16474–16475.

121. Park, Y. K.; Choi, S. B.; Kim, H.; Kim, K.; Won, B. -H.; Choi, K.; Choi, J. -S.; Ahn, W. -S.; Won, N.; Kim, S.; Jung, D. H.; Choi, S. -H.; Kim, G. -H.; Cha, S. -S.; Jhon, Y. H.; Yang, J. K. Crystal structure and guest uptake of a mesoporous metal-organic framework containing cages of 3.9 and 4.7 nm in diameter. *Angew. Chem. Int. Ed.* **2007**, *46*, 8230–8233.

122. Fang, Q. -R.; Zhu, G. -S.; Jin, Z.; Ji, Y. -Y.; Ye, J. -W.; Xue, M.; Yang, H.; Wang, Y.; Qiu, S. -L. Mesoporous metal-organic framework with rare etb topology for hydrogen storage and dye assembly. *Angew. Chem. Int. Ed.* **2007**, *46*, 6638–6642.

123. Rowsell, J. L. C.; Yaghi, O. M. Effects of functionalization, catenation, and variation of the metal oxide and organic linking units on the low-pressure hydrogen adsorption properties of metal-organic frameworks. *J. Am. Chem. Soc.* **2006**, *128*, 1304–1315.

124. Lin, H.; Jia, J.; Hubberstey, P.; Shroeder, M.; Champness, N. R. Hydrogen storage in metal-organic frameworks. *CrystEngComm* **2007**, *9*, 438–448.

125. Kondo, M.; Okubo, T.; Asami, A.; Noro, S. -I.; Yoshitomi, T.; Kitagawa, S.; Ishii, T.; Matsuzaka, H.; Seki, K. Rational synthesis of stable channel-like cavities with methane gas adsorption properties: $[\{Cu_2(pzdc)_2(L)\}_n]$ (pzdc = pyrazine-2,3-dicarboxylate; L = a pillar ligand). *Angew. Chem. Int. Ed.* **1999**, *38*, 140–143.

126. Kitaura, R.; Fujimoto, K.; Noro, S. -I.; Kondo, M.; Kitagawa, S. A pillared-layer coordination polymer network displaying hysteretic sorption: $[Cu_2(pzdc)_2(dpyg)]_n$ (pzdc = pyrazine-2,3-dicarboxylate; dpyg = 1,2-di(4-pyridyl)glycol). *Angew. Chem. Int. Ed.* **2002**, *41*, 133–135.

127. Evans, O. R.; Lin, W. Crystal engineering of NLO materials based on metal-organic coordination networks. *Acc. Chem. Res.* **2002**, *35*, 511–522.

128. Chun, H.; Dybtsev, D. N.; Kim, H.; Kim, K. Synthesis, X-ray crystal structures, and gas sorption properties of pillared square grid nets based on paddle-wheel motifs: implications for hydrogen storage in porous materials. *Chem. Eur. J.* **2005**, *11*, 3521–3529.

129. Ma, B. -Q.; Mulfort, K. L.; Hupp, J. T. Microporous pillared paddle-wheel frameworks based on mixed-ligand coordination of zinc ions. *Inorg. Chem.* **2005**, *44*, 4912–4914.

130. Chun, H.; Moon, J. Discovery, synthesis, and characterization of an isomeric coordination polymer with pillared Kagome net topology. *Inorg. Chem.* **2007**, *46*, 4371–4373.

131. Eubank, J. F.; Walsh, R. D.; Eddaoudi, M. Terminal co-ligand directed synthesis of a neutral, non-interpenetrated (10,3)-*a* metal-organic framework. *Chem. Commun.* **2005**, 2095–2097.

132. Jia, J.; Lin, X.; Wilson, C.; Blake, A. J.; Champness, N. R.; Hubberstey, P.; Walker, G.; Cussen, E. J.; Schröder, M. Twelve-connected porous metal-organic frameworks with high $H_2$ adsorption. *Chem. Commun.* **2007**, 840–842.

133. Okubo, T.; Kondo, M.; Kitagawa, S. Synthesis, structure, and magnetic properties of one-dimensional copper(II) coordination polymer, {[Cu(pyrazine-2,3-dicarboxylate)-$(H_2O)_2$]·$2H_2O$}$_n$. *Synth. Met.* **1997**, *85*, 1661–1662.

134. Brant, J. A.; Liu, Y.; Sava, D. F.; Beauchamp, D.; Eddaoudi, M. Single-metal-ion-based molecular building blocks (MBBs) approach to the design and synthesis of metal-organic assemblies. *J. Mol. Struct.* **2006**, *796*, 160–164.

135. Eddaoudi, M.; Eubank, J. F. Periodic nanostructures based on metal-organic frameworks (MOFs): en route to zeolite-like metal-organic frameworks (ZMOFs). In *Organic Nanostructures*; Atwood, J. L.; Steed, J.W.,Eds., Wiley-VCH Verlag GmbH & Co KGaA: Weinheim, 2008; pp. 251–276.

136. Liu, Y.; Kravtsov V.Ch.; Walsh R. D.; Poddar, P.; Hariharan, S.; Eddaoudi, M. Directed assembly of metal-organic cubes from deliberately predesigned molecular building blocks. *Chem. Commun.* **2004**, 2806–2807.

137. Liu, Y.; Kravtsov VCh, Beauchamp D. A.; Eubank, J. F.; Eddaoudi, M. 4-Connected metal-organic assemblies mediated via heterochelation and bridging of single metal ions: Kagomé lattice and the $M_6L_{12}$ octahedron. *J. Am. Chem. Soc.* **2005**, *127*, 7266–7267.

138. Zhang, J. P.; Horike, S.; Kitagawa, S. A flexible porous coordination polymer functionalized by unsaturated metal clusters. *Angew. Chem. Int. Ed.* **2007**, *46*, 889–892.

139. He, J.; Yin, Y.; Wu, T.; Li, D.; Huang, X. C. Design and solvothermal synthesis of luminescent copper(I)-pyrazolate coordination oligomer and polymer frameworks. *Chem. Commun.* **2006**, 2845–2847.

140. Dincă, M.; Yu, A. F.; Long, J. R. Microporous metal-organic frameworks incorporating 1,4-benzeneditetrazolate: syntheses, structures, and hydrogen storage properties. *J. Am. Chem. Soc.* **2006**, *128*, 8904–8913.

141. Dincă, M.; Dailly, A.; Liu, Y.; Brown, C. M.; Neumann, D. A.; Long, J. R. Hydrogen storage in a microporous metal-organic framework with exposed $Mn^{2+}$ coordination sites. *J. Am. Chem. Soc.* **2006**, *128*, 16876–16883.

142. Dincă, M.; Han, W. S.; Liu, Y.; Dailly, A.; Brown, C. M.; Long, J. R. Observation of $Cu^{2+}$-$H_2$ interactions in a fully desolvated sodalite-type metal-organic framework. *Angew. Chem. Int. Ed.* **2007**, *46*, 1419–1422.

143. Dincă, M.; Dailly, A.; Tsay, C.; Long, J. R. Expanded sodalite-type metal-organic frameworks: increased stability and $H_2$ adsorption through ligand-directed catenation. *Inorg. Chem.* **2008**, *47*, 11–13.

144. Jiang, C.; Yu, Z.; Wang, S.; Jiao, C.; Li, J.; Wang, Z.; Cui, Y. Rational design of metal-organic frameworks based on 5-(4-pyridyl)tetrazolate: from 2D grids to 3D porous networks. *Eur. J. Inorg. Chem.* **2004**, 3662–3667.

145. Luo, T.; Tsai, H.; Yang, S.; Liu, Y.; Yadav, R. D.; Su, C.; Ueng, C.; Lin, L.; Lu, K. Crystal engineering: toward intersecting channels from a neutral network with a bcu-type topology. *Angew. Chem. Int. Ed.* **2005**, *44*, 6063–6067.

146. Xiong, R.; Xue, X.; Zhao, H.; You, X.; Abrahams, B. F.; Xue, Z. Novel, acentric metal-organic coordination polymers from hydrothermal reactions involving in situ ligand synthesis. *Angew. Chem. Int. Ed.* **2002**, *41*, 3800–3803.

147. Xue, X.; Wang, X.; Wang, L.; Xiong, R.; Abrahams, B. F.; You, X.; Xue, Z.; Che, C-M. Hydrothermal preparation of novel Cd(II) coordination polymers employing 5-(4-pyridyl)tetrazolate as a bridging ligand. *Inorg. Chem.* **2002**, *41*, 6544–6546.

148. Li, J.; Tao, J.; Huanga, R.; Zhang, L. Poly[$\mu_4$-5-(3-carboxylatophenyl)-1*H*-tetrazolato-zinc(II)]. *Acta Crystallogr.* **2005**, *E61*, m984–m985.

149. Qu, Z. -R.; Zhao, H.; Wang, X. -S.; Li, Y. -H.; Song, Y. -M.; Liu, Y. -J.; Ye, Q.; Xiong, R. -G.; Abrahams, B. F.; Xue, Z. -L.; You, X. -Z. Homochiral Zn and Cd coordination polymers containing amino acid-tetrazole ligands. *Inorg. Chem.* **2003**, *42*, 7710–7712.

150. Wang, X. -S.; Huang, X. -F.; Xiong, R. -G. An unexpected intermediate or precipitate - novel 3D Cd-coordination polymer formed in the preparation of 5-substituted 1H-tetrazoles from nitrile in water. *Wuji Huaxue Xuebao* **2005**, *21*, 1020–1024.

151. Zhao, H.; Ye, Q.; Wu, Q.; Song, Y.; Liu, Y.; Xiong, R. A novel one-dimensional zinc coordination polymer, [zinc{(4,5-ditetrazoyl)-imidazole}{(1,10)-phenanthroline}-($H_2O$)]$_n$. *Z. Anorg. Allg. Chem.* **2004**, *630*, 1367–1370.

152. Wang, Z.; Kravtsov, V.; Zaworotko, M. J. Ternary nets formed by self-assembly of triangles, squares, and tetrahedra. *Angew. Chem. Int. Ed.* **2005**, *44*, 2877–2880.

153. Morris, J.; Noll, B. C.; Henderson, K. W. High-connectivity networks: characterization of the first uninodal 9-connected net and two topologically novel 7-connected nets. *Chem. Commun.* **2007**, 5191–5193.

154. Cairns, A. J.; Perman, J. A.; Wojtas, L.; Kravtsov VCh, Alkordi M. H.; Eddaoudi, M.; Zaworotko, M. J. Supermolecular building blocks (SBBs) and crystal design: 12-Connected open frameworks based on a molecular cubohemioctahedron. *J. Am. Chem. Soc.* **2008**, *130*, 1560–1561.

155. Nouar, F.; Eubank, J. F.; Bousquet, T.; Wojtas, L.; Zaworotko, M. J.; Eddaoudi, M. Supermolecular building blocks (SBBs) for the design and synthesis of highly porous metal-organic frameworks. *J. Am. Chem. Soc.* **2008**, *130*, 1833–1835.

156. Holden, A. *Shapes, Space, and Symmetry*; Columbia University Press: New York, 1971.

157. Johnson, N. Convex polyhedra with regular faces. *Can. J. Math.* **1966**, *18*, 169–200.

158. Moulton, B.; Lu, J.; Mondal, A.; Zaworotko, M. Nanoballs: nanoscale faceted polyhedra with large windows and cavities. *Chem. Commun.* **2001**, 863–864.

159. Eddaoudi, M.; Kim, J.; Wachter, J. B.; Chae, H. K.; O'Keeffe, M.; Yaghi, O. M. Porous metal-organic polyhedra: 25 Å Cuboctahedron constructed from 12 $Cu_2(CO_2)_4$ paddle-wheel building blocks. *J. Am. Chem. Soc.* **2001**, *123*, 4368–4369.

160. Sudik, A. C.; Millward, A. R.; Ockwig, N. W.; Cote, A.; Kim, J.; Yaghi, O. M. Design, synthesis, structure, and gas ($N_2$, Ar, $CO_2$, $CH_4$, and $H_2$) sorption properties of porous metal-organic tetrahedral and heterocuboidal polyhedra. *J. Am. Chem. Soc.* **2005**, *127*, 7110–7118.

161. Sudik, A. C.; Cote, A. P.; Wong-Foy, A. G.; O'Keeffe, M.; Yaghi, O. M. A metal-organic framework with a hierarchical system of pores and tetrahedral building blocks. *Angew. Chem. Int. Ed.* **2006**, *45*, 2528–2533.

162. McManus, G.; Wang, Z.; Zaworotko, M. J. Suprasupermolecular chemistry: infinite networks from nanoscale metal-organic building blocks. *Cryst. Growth Des.* **2004**, *4*, 11–13.

163. Zhou, H.; Lachgar, A. Octahedral metal clusters $[Nb_6Cl_{12}(CN)_6]^{4-}$ as molecular building blocks: from supramolecular assemblies to coordination polymers. *Eur. J. Inorg. Chem.* **2007**, 1053–1066.

164. Moon, D.; Kang, S.; Park, J.; Lee, K.; John, R. P.; Won, H.; Seong, G. H.; Kim, Y. S.; Kim, G. H.; Rhee, H.; Lah, M. S. Face-driven corner-linked octahedral nanocages: $M_6L_8$ cages formed by $C_3$-symmetric triangular facial ligands linked via $C_4$-symmetric square

tetratopic $Pd^{II}$ ions at truncated octahedron corners. *J. Am. Chem. Soc.* **2006**, *128*, 3530–3531.

165. Park, J.; Hong, S.; Moon, D.; Park, M.; Lee, K.; Kang, S.; Zou, Y.; John, R. P.; Kim, G. H.; Lah, M. S. Porous metal-organic frameworks based on metal-organic polyhedra with nanosized cavities as supramolecular building blocks: two-fold interpenetrating primitive cubic networks of $[Cu_6L_8]^{12+}$ nanocages. *Inorg. Chem.* **2007**, *46*, 10208–10213.
166. Perry, J. J. IV; Kravtsov, V.; McManus, G. J.; Zaworotko, M. J. Bottom up synthesis that does not start at the bottom: quadruple covalent cross-linking of nanoscale faceted polyhedra. *J. Am. Chem. Soc.* **2007**, *129*, 10076–10077.
167. Zou, Y.; Park, M.; Hong, S.; Lah, M. S. A designed metal-organic framework based on a metal-organic polyhedron. *Chem. Commun.* **2008**, 2340–2342.
168. Li, H.; Laine, A.; O'Keeffe, M.; Yaghi, O. M. Supertetrahedral sulfide crystals with giant cavities and channels. *Science* **1999**, *283*, 1145–1147.
169. Li, H.; Eddaoudi, M.; Laine, A.; O'Keeffe, M.; Yaghi, O. M. Noninterpenetrating indium sulfide supertetrahedral cristobalite framework. *J. Am. Chem. Soc.* **1999**, *121*, 6096–6097.
170. Cahill, C. L.; Ko, Y.; Parise, J. B. A novel 3-dimensional open framework sulfide based upon the $[In_{10}S_{20}]^{10-}$ supertetrahedron: DMA-InS-SB1. *Chem. Mater.* **1998**, *10*, 19–21.
171. Huang, X. -C.; Lin, Y. -Y.; Zhang, J. -P.; Chen, X. -M. Ligand-directed strategy for zeolite-type metal-organic frameworks: Zinc(II) imidazolates with unusual zeolitic topologies. *Angew. Chem. Int. Ed.* **2006**, *45*, 1557–1559.
172. Park, K. S.; Ni, Z.; Cote, A. P.; Choi, J. Y.; Huang, R.; Uribe-Romo, F. J.; Chae, H. K.; O'Keeffe, M.; Yaghi, O. M. Exceptional chemical and thermal stability of zeolitic imidazolate frameworks. *Proc. Natl Acad. Sci. USA* **2006**, *103*, 10186–10191.
173. Zaworotko, M. J. Crystal engineering of diamondoid networks. *Chem. Soc. Rev.* **1994**, *23*, 283–288.
174. Liu, Y.; Kravtsov V.Ch., Larsen R.; Eddaoudi, M. Molecular building blocks approach to the assembly of zeolite-like metal-organic frameworks (ZMOFs) with extra-large cavities. *Chem. Commun.* **2006**, 1488–1490.
175. Eddaoudi, M.; Eubank, J. F.; Liu, Y.; Kravtsov V.Ch., Larsen R. W.; Brant, J. A. Zeolites embrace metal-organic frameworks: building block approach to the design and synthesis of zeolite-like metal-organic frameworks (ZMOFs). *Stud. Surf. Sci. Catal.* **2007**, *170B*, 2021–2029 (from zeolites to porous MOF materials).
176. Sava, D. F.; Kravtsov V.Ch., Nouar F.; Wojtas, L.; Eubank, J. F.; Eddaoudi, M. Quest for zeolite-like metal-organic frameworks: on pyrimidinecarboxylate bis-chelating bridging ligands. *J. Am. Chem. Soc.* **2008**, *130*, 3768–3770.
177. Kaes, C.; Katz, A.; Hosseini, M. W. Bipyridine: the most widely used ligand. A review of molecules comprising at least two 2,2′-bipyridine units. *Chem. Rev.* **2000**, *100*, 3553–3590.
178. Navarro, J. A. R.; Lippert, B. Simple 1:1 and 1:2 complexes of metal ions with heterocycles as building blocks for discrete molecular as well as polymeric assemblies. *Coord. Chem. Rev.* **2001**, *222*, 219–250.
179. Tabares, L. C.; Navarro, J. A. R.; Salas, J. M. Cooperative guest inclusion by a zeolite analogue coordination polymer. Sorption behavior with gases and amine and group 1 metal salts. *J. Am. Chem. Soc.* **2001**, *123*, 383–387.

180. Banerjee, R.; Phan, A.; Wang, B.; Knobler, C.; Furukawa, H.; O'Keeffe, M.; Yaghi, O. M. High-throughput synthesis of zeolitic imidazolate frameworks and application to $CO_2$ capture. *Science* **2008**, *319*, 939–943.

181. Wang, B.; Cote, A. P.; Furukawa, H.; O'Keeffe, M.; Yaghi, O. M. Colossal cages in zeolitic imidazolate frameworks as selective carbon dioxide reservoirs. *Nature* **2008**, *453*, 207–211.

182. Davis, M. E. Ordered porous materials for emerging applications. *Nature* **2002**, *417*, 813–821.

183. Davis, M. E. The quest for extra-large pore, crystalline molecular sieves. *Chem. Eur. J.* **1997**, *3*, 1745–1750.

184. Cheetham, A. K.; Rao, C. N. R.; Feller, R. K. Structural diversity and chemical trends in hybrid inorganic-organic framework materials. *Chem. Commun.* **2006**, 4780–4795.

185. Evans, O. R.; Ngo, H. L.; Lin, W. Chiral porous solids based on lamellar lanthanide phosphonates. *J. Am. Chem. Soc.* **2001**, *123*, 10395–10396.

186. Sawaki, T.; Aoyama, Y. Immobilization of a soluble metal complex in an organic network. Remarkable catalytic performance of a porous dialkoxyzirconium polyphenoxide as a functional organic zeolite analogue. *J. Am. Chem. Soc.* **1999**, *121*, 4793–4798.

187. Gomez-Lor, B.; Gutierrez-Puebla, E.; Iglesias, M.; Monge, M. A.; Ruiz-Valero, C.; Snejko, N. $In_2(OH)_3(BDC)_{1.5}$ (BDC = 1,4-benzendicarboxylate): an In(III) supramolecular 3D framework with catalytic activity. *Inorg. Chem.* **2002**, *41*, 2429–2432.

188. Zou, Q.; Sakurai, H.; Xu, Q. Preparation, adsorption properties, and catalytic activity of 3D porous metal-organic frameworks composed of cubic building blocks and alkali-metal ions. *Angew. Chem. Int. Ed.* **2006**, *45*, 2542–2546.

189. Horcajada, P.; Serre, C.; Vallet-Regi, M.; Sebban, M.; Taulelle, F.; Férey, G. Metal-organic frameworks as efficient materials for drug delivery. *Angew. Chem. Int. Ed.* **2006**, *45*, 5974–5978.

190. Iwamoto, T. The Hofmann-type and related inclusion compounds. In *Inclusion Compounds: Structural Aspects of Inclusion Compounds Formed by Inorganic and Organometallic Host Lattices*; Atwood, J. L.; Davies, J. E. D.; MacNicol, D. D., Eds., Academic Press: London, 1984; pp. 29–57.

191. Millward, A. R.; Yaghi, O. M. Metal-organic frameworks with exceptionally high capacity for storage of carbon dioxide at room temperature. *J. Am. Chem. Soc.* **2005**, *127*, 17998–17999.

192. Hydrogen, Fuel Cells & Infrastructure Technologies Program: Multi-year research, Development and Demonstration, Plan., U.S., Department of Energy, February 2005, Chapter 3, http://www.eere.energy.gov/hydrogenandfuelcell/mypp/.

193. Basic Research Needs for the Hydrogen Economy, report of the Basic Energy Sciences workshop on Hydrogen, Production, Storage and Use, U. S. Department of Energy, May 13–15, **2005**, http://www.sc.doe.gov/bes/.

194. Fichtner, M. Nanotechnological aspects in materials for hydrogen storage. *Adv. Eng. Mater.* **2005**, *7*, 443–455.

195. Rowsell, J. L. C.; Yaghi, O. M. Metal-organic frameworks: a new class of porous materials. *Micro. Meso. Mater.* **2004**, *73*, 3–14.

196. Zhou, L. Progress and problems in hydrogen storage methods. *Renewable Sustain. Energy Rev.* **2005**, *9*, 395–408.

197. Barrett, S. Patent analysis identifies trends in fuel cell R&D. *Fuel Cell Bull.* **2005**, 12–13.

198. Rosi, N. L.; Eckert, J.; Eddaoudi, M.; Vodak, D. T.; Kim, J.; O'Keeffe, M.; Yaghi, O. M. Hydrogen storage in microporous metal-organic frameworks. *Science* **2003**, *300*, 1127–1129.

199. Jaroniek, M. M. R. *Physical Adsorption on Heterogeneous Solids*; Elsevier: Amsterdam, 1988.

200. Czepirski, L.; Jagiello, J. Virial-type thermal equation of gas-solid adsorption. *Chem. Eng. Sci.* **1989**, *44*, 797–801.

201. Li, Y.; Yang, R. T. Hydrogen storage in metal-organic and covalent-organic frameworks by spillover. *AIChE J.* **2008**, *54*, 269–279.

202. Collins, D.; Zhou, H. Hydrogen storage in metal-organic frameworks. *J. Mater. Chem.* **2007**, *17*, 3154–3160.

203. Hirscher, M.; Panella, B. Hydrogen storage in metal-organic frameworks. *Scripta Materialia 56*, **2007**, 809–812.

204. Isaeva, V. I.; Kustov, L. M.; Metal-organic framework compounds as new materials for hydrogen storage. *Rossiiskii Khimicheskii Zhurnal* **2006**, *50*, 56–72.

205. Thomas, K. Hydrogen adsorption and storage on porous materials. *Catal. Today* **2007**, *120*, 389–398.

206. Férey, G. Hybrid porous solids: past, present, future. *Chem. Soc. Rev.* **2008**, *37*, 191–214.

207. Li, Y.; Yang, R. T. Significantly enhanced hydrogen storage in metal-organic frameworks via spillover. *J. Am. Chem. Soc.* **2006**, *128*, 726–727.

208. Li, Y.; Yang, R. T. Hydrogen storage in metal-organic frameworks by bridged hydrogen spillover. *J. Am. Chem. Soc.* **2006**, *128*, 8136–8137.

209. Li, Y.; Yang, R. T.; Gas adsorption and storage in metal-organic framework MOF-177. *Langmuir* **2007**, *23*, 12937–12944.

210. Biradha, K.; Hongo, Y.; Fujita, M. Crystal-to-crystal sliding of 2D coordination layers triggered by guest exchange. *Angew. Chem. Int. Ed.* **2002**, *41*, 3395–3398.

211. Shmilovits, M.; Diskin-Posner, Y.; Vinodu, M.; Goldberg, I. Crystal engineering of "porphyrin sieves" based on coordination polymers of Pd- and Pt-tetra(4-carboxyphenyl) porphyrin. *Cryst. Growth Des.* **2003**, *3*, 855–863.

212. Shmilovits, M.; Vinodu, M.; Goldberg, I. Coordination polymers of tetra(4-carboxyphenyl)porphyrins sustained by tetrahedral zinc ion linkers. *Cryst. Growth Des.* **2004**, *4*, 633–638.

213. Kempe, R. Coordination polymers that contain accessible metal centres—synthesis and structure of a nickel cluster connected porphyrin network. *Z. Anorg. Allg. Chem.* **2005**, *631*, 1038–1040.

214. Kitaura, R.; Onoyama, G.; Sakamoto, H.; Matsuda, R.; Noro, S.; Kitagawa, S. Immobilization of a metallo Schiff base into a microporous coordination polymer. *Angew. Chem. Int. Ed.* **2004**, *43*, 2684–2687.

215. Cho, S.; Ma, B.; Nguyen, S.; Hupp, J. T.; Albrecht-Schmitt, T. E. A metal-organic framework material that functions as an enantioselective catalyst for olefin epoxidation. *Chem. Commun.* **2006**, 2563–2565.

216. Mulfort, K. L.; Hupp, J. T. Chemical reduction of metal-organic framework materials as a method to enhance gas uptake and binding. *J Am Chem Soc* **2007**, *129*, 9604–9605.

217. Qiu, L.; Xie, A.; Zhang, L. Encapsulation of catalysts in supramolecular porous frameworks: size- and shape-selective catalytic oxidation of phenols. *Adv. Mater.* **2005**, *17*, 689–692.

218. Brunel, D.; Lasperas, M. Design of chiral hybrid organic-inorganic mesoporous materials as enantioselective epoxidation and alkylation catalysts. *Nanostruct. Catal.* **2003**, 157–177.

219. Hutchings, G. Enantioselective reactions using modified microporous and mesoporous materials. *Surf. Chem. Catal.* **2002**, 241–274.

220. Kesanli, B.; Lin, W. Chiral porous coordination networks: rational design and applications in enantioselective processes. *Coord. Chem. Rev.* **2003**, *246*, 305–326.

221. Wu, D.; Hu, A.; Zhang, L.; Lin, W. A homochiral porous metal-organic framework for highly enantioselective heterogeneous asymmetric catalysis. *J. Am. Chem. Soc.* **2005**, *127*, 8940–8941.

222. Dybtsev, D. N.; Nuzhdin, A. L.; Chun, H.; Bryliakov, K. P.; Talsi, E.; Fedin, V.; Kim, K. A homochiral metal-organic material with permanent porosity, enantioselective sorption properties, and catalytic activity. *Angew. Chem. Int. Ed.* **2006**, *45*, 916–920.

223. Rieter, W.; Taylor, K.; Lin, W. Surface modification and functionalization of nanoscale metal-organic frameworks for controlled release and luminescence sensing. *J. Am. Chem. Soc.* **2007**, *129*, 9852–9853.

# 3

# TOPOLOGY AND INTERPENETRATION

STUART R. BATTEN
*School of Chemistry, Monash University, Australia*

## 3.1 INTRODUCTION

A very useful approach to both the design and analysis of crystal structures is the use of nets. Complicated structures can be reduced to simple networks to gain a better understanding of the way that the constituent components arrange themselves. It also provides a powerful way of controlling that arrangement and thus ultimately properties of the material overall.

A net is an infinite array of nodes, each of which is connected to three or more others by 2-connecting links. Links must be 2-connecting; if they connect three or more nodes then they themselves are nodes. Similarly, nodes are 3-connecting or higher; 2-connecting "nodes" are in fact links.

The net approach was first developed by Wells[1–3] to describe crystal structures, and was later applied to coordination polymer *design* by Robson and Hoskins.[4–6] In the network analysis of crystal structures it is just a matter of identifying the nodes in the network, and then analyzing their connectivity. Nodes can be metal ions, clusters, and/or ligands. Care, however, should be taken in identifying the nodes and connections. For example, a particular net will be generated when a cluster of metal ions and ligands is chosen as a node. A different net, however, may result if the individual metals and ligands of that cluster are instead assigned as the nodes. Thus the makeup of the nodes should always be clearly specified in a topological description.

*Metal-Organic Frameworks: Design and Application*, Edited by Leonard R. MacGillivray

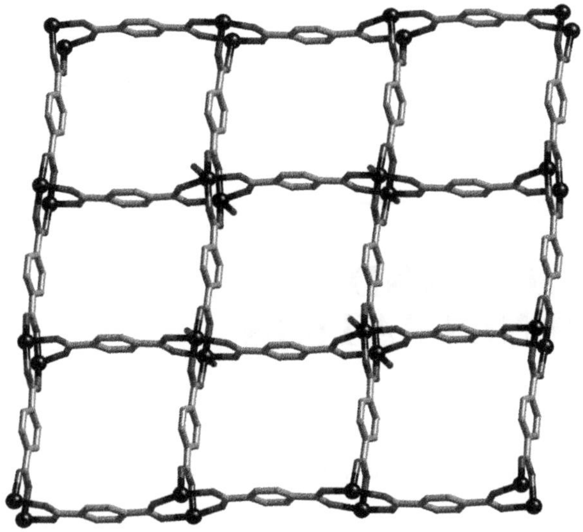

**FIGURE 3.1** A sheet structure containing 4-connecting $Zn_2(O_2C)_4$ secondary building units (SBUs).

Often, however, it is better to choose a cluster-like subunit of the framework as a node. These are sometimes known as secondary building units (SBUs),[7] particularly (but not exclusively) in relation to metal acetate clusters formed *in situ* during the assembly of the framework. These can often greatly simplify the description of the network. For example, the structure of $Zn(BDC)(H_2O)$.DMF, BDC = 1,4-benzenedicarboxylate contains zinc dimers linked by BDC anions into a 2D sheet (Figure 3.1).[8] By assigning $Zn_2(O_2C)_4$ moieties as nodes, which are linked by the aromatic backbones of the ligands, the simple and common (4,4) topology is achieved. If, however, the individual ligands and metals are chosen as nodes, a more complicated (and rare) binodal 4-connected net results.

Nonetheless, the nodes chosen should also be chemically sensible—for example, it is much better to describe a tetradentate ligand as a single 4-connecting node rather than two interconnected 3-connecting nodes (it is a single chemical species). These two different ways of describing the ligand will give different network topologies, and thus the topological description is dependent on the nodal choice (as it was for the previous example). Ultimately, the network description should correspond to the chemical structure of the coordination polymers, and must help simplify the description (or design) of the overall structure. If not, then this approach is of little benefit.

The design approach is similarly straightforward. The net targeted is deconstructed into nodes and links and then chemical equivalents are chosen and mixed. If, for example, one wished to design a diamond net, which is composed of interlinked tetrahedral nodes, then one could use tetrahedral metals (e.g., $Zn^{II}$, $Cd^{II}$, $Cu^{I}$) linked by linear ligands, tetrahedral ligands linked by 2-connecting metal ions (or metal complexes), or both tetrahedral ligands and metals. Thus reaction of $Zn^{II}$ or $Cd^{II}$

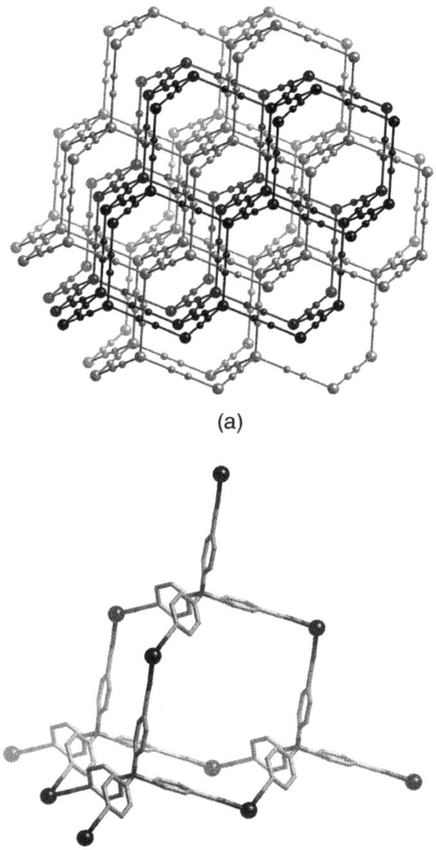


**FIGURE 3.2** (a) The two interpenetrating diamondoid nets of $Cd(CN)_2$, and (b) a large adamantane cavity in the diamond structure of $Cu^{I}[C(C_6H_4CN)_4]BF_4.xC_6H_5NO_2$.

with the linear cyanide ligand gives diamond like nets for $M(CN)_2$, M = Zn, Cd (Figure 3.2a).[5] Alternatively, the structure of $Cu^{I}[C(C_6H_4CN)_4]BF_4.xC_6H_5NO_2$ contains both tetrahedral metal and ligand nodes, and again a (very open) diamond network is formed[4] (Figure 3.2b).

It should be noted, as we will see below, that there are multiple possible networks for the same types of nodes. So for tetrahedral nodes, not only are diamond nets possible, but also other topologies such as Lonsdaleite. Nonetheless, in this case diamond is overwhelmingly the most common topology. By contrast, for 3-connecting and 4-connecting planar nodes there are numerous 2D and 3D possibilities, many of which are relatively common.

It is also important to emphasize that a network is a topological description and not a geometric one. As a result, the two networks shown in Figure 3.3, which are clearly geometrically different, are in fact topologically identical. One can be converted into

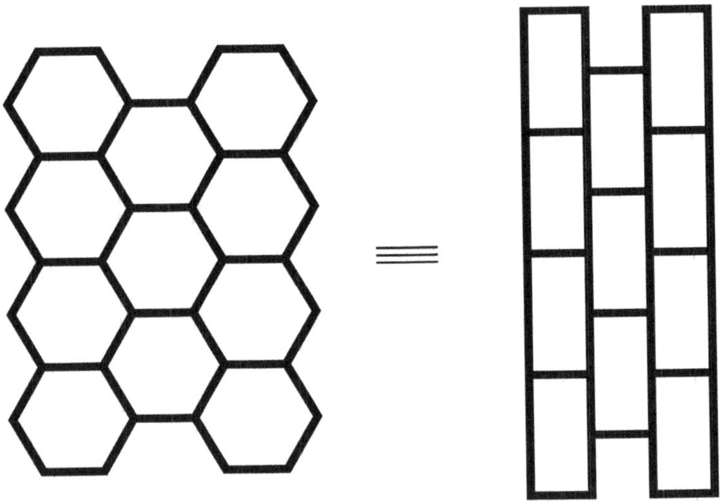

**FIGURE 3.3** Two *topologically* identical (but geometrically different) nets.

the other just by changing the geometry around the nodes, and without the breaking of any links. Similarly, from a topological viewpoint tetrahedral and square planar nodes are in fact identical—they are both 4-connecting nodes. In practice, however, the topologies that result from these geometrically different nodes are usually very different. Also, from a chemical viewpoint these geometries are obviously quite different with regard to the design of ligands and metal ions, and so we will make the (strictly incorrect) distinction here.

Furthermore, the local chemical geometry can be very different to the nodal geometry. For example, the structure of $Zn(dca)_2$, dca = dicyanamide, $N(CN)_2^-$ contains *tetrahedral* $Zn^{II}$ nodes linked by dca anions.[9] The dca bridges are kinked, and so the overall topology is of a square grid sheet with *square planar* 4-connected nodes. The geometry of the underlying net itself can also be very different to the "ideal," most symmetric version of it. The structure of Cu(CN)(4-cyanopyridine) contains bridging cyanide and 4-cyanopyridine ligands of very different lengths (5.04 and 9.45 Å, respectively). This gives a very distorted net, but nonetheless it is still the diamond net.

## 3.2 NOMENCLATURE

Once a structure has been simplified to a net, that net needs to be identified and classified. There are a number of pieces of information about the net that aid the identification, particularly in cases where the net is unusual or highly distorted from the most symmetrical version of the net.

The first of these is to identify the numbers of unique nodes and their connectivity. The number of nodes should be those that are *topologically* unique, not just chemically or crystallographically unique (which may be more). For example, the structure of

$Cu^I[C(C_6H_4CN)_4]BF_4.xC_6H_5NO_2$, shown in Figure 3.2(b), contains two chemically and crystallographically different nodes—the $Cu^I$ atom and the central C atom of the ligand. Nonetheless, both 4-connecting nodes are topologically equivalent, and the underlying net is that of the uninodal diamond net.

The next pieces of information about a net are the shortest circuits around each of the unique nodes. A shortest circuit is the number of nodes in the smallest loop in the net that can be constructed containing two given links from a given node. Thus, for the 2D hexagonal network shown in Figure 3.3 the shortest circuit for each pair of unique links around the 3-connecting node is a six-membered ring. Note that in most nets not all shortest circuits are of the same size—for some unique pairs of links the shortest circuit may be larger than for other pairs.

These shortest circuits can be represented mathematically by using Schläfli symbols. These list the numbers of nodes in each shortest circuit for each unique pairs of links, with superscripts to represent multiple occurrences of the same sized rings. For a $p$-connected net, there will be $p(p-1)/2$ unique pairs of links for which the shortest circuit must be given. Thus the net in Figure 3.3 has the Schläfli symbol $6^3$. Significantly, for this net all the shortest circuits are of the same size. Wells[1–3] designated these particular nets as *uniform nets*. Furthermore, all the nodes have the same connectivity, and thus it is a *Platonic uniform net*. For these types of nets the symbol $(n,p)$ is used, where $n$ is the number of nodes in the shortest circuits and $p$ is the connectivity of the nodes. Thus the net in Figure 3.3 is most commonly known as the (6,3) net. There are also a number of other mathematical properties of nets which can be calculated, including "long" Schläfli symbols (or vertex symbols); for an explanation of these the reader is directed elsewhere.[10–17]

The (6,3) net also illustrates another common feature of nets—the same net can be given different names. Wells's $(n,p)$ and Schläfli symbols are just two ways of naming nets. Furthermore, both these notations can, unfortunately, be the same for different nets—for example there are a number of (10,3) nets (labeled by Wells as (10,3)-a, (10,3)-b, (10,3)-c, etc.), and both the diamond and Lonsdaleite nets have the same $6^6$ Schläfli symbol. Thus nets are sometimes named uniquely after simple materials already known to have that topology, such as diamond, Lonsdaleite, rutile, PtS, α-Po, $SrAl_2$, sodalite, NbO, $CdSO_4$, and $ThSi_2$. Other nets, however, have no known simple prototype, and so a three-letter code system has been developed to give nets unique labels (analogous to that employed for zeolite nets).[18] Thus the (10,3)-b net of Wells can also be referred to as the $ThSi_2$ or **ths** net, and has the Schläfli symbol $10^3$.

Armed with this information about the net displayed by the compound in question, and the general nomenclature for nets, it is then a matter of matching the given net against those known and labeled (assuming the net is not new). There are a number of resources for nets, including the books by Wells,[1–3] O'Keeffe and Hyde,[17] Öhrström and Larsson,[19] and Batten, Neville and Turner[20] as well as a number of reviews.[6,16,21,22] Furthermore, the Reticular Chemistry Structure Research (RCSR) website is a valuable and extensive searchable database of nets.[18] Automated programs such as Systre,[23] OLEX,[24] and TOPOS[25] are also very useful aids in assigning network topologies. Some of the more common 2D and 3D nets, however, are briefly given below.

## 3.3 COMMON 2D NETS

A number of common 2D nets are given in Figures 3.3 and 3.4. By far the most common are (6,3) and (4,4) (Figures 3.3 and 3.4a) nets. Note that the (4,4) notation given by Wells and now widely used is not strictly correct—the shortest circuits formed by the *trans* links are six-membered, and thus the Schläfli symbol is $4^4.6^2$. Wells chose to ignore these pairs of links in deriving the (4,4) notation.

The (6,3) net should not be confused with the (3,6) net (Figure 3.4b); the later has six-connecting nodes and three-membered shortest circuits (again, only circuits from adjacent nodes have been considered, arbitrarily, in deriving this symbol). Similarly, neither net should be confused with a 3,6-connected net, such as that shown in Figure 3.4(c). This is a binodal net containing both 3-connecting and 6-connecting nodes. The 3-connected net shown in Figure 3.4(d) contains two types of shortest circuits—four-membered and eight-membered—to give the Schläfli symbol $4.8^2$. The net in Figure 3.4(e) contains two sorts of nodes in a 4:1 ratio, giving a Schläfli symbol of

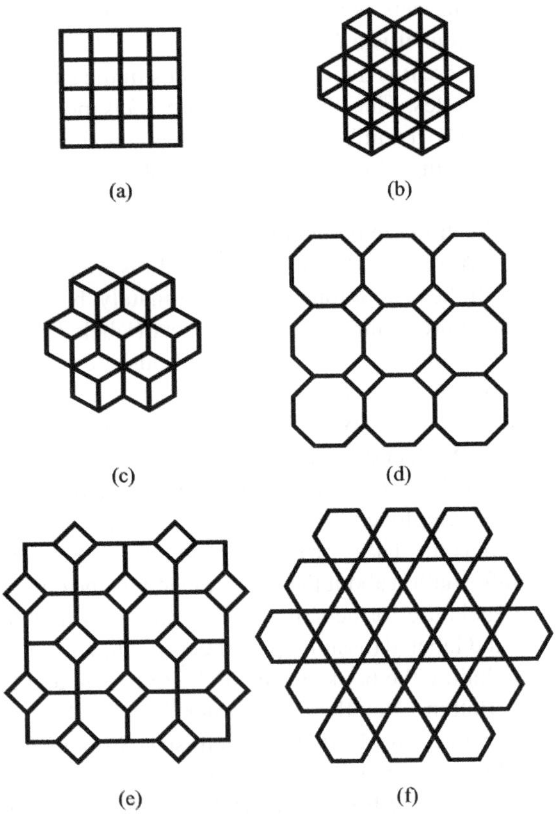

**FIGURE 3.4** A selection of 2D nets: (a) (4,4), (b) (3,6), (c) the 3,6-connected $(4^3)_2(4^6.6^6.8^3)$ net, (d) $4.8^2$, (e) $(4.6^2)_4(6^4.10^2)$, and (f) the Kagomé lattice.

$(4.6^2)_4(6^4.10^2)$. The Kagomé lattice (Figure 3.4f) is a particularly interesting net for magnetochemists, due to the formation of the three-membered rings. In a network with antiferromagnetic coupling between adjacent nodes, the three-membered rings can lead to spin frustration. While two corners of a triangle can align their magnetic spins antiparallel, the third corner cannot be antiparallel with *both* of the other two corners. Such spin frustration has indeed been observed in a coordination polymer with this topology.[26]

The 2D nets shown in Figures 3.3 and 3.4 are essentially planar in nature—that is they can be mapped onto a plane. There are a number of other 2D nets, however, which cannot be mapped onto a plane (without links lying on top of each other), but rather have a degree of "thickness". These can be created by cross-linking two or more layers. For example, the net shown in Figure 3.5(a) contains two layers of 1D rods in which the rods run at an inclined angle to those of the adjoining layer. These two layers are linked to create a 2D bilayer network. Such a topology has been observed in the structure of $Ag(bpea)L_{1/2}.H_2O$, bpea = 1,2-bis(-pyridyl)ethane, L = 4,4′-biphenylcarboxylate.[27]

Similarly, bilayers can be created by linking 2D nets rather than layers of 1D chains. A series of interesting bilayer structures containing pairs of interlinked (4,4) sheets have been observed for lanthanoid complexes of 4,4′-bipyridine-*N,N′*-dioxide.[28] These can have either 5-connecting, 6-connecting, or both 7- and 8-connecting nodes (Figure 3.5b–d). Cross-linking of (6,3) sheets can also occur; Figure 3.5(e) shows the 3,4-connected net formed in the structure of $Cu_4(dca)_4(4,4'\text{-bipy})_3(MeCN)_2$, 4,4′-bipy = 4,4′-bipyridine.[29] Finally, there are a few scattered reports of 2D structures containing three or more cross-linked layers, including a trilayer structure containing two outer (4,4) sheets and an inner layer of 1D rods,[30] and a pentalayer structure containing five interconnected layers of rods.[31]

## 3.4 COMMON 3D NETS

### 3.4.1 3-Connected Nets

Three-connected nets are of significant interest because, all things being equal, the lower the connectivity of the nodes the more open space available in the network. The most common 3-connected nets are the (10,3) nets (Figure 3.6), and in particular, (10,3)-a and (10,3)-b. The (10,3)-a net is unusual in that it is inherently chiral—the fourfold helices visible in the figure are all of the same hand. As a result, this net is of particular interest to those seeking chiral nets for applications such as enantioselective absorption and catalysis.[32] The (10,3)-b net is also interesting in that it can expand or collapse like a wine-rack by changing only the torsion angle around certain bonds.[33] The (10,3)-c and (10,3)-d nets are less common; (10,3)-d can be confused with (10,3)-a; however, in the (10,3)-d net adjoining fourfold helices are of opposite handedness.

There are few reports of other 3D 3-connected nets. The (12,3) net is interesting in that it is self-penetrating, as discussed later. The binodal nets LiGe, **lvt-a** and **nbo-a** are shown in Figure 3.7. The later can be generated from the NbO 4-connected net (vide infra) by replacing the 4-connecting nodes with pairs of 3-connecting nodes. This

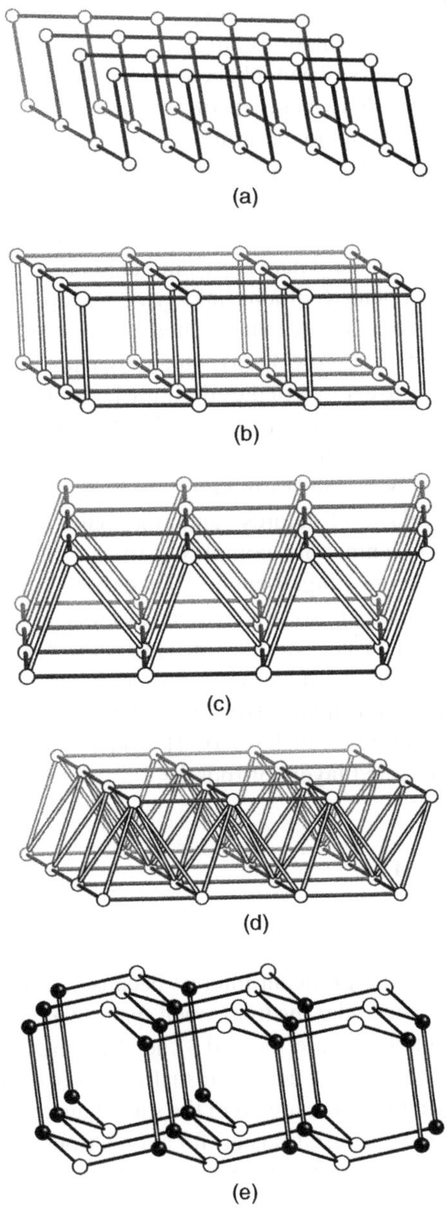


**FIGURE 3.5** Two-dimensional bilayer structures can be created by (a) cross-linking two layers of rods, or by cross-linking pairs of (4,4) sheets by (b) one link per node, (c) two links per node, or (d) four links per node, or by (e) linking pairs of (6,3) sheets.

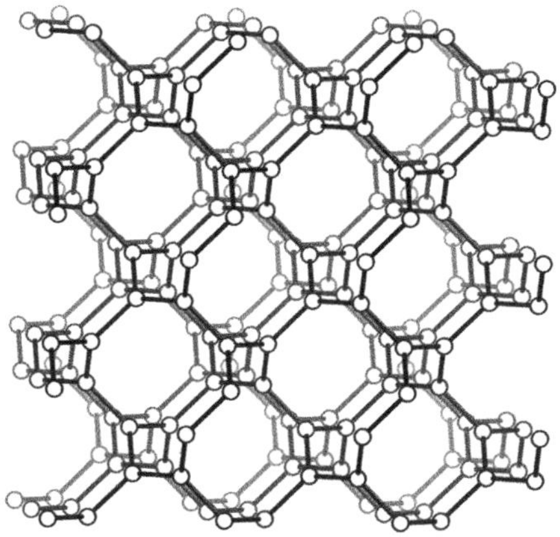

(a)

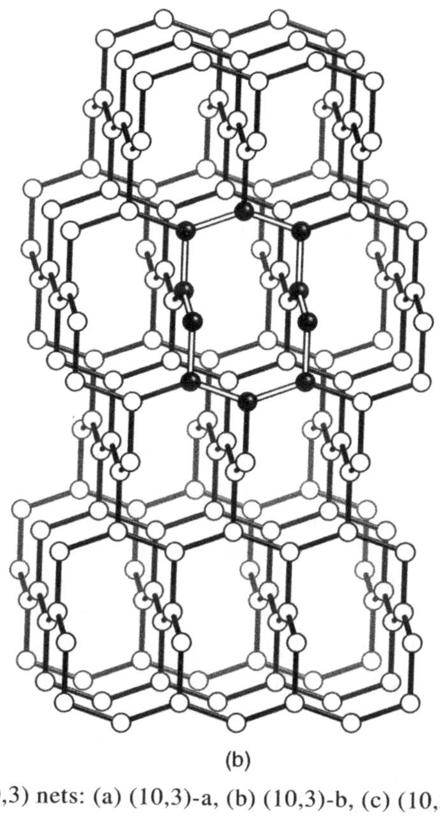

(b)

**FIGURE 3.6** Four (10,3) nets: (a) (10,3)-a, (b) (10,3)-b, (c) (10,3)-c, and (d) (10,3)-d.

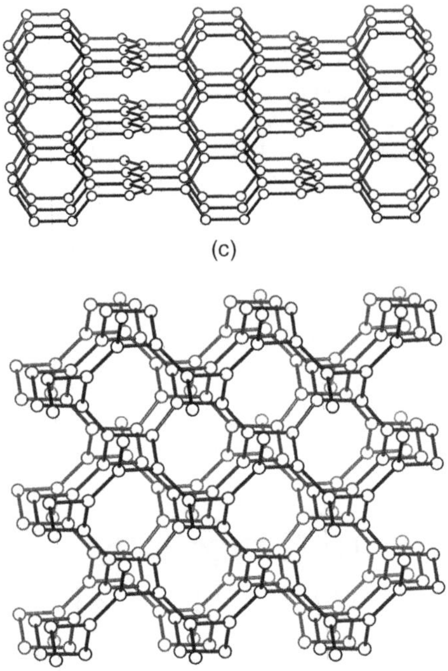

(c)

(d)

**FIGURE 3.6** (*Continued*).

replacement to generate a new net has been termed *decoration* (if the replacement nodes have the same connectivity as those they replace, it is termed *augmentation*).[16]

### 3.4.2 4-Connected Nets

The most common 4-connected 3D net, and indeed the most common 3D net overall, is the diamond net (Figure 3.8a). A feature of the net are the adamantane cavities, bounded by four six-membered rings, as highlighted in the figure. The Lonsdaleite net (Figure 3.8b) is, by contrast, very rare despite its simplicity. The difference to diamond can be seen in the nature of the cavities—in Lonsdaleite there are two sorts of cavities, one bounded by three six-membered rings, the other by five. Other common 4-connected nets with tetrahedral or pseudo-tetrahedral nodes are quartz (Figure 3.8c), which is chiral, $SrAl_2$ or **sra** (Figure 3.8d), which is the second most common 4-connected net, and the zeolitic sodalite net (Figure 3.8e).

There are also a number of reported 4-connected nets with planar nodes. In the most symmetrical version of the NbO net (Figure 3.9a) all adjacent nodes are perpendicular to each other. In the $CdSO_4$ net (Figure 3.9b) only half are perpendicular, with the others coplanar. Other nets with planar 4-connecting nodes include **lvt**, (8,4), **qzd** (or $7^5.9$), **usf**, and **mot** (Figure 3.9).

The last net (**mot**) contains two sorts of nodes, and there are a number of other 4-connected nets containing multiple nodes. Of particular interest are those containing

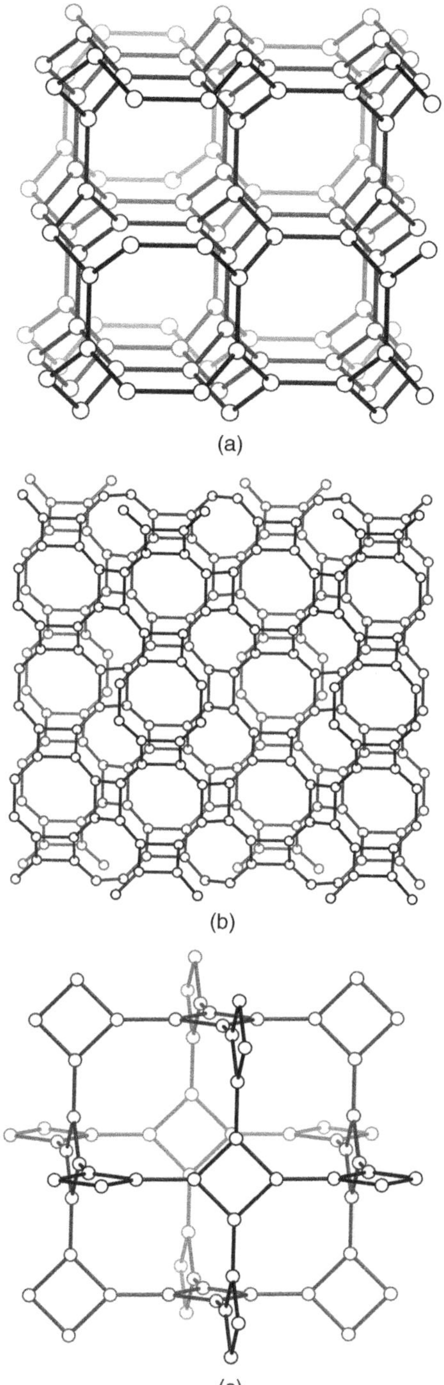


**FIGURE 3.7** The 3-connected nets (a) **lig**, (b) **lvt-a**, and (c) **nbo-a**.

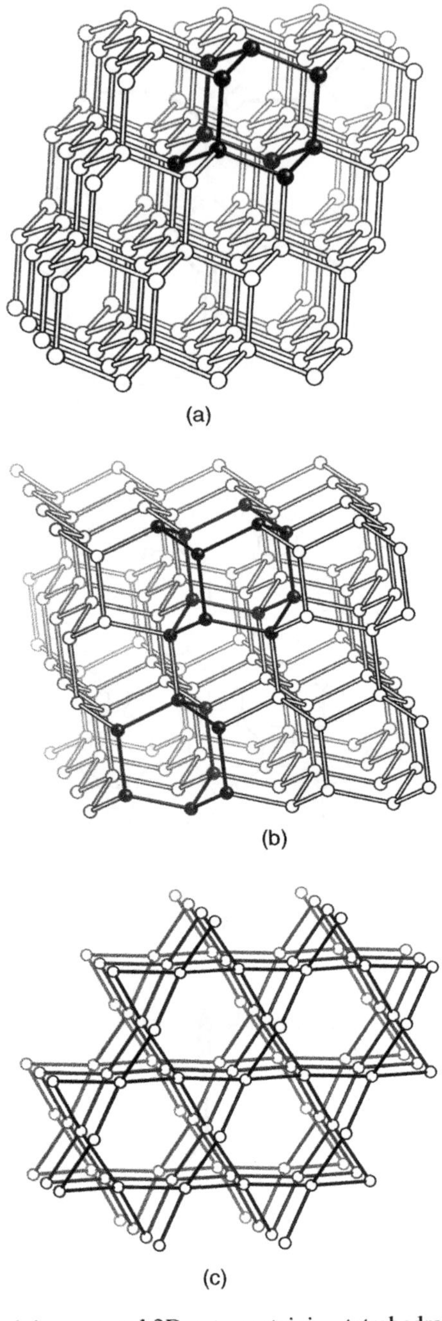


**FIGURE 3.8** Selected 4-connected 3D nets containing tetrahedral nodes: (a) diamond, (b) Lonsdaleite, (c) quartz, (d) $SrAl_2$, and (e) sodalite.

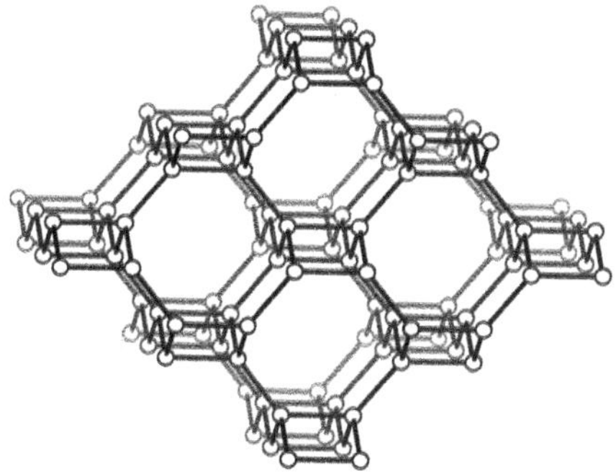

(d)

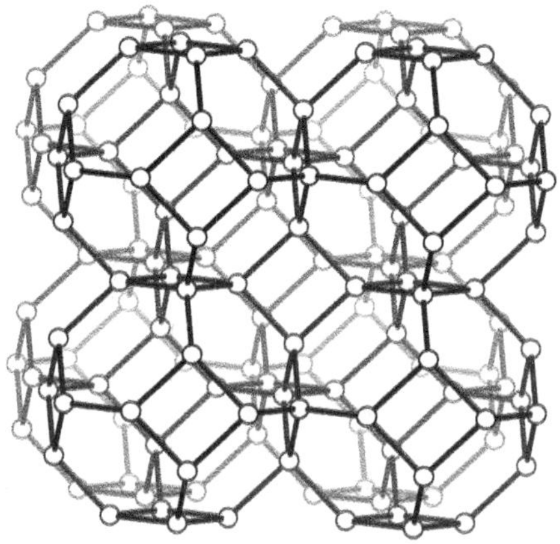

(e)

**FIGURE 3.8** (*Continued*).

both tetrahedral and square planar nodes. In the PtS net (Figure 3.10a) these nodes alternate (each tetrahedral node is connected to only square planar nodes, and vice versa), while in the moganite net there are links directly between the tetrahedral nodes (Figure 3.10b).

### 3.4.3 5-Connected Nets

There are only a few 5-connected nets. The simplest ways to generate these nets are to cross-link (6,3) sheets (boron nitride topology), or to link (4,4) sheets with links alternating between connecting to the sheet above or below (**sqp** topology).

### 3.4.4 6-Connected Nets

By far the most common 6-connected net is α-Po (Figure 3.11a). This net contains octahedral nodes. For trigonal prismatic nodes, the **acs** net can be formed

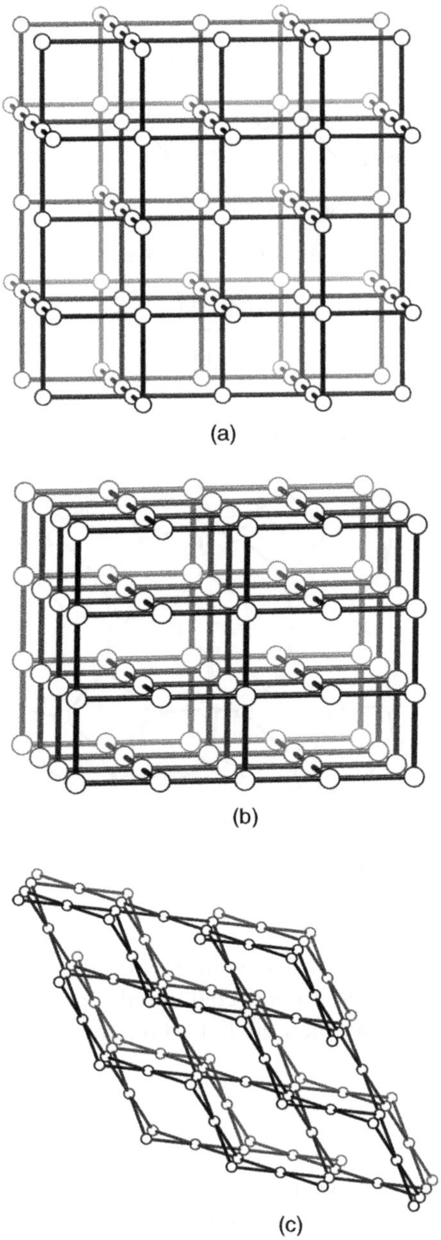


**FIGURE 3.9** Selected 4-connected 3D nets containing square planar nodes: (a) NbO, (b) CdS, (c) **lvt**, (d) (8,4), (e) **qzd**, (f) **usf**, and (g) **mot**.

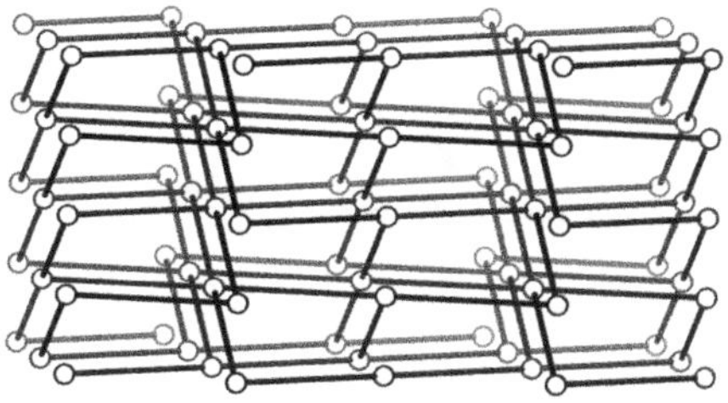

(d)

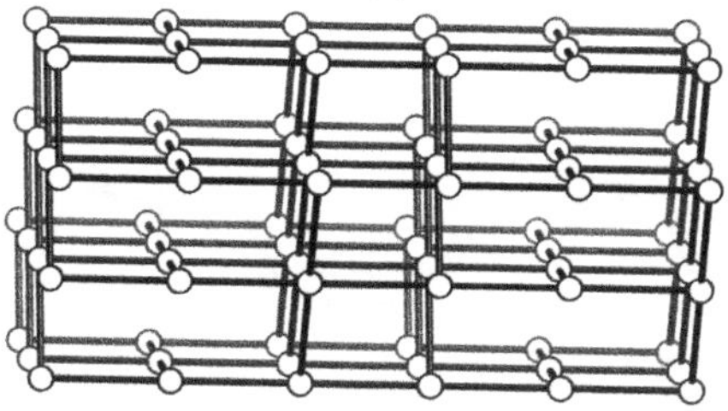

(e)

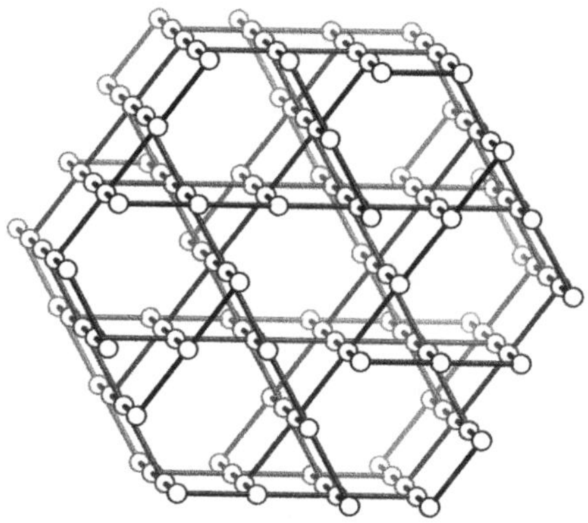

(f)

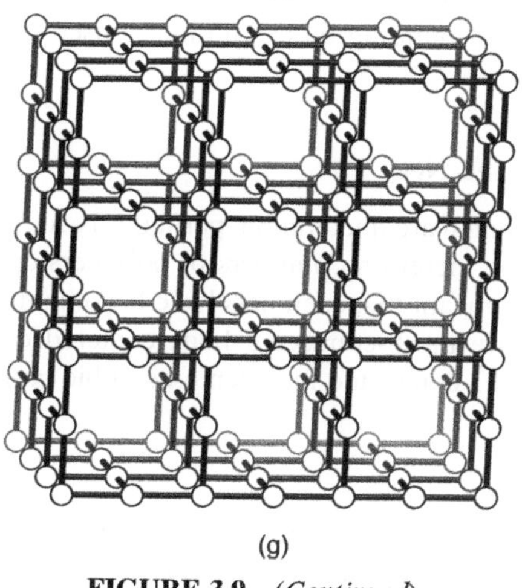

(g)

**FIGURE 3.9** (*Continued*).

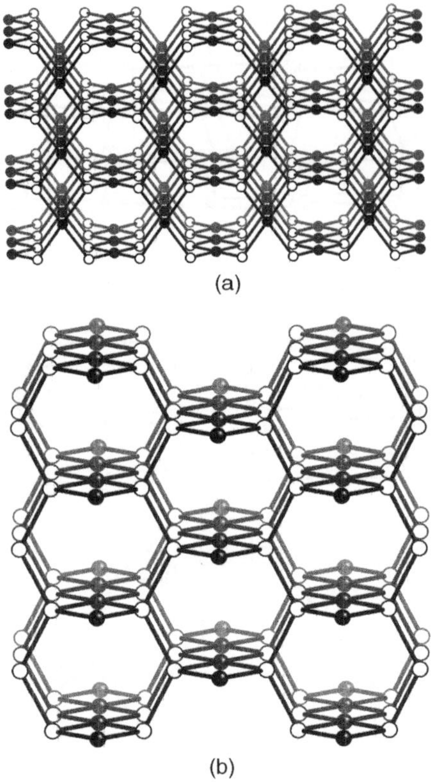


**FIGURE 3.10** Two binodal 4-connected nets: (a) PtS and (b) moganite.

(Figure 3.11b), while a 1:1 mixture of both geometries is found in the NiAs net (Figure 3.11c). These last two topologies are rather rare, mainly because the geometry is unusual for the sorts of building blocks normally used in coordination polymer construction.

### 3.4.5 Higher Connected Nets

Connectivities higher than 6 are uncommon, particularly for transition metal coordination polymers, but the increasing numbers of polymers containing lanthanide or metal cluster nodes has meant that such nets have become more relevant. The most fundamental of these are the 8-connected body-centered cubic lattice (CsCl net) (Figure 3.12a) and the 12-connected face-centered cubic lattice (Figure 3.12b).

### 3.4.6 Multinodal Nets

Nets can, of course, be constructed from more than one type of node, as we saw for 4-connected nets with both tetrahedral and square planar nodes. There are also

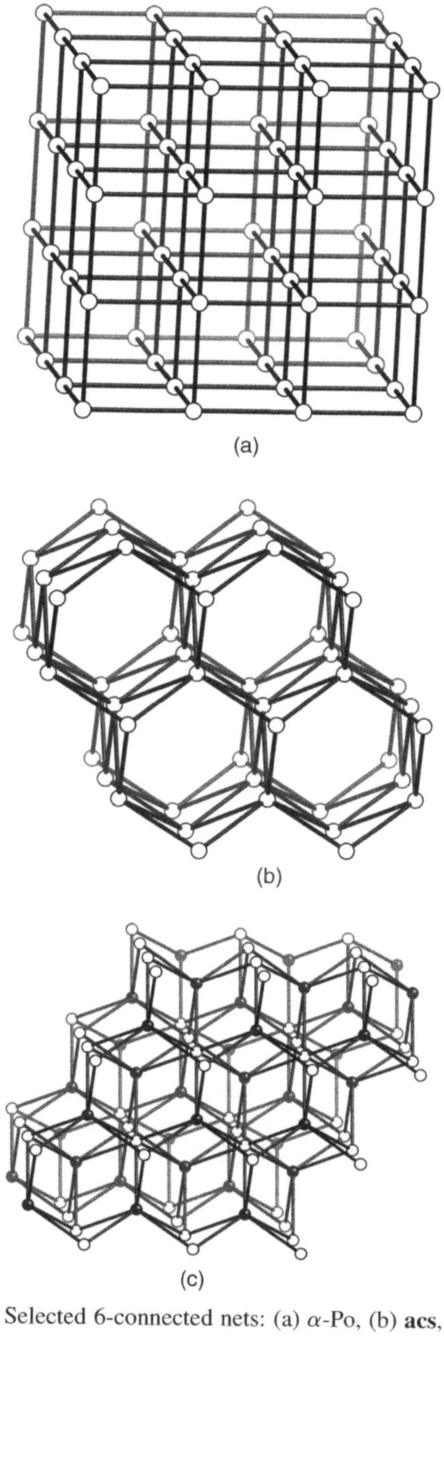


**FIGURE 3.11** Selected 6-connected nets: (a) $\alpha$-Po, (b) **acs**, and (c) NiAs.

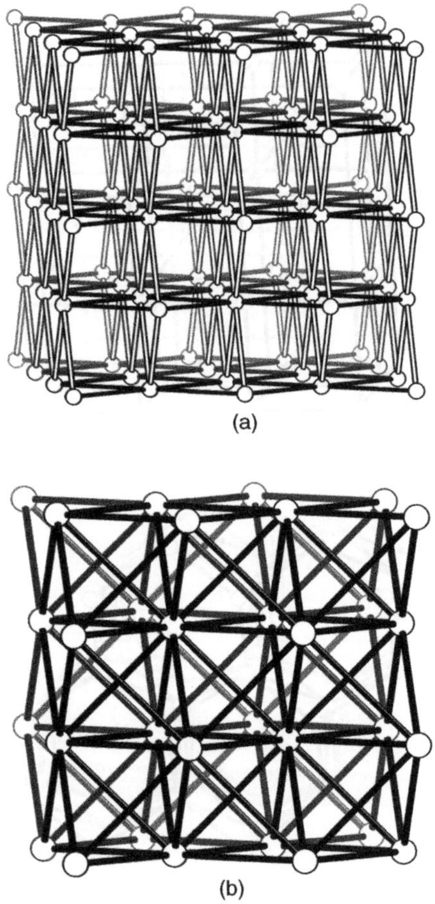


**FIGURE 3.12** (a) The 8-connected CsCl (or body-centered cubic) net and (b) the 12-connected face-centered cubic net.

numerous nets in which the nodes have different connectivities. Two closely related 3,4-connected nets are the boracite net (Figure 3.13a) and the twisted boracite net (Figure 3.13b). The former contains tetrahedral nodes while the later has square planar 4-connecting nodes. The $Pt_3O_4$ net also contains alternating square planar and trigonal nodes (Figure 3.13c). The structure of $\alpha$-$C_3N_4$ also has a 3,4-connected net, with tetrahedral 4-connecting nodes (Figure 3.13d). In all these nets, 3-connecting nodes are directly linked to only 4-connecting nodes, and vice versa. As an example of an alternative arrangement, the structure of InS has 4-connecting nodes in which one quarter of the links are to other 4-connecting nodes (Figure 3.13e).

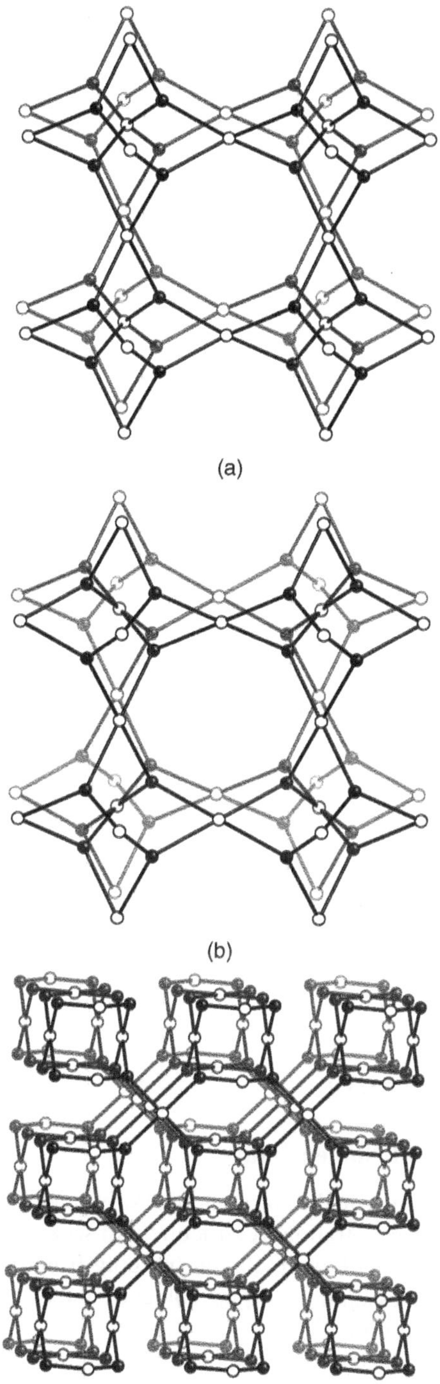

FIGURE 3.13 Selected 3,4-connected nets: (a) boracite, (b) twisted boracite, (c) $Pt_3O_4$, (d) $\alpha$-$C_3N_4$, and (e) InS.

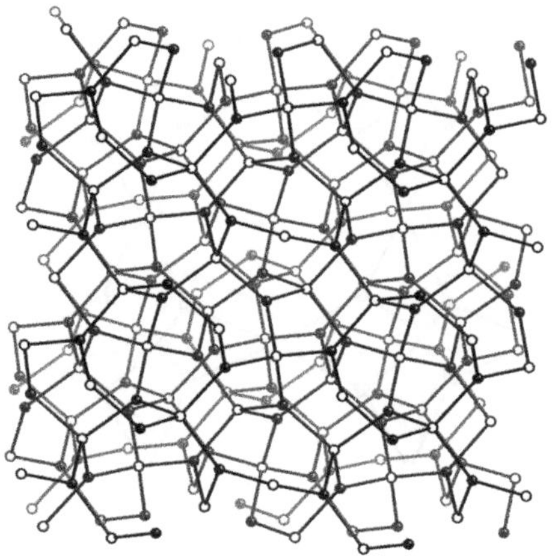

(d)

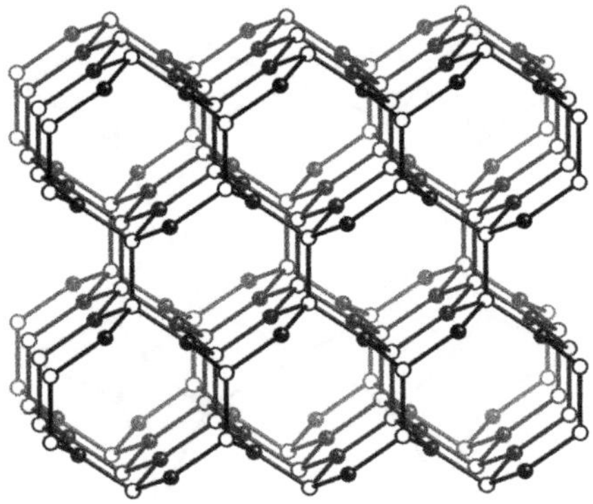

(e)

**FIGURE 3.13** (*Continued*).

## 3.5 INTERPENETRATION

### 3.5.1 Introduction and Nomenclature

The crystal engineer trying to make porous networks is inevitably going to come across the phenomenon of interpenetration. A structure will seek to fill available space as it forms, whether that is through intercalation (ranging from ordered guest species to completely disordered, essentially liquid solvent), interdigitation (for 1D or 2D networks), or interpenetration.[34]

Interpenetration is defined as occurring when two or more polymeric networks are not chemically bonded to each other but cannot be separated (in an imaginary, topological sense) without the breaking of bonds.[22] They are polymeric equivalents of molecular catenanes and rotaxanes. Thus, for example, the structure of [Mn $(dca)_2(H_2O)_2].H_2O$, which contains (4,4) sheets which stack to give channels

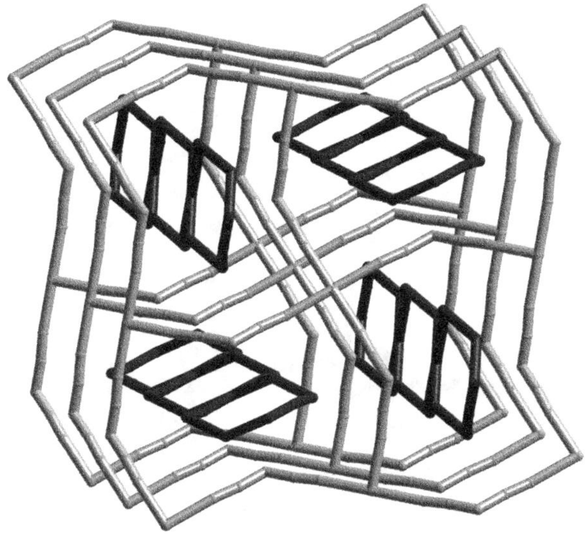

**FIGURE 3.14** One-dimensional nets intercalated within (but not interpenetrating) 2D sheets.

containing 1D chains of the same formula (Figure 3.14), is not interpenetrating.[35,36] This is because the chains could (in a theoretical sense) be pulled out of the channels without links in either net being broken.

While interpenetration can be an annoyance to those seeking maximum porosity, it does not always result in non-porous materials, and it can, in fact, also produce other interesting properties. For example, the structure of $Cu_3(tpt)_4(ClO_4)_3$, tpt = 2,4,6-tri(4-pyridyl)-1,3,5-triazine still contains 35% void space despite the twofold interpenetration of the boracite networks.[37] Interpenetration can even be beneficial to porous networks, stabilizing structures that would otherwise likely collapse upon removal of solvent.[38] Structures with variable sized pores due to the movement of interpenetrating nets relative to each other can produce interesting materials which show properties such as spin cross-over,[39] counterion controlled porosity,[40] or pressure-related hysteresis of gas sorption.[41] In one material the interpenetrating nets produce large, completely sealed-off cavities (Figure 3.15),[42] while other interpenetrating structures with interesting magnetic or electronic properties have also been reported.[43–47]

We described earlier the importance of describing the topology of coordination polymers. In interpenetrating nets, however, it is also important to describe the *topology of interpenetration*; that is, the way the nets interpenetrate. This is analogous to molecular structures where not only the geometry of the individual molecules is described, but also the packing arrangement of those molecules.

This is particularly important as nets can interpenetrate in different ways. For example, consider the two pairs of interpenetrating nets in Figure 3.16. In both arrangements the sheets have (4,4) topology, but the way they interpenetrate each other is clearly different.

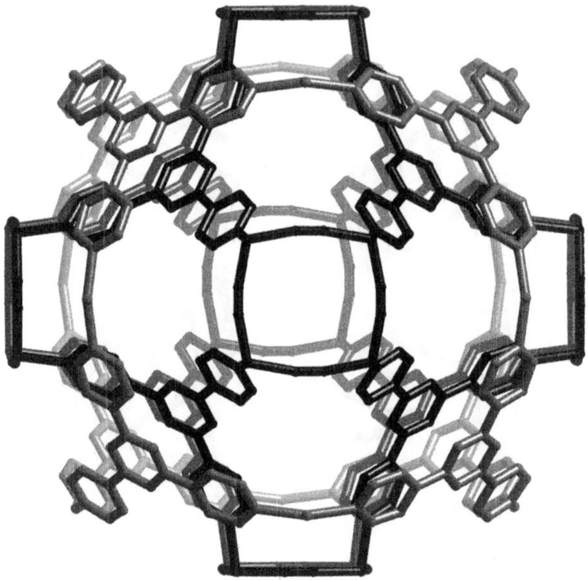

**FIGURE 3.15** A single closed-off cavity created by the interpenetration of two nets.

There are, today, many, many examples of interpenetrating structures. Thus we will not discuss all the various possible permutations here, but rather point the reader to a series of notable reviews.[20,22,48–52] Instead we will highlight just a few of the interesting topological considerations.

### 3.5.2 1D and 2D Interpenetration

There are two important topological properties of interpenetrating 1D and 2D networks. Firstly, the nets can interpenetrate such that they are all parallel (either their directions of propagation (1D nets) or their mean planes (2D nets)) or inclined in two or more directions. Secondly, the interpenetration can produce entanglements that are either of the same dimensionality of the nets or higher. Both these considerations lead to different topologies of interpenetration. To reflect this, a notation has been developed[22,48] which takes the form

$$m\mathrm{D}(/n\mathrm{D}) \rightarrow p\mathrm{D}\ \text{parallel/inclined interpenetration}$$

where $m$D is the dimension of the interpenetrating nets, $n$D is the dimension of the second type of net (if applicable), and $p$D is the dimension of the overall entanglement ($m,n \geq p$D).

For example, consider the two interpenetrating 1D nets in Figure 3.17(a). The nets are entangled in pairs such that the mean directions of propagation are parallel, and the nets interpenetrate such that the overall entanglement is still 1D. Thus this is an

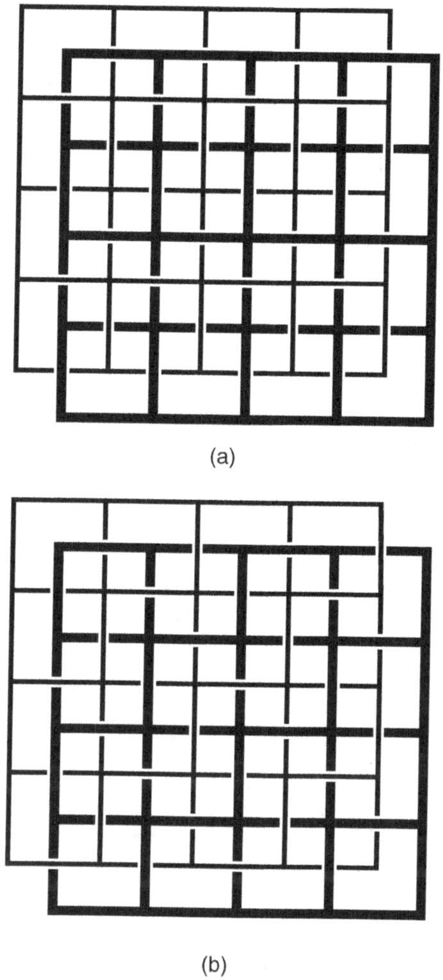

FIGURE 3.16 Two topologically different modes of interpenetration for (4,4) sheets.

example of 1D → 1D parallel interpenetration. In Figure 3.17(b), the 1D nets are also parallel, but their mean directions of propagation are much more offset, and thus an overall 2D entanglement is generated (1D → 2D parallel interpenetration). Figure 3.17(c) also shows the interpenetration of 1D nets; however, in this case there are two sets of nets. The mean directions of propagations between the two sets are inclined to each other, and an overall 3D entanglement is produced (1D → 3D inclined interpenetration). Other entanglement topologies possible are 1D → 3D parallel and 1D → 2D inclined.

For 2D nets there are three possibilities: 2D → 2D parallel, 2D → 3D parallel and 2D inclined interpenetration (the "→3D" in the last case is redundant—inclined interpenetration of 2D nets must give a 3D entanglement). The 2D → 2D

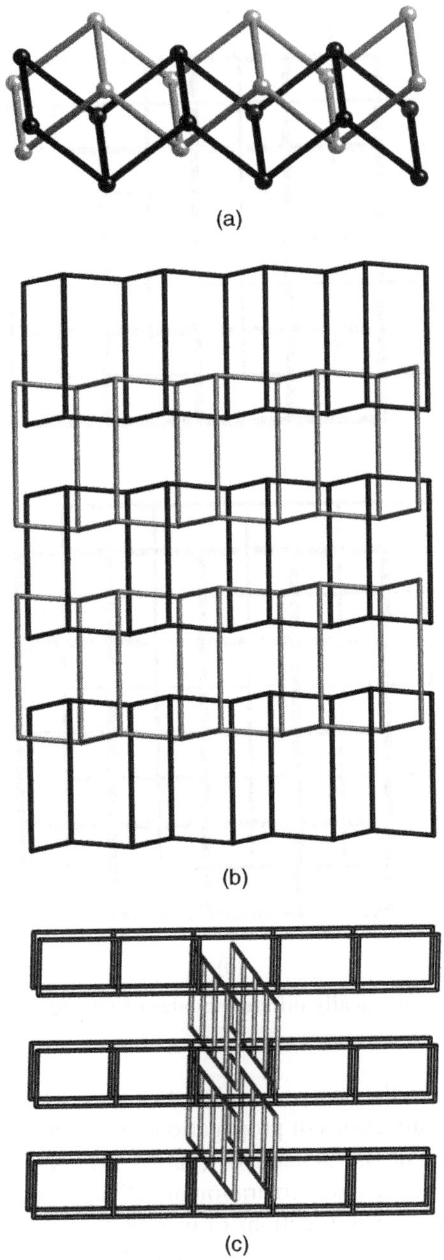

(a) (b) (c)

**FIGURE 3.17** Different modes of interpenetration for 1D nets: (a) 1D → 1D parallel interpenetration, (b) 1D → 2D parallel interpenetration, and (c) 1D → 3D inclined interpenetration.

**FIGURE 3.18** Twofold 2D → 2D parallel interpenetration of (6,3) sheets in Ag(tcm).

interpenetration of (6,3) sheets in the structure of Ag(tcm), tcm = tricyanomethanide, $C(CN)_3^-$ is shown in Figure 3.18.[53–55] Note the undulating nature of each sheet, and that the mean planes of each net are not only parallel, but also coincident—that is they have the same mean plane. Figure 3.16 showed earlier (schematically) two different interpenetration topologies for (4,4) nets showing 2D → 2D parallel interpenetration.

For 2D → 3D parallel interpenetration, the mean planes of the interpenetrating nets are parallel but no longer coincident. There are two possible motifs that are normally observed for the sheets displaying this mode of interpenetration. The first is that the sheets are highly undulating, as in the structure of $[Cu(bpee)_{1.5}(PPh_3)]PF_6 \cdot 1.5CH_2Cl_2$, bpee = 1,2-bis(4-pyridine)ethene,[56] shown in Figure 3.19(a). The second possibility is that the sheets have a degree of "thickness," such as in the bilayer structure of $Cu_4(dca)_4(4,4'\text{-bipy})_3(MeCN)_2$ (Figure 3.19b).[57]

A particularly unusual way for 2D parallel nets to interpenetrate is Borromean interpenetration.[50] Borromean rings (Figure 3.20a) are discrete rings that are entangled in such a way that any two rings can be separated in the absence of the third.

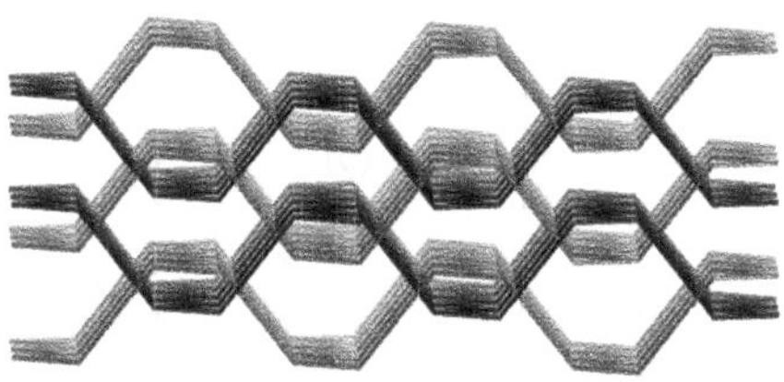

(a)

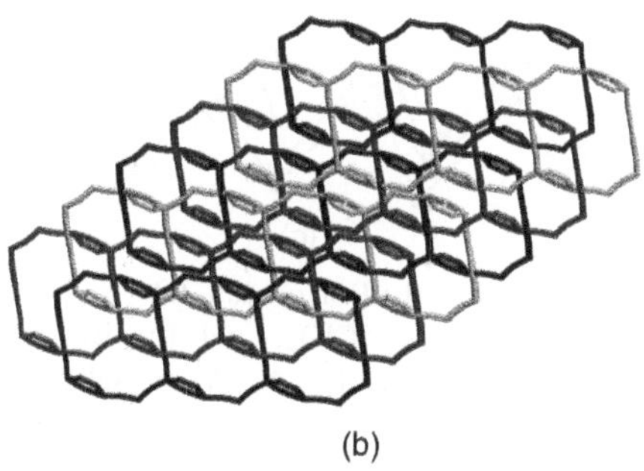

(b)

**FIGURE 3.19** (a) 2D → 3D parallel interpenetration of corrugated (6,3) sheets in the structure of $[Cu(bpee)_{1.5}(PPh_3)]PF_6 \cdot 1.5CH_2Cl_2$ and (b) 2D → 3D parallel interpenetration of "thick" 2D sheets in the structure of $Cu_4(dca)_4(4,4'\text{-bipy})_3(MeCN)_2$.

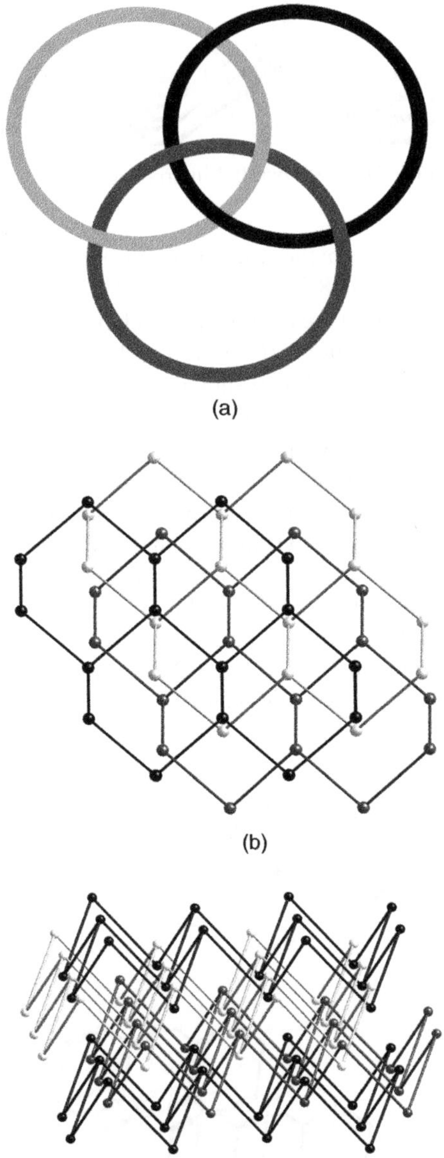


**FIGURE 3.20** (a) Borromean rings, (b) Borromean 2D → 2D parallel interpenetration in the structure of $[Cu(tmeda)_2\{Ag(CN)_2\}_3]ClO_4$, and (c) Borromean 2D → 3D parallel interpenetration observed in $[Ag_2L_3]X_2$, L = *N*, *N′*-bis(salicylidene)-1,4-diaminobutane, X = $NO_3$, $ClO_4$.

That is, any two rings are not themselves entangled, but the addition of the third produces an arrangement in which the rings cannot be separated without breaking one. A similar situation can exist for parallel 2D nets (for both 2D and 3D overall entanglements). The structure of $[Cu(tmeda)_2\{Ag(CN)_2\}_3]ClO_4$, tmeda = *N,N,N′,N′*-tetramethylethylenediamine contains discrete layers of threefold interpenetrating (6,3) nets which can be defined.[58] These sheets interpenetrate such that any two are only made inseparable by the presence of the third (Figure 3.20b). In this structure the overall entanglement is 2D, but a similar situation can exist for 2D → 3D interpenetration, such as is found in the structure of $[Ag_2L_3]X_2$, L = *N,N′*-bis(salicylidene)-1,4-diaminobutane, X = $NO_3$, $ClO_4$ (Figure 3.20c).[59]

The other common interpenetration motif for 2D nets is inclined interpenetration, which generates 3D entanglements. Like the different topologies for parallel interpenetration shown in Figure 3.16, there are different topologies of interpenetration for inclined sheets. Three different ways for inclined (4,4) sheets to interpenetrate are shown in Figure 3.21; these have been classed as parallel/parallel, diagonal/diagonal, and parallel/diagonal interpenetration.[60] Usually the interpenetration occurs between two stacks of sheets which are inclined to each other. Occasionally, however, higher numbers of inclined stacks interpenetrate. For example, in the structure of $Co_2(azpy)_3(NO_3)_4.Me_2CO.2H_2O$, azpy = *trans*-4,4′-azopyridine there are *four* sets of stacked (6,3) sheets which are all mutually inclined to each other and interpenetrate.[61]

Finally, the structure of $[AgL_2]SbF_6$, L = 3-cyanophenyl 4-cyanobenzoate is worthy of mention.[62] It contains layers of doubly interpenetrating (4,4) sheets. Remarkably, these double layers then show inclined interpenetration, and thus the overall interpenetration topology could be classed as 2D → 2D parallel → 3D inclined interpenetration.

### 3.5.3 3D Interpenetration

There are numerous examples of interpenetrating 3D networks, the most common of which are the diamond and $\alpha$-Po nets. We shall start, however, with the (10,3)-a net. As mentioned earlier, this net is inherently chiral. This leads to the possibility of a (10,3)-a net being interpenetrated by a net of either the same handedness (homochiral interpenetration), or of the opposite hand (heterochiral or racemic interpenetration). These two possibilities for doubly interpenetrating systems are shown in Figure 3.22; the chirality of the nets can be noted by the handedness of the fourfold helices apparent from this view. Both situations have been observed. For example, $Ni_3(btc)_2(py)_6(eg)_6.{\sim}3eg.{\sim}4H_2O$, btc = 1,3,5-benzenetricarboxylate, py = pyridine, eg = ethylene glycol contains four interpenetrating nets all of the same hand,[63] while $Zn(tpt)_{2/3}(SiF_6)(H_2O)_2(MeOH)$ contains eight interpenetrating nets, four of each hand to give an overall racemate.[64]

Despite this example, differing interpenetration topologies for a given net topology are unusual for 3D nets. Usually there is a preferred ("normal") interpenetration topology, like that shown for diamond nets such as $Cd(CN)_2$ (Figure 3.2a). Occasionally, however, unusual motifs are observed. For the diamond net, the

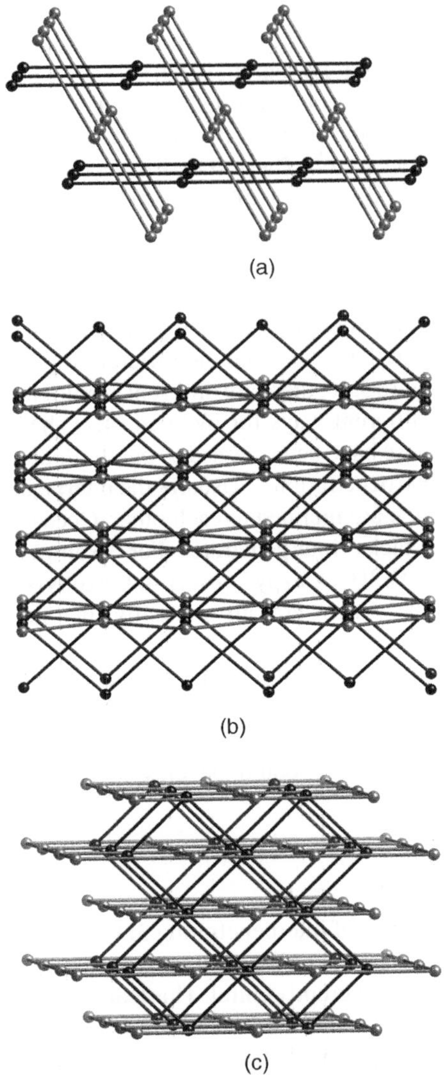

**FIGURE 3.21** Different interpenetration topologies for inclined (4,4) sheets: (a) parallel/parallel, (b) diagonal/diagonal, and (c) parallel/diagonal.

structure of $\beta$-Cu(dca)(bpee) is a good illustrative example.[29] The structure contains five interpenetrating nets, and a convenient way to illustrate the different topology of interpenetration in this case is to examine the relationship between the adamantane cavities of the five nets. This can be done by selecting one adamantane cavity of one of the nets, and then selecting a six-membered window of that cavity. This can then be used to uniquely define adamantane cavities of the four other nets. Two of those adamantane cavities will be related to that of the chosen net in the way shown in Figure 3.23(a). Each window of the first net's cavity is penetrated by rods of the other

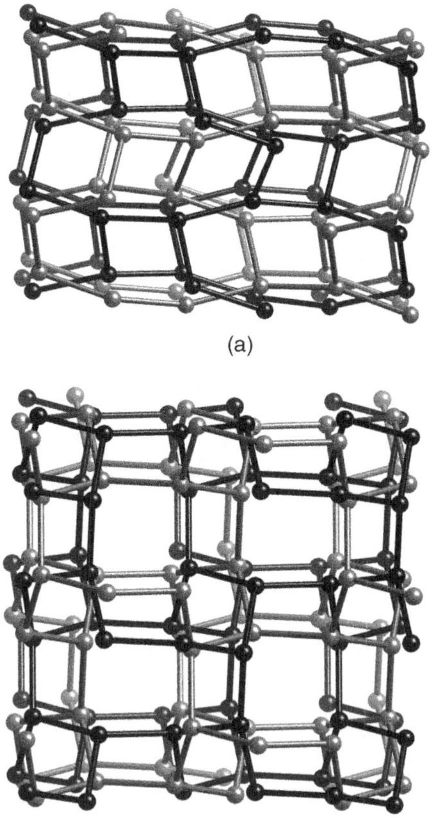

(a)

(b)

**FIGURE 3.22** (a) Heterochiral and (b) homochiral interpenetration of two (10,3)-a nets.

net's adamantane cage (plus the link projecting from the "top"). Almost all diamond nets interpenetrate in such a way that the cavities of different nets are related to each other in this way. Nets 4 and 5, however, display different relationships. For example, the equivalent cage in net 4 is shown in Figure 3.23(b). Note that one of the windows of the first cage is not penetrated at all, while another window has *two* rods passing through it. Thus this compound shows "abnormal" interpenetration. The five interpenetrating nets in this structure are in fact equivalent, so these motifs are repeated throughout the structure with the different nets swapping roles.

Similarly, there is a normal mode of interpenetration for $\alpha$-Po nets, whereby each cavity contains a node of the other net, and each cubic cavity catenates with eight cavities of the other net (Figure 3.24a). An unusual abnormal mode of interpenetration is found in the structure of $[Mn(bpea)(H_2O)_4](ClO_4)_2(bpea)_4$, bpea = 1,2-bis(4-pyridyl)ethane, in which the $\alpha$-Po nets are defined by both coordinative and hydrogen bonding.[65] In this interpenetration topology each cavity catenates with 10 cavities of the other net (Figure 3.24b).

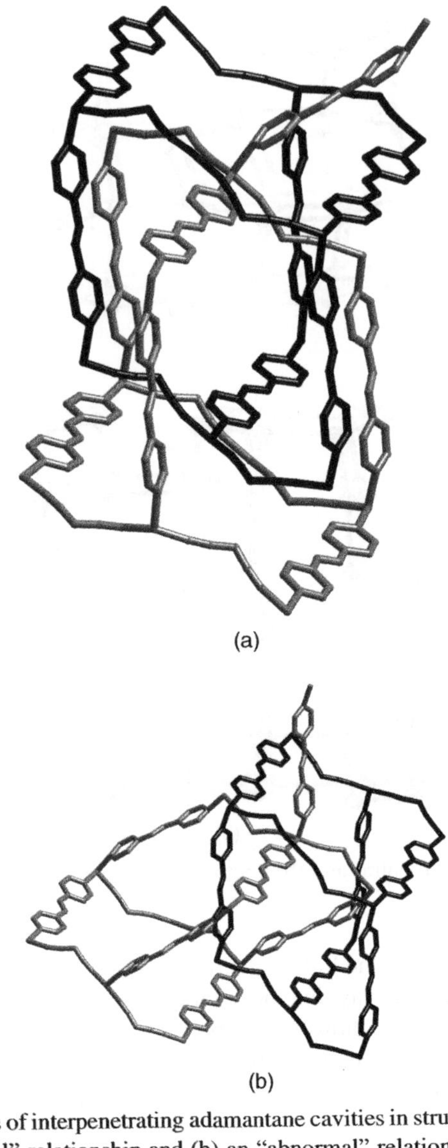

(a)

(b)

**FIGURE 3.23** Pairs of interpenetrating adamantane cavities in structure of $\beta$-Cu(dca)(bpee) showing (a) a "normal" relationship and (b) an "abnormal" relationship.

### 3.5.4 Heterogeneous Interpenetration

The examples of interpenetration discussed above are largely homogeneous in nature, with interpenetrating nets being chemically and topologically identical (and in many cases, related by symmetry). However, it is possible for chemically and/or topologically (even dimensionally) different nets to interpenetrate (although it is less common than homogenous interpenetration). For example, $K_2[PdSe_{10}]$ contains two

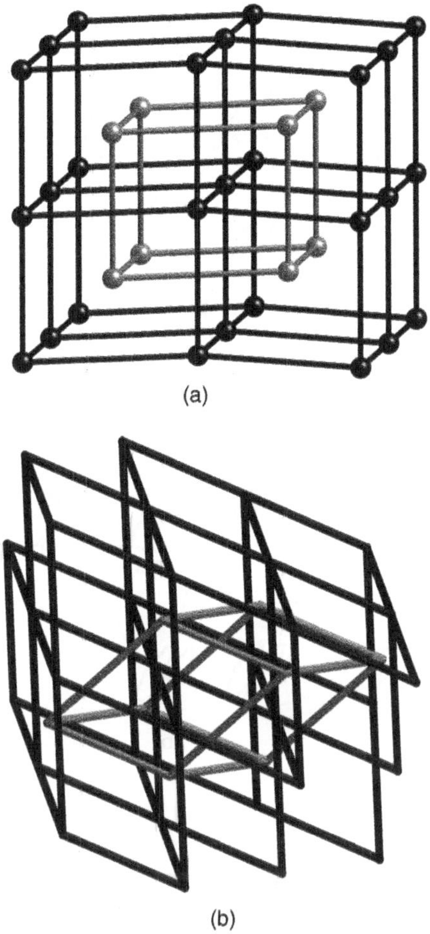


**FIGURE 3.24** Interpenetration of α-Po nets: (a) two nets showing the "normal" interpenetration mode and (b) the "abnormal" twofold interpenetration seen for $[Mn(bpea)(H_2O)_4](ClO_4)_2(bpea)_4$.

interpenetrating diamond nets.[66] In one net, the Pd atoms are connected by $Se_4^{2-}$ bridges, while the other contains $Se_6^{2-}$ bridges. Necessarily the bridging distances are the same, a constraint which is negotiated by the flexible, twisted nature of the bridges. The structure of $[Ni(azpy)_2(NO_3)_2]_2[Ni_2(azpy)_3(NO_3)_4].4CH_2Cl_2$ contains 2D nets displaying inclined interpenetration.[67] Not only are the interpenetrating nets chemically different, but also one type has (4,4) topology and the other type has (6,3) topology.

Interpenetration of nets of different dimensions is rare but topologically interesting. The structure of $[Cu_5(bpp)_8(SO_4)_4(EtOH)(H_2O)_5](SO_4).EtOH.25.5H_2O$, bpp = 1,3-bis(4-pyridyl)propane contains 1D chains interpenetrating 2D sheets in an inclined fashion (1D/2D → 3D inclined interpenetration; Figure 3.25a).[68] $[Co(bix)_2(H_2O)_2]$

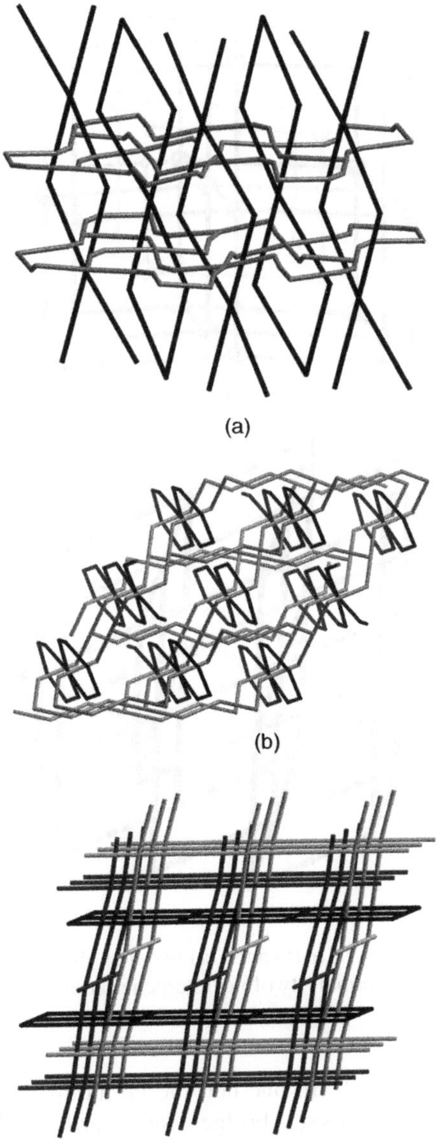


**FIGURE 3.25** Interpenetration of different dimensionality nets: (a) 1D/2D → 3D interpenetration seen for $[Cu_5(bpp)_8(SO_4)_4(EtOH)(H_2O)_5](SO_4).EtOH.25.5H_2O$, (b) 1D/3D interpenetration seen for $[Co(bix)_2(H_2O)_2](SO_4).7H_2O$, and (c) 2D/3D interpenetration seen for $[Co(mppe)_2(NCS)_2].2[Co(mppe)_2(NCS)_2].5MeOH$.

$(SO_4).7H_2O$, bix = 1,4-bis(imidazole-1-ylmethyl)benzene shows interpenetration of 1D chains and 3D $CdSO_4$ nets (1D/3D interpenetration; Figure 3.25b),[69] while [Co $(mppe)_2(NCS)_2].2[Co(mppe)_2(NCS)_2]$.5MeOH, mppe = 1-methyl-1′-(4-pyridyl)-2-4-(pyrimidyl)ethylene has two $CdSO_4$ 3D nets interpenetrating 2D (4,4) sheets (2D/3D interpenetration; Figure 3.25c).[70]

### 3.5.5 Self-Penetration

If interpenetrating nets are polymeric equivalents of molecular catenanes and rotaxanes, the polymeric equivalent of a knot is a self-penetrating network. In these structures, shortest circuits are penetrated by rods of the same network. The qualification that it is *shortest circuits* that must be penetrated is necessary as it would otherwise be possible to define large enough rings and rods in most 3D nets to produce these rotaxane-like interactions.

The relationship between interpenetrating nets, self-penetrating nets, and single nets can be illustrated by a series of related structures containing the dca and tcm anions. The structures of $M(dca)_2$, M = $Cr^{II}$, $Mn^{II}$, $Fe^{II}$, $Co^{II}$, $Ni^{II}$, $Cu^{II}$ contain single rutile-related networks, with 3-connecting dca ligands and 6-connecting metal ions (Figure 3.26a).[71–73] The tcm anion is larger than dca, and so the structures of $M(tcm)_2$, M = Cr, Mn, Fe, Co, Ni, Cu, Zn, Cd, Hg contain two interpenetrating rutile nets (Figure 3.26b).[74,75] When both ligands are used in the one compound (i.e. M(dca)(tcm)) a structural compromise is reached—a single network is produced which is different but closely related to rutile (it has the same Schläfli symbol, $(4.6^2)_2(4^2.6^{10}8^3)$) (Figure 3.26c).[76] Furthermore, in the $M(dca)_2$ structures the interpenetration is such that the six-membered shortest circuits are penetrated by rods of the other net (note that the four-membered shortest circuits are not penetrated). For the self-penetrating

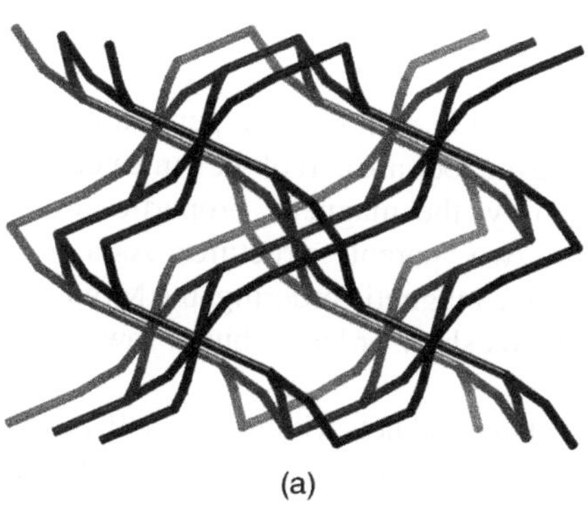

(a)

**FIGURE 3.26** Three related structures: (a) the single rutile net of $M(dca)_2$, (b) the two interpenetrating nets of $M(tcm)_2$, and (c) the single self-penetrating net of M(dca)(tcm).

(b)

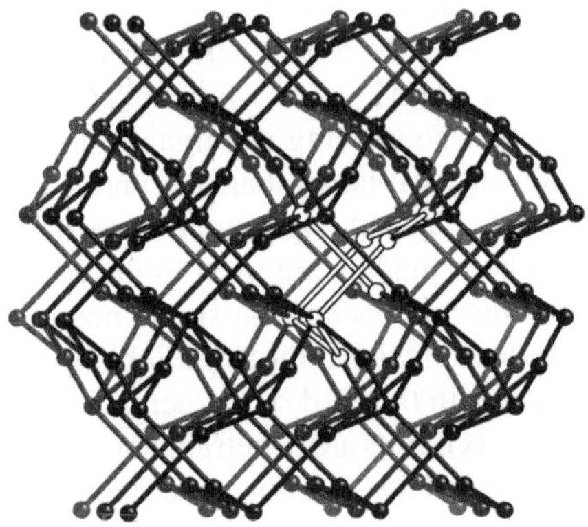

(c)

**FIGURE 3.26** (*Continued*).

M(dca)(tcm), a similar situation arises—rods penetrate the six-membered shortest circuits, except in this case the rods are part of the same network as the shortest circuits they penetrate. Interestingly, the magnetic properties of M(dca)(tcm) are also a compromise between the two "parent" structures. As a general trend, the $M(dca)_2$ compounds show long-range magnetic ordering, the $M(tcm)_2$ compounds do not, and the M(dca)(tcm) compounds show ordering but at lower temperatures than their M $(dca)_2$ equivalents.

There are also two fundamental nets which show self-penetration. In the (12,3) net, found in the structure of $Ni(tpt)(NO_3)_2$,[77] the 12-membered shortest circuits form catenane-like motifs (Figure 3.27a). Similarly, the (8,4) net (Figure 3.27b), which contains planar 4-connecting nodes, has catenane-like motifs between the eight-membered shortest circuits.[78]

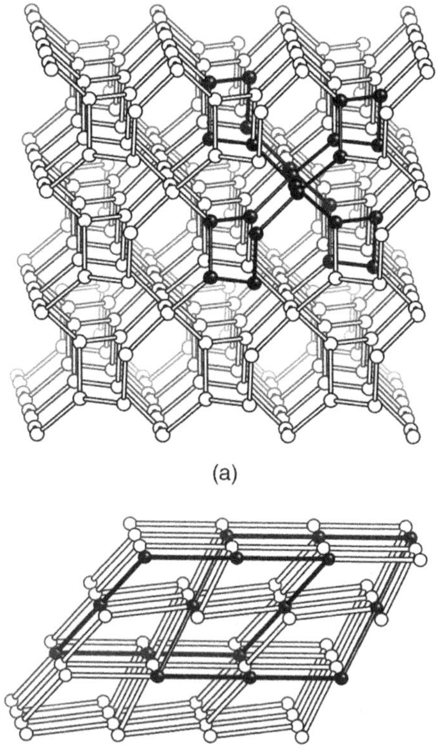

FIGURE 3.27 The self-penetrating (a) (12,3) (undistorted) and (b) (8,4) nets.

## REFERENCES


1. Wells, A. F. *Three-dimensional Nets and Polyhedra*. Wiley-Interscience: New York, 1977.
2. Wells, A. F. *Further Studies of Three-dimensional Nets. ACA Monograph No. 8*. American Crystallographic Association: New York, 1979.
3. Wells, A. F. *Structural Inorganic Chemistry*. 5th edition. Oxford University Press: Oxford, 1984.
4. Hoskins, B. F.; Robson, R. Infinite polymeric frameworks consisting of three dimensionally linked rod-like segments. *J. Am. Chem. Soc.* **1989**, *111*, 5962.
5. Hoskins, B. F.; Robson, R. Design and construction of a new class of scaffolding-like materials comprising infinite polymeric frameworks of 3D-linked molecular rods. A reappraisal of the $Zn(CN)_2$ and $Cd(CN)_2$ structures and the synthesis and structure of the diamond-related frameworks $[N(CH_3)_4][Cu^{I}Zn^{II}(CN)_4]$ and $Cu^{I}$[4,4′,4″,4‴-tetracyano-tetraphenylmethane]$BF_4.xC_6H_5NO_2$. *J. Am. Chem. Soc.* **1990**, *112*, 1546.
6. Robson, R. A net-based approach to coordination polymers. *J. Chem. Soc., Dalton Trans.* **2000**, 3735.

7. Kim, J.; Chen, B.; Reineke, T. M.; Li, H.; Eddaoudi, M.; Moler, D. B.; O'Keeffe, M.; Yaghi, O. M. Assembly of metal-organic frameworks from large organic and inorganic secondary building units: new examples and simplifying principles for complex structures. *J. Am. Chem. Soc.* **2001**, *123*, 8239.
8. Li, H.; Eddaoudi, M.; Groy, T. L.; Yaghi, O. M. Establishing microporosity in open metal-organic frameworks: gas sorption isotherms for Zn(BDC) (BDC = 1,4-benzenedicarboxylate). *J. Am. Chem. Soc.* **1998**, *120*, 8571.
9. Jensen, P.; Batten, S. R.; Fallon, G. D.; Moubaraki, B.; Murray, K. S.; Price, D. J. Structural isomers of $M(dca)_2$ molecule-based magnets. Crystal structure of tetrahedrally coordinated sheet-like β-$Zn(dca)_2$ and β-Co/$Zn(dca)_2$, and the octahedrally coordinated rutile-like α-$Co(dca)_2$, where $dca^-$ = dicyanamide, $N(CN)_2^-$, and magnetism of β-$Co(dca)_2$. *Chem. Commun.* **1999**, 177.
10. Yaghi, O. M.; O'Keeffe, M.; Ockwig, N. W.; Chae, H. K.; Eddaoudi, M.; Kim, J. Reticular synthesis and the design of new materials. *Nature* **2003**, *423*, 705.
11. Friedrichs, O. D.; O'Keeffe, M.; Yaghi, O. M. Three-periodic nets and tilings: regular and quasiregular nets. *Acta Crystallogr., Sect. A* **2003**, *59*, 22.
12. Friedrichs, O. D.; O'Keeffe, M.; Yaghi, O. M. Three-periodic nets and tilings: semiregular nets. *Acta Crystallogr., Sect. A* **2003**, *59*, 515.
13. Bonneau, C.; Friedrichs, O. D.; O'Keeffe, M.; Yaghi, O. M. Three-periodic nets and tilings: minimal nets. *Acta Crystallogr., Sect. A* **2004**, *60*, 517.
14. Delgado-Friedrichs, O.; O'Keeffe, M. Crystal nets as graphs: terminology and definitions. *J. Solid State Chem.* **2005**, *178*, 2480.
15. Delgado-Friedrichs, O.; Foster, M. D.; O'Keeffe, M.; Proserpio, D. M.; Treacy, M. M. J.; Yaghi, O. M. What do we know about three-periodic nets? *J. Solid State Chem.* **2005**, *178*, 2533.
16. O'Keeffe, M.; Eddaoudi, M.; Li, H.; Reineke, T.; Yaghi, O. M. Frameworks for extended solids: Geometrical design principles. *J. Solid State Chem.* **2000**, *152*, 3.
17. O'Keeffe, M.; Hyde, B. G. *Crystal Structures. I. Patterns and Symmetry, Mineralogical Society of America Monograph*. Mineralogical Society of America: Washington, 1996.
18. Ockwig, N. W.; Delgado-Friedrichs, O.; O'Keeffe, M.; Yaghi, O. M. Reticular chemistry: occurrence and taxonomy of nets and grammar for the design of frameworks. *Acc. Chem. Res.* **2005**, *38*, 176. http://rcsr.anu.edu.au [Accessed on March 18, 2008].
19. Öhrström, L.; Larsson, K. *Molecule-Based Materials, The Structural Network Approach*. Elsevier: Amsterdam, 2005.
20. Batten, S. R.; Neville, S. M.; Turner, D. R. *Coordination Polymers. Design, Analysis and Application*. The Royal Society of Chemistry: Cambridge, 2008.
21. Öhrström, L.; Larsson, K. What kinds of three-dimensional nets are possible with tris-chelated metal complexes as building blocks? *J. Chem. Soc. Dalton Trans.* **2004**, 347.
22. Batten, S. R.; Robson, R. Interpenetrating nets: ordered, periodic entanglement. *Angew. Chem. Int. Ed.* **1998**, *37*, 1460.
23. Delgado-Friedrichs, O.; O'Keeffe, M. Identification of and symmetry computation for crystal nets. *Acta Crystallogr., Sect. A* **2003**, *59*, 351. http://gavrog.org [Accessed on March 18, 2008].
24. Dolomanov, O. V.; Blake, A. J.; Champness, N. R.; Schröder, M. OLEX: new software for visualization and analysis of extended crystal systems. *J. Appl. Cryst.* **2003**, *36*, 1283.

http://www.ccp14.ac.uk/ccp/web-mirrors/lcells/olex_index.htm [Accessed on March 18, 2008].

25. Blatov, V. A.; Shevchenko, A. P.; Serezhkin, V. N. TOPOS3.2: a new version of the program package for multipurpose crystal-chemical analysis. *J. Appl. Cryst.* **2000**, *33*, 1193.

26. Moulton, B.; Lu, J.; Hajndl, R.; Hariharan, S.; Zaworotko, M. J. Crystal engineering of a nanoscale Kagomé lattice. *Angew. Chem. Int. Ed.* **2002**, *41*, 2821.

27. Fu, Z.-Y.; Wu, X.-T.; Dai, J.-C.; Hu, S.-M.; Du, W.-X. A novel mixed-ligand molecular bilayer generated by self-assembly of "T-shaped" moieties, displaying an unusual entanglement. *New J. Chem.* **2002**, *26*, 978.

28. Hill, R. J.; Long, D.-L.; Champness, N. R.; Hubberstey, P.; Schröder, M. New approaches to the analysis of high connectivity materials: design frameworks based on $4^4$- and $6^3$-subnet tectons. *Acc. Chem. Res.* **2005**, *38*, 337.

29. Batten, S. R.; Harris, A. R.; Jensen, P.; Murray, K. S.; Ziebell, A. Copper(I) dicyanamide coordination polymers: ladders, sheets, layers, diamond-like networks and unusual interpenetration. *J. Chem. Soc., Dalton Trans.* **2000**, 3829.

30. Lu, J. Y.; Babb, A. M. The first triple-layer 2-D coordination polymer: $[Cu_3(bpen)(IN)_6(H_2O)_2]$. *Inorg. Chem.* **2001**, *40*, 3261.

31. Carlucci, L.; Ciani, G.; Proserpio, D. M.; Rizzato, S. Chiral packing of chiral quintuple layers polycatenated to give a three-dimensional network in the coordination polymer $[Co_5(bpe)_9(H_2O)_8(SO_4)_4](SO_4)\cdot 14H_2O$ [bpe = 1,2-bis(4-pyridyl)ethane]. *Chem. Commun.* **2000**, 1319.

32. Bradshaw, D.; Claridge, J. B.; Cussen, E. J.; Prior, T. J.; Rosseinsky, M. J. Design, chirality, and flexibility in nanoporous molecule-based materials. *Acc. Chem. Res.* **2005**, *38*, 273.

33. Biradha, K.; Fujita, M. A springlike 3D-coordination network that shrinks or swells in a crystal-to-crystal manner upon guest removal or readsorpion. *Angew. Chem. Int. Ed.* **2002**, *41*, 3392.

34. Batten, S. R.; Hoskins, B. F.; Robson, R. Interdigitation, interpenetration and intercalation in layered cuprous tricyanomethanide derivatives. *Chem. Eur. J.* **2000**, *6*, 156.

35. Murray, K. S.; Batten, S. R.; Moubaraki, B.; Price, D. J.; Robson, R. Molecular magnetism in manganese dicyanamide extended network structures. *Mol. Cryst. Liq. Cryst.* **1999**, *335*, 313.

36. Batten, S. R.; Jensen, P.; Kepert, C. J.; Kurmoo, M.; Moubaraki, B.; Murray, K. S.; Price, D. J. Syntheses, structures and magnetism of $\alpha$-$Mn(dca)_2$, $Mn(dca)_2(H_2O)_2.H_2O$, $Mn(dca)_2(C_2H_5OH)_2.(CH_3)_2CO$, $Fe(dca)_2(CH_3OH)_2$ and $Mn(dca)_2(L)_2$, where L = pyridine, $CH_3OH$, DMF and $dca^-$ = dicyanamide, $N(CN)_2^-$. *J. Chem. Soc., Dalton Trans.* **1999**, 2987.

37. Abrahams, B. F.; Batten, S. R.; Hamit, H.; Hoskins, B. F.; Robson, R. A cubic 3,4-connected net with large cavities in solvated $[Cu_3(tpt)_4](ClO_4)_3$ (tpt = 2, 4, 6-Tri(4-pyridyl)-1, 3, 5-triazine). *Angew. Chem. Int. Ed. Engl.* **1996**, *35*, 1690.

38. Chen, B.; Eddaoudi, M.; Hyde, S. T.; O'Keeffe, M.; Yaghi, O. M. Interwoven metal-organic framework on a periodic minimal surface with extra-large pores. *Science* **2001**, *291*, 1021.

39. Halder, G. J.; Kepert, C. J.; Moubaraki, B.; Murray, K. S.; Cashion, J. D. Guest-dependent spin cross-over in a nanoporous molecular framework material. *Science* **2002**, *298*, 1762.

40. Majl, T.; Matsuda, R.; Kitagawa, S. A flexible interpenetrating coordination framework with a bimodal porous functionality. *Nat. Mater.* **2007**, *6*, 142.

41. Kitaura, R.; Seki, K.; Akiyama, G.; Kitagawa, S. Porous coordination-polymer crystals with gated channels specific for supercritical gases. *Angew. Chem. Int. Ed.* **2003**, *42*, 428.

42. Batten, S. R.; Hoskins, B. F.; Robson, R. Two interpenetrating 3D networks which generate spacious sealed-off compartments enclosing of the order of 20 solvent molecules in the structures of $Zn(CN)(NO_3)(tpt)_{2/3}$.solv (tpt = 2,4,6-tri(4-pyridyl)-1,3,5-triazine, solv = ~3/4$C_2H_2Cl_4$. 3/4$CH_3OH$ or ~3/2$CHCl_3$. 1/3$CH_3OH$). *J. Am. Chem. Soc.* **1995**, *117*, 5385.

43. Saha, M. K.; Morón, M. C.; Palacio, F.; Bernal, I. A new bimetallicintercalated 3D assembly magnet $[\{(323)Ni\}_3\{Fe^{III}(CN)_6\}_2]_n.12nH_2O$ (323 = *N,N′*-bis(3-aminopropyl) ethylenediamine): an unprecedented concomitant presence of meridional and facial arrangement of ferricyanide anion. *Inorg. Chem.* **2005**, *44*, 1354.

44. Miller, J. S.; Vos, T. E.; Shum, W. W. Interpenetrating-lattice-structured magnets exhibiting anomalous magnetic properties. *Adv. Mater.* **2005**, *17*, 2251.

45. Ermer, O. Sevenfold diamond structure and conductivity of copper dicyanoduinonediimines $Cu(DCNQI)_2$. *Adv. Mater.* **1991**, *3*, 608.

46. Sinzger, K.; Hunig, S.; Jopp, M.; Bauer, D.; Bietsch, W.; von Schutz, J. U.; Wulf, H. C.; Kremer, R. K.; Metzenthin, T.; Bau, R.; Khan, S. I.; Lindbaum, A.; Lengauer, C. L.; Tillmanns, E. The organic metal $(Me_2\text{-}DCNQI)_2Cu$: dramatic changes in solid-state properties and crystal structure due to secondary deuterium effects. *J. Am. Chem. Soc.* **1993**, *115*, 7696.

47. Aumuller, A.; Erk, P.; Klebe, G.; Hunig, S.; von Schutz, J. U.; Werner, H.-P. A radical anion salt of 2,5-dimethyl-*N,N′*-dicyanoquinonediimine with extremely high electrical conductivity. *Angew. Chem. Int. Ed. Engl.* **1986**, *25*, 740.

48. Batten, S. R. Topology of interpenetration. *CrystEngComm* **2001**, *3*, 67. http://www.chem.monash.edu.au/staff/sbatten/interpen/ [Accessed on March 18, 2008].

49. Blatov, V. A.; Carlucci, L.; Ciani, G.; Proserpio, D. M. Interpenetrating metal-organic and inorganic 3D networks: a computer-aided systematic investigation. Part I. Analysis of the Cambridge structural database. *CrystEngComm* **2004**, *6*, 377.

50. Carlucci, L.; Ciani, G.; Proserpio, D. M. Borromean links and other non-conventional links in 'polycatenated' coordination polymers: re-examination of some puzzling networks. *CrystEngComm* **2003**, *5*, 269.

51. Carlucci, L.; Ciani, G.; Proserpio, D. M. Polycatenation, polythreading and polyknotting in coordination network chemistry. *Coord. Chem. Rev.* **2003**, *246*, 247.

52. Baburin, I. A.; Blatov, V. A.; Carlucci, L.; Ciani, L.; Proserpio, D. M. Interpenetrating metal-organic and inorganic 3D networks: a computer-aided systematic investigation. Part II [1]. Analysis of the inorganic crystal structure database (ICSD). *J. Solid State Chem.* **2005**, *178*, 2452.

53. Konnert, J.; Britton, D. The crystal structure of $AgC(CN)_3$. *Inorg. Chem.* **1996**, *5*, 1193.

54. Batten, S. R.; Hoskins, B. F.; Robson, R. The Structures of [Ag(tcm)], $[Ag(tcm)(phz)_{1/2}]$ and [Ag(tcm)(pyz)] (tcm = tricyanomethanide, $C(CN)_3^-$, phz = phenazine, pyz = pyrazine). *New J. Chem.* **1998**, *22*, 173.

55. Abrahams, B. F.; Batten, S. R.; Hoskins, B. F.; Robson, R. $AgC(CN)_3$ based coordination polymers. *Inorg. Chem.* **2003**, *42*, 2654.

56. Knaust, J. M.; Lopez, S.; Keller, S. W. Dimensional control of Cu(I)-bis(4-pyridyl)ethylene coordination networks. *Inorg. Chim. Acta* **2001**, *324*, 81.

57. Carlucci, L.; Ciani, G.; Proserpio, D. M. Three-dimensional architectures of intertwined planar coordination polymers: the first case of interpenetration involving two different bidimensional polymeric motifs. *New J. Chem.* **1998**, *22*, 1319.

58. Leznoff, D. B.; Xue, B.-Y.; Batchelor, R. J.; Einstein, F. W. B.; Patrick, B. O. Gold-gold interactions as crystal engineering design elements in heterobimetallic coordination polymers. *Inorg. Chem.* **2001**, *40*, 6026.

59. Tong, M. L.; Chen, X.-M.; Ye, B.-H.; Ji, L.-N. Self-assembled three-dimensional coordination polymers with unusual ligand-unsupported Ag-Ag bonds: syntheses, structures, and luminescent properties. *Angew. Chem. Int. Ed.* **1999**, *38*, 2237.

60. Zaworotko, M. J. Superstructural diversity in two dimensions: crystal engineering of laminated solids. *Chem. Commun.* **2001**, 1.

61. Kondo, M.; Shimamura, M.; Noro, S.; Minakoshi, S.; Asami, A.; Seki, K.; Kitagawa, S. Microporous materials constructed from the interpenetrated coordination networks. Structures and methane adsorption properties. *Chem. Mater.* **2000**, *12*, 1288.

62. Hirsch, K. A.; Wilson, S. R.; Moore, J. S. Crystallization of 3-cyanophenyl 4-cyanobenzoate with $AgSbF_6$: a polar coordination network based on the crisscrossing of intertwined helices. *Chem. Commun.* **1998**, 13.

63. Kepert, C. J.; Rosseinsky, M. J. A porous chiral framework of coordinated 1,3,5-benzenetricarboxylate: quadruple interpenetration of the (10,3)-a network. *Chem. Commun.* **1998**, 31.

64. Abrahams, B. F.; Batten, S. R.; Hamit, H.; Hoskins, B. F.; Robson, R. A Wellsian '3D racemate': eight interpenetrating, enantiomorphic (10,3)-a nets, four right handed and four left handed. *Chem. Commun.* **1996**, 1313.

65. Hong, C. S.; Son, S.-K.; Lee, Y. S.; Jun, M.-J.; Do, Y. High-dimensional manganese(II) compounds with noncovalent and/or covalent bonds derived from flexible ligands: self-assembly and structural transformation. *Inorg. Chem.* **1999**, *38*, 5602.

66. Kim, K. W.; Kanatzidis, M. G. Hydrothermal synthesis of $K_2PdSe_{10}$. Coexistence of two large interpenetrating three-dimensional frameworks of $[Pd(Se_4)_2]^{2-}$ and $[Pd(Se_6)_2]^{2-}$. *J. Am. Chem. Soc.* **1992**, *114*, 4878.

67. Plater, M. J.; Foreman MRStJ.; Gelbrich T.; Coles, S. J.; Hursthouse, M. B. Synthesis and characterisation of infinite co-ordination networks from flexible dipyridyl ligands and cadmium salts. *J. Chem. Soc., Dalton Trans.* **2000**, 3065.

68. Carlucci, L.; Ciani, G.; Moret, M.; Proserpio, D. M.; Rizzato, S. Polymeric layers catenated by ribbons of rings in a three-dimensional self-assembled architecture: a nanoporous network with spongelike behavior. *Angew. Chem. Int. Ed.* **2000**, *39*, 1506.

69. Carlucci, L.; Ciani, G.; Proserpio, D. M. A new type of entanglement involving one-dimensional ribbons of rings catenated to a three-dimensional network in the nanoporous structure of $[Co(bix)_2(H_2O)_2](SO_4)\cdot 7H_2O$ [bix = 1,4-bis(imidazol-1-ylmethyl)benzene]. *Chem. Commun.* **2004**, 380.

70. Shin, D. M.; Lee, I. S.; Chung, Y. K.; Lah, M. S. Coordination polymers based on square planar Co(II) node and linear spacer: solvent-dependent pseudo-polymorphism and an unprecedented interpenetrating structure containing both 2D and 3D topological isomers. *Chem. Commun.* **2003**, 1036.

71. Batten, S. R.; Jensen, P.; Moubaraki, B.; Murray, K. S.; Robson, R. Structure and molecular magnetism of the rutile-related compounds $M(dca)_2$, M = $Co^{II}$, $Ni^{II}$, $Cu^{II}$, dca = dicyanamide, $N(CN)_2^-$. *Chem. Commun.* **1998**, 439.

72. Kurmoo, M.; Kepert, C. J. Hard magnets based on transition metal complexes with the dicyanamide anion, $\{N(CN)_2\}^-$. *New J. Chem.* **1998**, *22*, 1515.
73. Manson, J. L.; Kmety, C. R.; Huang, Q.-Z.; Lynn, J. W.; Bendele, G. M.; Pagola, S.; Stephens, P. W.; Liable-Sands, L. M.; Rheingold, A. L.; Epstein, A. J.; Miller, J. S. Structure and magnetic ordering of $M^{II}[N(CN)_2]_2$ (M = Co, Ni). *Chem. Mater.* **1998**, *10*, 2552.
74. Batten, S. R.; Hoskins, B. F.; Robson, R. 3D knitting patterns. Two independent, interpenetrating rutile-related infinite frameworks in the structure of $Zn[C(CN)_3]_2$. *J. Chem. Soc., Chem. Commun.* **1991**, 445.
75. Batten, S. R.; Hoskins, B. F.; Moubaraki, B.; Murray, K. S.; Robson, R. Crystal structures and magnetic properties of the interpenetrating rutile-related compounds $M(tcm)_2$ [M = octahedral, divalent metal; $tcm^-$ = tricyanomethanide, $C(CN)_3^-$] and the sheet structures of $[M(tcm)_2(EtOH)_2]$ (M = Co or Ni). *J. Chem. Soc., Dalton Trans.* **1999**, 2977.
76. Jensen, P.; Price, D. J.; Batten, S. R.; Moubaraki, B.; Murray, K. S. Self-penetration—a structural compromise between single networks and interpenetration: magnetic properties and crystal structures of $[Mn(dca)_2(H_2O)]$ and [M(dca)(tcm)], M = Co, Ni, Cu, dca = Dicyanamide, $N(CN)_2^-$, tcm = tricyanomethanide, $C(CN)_3^-$. *Chem. Eur. J.* **2000**, *6*, 3186.
77. Abrahams, B. F.; Batten, S. R.; Grannas, M. J.; Hamit, H.; Hoskins, B. F.; Robson, R. Ni $(tpt)(NO_3)_2$—a three-dimensional network with the exceptional (12,3) topology: a self-entangled single net. *Angew. Chem. Int. Ed.* **1999**, *38*, 1475.
78. Tong, M.-L.; Chen, X.-M.; Batten, S. R. A new self-penetrating uniform net, (8,4) (or $8^6$), containing planar 4-connecting nodes. *J. Am. Chem. Soc.* **2003**, *125*, 16170.

# 4

# HIGHLY CONNECTED METAL-ORGANIC FRAMEWORKS

PETER HUBBERSTEY, XIANG LIN, NEIL R. CHAMPNESS, AND MARTIN SCHRÖDER
*School of Chemistry, The University of Nottingham, University Park, Nottingham, UK*

## 4.1 INTRODUCTION

The study of metal–ligand coordination polymers, also called metal-organic frameworks (MOFs), is a field of research that has rapidly risen to the forefront of modern chemistry.[1] The area has grown from the initial, and now seminal, reports of Robson[2] toward advanced design strategies,[3–5] structural appreciation, and topological analysis[6,7] and has afforded a range of fascinating materials properties.[8–13] The field of coordination frameworks thus brings together diverse scientific disciplines combining synthetic methodology and structural analysis with the preparation of new generations of multi-functional materials. The research area has developed apace from its origins in coordination and supramolecular chemistry, and now embraces materials and separation science, theory, electronics, magnetism, and catalysis. Underpinning all of this research is the primary requisite for the designed synthesis and understanding of new structural materials. The basic principles that lie at the heart of designing such materials are based upon the original concept outlined by Robson, the *building-block* methodology.[2] This chapter will describe how the building-block approach can be used to design and synthesize metal-organic frameworks of high-connectivity. In particular, the chapter will discuss frameworks in which at least one of the nodes within the network adopts a connectivity greater than 6, that is, where a node is connected to more than six nearest neighbor nodes.

*Metal-Organic Frameworks: Design and Application*, Edited by Leonard R. MacGillivray

### 4.1.1 The Building-Block Methodology

Structural design of coordination frameworks can be achieved using a simple approach known as the building-block methodology. The basic principle of this design strategy as proposed by Robson,[2] relies upon the assembly of molecular and/or ionic moieties into extended arrays using complementary interactions between components. This strategy relies upon the use of divergent building blocks, that is, those that encourage the formation of extended arrays rather than discrete or molecular species. In its most basic form, the building-block methodology provides an excellent starting point for any coordination framework design. In essence by linking building blocks (nodes) through simple linear rods (connecting units or spacers) in a modular approach, the geometric and chemical properties of the nodes and spacers are expressed through three-dimensional space (Scheme 4.1).

The coordination chemistry of metal cations, especially of the late *d*-block transition series, is now very well developed and understood in terms of preferred

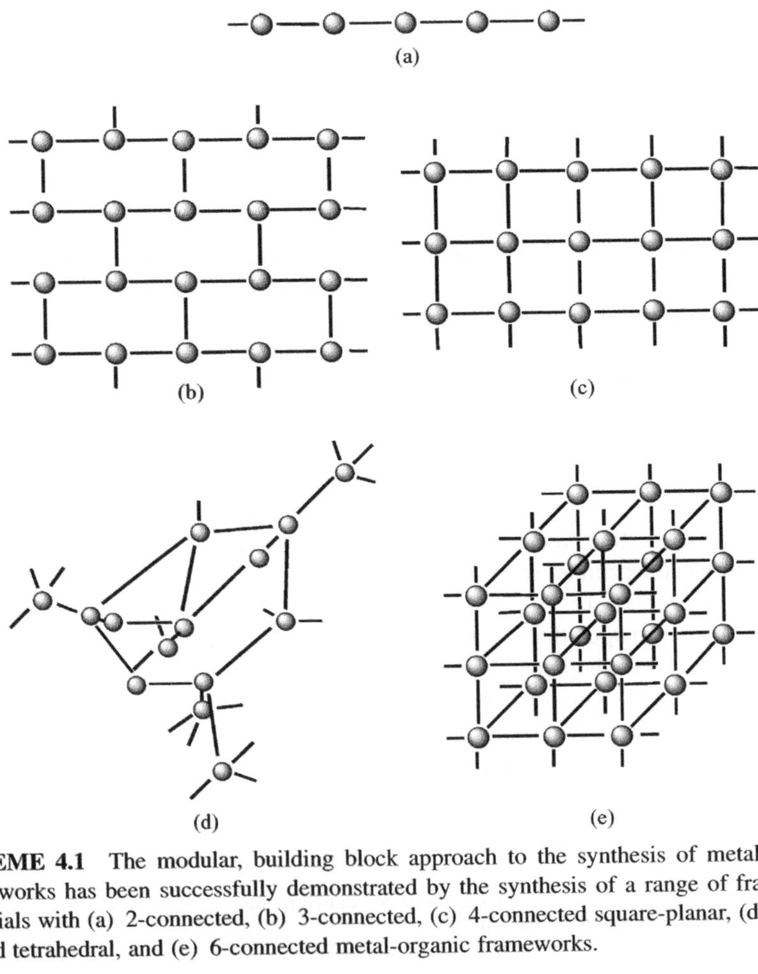

**SCHEME 4.1** The modular, building block approach to the synthesis of metal-organic frameworks has been successfully demonstrated by the synthesis of a range of framework materials with (a) 2-connected, (b) 3-connected, (c) 4-connected square-planar, (d) 4-connected tetrahedral, and (e) 6-connected metal-organic frameworks.

ligands, donor types, geometries, and stereochemistry. Metal cations that prefer linear coordination such as Ag(I) can be linked through simple connecting ligand spacers to afford a one-dimensional chain structure (Scheme 4.1a).[14] Two-dimensional sheet constructions can be built using trigonal (Scheme 4.1b)[15] or square-planar metal center nodes (Scheme 4.1c),[16] while three-dimensional architectures can be achieved by linking tetrahedral (Scheme 4.1d)[17,18] or octahedral (Scheme 4.1e)[19] metal centers with the same linear ligand spacers. Perhaps the most commonly studied example of a three-dimensional coordination framework is that constructed from tetrahedral metal centers (e.g., Cu(I)), and linear bridging ligands such as 4,4′-bipyridine[17] or *trans*-1, 2-bis(4-pyridyl)ethene.[18] In these materials the tetrahedral metal nodes are linked such that the geometry of the cation is expressed in three dimensions to generate a diamond-like structure. Thus, by considering and varying the metal geometry alone, a wide variety of different networks can be generated using relatively simple linear (i.e., 2-connected) ligand spacers. Higher complexity can be introduced by the incorporation of organic spacers with higher connectivity than 2, and this leads to an even greater variety of network types and structural patterning.[3]

The vast majority of metal-organic frameworks have a connectivity of 6 or lower, with 4- and 6-connected nets being significantly more common than 5-connected framework structures. Frameworks with connectivity of greater than 6 are far less common, reflecting the difficulty in designing such structures and the comparatively limited number of building blocks with connectivities greater than 6, the most common stereochemistry for a *d*-block transition ion node being octahedral, six-coordination. In identifying suitable cationic building blocks with connectivities greater than 6, two major approaches have been taken. One methodology is to use individual metal centers with high coordination numbers. This approach is limited to the exploitation of lanthanide *f*-block cations where higher coordination numbers of 8–12 are common. Alternatively, clusters of metal centers offering multiple coordination sites to enhance connectivity between multinuclear nodes can be considered. Typically, such clusters are constructed from *d*-block transition metal cations. This chapter will deal with these two approaches in turn and in detail.

Common to both methodologies are the bridging ligands. More than 25 different ligand spacers and anions have been utilized in generating the highly connected frameworks discussed in this chapter. Shown schematically in Scheme 4.2, they have been classified by their connectivity and their anionic charge. Thus ligands $L^{20X}$ (Scheme 4.2a) are 2-connecting neutral molecules, while ligands $L^{21X}$ (Scheme 4.2b) and $L^{22X}$ (Scheme 4.2c) are 2-connecting mono- and di-anions, respectively. Ligands $L^{31X}$ (Scheme 4.2d), $L^{32X}$ (Scheme 4.2d) and, $L^{33X}$ (Scheme 4.2d) are 3-connecting mono-, di- and tri-anions, respectively. Ligands $L^{40X}$ (Scheme 4.2e) are 4-connecting neutral molecules, and ligands $L^{42X}$ (Scheme 4.2e) and $L^{44X}$ (Scheme 4.2e) are 4-connecting di- and tetra-anions, respectively. The donor groups in these neutral molecular spacers are predominantly *O*-centers of *N*-oxides compatible with the hard Lewis acid nodes comprising *f*-block lanthanide cations, while *N*-donors from heterocyclic aromatic rings have been utilized with softer Lewis acid *d*-block metals. Anions used in these systems are predominantly *O*-donors from carboxylate or phenolate moieties. Whereas the heterocyclic *N*-donors comprise a single lone pair of

(a)

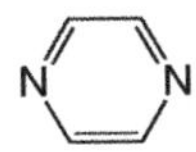

$L^{20A}$ Pyrazine

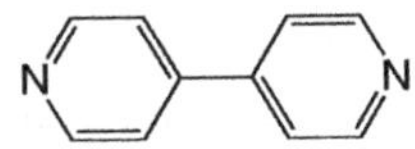

$L^{20B}$ 4,4'-Bipyridine

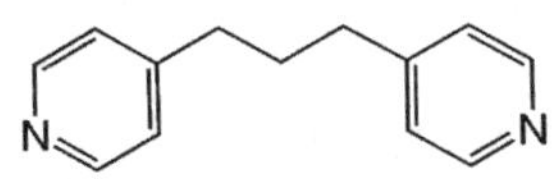

$L^{20C}$ 1,3-Bis(pyridin-4-yl)propane

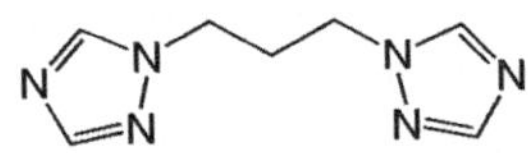

$L^{20D}$ 1,3-Bis(1,2,4-triazol-1-yl)propane

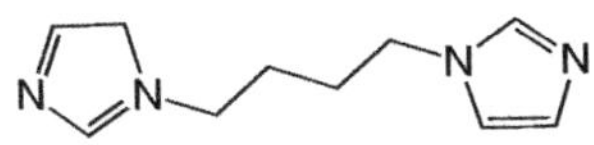

$L^{20E}$ 1,4-Bis(imidazol-1-yl)butane

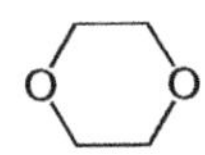

$L^{20F}$ Dioxane

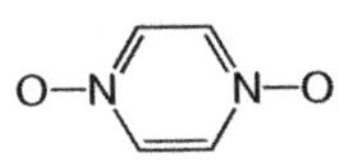

$L^{20G}$ Pyrazine-*N,N'*-dioxide

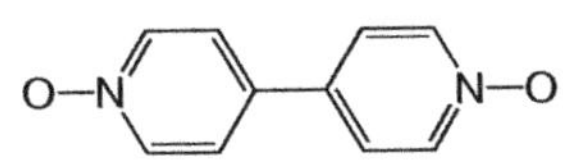

$L^{20H}$ 4,4'-Bipyridine-*N,N'*-dioxide

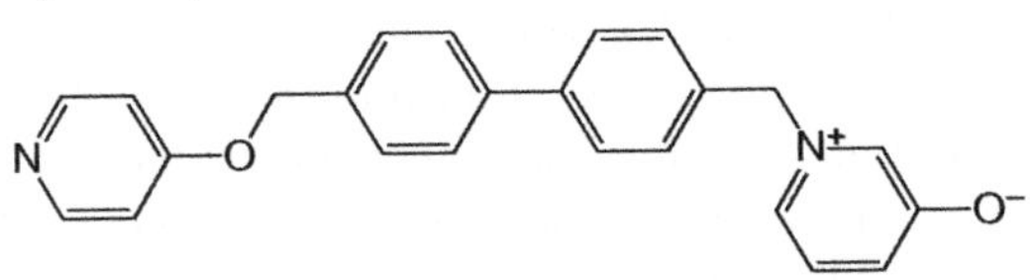

$L^{20I}$ 1-((4'-((pyridin-3-yloxy)methyl)biphenyl-4-yl) methylpyridinium-3-olate

(b)

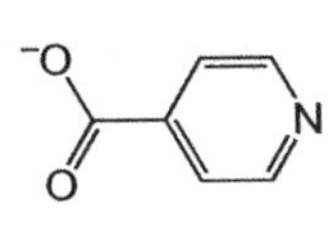

$L^{21A}$ Pyridine-4-carboxylate (Isonicotinate)

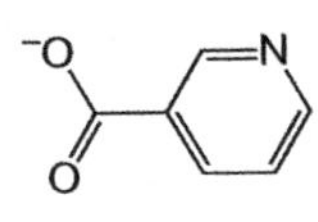

$L^{21B}$ Pyridine-3-carboxylate (Nicotinate)

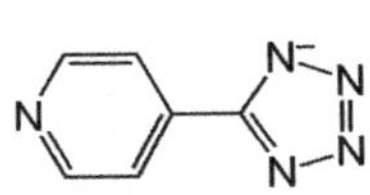

$L^{21C}$ Pyridine-4-tetrazolate

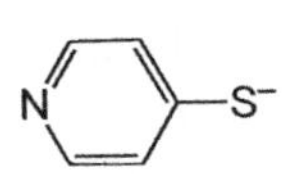

$L^{21D}$ 4-Pyridinethiolate

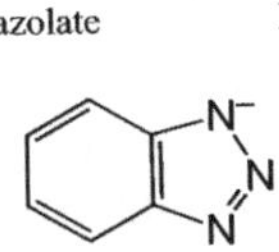

$L^{21E}$ Benzotriazolate

**SCHEME 4.2** (a) Bridging spacers with twofold connectivity. (b) Bridging mono-anions with twofold connectivity. (c) Bridging di-anions with twofold connectivity. (d) Bridging mono-, di- and tri-anions with threefold connectivity. (e) Bridging molecules, di- and tetra-anions with fourfold connectivity.

(c)

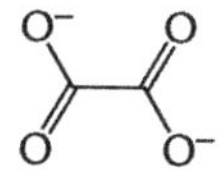

$L^{22A}$ Oxalate

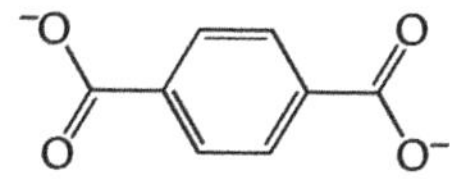

$L^{22B}$ Benzene-1,4-dicarboxylate (Terephthalate)

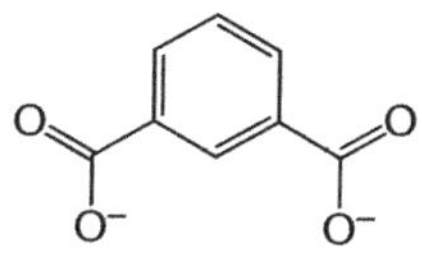

$L^{22C}$ Benzene-1,3-dicarboxylate (Isophthalate)

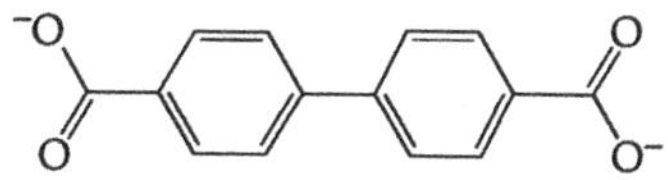

$L^{22D}$ Biphenyl-4,4'-dicarboxylate

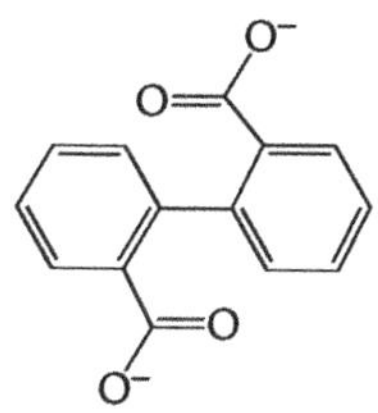

$L^{22E}$ Biphenyl-2,2'-dicarboxylate

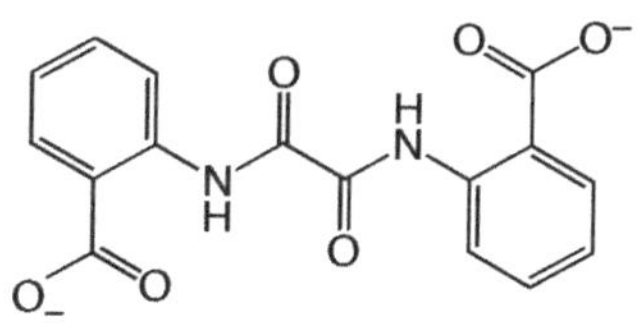

$L^{22F}$ *N,N′*-oxamidobis(benzene-2-carboxylate)

(d)

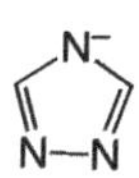

$L^{31A}$ Triazolate

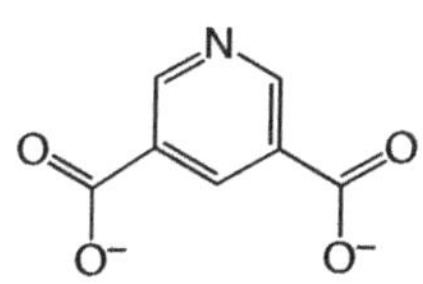

$L^{32A}$ Pyridine-3,5-dicarboxylate

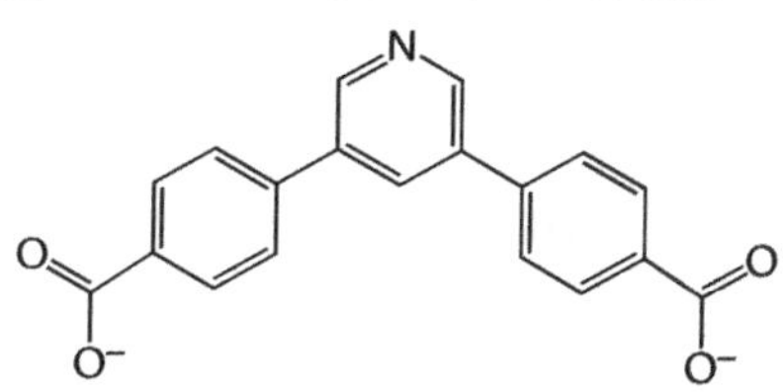

$L^{32B}$ Pyridine-3,5-bis(benzene-4-carboxylate)

**SCHEME 4.2** (*Continued*).

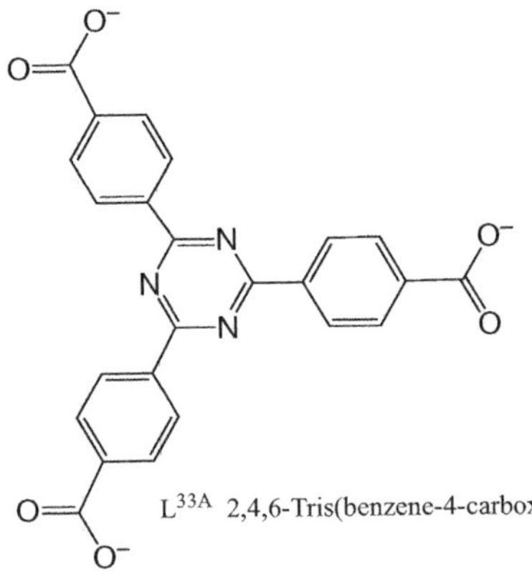

$L^{33A}$ 2,4,6-Tris(benzene-4-carboxylato)-*s*-triazine

(e)

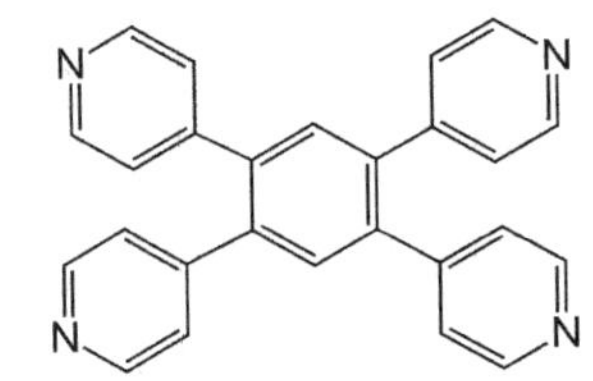

$L^{40A}$ 1,2,4,5-Tetrakis(pyridin-4-yl)benzene

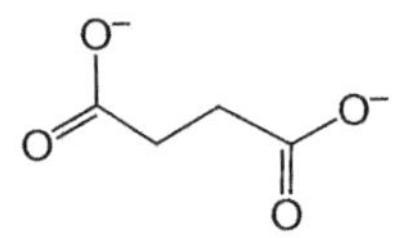

$L^{42A}$ Succinate

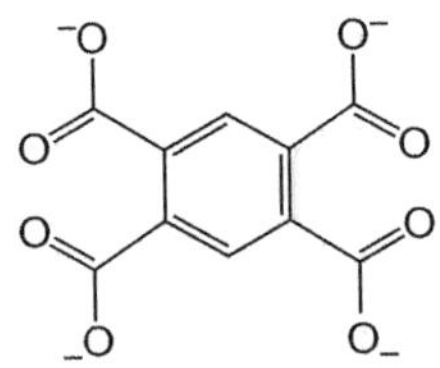

$L^{44A}$ Tetrakis(benzene-1,2,4,5-carboxylate

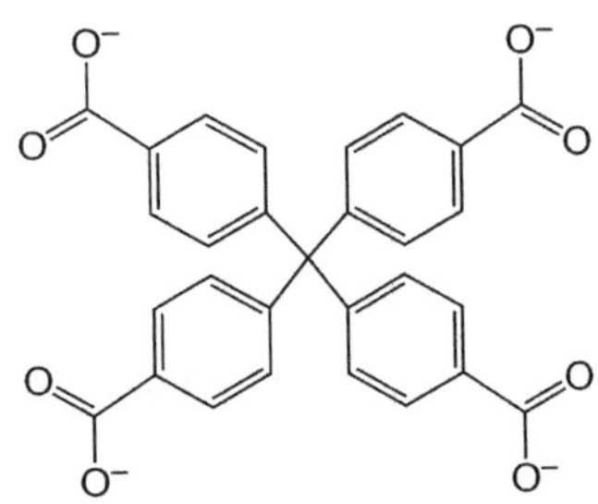

$L^{44B}$ Tetrakis(benzene-4-carboxylato)methane

**SCHEME 4.2** (*Continued*).

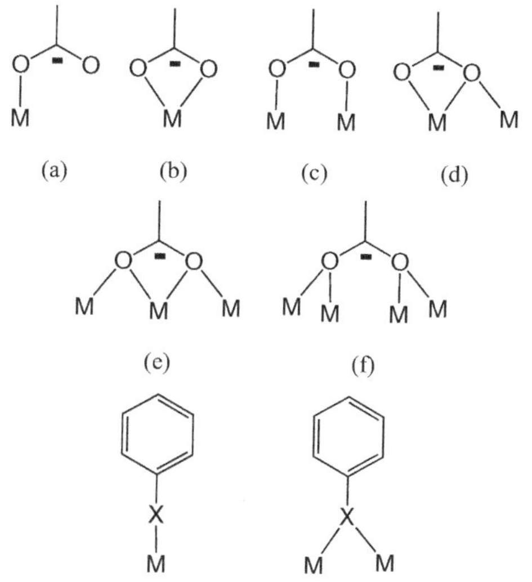

**SCHEME 4.3** Coordination modes of the carboxylate anion (a)–(f) and of the phenolate/thiophenolate anion (g)–(h).

electrons, the *O*-donors have at least two lone pairs and can therefore support cluster formation by bridging metal cations. Carboxylate anions also have the potential to form clusters owing to their bidenticity. Potential coordination modes of the carboxylate and phenolate moieties are illustrated in Scheme 4.3.

## 4.2 METAL CATIONS AS HIGHLY CONNECTED NODES

Any strategy using individual metal centers as highly connected nodes is limited by the available coordination numbers of the metal cations. Thus to target nets and nodes with connectivities greater than 6 the clearest strategy is to exploit the high coordination numbers of *f*-block cations. Lanthanide coordination frameworks with 4,4′-bipyridine-*N*,*N*′-dioxide ($L^{20H}$) and related pyridine *N*-oxide ligands (e.g., $L^{20G}$) have been studied and confirm the successful application of this strategy. $L^{20H}$ gives excellent hard acid/hard base complementarity between the lanthanide cation and the *N*-oxide donors, while the reduced steric footprint of this ligand compared with 4,4′-bipyridine when bound to a metal center avoids crowding at the metal center and encourages high coordination number and connectivity. $L^{20H}$ can bridge metal centers in both *anti*- and *syn*-conformations and can form double as well as single bridges. $L^{20H}$ acts most commonly as a $\mu_2$-bridge but exceptions are known in which a $\mu_3$- or $\mu_4$-bridging modes are adopted.[20–23] Although lanthanide coordination frameworks with 3-,[23] 4-[24] and 6-connected[25] topologies can be prepared, it can also be seen from the following selected examples that combining mixtures of Ln(III) centers and $L^{20H}$ successfully generates complex frameworks based on high coordination numbers and connectivity. The examples highlighted (Table 4.1) give an overview of the coordination

**TABLE 4.1 Frameworks Based on Individual Lanthanide(III) Nodes**

| Framework | | Metal Center | Bridging Ligand | Connectivity | Topology | Ref. |
|---|---|---|---|---|---|---|
| **7-Connected** | | | | | | |
| $\{[La(L^{20H})_4]\cdot BPh_4\cdot 3ClO_4\cdot 2.75MeOH\}_\infty$ | **7A** | $La(Opy)_8$ | $L^{20H}$ | (7,2) | $4^{17}.6^4$ | 27 |
| $\{[Pr(L^{21B})(L^{22E})]\}_\infty$ | **7B** | $Pr(O_2CR)_6(py)$ | $L^{21B}$; $L^{22E}$ | (7,2) | $3^6.4^{13}.6^2$ | 26 |
| $\{[La_2(L^{20H})_{7.3}(MeOH)(H_2O)_{0.4}]\cdot 6CF_3SO_3\}_\infty$ | **7C** | $La(Opy)_7(MeOH)$ | $L^{20H}$ | (7,2) | $3^{12}.4^{14}.6^{2*}$ | 28 |
| **8-Connected** | | | | | | |
| $\{[La(L^{20H})_4](CF_3SO_3)_3\cdot 4.2MeOH\}_\infty$ | **8A** | $La(Opy)_8$ | $L^{20H}$ | (8,2) | $4^{24}.6^4$ | 27 |
| $\{[Ln(L^{20G})_4](ClO_4)_3\}_\infty$, (Ln = La, Ce, Pr, Sm, Eu, Gd, Tb or Y) | **8B** | $Ln(Opy)_8$ | $L^{20G}$ | (8,2) | $4^{24}.6^4$ | 29 |
| $\{[La(L^{20H})_4](ClO_4)_3\cdot MeOH\cdot 4CHCl_2CHCl_2\}_\infty$ | **8M** | $La(Opy)_8$ | $L^{20H}$ | (8,2) | $3^3.4^{15}.5^8.6^2$ | 28 |
| $\{[Ln(L^{20H})_4](CF_3SO_3)_3\cdot 4MeOH\cdot C_2Cl_4\}_\infty$ (Ln = Yb or Eu). | **8R** | $Ln(Opy)_8$ | $L^{20H}$ | (8,8,2) | $(3^5.4^{14}.5^9)(3^5.4^{13}.5.^{10})_2$ | 28 |
| $\{[La_2(L^{20H})_{7.3}(MeOH)(H_2O)_{0.4}]\cdot 6CF_3SO_3\}_\infty$ | **7C** | $La(Opy)_8$ | $L^{20H}$ | (8,2) | $3^{12}.4^{14}.6^2$ | 28 |

*binodal system based upon 7- and 8-connected nodes.

frameworks that have been identified thus far using Ln(III) cations in combination with pyridine *N*-oxide ligands. The only example included in Table 4.1 which has bridging ligands different from pyridine *N*-oxides is $\{[Pr(L^{21B})(L^{22E})]\}_\infty$[26] in which a mixture of biphenyl-2,2′-dicarboxylate and nicotinate link Pr(III) centers.

### 4.2.1 7-Connected Frameworks

Framework structures containing 7-connected nodes are extremely rare in materials chemistry and it is difficult to envisage how this would be achieved with coordination frameworks if a lanthanide-based system were not used. Thus far, only three examples of 7-connected frameworks that uniquely form from an individual node have been reported. The only one that adopts a 3D construction is that in $\{[La(L^{20H})_4]\cdot BPh_4\cdot 3ClO_4\cdot 2.75MeOH\}_\infty$ **7A**.[27] The other two form 2D bilayers, $\{[Pr(L^{21B})(L^{22E})]\}_\infty$ **7B**[26] and $\{[La_2(L^{20H})_{7.3}(MeOH)(H_2O)_{0.4}]\cdot 6CF_3SO_3\}_\infty$ **7C**,[28] of which the latter is binodal (i.e., there are two chemically distinct $Ln^{3+}$ based nodes, one 7-connected, the other 8-connected within the lattice).

In the 3D framework formed by **7A**[27] each La(III) center is coordinated by oxygen atoms from eight separate $L^{20H}$ molecules, six of which form single bridges to six neighboring La(III) centers, while the remaining two form a double bridge to a single neighboring La(III) center producing a network structure that comprises intersecting $4^4$ and $6^3$ nets. The topology of the resulting 3D framework ($4^{16}.6^5$) is discussed in a later section.

The bilayer formed by **7B**[26] is the only example of a highly connected individual metal cation based framework with bridging ligands other than *N*-oxides. One biphenyl-2,2′-dicarboxylate ($L^{22E}$) uses both carboxylate moieties (mode (c) and mode (d); Scheme 4.3) to link four Pr(III) centers, while one nicotinate ($L^{21B}$) uses its carboxylate moiety (mode (c); Scheme 4.3) and *N*-donor pyridine to link three Pr(III) centers. Four of the seven bridges form a $4^4$-grid while the other three link the grids to form a 2D bilayer structure with $3^6.4^{13}.6^2$ topology as shown in Figure 4.1(a). It should be noted that it is also possible to consider the structure to be formed from a parallel series of carboxylate-bridged polymeric chains of Pr(III) cations linked by the nicotinate anions to form a $4^4$-grid, two of which are bridged by biphenyl-2,2′-dicarboxylate anions to form the bilayer structure (Figure 4.1a).

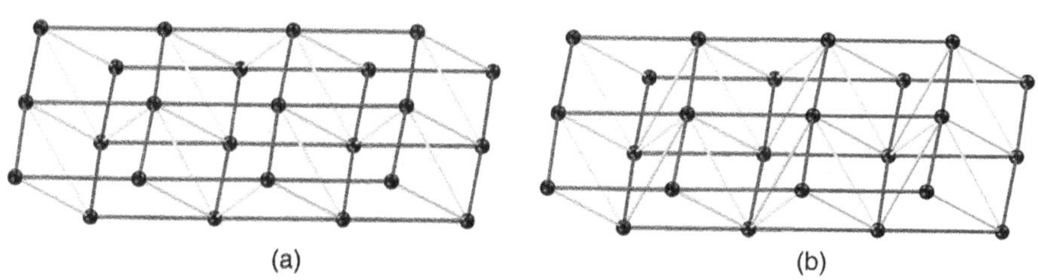


**FIGURE 4.1** Schematic representations of the bilayers formed by frameworks containing 7- and 8-connected nodes. (a) 7-connected bilayer of topology $3^6.4^{13}.6^2$. (b) 8-connected bilayer of topology $3^{12}.4^{14}.6^2$.

The binodal bilayer formed by **7C**[28] comprises two independent La(III) centers, one surrounded by eight $L^{20H}$ molecules while the other is ligated by seven $L^{20H}$ molecules and a MeOH molecule. Both nodes adopt tetragonal anti-prismatic coordination geometries with four of the bridging ligands used to generate a $4^4$-net. The remaining $L^{20H}$ molecules are all located on the same side of the net linking via three- and fourfold bridges from the 7- and 8-connected centers, respectively, to a second $4^4$-net to give a bilayer motif. The idealized network based on the 8-connected network as shown in Figure 4.1(b) has $3^{12}.4^{14}.6^2$ topology.

### 4.2.2 8-Connected Frameworks

Networks derived from 8-connected lanthanide nodes have been prepared as both 3D frameworks[27–29] and, as discussed above, as bilayer[28] network structures.

Three distinct groups of 3D frameworks with 8-connected topologies have been identified, (a) $\{[La(L^{20H})_4](CF_3SO_3)_3{\cdot}4.2MeOH\}_\infty$[27] and $\{[Ln(L^{20G})_4](ClO_4)_3\}_\infty$, (Ln = La, Ce, Pr, Sm, Eu, Gd, Tb or Y)[29]; (b) $\{[La(L^{20H})_4](ClO_4)_3{\cdot}MeOH{\cdot}4CHCl_2CHCl_2\}_\infty$;[28] and (c) $\{[Ln(L^{20H})_4](CF_3SO_3)_3{\cdot}4MeOH{\cdot}C_2Cl_4\}_\infty$ (Ln = Yb or Eu).[28] In all three compound groups, the lanthanides are coordinated exclusively by ligands $L^{20H}$ or $L^{20G}$ that act as bridges between adjacent lanthanide nodes in the extended framework structures. It is also worth noting that in all three structural types the interlayer bridging $L^{20H}$ or $L^{20G}$ molecules are located equally on opposite sides of the metal center coordination polyhedron.

The structure observed for $\{[La(L^{20H})_4]{\cdot}3CF_3SO_3{\cdot}4.2MeOH\}_\infty$**8A**[27] and $\{[Ln(L^{20G})_4]{\cdot}3ClO_4\}_\infty$, (Ln = La, Ce, Pr, Sm, Eu, Gd, Tb or Y) **8B**[29] (Table 4.1) is relatively simple. Each La(III) center is coordinated in a compressed square anti-prismatic geometry by eight $L^{20H}$ or $L^{20G}$ molecules which bridge to eight different neighboring La(III) centers forming a 3D body-centered cubic (CsCl-like) framework, the topology of which ($4^{24}.6^4$) is discussed in a later section.

The structures of $\{[La(L^{20H})_4]{\cdot}3ClO_4{\cdot}MeOH{\cdot}4CHCl_2CHCl_2\}_\infty$**8O**[28] and $\{[Ln(L^{20H})_4]{\cdot}3CF_3SO_3{\cdot}4MeOH{\cdot}C_2Cl_4\}_\infty$ (M = Yb or Eu) **8R**[28] are far more complex than the simple CsCl-like net. Although the asymmetric unit of **8O** contains only one unique La(III) center that of **8R** is bi-nodal. Unlike the situation for the binodal bilayer **7C**,[28] both metal centers are 8-connecting. In both 3D architectures, each type of Ln(III) center adopts a homoleptic $LaO_8$ coordination environment and is linked to eight adjacent metal centers to form 3D 8-connected framework structures which are *not* of the familiar CsCl-type. The topologies of these structures are discussed in a later section.

## 4.3 METAL CLUSTERS AS HIGHLY CONNECTED NODES

Although highly connected frameworks can be constructed using single metal centers, this approach is largely limited to *f*-block metal cations. Although metal centers of the *d*-block series are severely limited in terms of targeting individual metal centers as nodes in highly connected frameworks, it is possible to exploit metal clusters constructed from multiple transition metal centers. Such clusters are often formed *in situ* during the reaction of metal salts and spacer ligands, and have been termed "secondary building units," reflecting their role in the formation and structure of the resulting coordination framework material.[30] For example the $[Zn_4(O)(O_2CR)_6]$ cluster which forms the basis of MOF-5[30] can be regarded as a "secondary building unit" and is formed *in situ* during the reaction of Zn(II) salts with carboxylate ligands. It comprises four Zn(II) cations tetrahedrally arranged around an $O^{2-}$ anion with six carboxylate moieties located along each of the edges of the $Zn_4$ tetrahedron. When the six carboxylate moieties form part of a bridging ligand their octahedral disposition around the cluster leads to a six connected α-Po primitive cubic (**pcu**) network of $4^{12}.6^3$ topology. Since networks with higher connectivity than 6 form the subject of this review, MOF-5 and its analogues will not be considered further here.

More than 20 highly connected frameworks based on cluster nodes have been reported and a representative list is given in Table 4.2. These materials are highly disparate both in cluster identity and choice of bridging ligand, and many of them have been formed serendipitously reflecting the difficulty and challenges of predictive synthesis in this area. However, several metal cations recur; these include Co(II), Cu(II), Zn(II), and Cd(II). There are also examples containing K(I), Rb(I), Fe(III), Ni(II), La(III), and Pb(II). Bridging ligands are primarily 2-connecting but there are some examples with 3- or 4-connectivity giving rise to more complex network topologies. The so-called "secondary building units," which range from binuclear to dodecanuclear metal clusters, are summarized in Schemes 4.4–4.9.

### 4.3.1 Polynuclear Metal Cluster Nodes

***4.3.1.1 Binuclear Nodes*** Examples of frameworks incorporating binuclear nodes include $[M_2(\mu_2\text{-OPh})_2(\text{diox})_7]$ (M = K, Rb) in **7E**,[31] $[Cu_2(py)_4(\mu_{1,2}\text{-}O_2CR)_2(O_2CR)_2]$ in **8F**,[32] $[Co_2(im)_6(\mu_2\text{-}V_2O_6)_2]$ in **8G**,[33] and $[Ln_2(L^{22A})_4(\mu_{1,2}\text{-}O_2CR)_4]$ in **8P**.[34] With the exception of $[K_2(\mu_2\text{-OPh})_2(\text{diox})_7]$,[31] which has sevenfold connectivity, all these nodes have eightfold connectivity. The binuclear core $[M_2(\mu_2\text{-OPh})_2(\text{diox})_7]$ (Scheme 4.4a)[31] comprises two M(I) centers (M = K, Rb), one of which has three and the other four pendant $L^{20F}$ bridging molecules linked by two 4-chloro-2,6-dimethylphenoxide anions. The seven bridging molecules of $L^{20F}$ give the cluster its sevenfold connectivity. The simplest of the 8-connecting clusters is $[Cu_2(py)_4(\mu_{1,2}\text{-}O_2CR)_2(O_2CR)_2]$ (Scheme 4.4b),[32] in which pairs of five-coordinate Cu(II) centers, each of which is terminally coordinated by two *N*-donor pyridines (one from $L^{21A}$ and one from $L^{21B}$) and one monodentate carboxylate (mode (a); Scheme 4.3) from $L^{21A}$, are linked by two bridging carboxylates (mode (c); Scheme 4.3) from $L^{21B}$. Both $L^{21A}$ and $L^{21B}$ anions link to adjacent clusters to give eightfold connectivity. For $[Co_2(im)_6(\mu_{1,2}\text{-}V_2O_6)_2]$ (Scheme 4.4c),[33] pairs of six-coordinate Co(II) centers are linked through $[V_4O_{12}]^{4-}$ anions to form an inorganic chain of composition $Co_2V_4O_{12}$. The coordination geometry at each Co(II) center is completed by three *N*-donor imidazole moieties from $L^{20E}$ to give an overall eightfold connectivity. For $[Ln_2(ox)_4(\mu_{1,2}\text{-}O_2CR)_4]$ (Scheme 4.4d),[34] the ninefold coordination geometry of each Ln(III) center is generated by the carboxylate moieties of four $[Cu(L^{21A})_2]$ bridges, two of which adopt mode (c) (Scheme 4.3) and two mode (d) (Scheme 4.3) and two

**TABLE 4.2 Frameworks Based on Metal Cluster Nodes**

| Framework | | Metal Cluster Node | Scheme | Bridging Ligand | Connectivity | Topology | Ref. |
|---|---|---|---|---|---|---|---|
| **7-Connected** | | | | | | | |
| $\{[K_5(Ph^B)_5(L^{20F})_5]\}_\infty$ | **7D** | $K_5(\mu_2\text{-OPh})(\mu_3\text{-OPh})_4(diox)_9$ | 4.7d | $L^{20F}$ | (7,2) | $4^{15}.6^6$ | 31 |
| $\{[M_2(Ph^A)_2(L^{20F})_{3.5}]\}_\infty$ (M = K, Rb) | **7E** | $M_2(\mu_2\text{-OPh})_2(diox)_7$ (M = K, Rb) | 4.4a | $L^{20F}$ | (7,2) | $3^3.4^{12}.5^5.6$ | 31 |
| **8-Connected** | | | | | | | |
| $\{[Cu_3Cl_2(OH_2)_2(L^{21C})_4]\cdot 4dmf\cdot 5H_2O\}_\infty$ | **8C** | $Cu_3(\mu_2\text{-Cl})_2(OH_2)_2(py)_4(\mu_{1,2}\text{-tz})_4$ | 4.5e | $L^{21C}$ | (8,2) | $4^{24}.6^4$ | 39 |
| $\{[Cd_5Cl_2(OAc)_2(OH_2)_2(L^{31A})_6]\cdot 2H_2O\}_\infty$ | **8D** | $Cd_5(\mu_2\text{-Cl})_2(\mu_2\text{-OAc})_2(OH_2)_2(\mu_{1,2}\text{-tr})_6(tr)_6$ | 4.7c | $L^{31A}$ | (8,2) 12 → 8 | $4^{24}.6^4$ | 40 |
| $\{[Pb_8(O)_2(OH_2)_2(L^{22C})_6\}_\infty$ | **8E** | $Pb_8(\mu_4\text{-O})_2(OH_2)_2(\mu_{1,2}\text{-O}_2CR)_{10}(O_2CR)_2$ | 4.9c | $L^{22C}$ | (8,2) 12 → 8 | $4^{24}.6^4$ | 51 |
| $\{[Cu_2(L^{21A})_2(L^{21B})_2\}_\infty$ | **8F** | $Cu_2(py)_4(\mu_{1,2}\text{-O}_2CR)_2(O_2CR)_2$ | 4.4b | $L^{21A}$; $L^{21B}$ | (8,2,2) | $4^{24}.6^4$ | 32 |
| $\{[Co_2(V_4O_{12})(L^{20E})_3]\cdot 4H_2O\}_\infty$ | **8G** | $Co_2(im)_6(\mu_{1,2}\text{-}V_2O_6)_2$ | 4.4c | $[V_4O_{12}]$; $L^{20E}$ | (8,2,2) | $4^{24}.5.6^3$ | 33 |
| $\{[Zn_5(OH)_2(phen)_2(L^{22B})_4]\}_\infty$ | **8H** | $Zn_5(\mu_3\text{-OH})_2(phen)_2(\mu_{1,2}\text{-O}_2CR)_6(O_2CR)_2$ | 4.7a | $L^{22B}$ | (8,2) | $4^{24}.5.6^3$ | 47 |
| $\{[Zn_5(OH)_2(L^{21E})_2(L^{22B})_3]\}_\infty$ | **8I** | $Zn_5(\mu_3\text{-OH})_2(bt)_4(\mu_{1,2}\text{-O}_2CR)_6$ | 4.7b | $L^{21E}$; $L^{22B}$ | (8,2,2) 10 → 8 | $4^{24}.5.6^3$ | 48 |
| $\{[Cd_3(OH_2)_2(L^{20D})(L^{22B})_3]\}_\infty$ | **8J** | $Cd_3(OH_2)_2(tr)_2(\mu_{1,2}\text{-O}_2CR)_6$ | 4.5c | $L^{22B}$; $L^{20D}$ | (8,2,2) | $4^{20}.6^8$ | 37 |
| $\{[Cu_4(V_4O_{12})(L^{20C})_4]\cdot 3H_2O\}_\infty$ | **8K** | $Cu_4(py)_8(V_4O_{12})$ | 4.6c | $L^{20C}$ | (8,2) | $4^{24}.6^4$ | 46 |
| $\{[Cu_4(OH)_2(L^{21A})_4(L^{22B})]\}_\infty$ | **8L** | $Cu_4(\mu_3\text{-OH})_2(py)_4(\mu_{1,2}\text{-O}_2CR)_4(O_2CR)_2$ | 4.6e | $L^{22B}$; $L^{21A}$ | (8,2,2) 10 → 8 | $3^5.4^{11}.5^8.6^4$ | 32 |
| $\{[Zn_3(L^{20B})(L^{22D})_3]\cdot 2dmf\cdot 4H_2O\}_\infty$ | **8N** | $Zn_3(py)_2(\mu_{1,2}\text{-O}_2CR)_6$ | 4.5a | $L^{22D}$; $L^{20B}$ | (8,2,2) | $3^6.4^{18}.5^3.6$ | 36 |
| $\{[Co_3(L^{20B})(L^{22D})_3]\cdot 4dmf\cdot H_2O\}_\infty$ | **8O** | $Co_3(py)_2(\mu_{1,2}\text{-O}_2CR)_6$ | 4.5b | $L^{22D}$; $L^{20B}$ | (8,2,2) | $3^6.4^{18}.5^3.6$ | 35 |

| | | | | | | | |
|---|---|---|---|---|---|---|---|
| $\{[Ln_2(L^{22A})_2]\cdot 2[Cu(L^{21A})_2]\cdot 2H_2O\}_\infty$ (Ln = La, Pr or Nd) | **8P** | $Ln_2(ox)_4(\mu_{1,2}\text{-}O_2CR)_4$ | 4.4d | $[Cu(L^{21A})_2]$; $L^{22A}$ | (8,2,2) | $3^6.4^{18}.5^3.6$ | 34 |
| $\{[Co_4(OH)_2(L^{20B})_2(L^{22F})_3]\cdot L^{20B}\}_\infty$ | **8Q** | $Co_4(\mu_3\text{-}OH)_2(py)_4(\mu_{1,2}\text{-}O_2CR)_4(O_2CR)_2$ | 4.6d | $L^{20B}$; $L^{22F}$ | (8,2,2) 10 → 8 | $3^6.4^{14}.5^4.6^4$ | 45 |
| $\{H_2[Co_4(O)(L^{33A})_{8/3}]\}_\infty$ | **8S** | $Co_4(\mu_4\text{-}O)((\mu_{1,2}\text{-}O_2CR)_8$ | 4.6b | $L^{33A}$ | (8,3) | $(4^3)\ (4^8.6^4.8^{12}.10^4)$ | 44 |
| $\{[Cd_4(dmf)_4(L^{44B})_2]\cdot 4dmf\cdot 4H_2O\}_\infty$ | **8T** | $Cd_4(dmf)_4(\mu_{1,2}\text{-}O_2CR)_8$ | 4.6a | $L^{44B}$ | (8,4) | $(4^6)(4^{12}.6^{12}.8^4)$ | 43 |
| $\{[Cd_8(SO_4)(SPh)_{12}(L^{40A})_2]\cdot 2HSO_4\cdot 4H_2O$ | **8U** | $Cd_8(\mu_4\text{-}SO_4)(\mu_2\text{-}SPh)_{12}(py)_8$ | 4.9b | $L^{40A}$ | (8,4) | $(4^4.6^2)_4(4^{16}.6^{12})$ | 50 |
| $\{[Cd_3Cl_2(L^{31A})_2(L^{42A})_2]\}_\infty$ | **8V** | $Cd_3(\mu_2\text{-}Cl)_2(\mu_{1,2}\text{-}tr)_2(tr)_2(O_2CR)_6$ | 4.5f | $L^{31A}$; $L^{42A}$ | (8,4) | $(4^5.6)(4^{12}.6^{16})$ | 40 |
| $\{[Zn_3(L^{20I})_2(L^{22B})_3][Zn_2(L^{22B})_2]\cdot 2H_2O\}_\infty$ | **8W** | $Zn_3(\mu_2\text{-}Opy)_2(\mu_{1,2}\text{-}O_2CR)_4(O_2CR)_2$ | 4.5d | $L^{20I}$; $L^{22B}$ | (8,6) | $(4^{12}.5.6^2)(4^{20}.5^2.6^6)$ | 38 |
| **9-Connected** | | | | | | | |
| $\{[Rb_6(Ph^C)_6(L^{20F})_{4.5}]\}_\infty$ | **9A** | $Rb_6(\mu_3\text{-}OPh)_6(diox)_9$ | 4.8b | $L^{20F}$ | (9,2) | $3^6.4^{22}.5^8$ | |
| $\{[Co_3(OH)(L^{32A})_3]\cdot H_2O\}_\infty$ | **9B** | $Co_3(\mu_3\text{-}OH)(py)_3(\mu_{1,2}\text{-}O_2CR)_6$ | 4.5g | $L^{32A}$ | (9,3) | $(4^2.6)_3\ (4^6.6^{21}.8^9)$ | 41 |
| $\{[Fe_3(O)(L^{32B})_3]\cdot n(solv)\}_\infty$ | **9C** | $Fe_3(\mu_3\text{-}O)(py)_3(\mu_{1,2}\text{-}O_2CR)_6$ | 4.5g | $L^{32B}$ | (9,3) | $(4^2.6)_3\ (4^6.6^{21}.8^9)$ | 42 |
| $\{[Ni_3(OH)(L^{32B})_3]\cdot n(solv)\}_\infty$ | **9D** | $Ni_3(\mu_3\text{-}OH)(py)_3(\mu_{1,2}\text{-}O_2CR)_6$ | 4.5g | $L^{32B}$ | (9,3) | $(4^2.6)_3\ (4^6.6^{21}.8^9)$ | 42 |
| **10-Connected** | | | | | | | |
| $\{[Zn_7(OH_2)_6(L^{31A})_6(L^{44A})_2]\cdot 8H_2O\}_\infty$ | **10A** | $Zn_7(OH_2)_6(\mu_{1,2,3}\text{-}tr)_6(O_2CR)_8$ | 4.9a | $L^{31A}$; $L^{44A}$ | (10,2,4) | | 40 |
| **12-Connected** | | | | | | | |
| $\{[Cu_3(L^{21D})_2(CN)]\}_\infty$ | **12A** | $Cu_6(\mu_3\text{-}Spy)_4(pyS)_4(CN)_4$ | 4.8a | $CN^-$; $L^{21D}$ | (12,2,2,2) | $3^{24}.4^{36}.5^6$ | 49 |
| $\{[Cu_{12}(SMe)_6(CN)_6]\cdot 2H_2O\}_\infty$ | **12B** | $Cu_{12}(\mu_4\text{-}SMe)_6(CN)_{12}$ | 4.9d | $CN^-$ | (12,2) | $3^{24}.4^{36}.5^6$ | 52 |

*Abbreviations*: bt: benztriazole; dmf: dimethylformamide; im: Imidazole; $OAc^-$: acetate; $OPh^-$: aryloxide; $ox^{2-}$: oxalte; phen: 1,10- phenanthroline; py: pyridine; $pyS^-$: pyridine-4-carboxylate; $RCO_2^-$: carboxylate; ta: 1,2,3-triazole; tr: 1,2,4-triazole; tz: tetrazole.

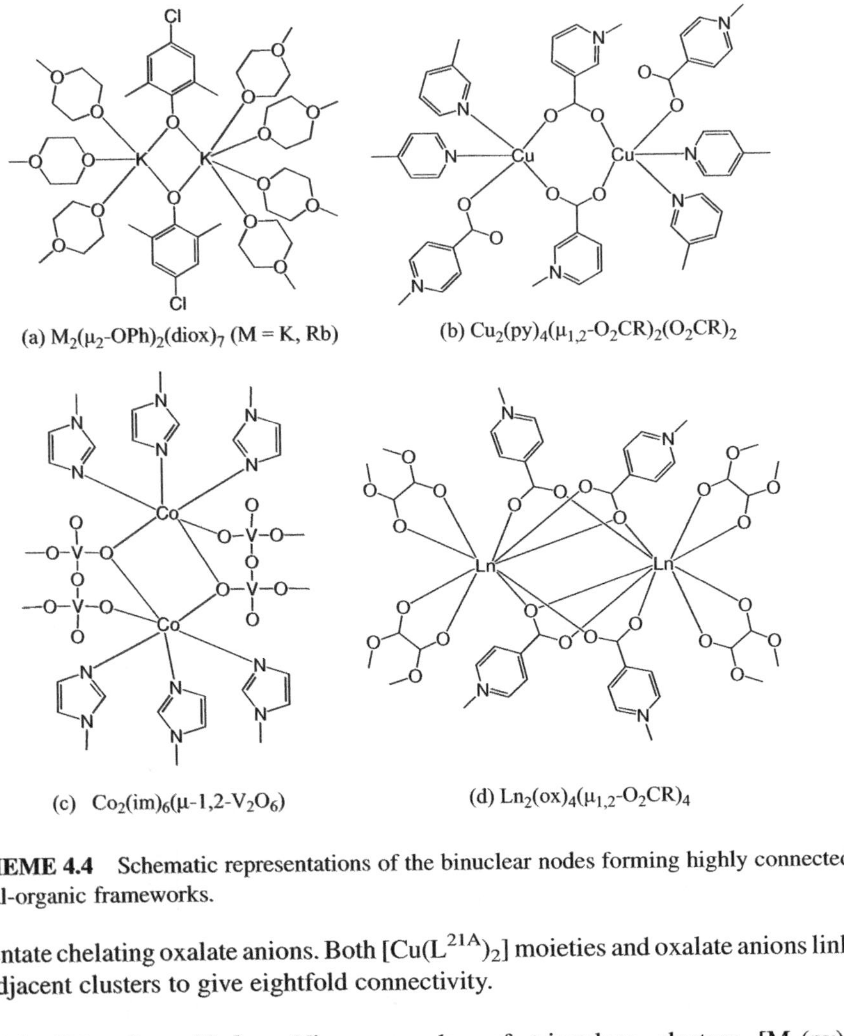

(a) $M_2(\mu_2\text{-OPh})_2(\text{diox})_7$ (M = K, Rb)

(b) $Cu_2(\text{py})_4(\mu_{1,2}\text{-}O_2CR)_2(O_2CR)_2$

(c) $Co_2(\text{im})_6(\mu\text{-}1,2\text{-}V_2O_6)$

(d) $Ln_2(\text{ox})_4(\mu_{1,2}\text{-}O_2CR)_4$

**SCHEME 4.4** Schematic representations of the binuclear nodes forming highly connected metal-organic frameworks.

bidentate chelating oxalate anions. Both $[Cu(L^{21A})_2]$ moieties and oxalate anions link to adjacent clusters to give eightfold connectivity.

***4.3.1.2 Trinuclear Nodes*** Nine examples of trinuclear clusters $[M_3(\text{py})_2(\mu_{1,2}\text{-}CO_2)_6]$ (M = Co in **8O**,[35] Zn in **8N**[36]), $[Cd_3(OH_2)_2(\text{tr})_2(\mu_{1,2}\text{-}O_2CR)_6$ in **8J**,[37] $[Zn_3(\mu_2\text{-Opy})_2(\mu_{1,2}\text{-}O_2CR)_4(O_2CR)_2]$ in **8W**,[38] $[Cu_3(\mu_2\text{-}Cl)_2(OH_2)_2(\text{py})_4(\mu_{1,2}\text{-tz})_4]$ in **8C**,[39] $[Cd_3(\mu_2\text{-}Cl)_2(\mu_{1,2}\text{-tr})_2(\text{tr})_2(O_2CR)_6]$ in **8V**,[40] and $[M_3(\mu_3\text{-}O)(\text{py})_3(\mu_{1,2}\text{-}O_2CR)_6]$ (M = Fe in **9C**,[41] Co in **9B**,[42] Ni in **9D**[41]) have been reported. The first six clusters incorporate three metal atoms in a chain to give 8-connected networks, while the remaining three, in which the three metal centers adopt a triangular arrangement, give rise to 9-connected networks. The clusters $[M_3(\text{py})_2(\mu_{1,2}\text{-}O_2CR)_6]$ (M = Co,[35] Zn[36]) and $[Cd_3(OH_2)_2(\text{tr})_2(\mu_{1,2}\text{-}O_2CR)_6]$[37] (Scheme 4.5a–c) differ solely in the coordination of the terminal metal cations; fourfold for Zn(II), fivefold for Co(II) and sixfold for Cd(II). The central metal cation is sixfold in all cases with each terminal-central pair of cations linked by three carboxylate moieties. To generate a fourfold Zn(II) geometry, coordination mode (c) (Scheme 4.3) is adopted by all three carboxylate moieties, while for the fivefold Co(II) and sixfold Cd(II) geometries, coordination mode (c) (Scheme 4.3) is adopted by two and mode (d) (Scheme 4.3) by one carboxylate moiety. The coordination geometries of the terminal cations are completed by pyridine or 1,2,4-triazolate nitrogen donors and in the case of Cd(II) a water molecule. In all three examples, the bridging ligands providing the six carboxylate anions ($L^{22B}$ or $L^{22D}$) and two *N*-donor moieties ($L^{20B}$ or $L^{20D}$) link to adjacent clusters to give eightfold connectivity.

$[Zn_3(\mu_2\text{-Opy})_2(\mu_{1,2}\text{-}O_2CR)_4(O_2CR)_2]$ (Scheme 4.5d)[38] also has four-coordinate terminal and six-coordinate central Zn(II) cations. In this case, however, one of the carboxylates bridging each terminal-central pair of cations is replaced by a phenolate oxygen donor, which adopts coordination mode (h) (Scheme 4.3), while the *N*-donor pyridine on the terminal Zn(II) cation is replaced by a carboxylate coordinating via mode (a) (Scheme 4.3). The bridging ligands providing the six carboxylate moieties ($L^{22B}$) and two phenoxide *O*-donors ($L^{20I}$) link to adjacent clusters to give eightfold connectivity.

The remaining two chain trinuclear clusters, which give rise to eightfold connectivity, have either one $\mu_2$-chloride anion and two $\mu_{1,2}$-tetrazolate moieties (from $L^{21C}$) as in $[Cu_3(\mu_2\text{-}Cl)_2(OH_2)_2(py)_4(\mu_{1,2}\text{-tz})_4]$ (Scheme 4.5e),[39] or one $\mu_2$-chloride, one $\mu_{1,2}$-triazolate, and one $\mu_{1,2}$-succinate as in $[Cd_3(\mu_2\text{-}Cl)_2(\mu_{1,2}\text{-tr})_2(tr)_2(O_2CR)_6]$ (Scheme 4.5f)[40] bridging each terminal–central pair of cations. The sixfold coordination of the terminal Cu(II) centers is completed by two pyridine moieties (from $L^{21C}$) and one water molecule, while that of the terminal Cd(II) centers is completed by one bidentate succinate carboxylate moiety, which adopts coordination mode (b) (Scheme 4.3), and one triazolate moiety.

The trinuclear clusters in $[M_3(\mu_3\text{-}O)(py)_3(\mu_{1,2}\text{-}O_2CR)_6]$ (M = Fe,[41] Co,[42] Ni[41]) Scheme 4.5g) comprise three metal cations surrounding, in an approximately planar fashion, a $\mu_3$-oxygen atom assigned as $O^{2-}$ for M = Fe and $OH^-$ for M = Co or Ni, with two carboxylate moieties, adopting coordination mode (c) (Scheme 4.3), linking each pair of metal cations. In all three cases, the cations are mixed valence $M^{2+}/M^{3+}$ species. The sixfold coordination geometries of the metal centers are completed by pyridine *N*-donors from $L^{32A}$ or $L^{32B}$ giving the clusters ninefold connectivity.

***4.3.1.3 Tetranuclear Clusters*** The five examples of tetranuclear clusters reported in the literature differ in the number and type of bridging species at their center: $[Cd_4(dmf)_4(\mu_{1,2}\text{-}O_2CR)_8]$ in **8T**[43] has none, $[Co_4(\mu_4\text{-}O)(\mu_{1,2}\text{-}O_2CR)_8]$ in **8S**[44] has one $\mu_4$-oxide, $[Co_4(\mu_3\text{-}OH)_2(py)_4(\mu_{1,2}\text{-}O_2CR)_4(O_2CR)_2]$ in **8Q**[45] and $[Cu_4(\mu_3\text{-}OH)_2(py)_4(\mu_{1,2}\text{-}O_2CR)_4(O_2CR)_2]$ in **8L**[32] have two $\mu_3$-hydroxides, while $[Cu_4(V_4O_{12})(py)_8]$ in **8K**[46] incorporates a tetranuclear $[V_4O_{12}]^{4-}$ anion. The cluster in $[Cd_4(dmf)_4(\mu_{1,2}\text{-}O_2CR)_8]$ (Scheme 4.6a),[43] which is arranged in a square-plane, is generated by pairs of carboxylate moieties from $L^{44B}$ linking pairs of Cd(II) centers. By adopting coordination mode (d), the carboxylate moieties occupy six positions of the sevenfold coordination geometries of the Cd(II) centers, with a dmf molecule located at the seventh position. The bridging carboxylate moieties serve to give an eightfold

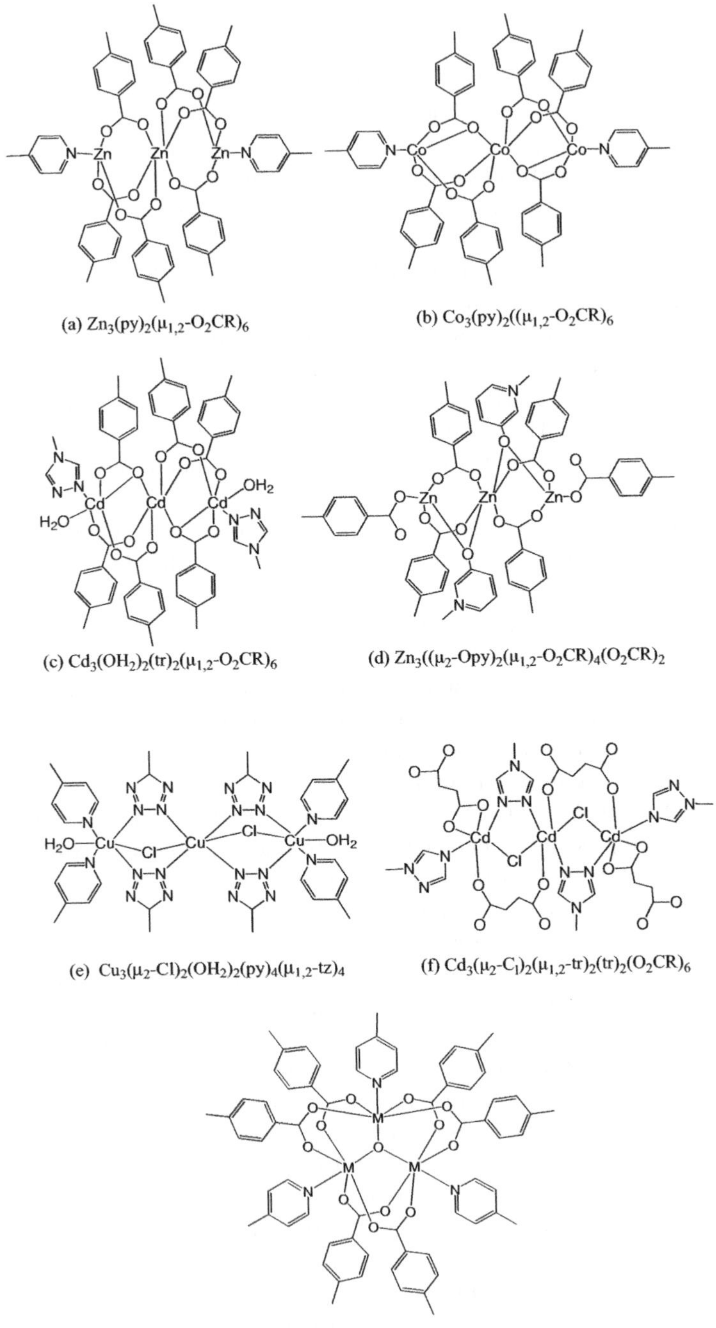

(a) $Zn_3(py)_2(\mu_{1,2}\text{-}O_2CR)_6$

(b) $Co_3(py)_2((\mu_{1,2}\text{-}O_2CR)_6$

(c) $Cd_3(OH_2)_2(tr)_2(\mu_{1,2}\text{-}O_2CR)_6$

(d) $Zn_3((\mu_2\text{-}Opy)_2(\mu_{1,2}\text{-}O_2CR)_4(O_2CR)_2$

(e) $Cu_3(\mu_2\text{-}Cl)_2(OH_2)_2(py)_4(\mu_{1,2}\text{-}tz)_4$

(f) $Cd_3(\mu_2\text{-}Cl)_2(\mu_{1,2}\text{-}tr)_2(tr)_2(O_2CR)_6$

(g) $M_3(\mu_3\text{-}O)(py)_3(\mu_{1,2}\text{-}O_2CR)_6$

**SCHEME 4.5** Schematic representations of the trinuclear nodes forming highly connected metal-organic frameworks.

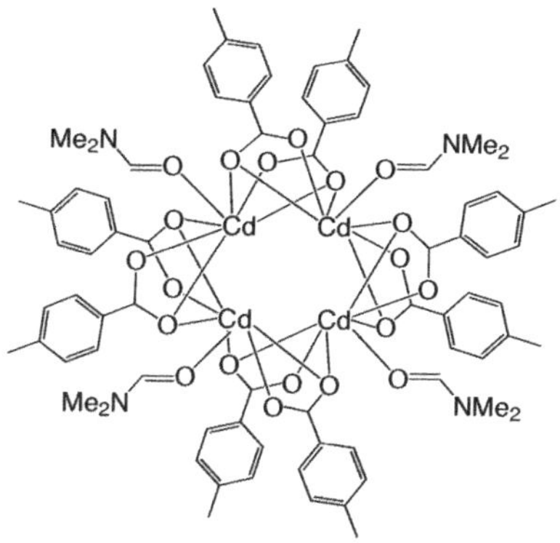

(a) $Cd_4(dmf)_4(\mu_{1,2}\text{-}O_2CR)_8$

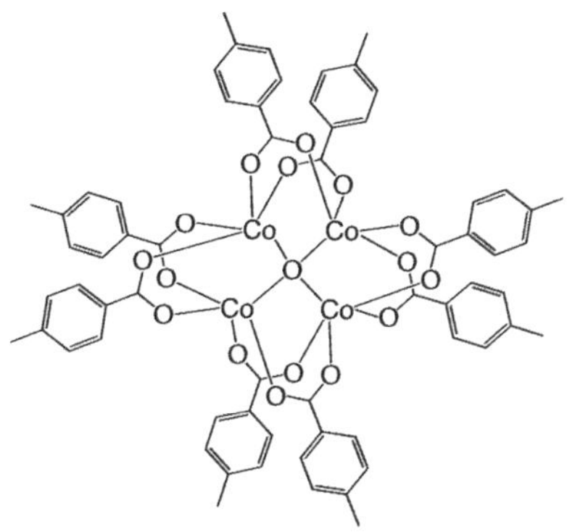

(b) $Co_4(\mu_4\text{-}O)(\mu_{1,2}\text{-}O_2CR)_8$

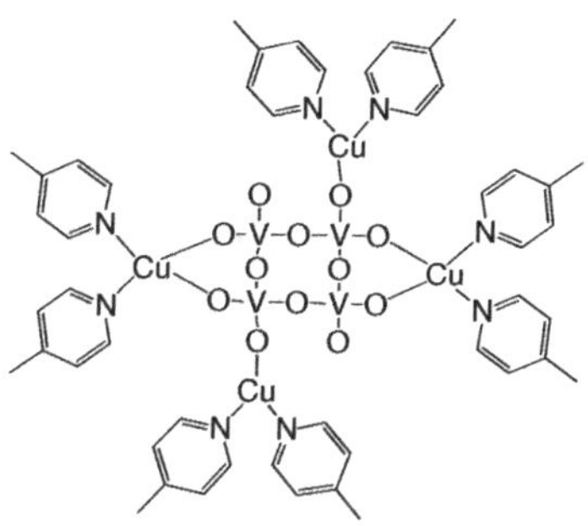

(c) $Cu_4(py)_8(V_4O_{12})$

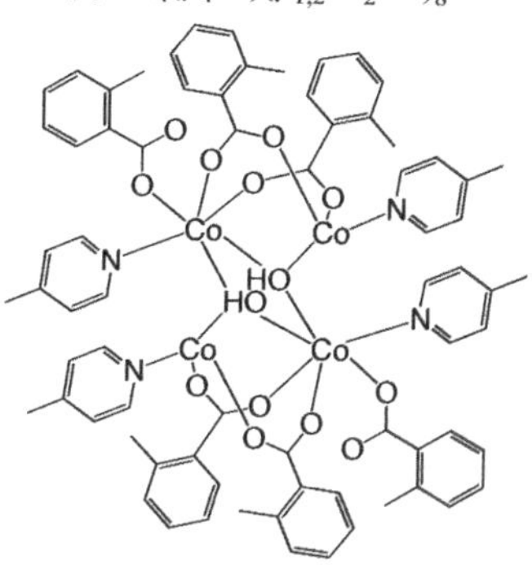

(d) $Co_4(\mu_3\text{-}OH)_2(py)_4(\mu_{1,2}\text{-}O_2CR)_4(O_2CR)_2$

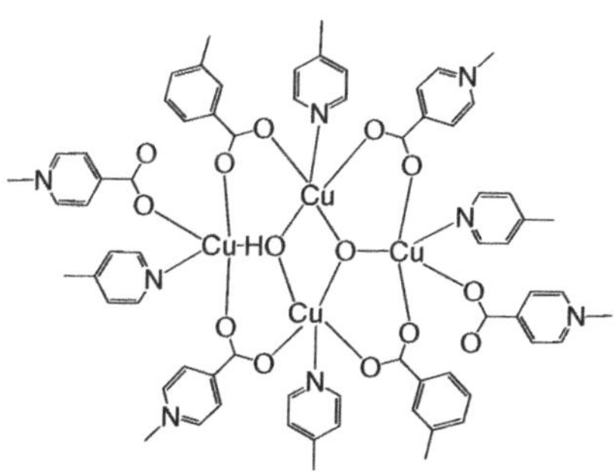

(e) $Cu_4(\mu_3\text{-}OH)_2(py)_4(\mu_{1,2}\text{-}O_2CR)_4(O_2CR)_2$

**SCHEME 4.6** Schematic representations of the tetranuclear nodes forming highly connected metal-organic frameworks.

connectivity to the cluster. In $[Co_4(\mu_4\text{-}O)(\mu_{1,2}\text{-}O_2CR)_8]$ (Scheme 4.6b)[44] the four Co (II) centers are disposed in a square-plane around a $\mu_4$-oxide with pairs of carboxylate moieties adopting coordination mode (c) (Scheme 4.3) and linking pairs of metal centers. The eight coordinated carboxylate moieties from $L^{33A}$ give eightfold connectivity to the cluster. In $[Cu_4(V_4O_{12})(py)_8]$ (Scheme 4.6c),[46] the eightfold connectivity is generated by each Cu(I) center binding two *N*-donor pyridines from the bridging $L^{20C}$ ligands. The three- and four-coordinate geometries of the Cu(I) centers are completed by either one or two links to oxygen atoms of the central $[V_4O_{12}]^{4-}$ unit.

The two $\mu_3$-hydroxide-centered tetranuclear clusters, $[Co_4(\mu_3\text{-}OH)_2(py)_4(\mu_{1,2}\text{-}O_2CR)_4(O_2CR)_2]$[45] and $[Cu_4(\mu_3\text{-}OH)_2(py)_4(\mu_{1,2}\text{-}O_2CR)_4(O_2CR)_2]$,[32] differ from the three previously described tetra-nuclear clusters in that a total of 10 bridging ligands radiate from the cluster. Overall, however, the clusters act as 8-connecting nodes as two pairs of bridging ligands form double bridges to adjacent clusters. In $[Co_4(\mu_3\text{-}OH)_2(py)_4(\mu_{1,2}\text{-}O_2CR)_4(O_2CR)_2]$ (Scheme 4.6d)[45] two $\mu_3\text{-}OH^-$ units link two binuclear $[Co_2(py)_2(O_2CR)_3]$ moieties in which the Co(II) centers are bridged by two carboxylates (mode (c); Scheme 4.3). One Co(II) center is six-coordinate binding to two bridging hydroxides and a single carboxylate terminally (mode (a); Scheme 4.3) and a single *N*-donor pyridine, while the other Co(II) center is four-coordinate and binding to one bridging hydroxide and a single terminal *N*-donor pyridine to give the cluster 10 bridging molecules; four $L^{20B}$ and six $L^{22F}$. In $[Cu_4(\mu_3\text{-}OH)_2(py)_4(\mu_{1,2}\text{-}O_2CR)_4(O_2CR)_2]$ (Scheme 4.6e)[32] the two $\mu_3\text{-}OH^-$ together with four carboxylate moieties (mode (c); Scheme 4.3) link the four coplanar Cu(II) centers to form an approximate parallelogram. Thus, two of the Cu(II) centers have terminal *N*-donor pyridine and monodentate carboxylate (mode (a); Scheme 4.3) as well as the two carboxylate moieties (mode (c); Scheme 4.3) and one bridging $OH^-$ in their five-coordinate geometry, while the other two have a single terminal *N*-donor pyridine as well as the two carboxylate moieties (mode (c); Scheme 4.3) and two bridging $OH^-$ in their five-coordinate geometry. Thus overall, each cluster has a total of 10 bridging ligands: eight $L^{21A}$ and two $L^{22B}$.

***4.3.1.4 Pentanuclear Clusters*** Of the pentanuclear clusters used for generation of high connectivity coordination polymers, $[Zn_5(\mu_3\text{-}OH)_2(phen)_2(\mu_{1,2}\text{-}O_2CR)_6(O_2CR)_2]$ in **8H**,[47] $[Zn_5(\mu_3\text{-}OH)_2(bt)_4(\mu_{1,2}\text{-}O_2CR)_6]$ in **8I**,[48] $[Cd_5(\mu_2\text{-}Cl)_2(\mu_2\text{-}OAc)_2(OH_2)_2(\mu_{1,2}\text{-}tr)_6(tr)_6]$ in **8D**,[40] and $[K_5(\mu_2\text{-}OPh)(\mu_3\text{-}OPh)_4(diox)_9]$ in **7D**,[31] the two Zn(II) clusters have similar cores. The five Zn(II) centers are disposed as two trigonal planar centers around $\mu_3$-bridging hydroxides with a common Zn(II) center. Each pair of Zn(II) centers is linked by a single carboxylate moiety (mode (c); Scheme 4.3) from $L^{22B}$ giving the central Zn(II) cation sixfold coordination. The peripheral Zn(II) centers have additionally either a bidentate 1,10-phenanthroline ligand as in $[Zn_5(\mu_3\text{-}OH)_2(phen)_2(\mu_{1,2}\text{-}O_2CR)_6(O_2CR)_2]$ (Scheme 4.7a),[47] or a single 1,2,3-triazolate moiety from bridging $L^{21E}$ as in $[Zn_5(\mu_3\text{-}OH)_2(bt)_4(\mu_{1,2}\text{-}O_2CR)_6]$ (Scheme 4.7b)[48] giving five- or fourfold coordination, respectively. Thus, whereas the former cluster has eight $L^{22B}$ bridging ligands, the latter has six $L^{22B}$ and four $L^{21E}$ bridging ligands. Nonetheless, both clusters act as 8-connecting nodes as four of the $L^{21E}$ ligands in the latter example form double bridges.

The pentanuclear cluster of $[Cd_5(\mu_2\text{-}Cl)_2(\mu_2\text{-}OAc)_2(OH_2)_2(\mu_{1,2}\text{-}tr)_6(tr)_6]$ (Scheme 4.7c)[40] is based on a centrosymmetric chain of five Cd(II) atoms. The central Cd(II) center is linked to its neighbors by one $\mu_{1,2}$-triazolate and one $\mu_2$-chloride, while the terminal ones are attached to their neighbors by two $\mu_{1,2}$-triazolates

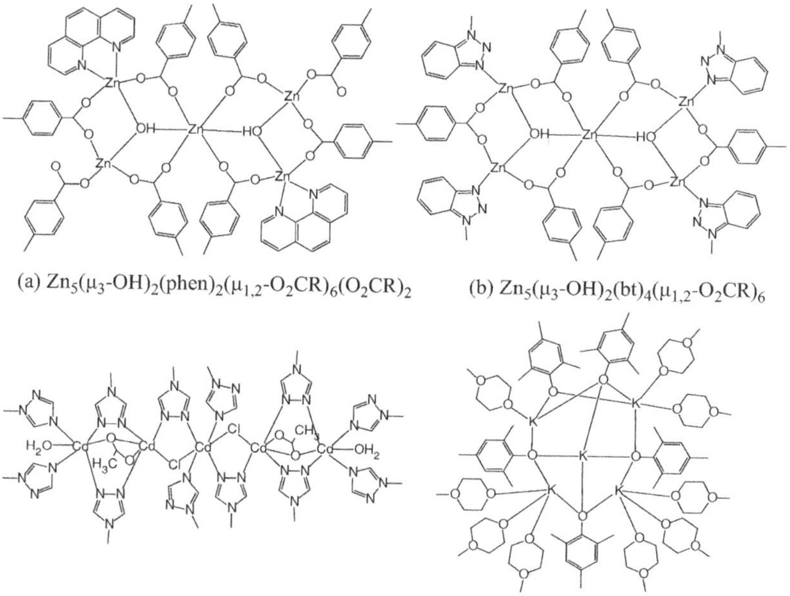

(a) $Zn_5(\mu_3\text{-}OH)_2(phen)_2(\mu_{1,2}\text{-}O_2CR)_6(O_2CR)_2$ (b) $Zn_5(\mu_3\text{-}OH)_2(bt)_4(\mu_{1,2}\text{-}O_2CR)_6$

(c) $Cd_5(\mu_2\text{-}Cl)_2(\mu_2\text{-}OAc)_2(OH_2)_2(\mu_{1,2}\text{-}tr)_6(tr)_6$ (d) $K_5(\mu_2\text{-}OPh)(\mu_3\text{-}OPh)_4(diox)_9$

**SCHEME 4.7** Schematic representations of the pentanuclear nodes forming the highly connected metal-organic frameworks.

and one acetate which adopts coordination mode (d) (Scheme 4.3). As a result both the central Cd(II) center and its neighbors are six-coordinate. The terminal Cd(II) centers are also six-coordinate, the other three coordination positions being occupied by two triazolate moieties and one water molecule. Hence, twelve $L^{31A}$ bridging ligands radiate from the cluster. Nonetheless, the cluster acts as an 8-connected node since eight $L^{31A}$ molecules act in concert as four double bridges.

The core of the pentanuclear cluster $[K_5(\mu_2\text{-}OPh)(\mu_3\text{-}OPh)_4(diox)_9]$ (Scheme 4.7d)[31] is best described as an eight-membered $K_4O_4$ ring comprising four K(I) cations and the oxygens of four 2,4,6-trimethylphenoxide anions with a potassium 2,4,6-trimethylphenoxide at its center. Of the four outer $K^+$ cations, two have three, one has two and the fourth has a single pendant dioxane ($L^{20F}$) bridging molecule; the central K(I) cation is coordinated solely by four 2,4,6-trimethylphenoxide anions. Of the nine dioxane ($L^{20F}$) molecules attached to the cluster, one is non-bridging, two form a double bridge to a second cluster while the other six act as single bridges, giving the system sevenfold connectivity.

***4.3.1.5 Hexanuclear Clusters*** Two hexanuclear clusters have been identified for this review; $[Cu_6(\mu_3\text{-}Spy)_4(pyS)_4(CN)_4]$ in **12A**[49] and $[Rb_6(\mu_3\text{-}OPh)_6(diox)_9]$ in **9A**.[31] The cluster in the former (Scheme 4.8a)[49] is based upon four $\mu_3$-thiolate sulfur donors from $L^{21D}$ arranged in a tetrahedron on the edges of which are located six Cu(I) centers,

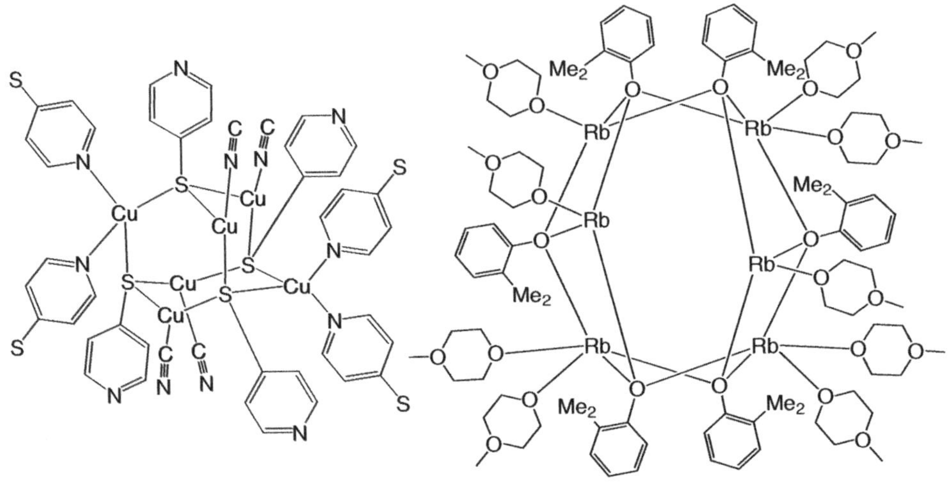

(a) $Cu_6(\mu_3\text{-}Spy)_4(pyS)_4(CN)_4$

(b) $Rb_6(\mu_3\text{-}OPh)_6(diox)_9$

**SCHEME 4.8** Schematic representations of the hexanuclear nodes forming highly connected metal-organic frameworks.

four of which have pendant bridging cyanide anions and two of which have pendant pyridine donors from $L^{21D}$. The eight $L^{21D}$ and four cyanide ligands render the cluster a 12-connected node. The core of $[Rb_6(\mu_3\text{-}OPh)_6(diox)_9]$ (Scheme 4.8b)[31] comprises six Rb(I) cations linked by six 2-iso-propylphenoxide anions with nine pendant dioxane ($L^{20F}$) molecules, all of which bridge to unique neighboring clusters to give a 9-connected uninodal framework.

***4.3.1.6 Heptanuclear Clusters*** The Zn(II) centers in the heptanuclear cluster $[Zn_7(OH_2)_6((\mu_{1,2,3}\text{-}tr)_6(O_2CR)_8]$ (Scheme 4.9a) in **10A**[40] are arranged in three groups: a chain of three surrounded by two centrosymmetrically related peripheral pairs, one on each side. The central Zn(II) center of the trinuclear chain is located on the crystallographic center of inversion. In the chain, the Zn(II) centers are linked through six $\mu_{1,2,3}$-1,2,4-triazolate anions giving the central Zn(II) sixfold coordination and providing three ligating atoms for the terminal Zn(II), the sixfold coordination geometries of which are completed by three water molecules. Of the six bridging triazolate anions, four (two on each side) act as monodentate donors to the other four Zn(II) centers, each pair of which are held in place by 1,3-located carboxylate moieties from $L^{44A}$ molecules. The other two link, in a similar fashion, to Zn(II) centers in adjacent clusters thus generating a chain of triazolate bridged Zn(II) centers. The coordination geometries of the peripheral Zn(II) centers are different. The four coordinate geometry of the one bridged by an extra triazolate is completed by one carboxylate (mode (a); Scheme 4.3) while the five-coordinate geometry of the other is completed by one monodentate (mode (a); Scheme 4.3) and one bidentate (mode (b); Scheme 4.3) carboxylate. Although chains of tetrazolate-bridged Zn(II) centers exist within the structure, each heptanuclear cluster may be considered to be attached to adjacent clusters through eight $L^{44A}$ molecules and two triazolate anions, and hence to act as a 10-connected node.

*4.3.1.7 Octanuclear Clusters* In $[Cd_8(\mu_4\text{-}SO_4)(\mu_2\text{-}SPh)_{12}(py)_8]$ (Scheme 4.9b) in **8U**,[50] eight Cd(II) centers are arranged at the corners of a cube, at the center of which is located a sulfate anion, with 12 $\mu_2$-SPh ligands bridging each pair of Cd(II) centers along the cube edges. Each Cd(II) center has a pyridine *N*-donor from separate $L^{40A}$ bridges, giving fivefold coordination and eightfold connectivity. The octanuclear Pb (II) cluster, $[Pb_8(\mu_4\text{-}O)_2(OH_2)_2(\mu_{1,2}\text{-}O_2CR)_{10}(O_2CR)_4]$ (Scheme 4.9c)[51] in **8E**

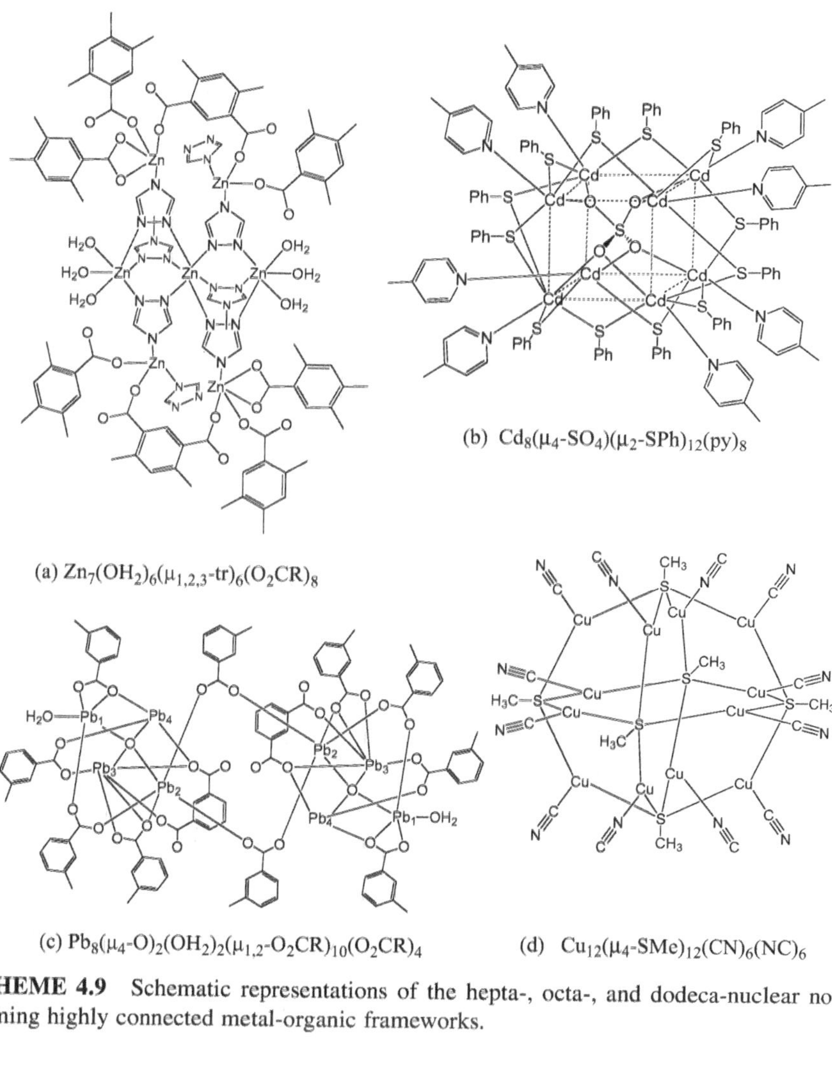

(a) $Zn_7(OH_2)_6(\mu_{1,2,3}\text{-}tr)_6(O_2CR)_8$

(b) $Cd_8(\mu_4\text{-}SO_4)(\mu_2\text{-}SPh)_{12}(py)_8$

(c) $Pb_8(\mu_4\text{-}O)_2(OH_2)_2(\mu_{1,2}\text{-}O_2CR)_{10}(O_2CR)_4$

(d) $Cu_{12}(\mu_4\text{-}SMe)_{12}(CN)_6(NC)_6$

**SCHEME 4.9** Schematic representations of the hepta-, octa-, and dodeca-nuclear nodes forming highly connected metal-organic frameworks.

comprises two crystallographically centrosymmetric $[Pb_4(\mu_4\text{-}O)(OH_2)(\mu_{1,2}\text{-}O_2CR)_4)(O_2CR)_2]$ cores bridged by two mode (c) (Scheme 4.3) carboxylates from $L^{22C}$ anions.

*4.3.1.8 **Dodecanuclear Clusters*** The basic unit of $[Cu_{12}(\mu_4\text{-}SMe)_6(CN)_{12}]$ (Scheme 4.9d) from **12B**[52] is a cube-octahedral Cu(I) cluster, each square face of which is bridged by a $\mu_4$-thiolate sulfur atom of a methyl thiolate anion. Attached to each three-coordinate Cu(I) center is a cyanide anion, giving the cluster 12-fold connectivity through the 12 $CN^-$ anions.

## 4.4 FRAMEWORK TOPOLOGIES

The topological description of the metal-organic frameworks depends not only on the connectivity of the metal-containing center, which, as discussed earlier, can be an individual metal cation or polynuclear cluster, but also the connectivity of the linker. In most cases the bridging spacer is a simple 2-connecting species, either linear or bent, in which case the framework can be viewed, for the purposes of this article, as uninodal and defined by the metal/metal cluster center alone. Incorporation of 3- and 4-connecting ligand linkers introduces the complication that both metal centers and ligand spacers can each be regarded as separate nodes. The overall structure can be then described in binodal terms by considering the connectivity of both the metal and ligand centers separately. In addition, materials incorporating metal nodes having different connectivities (such as **7C** and **8R**) are also regarded as multinodal.

The following discussion is ordered in sequence of increasing ligand linker connectivity within a sequence of increasing metal node connectivity. Although the vast majority of highly connected cationic metal nodes are 8-connected, there are several frameworks reported based upon 7-, 9-, 10- and 12-connected moieties.

### 4.4.1 7-Connected Cationic Frameworks

All three 3D frameworks based on 7-connected cationic nodes, $\{[La(L^{20H})_4]\cdot BPh_4\cdot 3ClO_4\cdot 2.75MeOH\}_\infty$**7A**,[27] $\{[K_5(Ph^B)_5(L^{20F})_5]\}_\infty$**7D**[31] and $\{[M_2(Ph^A)_2(L^{20F})_{3.5}]\}_\infty$ (M = K, Rb) **7E**[31] are uninodal. Whereas the nodes in **7A** are individual lanthanide cations, those in **7D** and **7E** are pentanuclear K(I) clusters (Scheme 4.7d) and binuclear K(I) or Rb(I) clusters (Scheme 4.4a), respectively. The bridging ligands are exclusively $L^{20H}$, $L^{20F}$ and $L^{20F}$ molecules, respectively. The frameworks of all three compounds comprise intersecting $4^4$- and $3^6$-subnet tectons[53] which differ in the way the subnet tectons intersect. The Schläfli symbols for the three frameworks are $4^{17}.6^4$ (Table 4.1; Figure 4.2a), $4^{15}.6^6$ (Table 4.2; Figure 4.2b) and $3^3.4^{12}.5^5.6$ (Table 4.2; Figure 4.2c), respectively. In **7A** and **7D**, the $6^3$-subnets intersect perpendicularly the $4^4$-subnets across the diagonal of the $4^4$ ring. In **7A** the $6^3$-subnets are perfectly eclipsed (Figure 4.2a), while in **7D** they are offset from each other beginning and ending at different $4^4$-subnet layers (Figure 4.2b). In **7E**, the $6^3$-subnets intersect perpendicularly the $4^4$-subnets across an edge of the $4^4$ ring (Figure 4.2c).

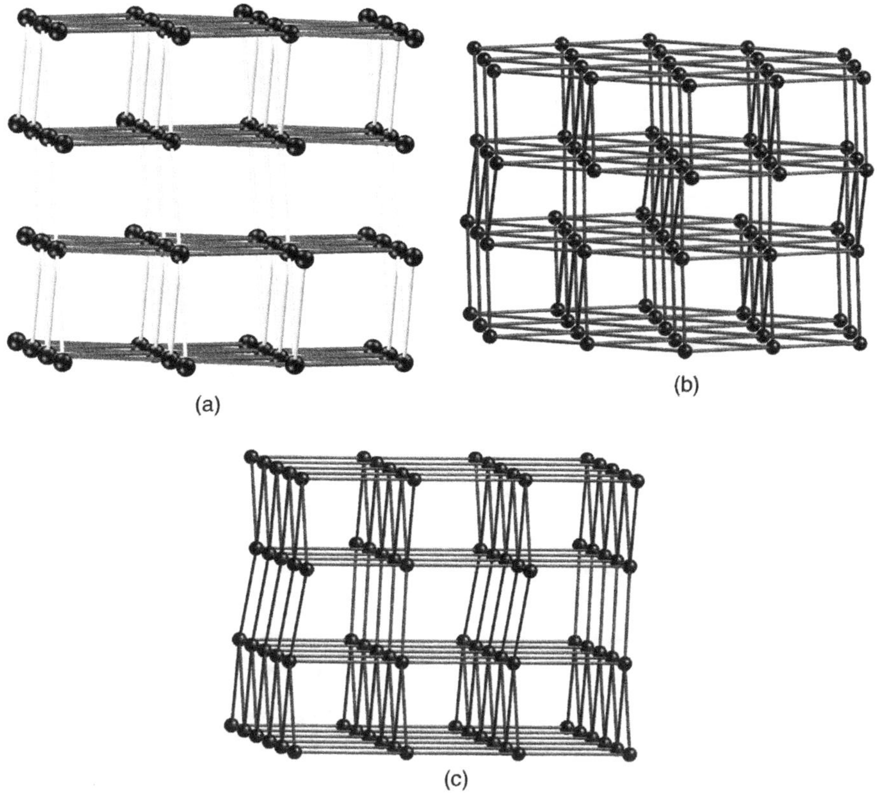


**FIGURE 4.2** Schematic representations of structures of frameworks containing 7-connected nodes. (a) $4^{17}.6^4$ **7A**;[27] (b) $4^{15}.6^6$-**7D**;[31] (c) $3^3.4^{12}.5^5.6$-**7E**.[31]

### 4.4.2 8-Connected Cationic Frameworks

The majority of networks involving 8-connected clusters are uninodal in the sense that they incorporate either linear or bent, 2-connecting organic molecules or anions (**8A–8R**). However, one of these (**8R**) contains two crystallographically distinct Ln(III) nodes bridged by $L^{20H}$ molecules and hence is strictly binodal. There is also one binodal framework (**8S**) with a triangular 3-connecting anion, $L^{33A}$, three (**8T–8V**) with 4-connecting ligands, $L^{40A}$, $L^{42A}$ or $L^{44B}$, and one (**8W**) with the octahedrally disposed 6-connecting cluster, $[Zn_2(O_2CR)_4(py)_2]$.

The most common 8-connected topology is the $4^{24}.6^4$ body centered cubic, CsCl-like, lattice (Tables 4.1 and 4.2; Figure 4.3a). It is adopted by two lanthanide-based networks $\{[La(L^{20H})_4]\cdot 3CF_3SO_3\cdot 4.2MeOH\}_\infty$**8A**[27] and $\{[Ln(L^{20G})_4]\cdot 3ClO_4\}_\infty$, (Ln = La, Ce, Pr, Sm, Eu, Gd, Tb or Y) **8B**[29] and by four cluster-based compounds $\{[Cu_3Cl_2(OH_2)_2(L^{21C})_4]\cdot 4dmf\cdot 5H_2O\}_\infty$**8C**,[39] $\{[Cd_5Cl_2(OAc)_2(OH_2)_2(L^{31A})_6]\cdot 2H_2O\}_\infty$**8D**,[40] $\{[Pb_8(O)_2(OH_2)_2(L^{22C})_6\}_\infty$**8E**[51] and $\{[Cu_2(L^{21A})_2(L^{21B})_2\}_\infty$**8F**.[32] The clusters range in nuclearity from binuclear for **8F** (Scheme 4.4b), through

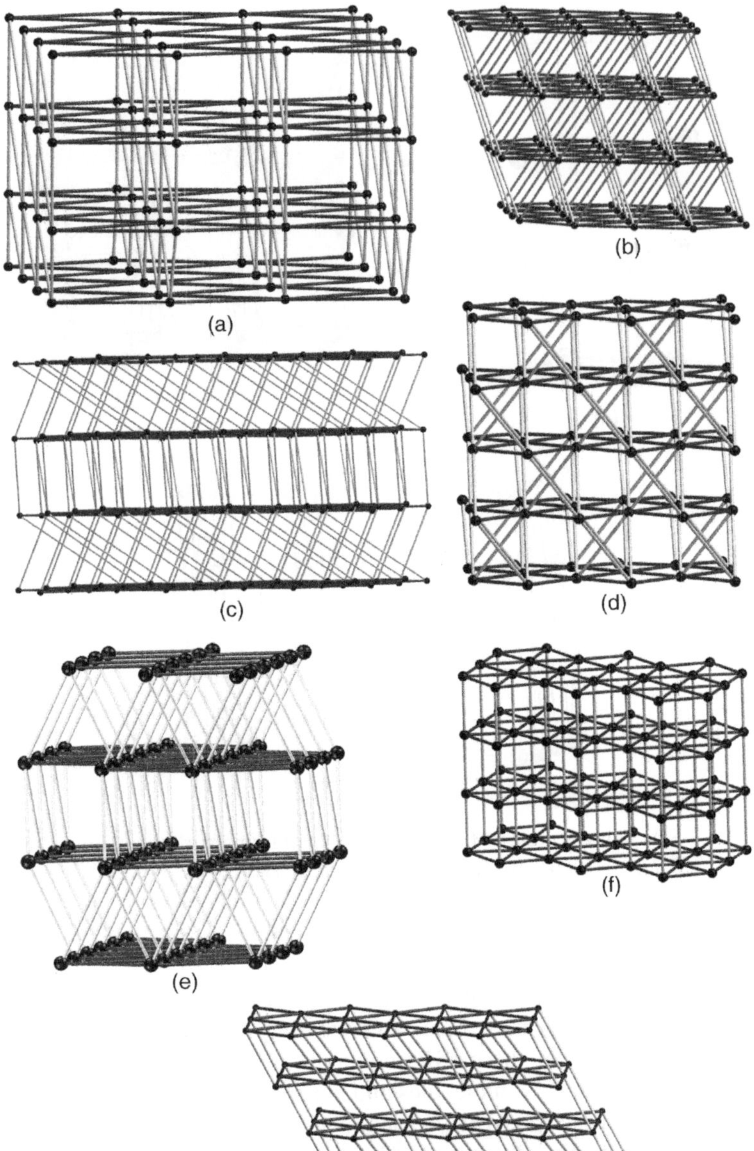


**FIGURE 4.3** Schematic representations of the structures adopted by 8-connected cationic nodes linked by 2-connecting organic bridges. (a) $4^{24}.6^{4}$-**8A**[27] **8B**,[28] **8C**,[39] **8D**,[40] **8E**,[51] and **8F**;[32] (b) $4^{24}.5.6^{3}$-**8G**,[33] **8H**,[47] **8I**;[48] (c) $4^{20}.6^{8}$-**8J**;[37] (d) $3^{5}.4^{11}.5^{8}.6^{4}$-**8L**.[32] (e) $3^{3}.4^{15}.5^{8}.6^{2}$-**8M**;[26] (f) $3^{6}.4^{18}.5^{3}.6$-**8N**,[36] **8O**,[35] and **8P**;[34] (g) $3^{6}.4^{14}.5^{4}.6^{4}$-**8Q**;[45]

trinuclear for **8C** (Scheme 4.5e) and pentanuclear for **8D** (Scheme 4.7c) to octanuclear for **8E** (Scheme 4.9c). In **8A–8E**, the bridging ligands are exclusively $L^{20H}$, $L^{20G}$, $L^{21C}$, $L^{31A}$ and $L^{22C}$, respectively. In **8F** there are equal numbers of two linking moieties, $L^{21A}$ and $L^{21B}$. Although they act as single connections in **8A–8C** and **8F** from eight-coordinate Ln(III) cations or eight-branched clusters, in **8D** and **8E**, 12-branched clusters form four single and four double connections to adjacent clusters to give the 8-connected $4^{24}.6^4$ topology (Figure 4.3a). This CsCl-like body centered cubic array can be described[53] as being formed from two series of $4^4$-subnet tectons aligned at 90° to each other. Alternatively, it can be described[53] as being formed from parallel $4^4$-subnet tectons with each center providing four co-planar links (in the form of two co-planar zigzag chains) to four different centers in the two adjacent nets (Figure 4.3a).

A derivative of the CsCl-like array is the $4^{24}.5.6^3$ topological architecture (Table 4.2; Figure 4.3b) adopted by $\{[Co_2(V_4O_{12})(L^{20E})_3]\cdot 4H_2O\}_\infty$**8G**[33] $\{[Zn_5(OH)_2(phen)_2(L^{22B})_4]\}_\infty$**8H**[47] and $\{[Zn_5(OH)_2(L^{21E})_2(L^{22B})_3]\}_\infty$**8I**.[48] Whereas the cluster in **8G** is binuclear (Scheme 4.4c), those in **8H** (Scheme 4.7a) and **8I** (Scheme 4.7b) are pentanuclear. Although the bridging ligands are exclusively $L^{22B}$ anions in **8H**, there are two linking moieties, $L^{20E}$ molecules and $[V_4O_{12}]^{4-}$ anions and $L^{21E}$ and $L^{22B}$ anions, respectively, in **8G** and **8I**. Although they act as single bridges in **8G** and **8H**, in **8I** 10-branched clusters form six single connections and two double bridges to adjacent clusters to give an overall 8-connected topology (Figure 4.3b). The coplanar zigzag chains linking the parallel $4^4$-subnet tectons, instead of bridging across the diagonal of a single window to give the $4^{24}.6^4$ topology of **8A–8F** (Figure 4.3a), bridge across the diagonal of two neighboring windows in **8G–8I** to give a $4^{24}.5.6^3$ topology (Figure 4.3b).

Similar topological architectures are adopted by $\{[Cd_3(OH_2)_2(L^{20D})(L^{22B})_3]\}_\infty$**8J**[37] and $\{[Cu_4(V_4O_{12})(L^{20C})_4]\cdot 3H_2O\}_\infty$**8K**.[46] In **8J**, the eight-branched trinuclear Cd(II) clusters (Scheme 4.5c) are bridged by $L^{20D}$ molecules and $L^{22B}$ anions in a 1:3 molar ratio. In **8K**, complex eight-branched $Cu_4V_4O_{12}$ clusters (Scheme 4.6c) are linked exclusively by $L^{20C}$ molecules. The zigzag chains linking the parallel $4^4$-subnets, instead of bridging across the diagonal of a single window to give the $4^{24}.6^4$ topology of **8A–8F** (Figure 4.3a), bridge across the diagonal of either six or eight neighboring windows in **8J** and **8K**, respectively. The authors report topologies of $4^{20}.6^8$ (Table 4.2; Figure 4.3c) for **8J** and $4^{24}.6^4$ for **8K.** Although the topology quoted for the latter is the same as that of the CsCl-like structure it is clearly architecturally different. The above four topologies are compared schematically in Scheme 4.10 which illustrates how pairs of organic molecules/anions, which form the zigzag chains linking the $4^4$-subnet tectons, bridge the windows of the square grids.

A $3^5.4^{11}.5^8.6^4$ topology is adopted by $\{[Cu_4(OH)_2(L^{21A})_4(L^{22B})]\}_\infty$**8L**[32] in which the 10-branched tetranuclear Cu(II) cluster (Scheme 4.6e) is linked by $L^{21A}$ and $L^{22B}$ anions in a 4:1 molar ratio. Two of the $L^{21A}$ anions form double bridges to give an eightfold connectivity in which parallel $4^4$-subnet tectons are linked by zigzag chains that translate along the direction of the diagonal of the $4^4$-subnet tecton but which bridge two edges of the grids (Figure 4.3d). Zigzag chains above and below each square-grid are linked to opposite corners, and the $4^4$-subnet tecton comprises single

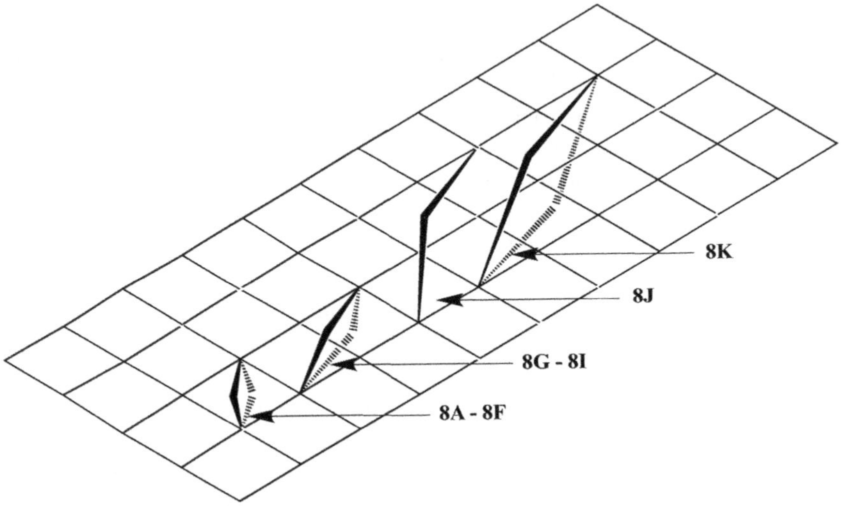


**SCHEME 4.10** Schematic representation of bridging modes adopted by the zigzag chains linking $4^4$-subnet tectons. For **8A–8F** (topology $4^{24}.6^4$) bridging occurs across the diagonal of a single window; for **8G–8I** (topology $4^{24}.5.6^3$) bridging across the diagonal of two neighboring windows is observed; for **8J** (topology $4^{20}.6^8$) bridging across the diagonal of six neighboring windows is observed; for **8K** (topology $4^{16}.6^{12}$) bridging across the diagonal of eight neighboring windows is observed. For all except **8J**, the zigzag chains above and below the grids are coplanar and hence propagate in the same direction (as shown). For **8J**, the zigzag chains above and below the grids are not coplanar and propagate in approximately perpendicular directions.

$L^{21A}$ bridges while the zigzag chains are formed from the $L^{21A}$ double bridges and bridging $L^{22B}$ anions.

A more complex variant of the CsCl-type network is that adopted by $\{[La(L^{20H})_4](ClO_4)_3 \cdot MeOH \cdot 4CHCl_2CHCl_2\}_\infty$ **8M**.[28] Whereas alternating $4^4$-subnet tectons in the CsCl-type structure are displaced only by translation (Figure 4.3a), in **8M** they are displaced by both translation and rotation by 61.5° to give a $3^3.4^{15}.5^8.6^2$ topology (Figure 4.3e). As a result, whereas the zigzag chains in the CsCl-type structure are distributed on both sides of the $4^4$-subnet tecton in a parallel fashion, bridging the diagonal of the square grid, in **8M** the zizag chains bridge the edges of one grid and the diagonal of the next (Figure 4.3e).

A totally different topology based on a $3^6$-subnet[53] is embraced by the recurrent 8-connected architecture of $\{[Zn_3(L^{20B})(L^{22D})_3] \cdot 2dmf \cdot 4H_2O\}_\infty$, **8N**[36] $\{[Co_3(L^{20B})_2(L^{22D})_3] \cdot 4dmf \cdot H_2O\}_\infty$ **8O**,[35] in both of which trinuclear Zn(II) (Scheme 4.5a) or Co(II) clusters (Scheme 4.5b) are bridged by $L^{20B}$ molecules and $L^{22D}$ anions in a 1:3 molar ratio, and $\{[Ln_2(L^{22A})_2] \cdot 2[Cu(L^{21A})_2] \cdot 2H_2O\}_\infty$ (Ln = La, Pr or Nd) **8P**[34] in which Ln(III) nodes are linked by an equimolar mixture of $L^{22A}$ anions and $[Cu(L^{21A})_2]$ moieties. Parallel $3^6$-subnet tectons are linked by mutually opposite connections to adjacent grids to give a $3^6.4^{18}.5^3.6$ topology (Table 4.2; Figure 4.3f). Consideration of the bridges within the $3^6$-subnet tectons in conjunction with the connections reveals three parallel $4^4$-subnets arranged perpendicularly to the $3^6$-subnet tectons at approximately 60° intervals. In **8N** and **8O**, the links in the $3^6$-subnet tectons are $L^{22F}$ dianions, while those bridging the nets are $L^{20B}$ molecules, whereas in **8P** the links in the $3^6$ subnet tectons are $L^{22A}$ dianions, while those bridging the subnet tectons are $[Cu(L^{21A})_2]$ moieties.

A fourth network based on parallel $3^6$-subnet tectons is adopted by $\{[Co_4(OH)_2(L^{20B})_2(L^{22F})_3]\cdot L^{20B}\}_\infty$**8Q**.[45] This system however differs from that adopted by **8N–8P** in that it has two different types of $4^4$-subnet tectons arranged perpendicularly to the $3^6$-subnet tectons to give an architecture with $3^6.4^{14}.5^4.6^4$ topology (Table 4.2; Figure 4.3g). Although the structure of **8Q** is based on a tetranuclear Co(II) cluster from which 10 bridging ligands radiate, four $L^{20B}$ molecules and six $L^{22F}$ anions (Scheme 4.6d), the latter form double bridges giving an overall eightfold connectivity.

The extended structure of $\{[Ln(L^{20H})_4](CF_3SO_3)_3\cdot 4MeOH\cdot C_2Cl_4\}_\infty$ (Ln = Yb, Eu) **8R** (Table 4.1)[28] comprises two crystallographically distinct 8-connected Ln(III) nodes in a stoichiometry of 1:2 linked by $L^{20H}$ molecules. Hence, although the complex contains 2-connecting $L^{20H}$ bridging molecules, it is strictly binodal.[53] Topologically, layers of puckered and irregular $6^3$-subnet tectons are perpendicularly connected via $L^{20H}$ molecules. For each six-membered ring, three $L^{20H}$ molecules connect to the subnet tecton above and three to the one below resulting in a 4-connected subnet of $SrAl_2$-structure. This arrangement is further intersected by two sets of $4^4$-subnet tectons, which meet at the 8-connected Ln(III) nodes that lie between the $6^3$ layers of the $SrAl_2$-subnet, to give a bimodal framework with an overall topology of $(3^5.4^{14}.5^9)(3^5.4^{13}.5^{10})_2$ (Table 4.1).

The (3,8) binodal network, $\{H_2[Co_4(O)(L^{33A})_{8/3}]\}_\infty$**8S**[44] is constructed by the linking of 8-connected tetranuclear Co(II) clusters (Scheme 4.6b) through triangular 3-connecting anion, $L^{33A}$ and has an overall topology of $(4^3)(4^8.6^4.8^{12}.10^4)$ (Table 4.2; Figure 4.4a).

Three (4,8)-connected binodal networks have been constructed using neutral molecules ($L^{40A}$), dianions ($L^{42A}$) and tetra-anions ($L^{44B}$) as fourfold connectors. $\{[Cd_4(dmf)_4(L^{44B})_2]\cdot 4dmf\cdot 4H_2O\}_\infty$**8T**[43] in which 8-connected tetranuclear Cd(II) clusters (Scheme 4.6a) are linked by $L^{44B}$ anions adopts the classical fluorite $CaF_2$ structure with $(4^6)(4^{12}.6^{12}.8^4)$ topology (Table 4.2; Figure 4.4b). Reaction of $[Me_4N]_4[Cd_{10}S_4(SPh)_{16}]$, pre-heated to 250°C for 3 h, with an aqueous mixture of 1,3-bis(pyridin-4-yl)propane sodium sulfate, lead to the *in situ* formation of the neutral 4-connector $L^{40A}$ and crystallization of the binodal (4,8) network, $\{[Cd_8(SO_4)(SPh)_{12}(L^{40A})_2]\cdot 2HSO_4\cdot 4H_2O$ **8U**.[50] Topological analysis identified the $(4^4.6^2)_4(4^{16}.6^{12})$ network (Table 4.2; Figure 4.4c) based on an octanuclear Cd(II) cluster (Scheme 4.9b). The trinuclear Cd(II) cluster nodes (Scheme 4.5f) and succinate anions in $\{[Cd_3Cl_2(L^{31A})_2(L^{42A})_2]\}_\infty$**8V**[40] act as eight- and fourfold connectors, respectively, to give a binodal (4,8) network with $(4^5.6)(4^{12}.6^{16})$ topology (Table 4.2).

One (6,8)-connected self-penetrating binodal network, $\{[Zn_3(L^{20I})(L^{22B})_3][Zn_2(L^{20I})(L^{22B})_2]\cdot 2H_2O\}_\infty$**8W**[38] has been reported. It comprises 6-connecting dinuclear Zn(II) and 8-connecting trinuclear Zn(II) clusters (Scheme 4.5a) which

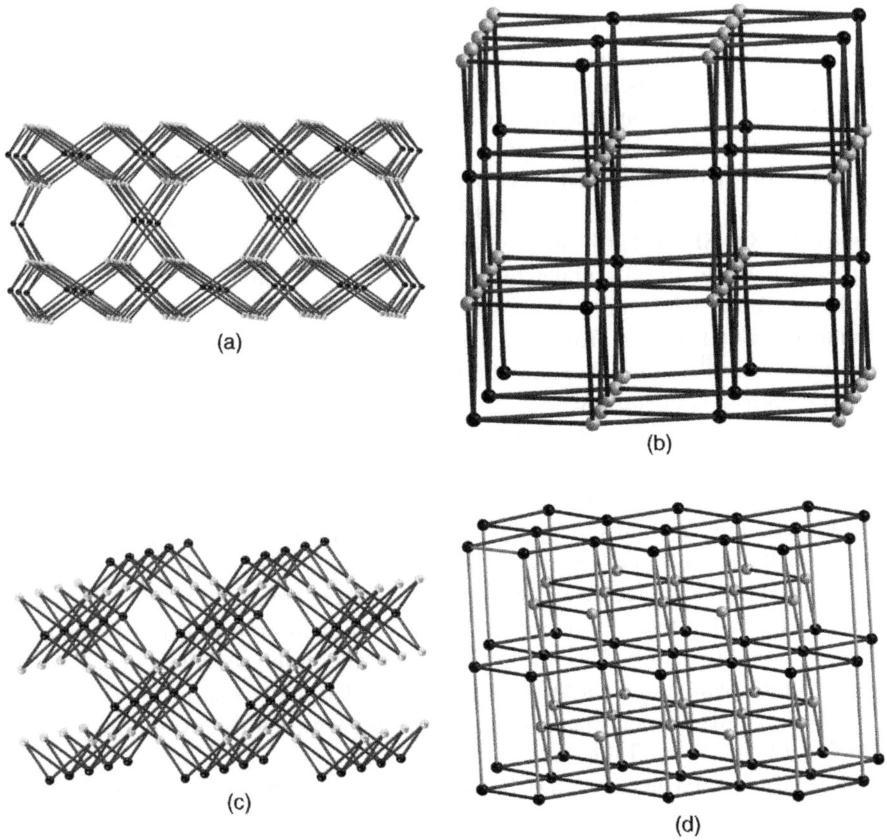


**FIGURE 4.4** Schematic representations of structures of frameworks containing 8-connected nodes linked by (a) 3-, (b–c) 4- and (d) 6-connecting organic bridges. (a) $(4^3)(4^8.6^4.8^{12}.10^4)$-**8S**;[44] (b) $(4^6)(4^{12}.6^{12}.8^4)$-**8T**;.[43] (c) $(4^4.6^2)_4(4^{16}.6^{12})$-**8U**;[50] (d) $(4^{12}.5.6^2)(4^{20}.5^2.6^6)$-**8W**.[38]

combine to form an architecture of $(4^{12}.5.6^2)(4^{20}.5^2.6^6)$ topology (Table 4.2; Figure 4.4d).

### 4.4.3 9-Connected Cationic Frameworks

Only one of the four frameworks reported with 9-connected nodes is uninodal; the other three are binodal. The uninodal 9-connected network formed by linking hexanuclear Rb(I) clusters (Scheme 4.8b) through dioxane bridges, $\{[Rb_6(Ph^C)_6(L^{20F})_{4.5}]\}_\infty$**9A**[31] has $3^6.4^{22}.5^8$ topology. The structure comprises parallel sheets of $4^4$-nets that are doubly intersected by $(3.4^3)(3^2.4^3)$-subnets at an angle of *ca.* 60° (Table 4.2; Figure 4.5a). The tiling pattern of the $(3.4^3)(3^2.4^3)$-subnets comprises rows of squares alternating with rows of mixed triangles and squares.

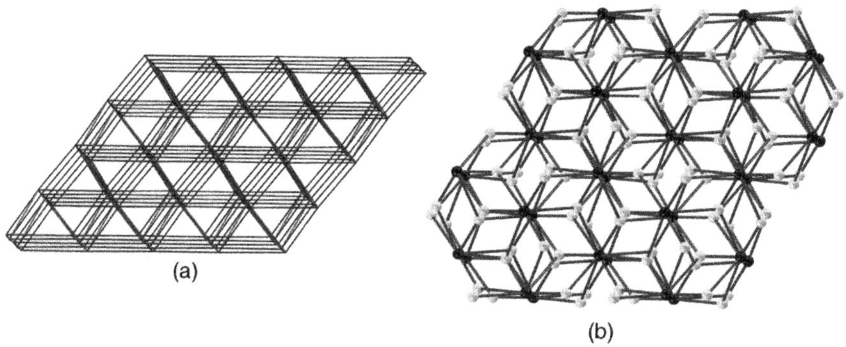


**FIGURE 4.5** Schematic representations of structures of frameworks containing 9-connected nodes. (a) $3^6.4^{22}.5^8$-**9A**[31]; (b) $(4^2.6)(4^6.6^{21}.8^9)$-**9B**,[41] **9C**,[42] **9D**.[42]

The three analogous binodal (9,3)-connected networks $\{[Co_3(OH)(L^{32A})_3]\cdot H_2O\}_\infty$**9B**[41] $\{[Fe_3(O)(L^{32B})_3]\cdot n(solv)\}_\infty$**9C**[42] and $\{[Ni_3(OH)(L^{32B})_3]\cdot n(solv)\}_\infty$**9D**[42] are topologically identical, adopting the $(4^2.6)_3$ $(4^6.6^{21}.8^9)$ arrangement (Table 4.2; Figure 4.5b). Trinuclear cationic clusters (Scheme 4.5g) act as 9-connecting nodes while the bridging anions, $L^{32A}$ or $L^{32B}$, act as 3-connecting spacers. The frameworks differ solely in the size of the bridging ligand, carboxylate units in $L^{32A}$ being replaced by benzene-4-carboxylate moieties in $L^{32B}$.

### 4.4.4 10-Connected Cationic Frameworks

The structure of $\{[Zn_7(OH_2)_6(L^{31A})_6(L^{44A})_2]\cdot 8H_2O\}_\infty$**10A**[40] (Table 4.2) is described by the authors as a novel 4,10-connected net comprising 4-connected $L^{44A}$ molecules and 10-connected heptanuclear Zn(II) clusters (Scheme 4.9a). They do acknowledge, however, that other descriptions are viable. For example, since two of the connectors contributing to the 10-fold connectivity of the $Zn_7$-clusters are $L^{31A}$ anions identical to those linking the Zn(II) centers in the presumed $Zn_7$-clusters (Scheme 4.9a), the Zn(II) cations and $L^{31A}$ anions can be considered to form part of infinite chains. The chains are linked by eight separate 4-connecting $L^{44A}$ molecules to give the 3D framework.

### 4.4.5 12-Connected Cationic Frameworks

Two compounds, $\{[Cu_3(L^{21D})_2(CN)]\}_\infty$**12A**[49] and $\{[Cu_{12}(SMe)_6(CN)_6]\cdot 2H_2O\}_\infty$**12B**,[52] adopt the face centered cubic (cubic close packed) 12-connected architecture (Table 4.2; Figure 4.6) with $3^{24}.4^{36}.5^6$ topology. They are both based upon Cu(I) architectures with bridging cyanide ligands. Whereas the former[49] is based on a hexanuclear Cu(I) cluster with four pendant cyanide anions and eight *N*-donor pyridines to give a 12-fold connectivity (Scheme 4.8a), the latter[52] is based on a dodecanuclear Cu(I) cluster with 12 pendant cyanide anions (Scheme 4.9d). Whereas the former framework is formed *in situ* from the hydrothermal reaction of copper acetate, (4-pyridylthio)acetic acid sodium thiocyanate and sodium hydroxide, the

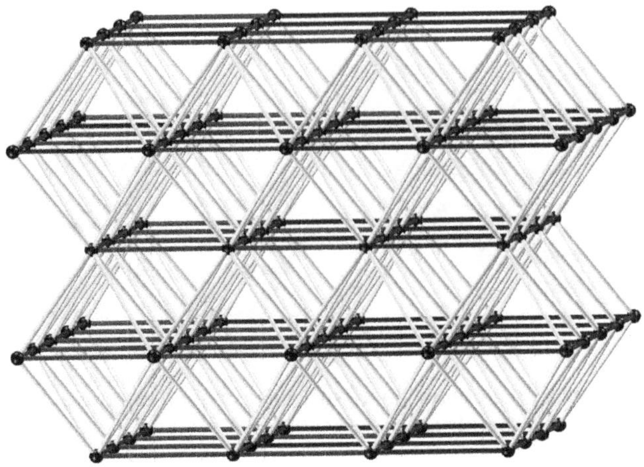

**FIGURE 4.6** Schematic representation of the $3^{24}.4^{36}.5^{6}$ topology adopted by 12-connected nodes in **12A**,[49] and **12B**.[52]

latter framework is formed *in situ* from a solution of hydrated copper acetate or nitrate and sodium thiocyanate in a mixture of methanol and acetonitrile under solvothermal (140°C) conditions.

## 4.5 CONCLUSIONS

In identifying suitable building blocks with connectivities greater than 6, two major approaches have been taken. The use of either individual metal centers with high coordination numbers or multi-nuclear metal clusters. Bridging ligands are primarily 2-connecting but there are examples of higher ligand connectivity giving rise to more complex network topologies.

Just seven highly connected frameworks based on individual metal centers have been identified; they are exclusively lanthanide-containing materials in which Ln(III) nodes adopt either seven- or eightfold coordination geometries to generate 7- or 8-connected frameworks. Many more highly connected frameworks based on metal clusters (so-called "secondary building units") have been identified. Ranging in nuclearity from bi- to dodeca-nuclear, these clusters generate architectures with connectivity from 7- to 12-fold. They are highly disparate both in choice of metal center and bridging ligand. Several metal cations recur; these include Co(II), Cu(II), Zn(II), and Cd(II). There are also examples containing K(I), Rb(I), Fe(III), Ni(II), La(III), and Pb(II). Some of the clusters are centered on $O^{2-}$, $OH^{-}$, $Cl^{-}$, or, in one case, $SO_4^{2-}$ anions but all have peripheral bridging species, including $\mu_{1,2}$-carboxylate, $\mu_2$-phenolate and $\mu_2$-thiolate anions. But for the fact that the clusters are limited in their composition to a small number of metal cations, central anions and peripheral bridging moieties, there seems to be little rational design behind their structural constitution, most being formed serendipitously. Although the majority of the bridges are single, some molecules form double bridges leading to a lower connectivity than the number of pendant bridging species radiating from the cluster.

A wide diversity of framework topologies has been identified. This is in no doubt due not only to the wide range of metal cluster nodes identified but also to the variety of bridging species used and the fact that many materials contain a mixture of bridges. There is, however, a clear trend within the 8-connected frameworks showing that those with just one type of linker tend to form more symmetrical architectures than those with two or more types. For example, five (**8A–8E**) of the six compounds that adopt the $4^{24}.6^{4}$ CsCl-like lattice (**8A–8F**) have just one linker while the other (**8F**) has two structurally very similar linkers, isonicotinate ($L^{21A}$) and nicotinate ($L^{21B}$). In contrast, only four (**8H**, **8M**, **8K**, and **8R**) of the remaining twelve 8-connected compounds (**8G–8R**), which adopt a variety of more complex topologies, have just one linker. As the relative sizes of the two linkers become more disparate, the inequality can be accommodated either by greater separation between the 2D subnet tectons, by bridging the diagonal of an increased number of grid windows (Scheme 4.10), or by formation of $3^{6}$-subnet tectons rather than $4^{4}$-subnet tectons (**8N**, **8O**, **8P** and **8Q**).

In view of the significant variety of potential metal cluster nodes and their geometries and the very many possible permutations of bridging molecules/anions, it is clear that there remains enormous potential for the future discovery of highly connected metal-organic frameworks.

## ACKNOWLEDGMENTS

We thank the EPSRC UK and the University of Nottingham for support. MS gratefully acknowledges receipt of a Royal Society Wolfson Merit Award and of a Royal Society Leverhulme Trust Senior Research Fellowship.

## REFERENCES

1. (a) Blake, A. J.; Champness, N. R.; Hubberstey, P.; Li, W.-S.; Withersby, M. A.; Schröder, M. *Coord. Chem. Rev.* **1999**, *183*, 117; (b) Khlobystov, A. N.; Blake, A. J.; Champness, N. R.; Lemenovskii, D. A.; Majouga, A. G.; Zyk, N. V.; Schröder, M. *Coord. Chem. Rev.* **2001**, *222*, 155; (c) Champness, N. R. *Dalton Trans.* **2006**, 877.
2. Hoskins, B. F.; Robson, R. *J. Am. Chem. Soc.* **1990**, *112*, 1546; Robson, R.; Abrahams, B. F.; Batten, S. R.; Gable, R. W.; Hoskins, B. F.; Liu, J. P. *ACS Symp. Ser.* **1992**, *499*, 256.
3. Eddaoudi, M.; Moler, D. B.; Li, H. L.; Chen, B. L.; Reineke, T. M.; O'Keeffe, M.; Yaghi, O. M. *Acc. Chem. Res.* **2001**, *34*, 319.
4. Champness, N. R.; Oxtoby, N. S. *Encyclopaedia of Nanoscience and Nanotechnology* 2004, 845.

5. Hosseini, M. W. *Acc. Chem. Res.* **2005**, *38*, 313.
6. Ockwig, N. W.; Delgado-Friedrichs, O.; O'Keeffe, M.; Yaghi, O. M. *Acc. Chem. Res.* **2005**, *38*, 176.
7. Hill, R. J.; Long, D.-L.; Champness, N. R.; Hubberstey, P.; Schröder, M. *Acc. Chem. Res.* **2005**, *38*, 337.
8. Rosi, N. L.; Eckert, J.; Eddaoudi, M.; Vodak, D. T.; Kim, J.; O'Keeffe, M.; Yaghi, O. M. *Science* **2003**, *300*, 1127.
9. Zhao, X. B.; Xiao, B.; Fletcher, A. J.; Thomas, K. M.; Bradshaw, D.; Rosseinsky, M. J. *Science* **2004**, *306*, 1012.
10. Kitaura, R.; Kitagawa, S.; Kubota, Y.; Kobayashi, T. C.; Kindo, K.; Mita, Y.; Matsuo, A.; Kobayashi, M.; Chang, H. C.; Ozawa, T. C.; Suzuki, M.; Sakata, M.; Takata, M. *Science* **2002**, *298*, 2358.
11. Maspoch, D.; Ruiz-Molina, D.; Wurst, K.; Domingo, N.; Cavallini, M.; Biscarini, F.; Tejada, J.; Rovira, C.; Veciana, J. *Nat. Mater.* **2003**, *2*, 190.
12. Coronado, E.; Galan-Mascaros, J. R.; Gomez-Garcia, C. J.; Laukhin, V. *Nature* **2000**, *408*, 447.
13. Millward, A. R.; Yaghi, O. M. *J. Am. Chem. Soc.* **2005**, *127*, 17998.
14. Withersby, M. A.; Blake, A. J.; Champness, N. R.; Hubberstey, P.; Li, W.-S.; Schröder, M. *Angew. Chem., Int. Ed. Engl.* **1997**, *36*, 2327.
15. Withersby, M. A.; Blake, A. J.; Champness, N. R.; Cooke, P. A.; Hubberstey, P.; Schröder, M. *New J. Chem.* **1999**, *23*, 573.
16. Long, D.-L.; Blake, A. J.; Champness, N. R.; Schröder, M. *Chem. Commun.* **2000**, 1369.
17. Macgillivray, L. R.; Subramanian, S.; Zaworotko, M. J. *J. Chem. Soc. Chem. Commun.* **1994**, 1325.
18. Blake, A. J.; Champness, N. R.; Chung, S. S. M.; Li, W.-S.; Schröder, M. *Chem. Commun.* **1997**, 1005.
19. Long, D.-L.; Hill, R. J.; Blake, A. J.; Champness, N. R.; Hubberstey, P.; Wilson, C.; Schröder, M. *Chem. Eur. J.* **2005**, *11*, 1384 and references therein.
20. (a) Shi, J.-M.; Xu, W.; Liu, Q.-Y.; Liu, F.-L.; Huang, Z.-L.; Lei, H.; Yu, W.-T.; Fang, Q. *Chem. Commun.* **2002**, 756; (b) Seward, C.; Hu, N.-X.; Wang, S. *J. Chem. Soc., Dalton Trans.* **2001**, *134;* (c) Liu, G.-F.; Qiao, Z.-P.; Wang, H.-Z.; Chen, X.-M.; Yang, G. *New J. Chem.* **2002**, *26*, 791.
21. Reineke, T. M.; Eddaoudi, M.; Fehr, M.; Kelley, D.; Yaghi, O. M. *J. Am. Chem. Soc.* **1999**, *121*, 1651.
22. (a) Kiritis, V.; Michaelides, A.; Skoulika, S.; Golhen, S.; Ouahab, L. *Inorg. Chem.* **1998**, *37*, 3407; (b) Reineke, T. M.; Eddaoudi, M.; O'Keeffe, M.; Yaghi, O. M. *Angew. Chem., Int. Ed. Engl.* **1999**, *38*, 2590; (c) Cao, R.; Sun, D.; Liang, Y.; Hong, M.; Tatsumi, K.; Shi, Q. *Inorg. Chem.* **2002**, *41*, 2087; (d) Hamilton, B. H.; Kelly, K. A.; Wagler, T. A.; Espe, M. P.; Ziegler, C. J. *Inorg. Chem.* **2002**, *41*, 4984; (e) Reineke, T. M.; Eddaoudi, M.; Moler, D.; O'Keeffe, M.; Yaghi, O. M. *J. Am. Chem. Soc.* **2000**, *122*, 4843.
23. Long, D.-L.; Blake, A. J.; Champness, N. R.; Wilson, C.; Schröder, M. *Chem. Eur. J.* **2002**, *8*, 2026.
24. Long, D.-L.; Blake, A. J.; Champness, N. R.; Schröder, M. *Chem. Commun.* **2000**, 1369.
25. Long, D.-L.; Hill, R. J.; Blake, A. J.; Champness, N. R.; Hubberstey, P.; Wilson, C.; Schröder, M. *Chem. Eur. J.* **2005**, *11*, 1384.

26. Ren, P.; Shi, W.; Cheng, P. *Inorg. Chem. Commun.* **2008**, *11*, 125.
27. Long, D.-L.; Blake, A. J.; Champness, N. R.; Wilson, C.; Schröder, M. *Angew. Chem., Int.* **2001**, *40*, 2444.
28. Long, D.-L.; Hill, R. J.; Blake, A. J.; Champness, N. R.; Hubberstey, P.; Proserpio, D. M.; Wilson, C.; Schröder, M. *Angew. Chem., Int.* **2004**, *43*, 1851.
29. Sun, H.-L.; Gao, S.; Ma, B.-Q.; Chang, F.; Fu, W.-F. *Microporous Mesoporous Mater.* **2004**, *73*, 89.
30. Yaghi, O. M.; O'Keeffe, M.; Ockwig, N. W.; Chae, H. K.; Eddaoudi, M.; Kim, J. *Nature* **2003**, *423*, 705.
31. Morris, J. J.; Noll, B. C.; Henderson, K. W. *CrystEngComm.* **2007**, 5191.
32. Zhang, J.; Kang, Y.; Zhang, J.; Li, Z.-J.; Qin, Y.-Y.; Yao, Y.-G. *Eur. J. Inorg. Chem.* **2006**, 2253.
33. Qi, Y.-F.; Xiao, D.; Wang, E.; Zhang, Z.; Wang, X. *Aust. J. Chem.* **2007**, *60*, 871.
34. Cheng, J.-W.; Zheng, S.-T.; Yang, G.-Y. *Dalton Trans.* **2007**, 4059.
35. Pan, L.; Liu, H.; Lei, X.; Huang, X.; Olson, D. H.; Turro, N. J.; Li, J. *Angew. Chem. Int.* **2003**, *42*, 542.
36. Fang, Q.; Shi, X.; Wu, G.; Tian, G.; Zhu, G.; Zhu, G.; Wang, R.; Qui, S. *J. Solid State Chem.* **2003**, *176*, 1.
37. Wang, X.-L.; Qin, C.; Wang, E.-B.; Su, Z.-M. *Chem. Eur. J.* **2006**, *12*, 2680.
38. Lan, Y.-Q.; Wang, X.-L.; Li, S.-L.; Su, Z.-M.; Shao, K.-Z.; Wang, E.-B. *Chem. Commun.* **2007**, 4863.
39. Luo, T.-T.; Tsai, H.-L.; Yang, S.-L.; Liu, Y.-H.; Yadav, R. D.; Su, C.-C.; Ueng, C.-H.; Lin, L.-G.; Lu, K.-L. *Angew. Chem. Int.* **2005**, *44*, 6063.
40. Zhai, Q.-G.; Lu, C.-Z.; Wu, X.-Y.; Batten, S. R. *Cryst. Growth Des.* **2007**, *7*, 2332.
41. Zhang, X.-M.; Zheng, Y.-Z.; Li, C.-R.; Zhang, W.-X.; Chen, X.-M. *Cryst. Growth Des.* **2007**, *7*, 980.
42. Jia, J.; Lin, X.; Wilson, C.; Blake, A. J.; Champness, N. R.; Hubberstey, P.; Walker, G.; Cussen, E. J.; Schröder, M. *Chem. Commun.* **2007**, 840.
43. Chun, H.; Kim, D.; Dybtsev, D. N.; Kim, K. *Angew. Chem. Int.* **2004**, *43*, 971.
44. Ma, S.; Zhou, H.-C. *J. Am. Chem. Soc.* **2007**, *128*, 11734.
45. Liu, C.-M.; Gao, S.; Zhang, D.-Q.; Zhu, D.-B. *Cryst. Growth Des.* **2007**, *7*, 1312.
46. Qu, X.; Xu, L.; Gao, G.; Li, F.; Yang, Y. *Inorg. Chem.* **2007**, *46*, 4775.
47. Wang, X.-L.; Qin, C.; Wang, E.-B.; Su, Z.-M.; Xua, L.; Batten, S. R. *Chem. Commun.* **2005**, 4789.
48. Yang, 'E.-C.; Zhao H.-K.; Ding, B.; Wang, X.-G.; Zhao, X.-J. *Cryst. Growth Des.* **2007**, *7*, 2009.
49. Zhang, X.-M.; Fang, R.-Q.; Wu, H.-S. *J. Am. Chem. Soc.* **2005**, *127*, 7670.
50. Zheng, N.; Bu, X.; Feng, P. *J. Am. Chem. Soc.* **2002**, *124*, 9688.
51. Yang, E.-C.; Li, J.; Ding, B.; Liang, Q.-Q.; Wang, X.-G.; Zhao, X.-J. *CrystEngComm.* **2008**, *10*, 158.
52. Li, D.; Wu, T.; Zhou, X.-P.; Zhou, R.; Huang, X.-C. *Angew. Chem. Int.* **2005**, *44*, 4175.
53. Hill, R. J.; Long, D.-L.; Champness, N. R.; Hubberstey, P.; Schröder, M. *Acc. Chem. Res.* **2005**, *38*, 337.

# 5

# SURFACE PORE ENGINEERING OF POROUS COORDINATION POLYMERS

SUJIT K. GHOSH[1,4] AND SUSUMU KITAGAWA[1,2,3]

[1]*Department of Synthetic Chemistry and Biological Chemistry, Graduate School of Engineering, Kyoto University, Nishikyo-ku, Kyoto, Japan*
[2]*Kitagawa Integrated Pore Project, Exploratory Research for Advanced Technology (ERATO), Japan Science and Technology Agency (JST), Shimogyo-ku, Kyoto, Japan*
[3]*Institute for Integrated Cell Material Sciences (iCeMS), Kyoto University, Sokyo-ku, Kyoto, Japan*
[4]*Department of Chemistry, Indian Institute of Science Education and Research (IISER), Pune, India*

## 5.1 INTRODUCTION

Porous coordination polymers (PCPs) or metal-organic frameworks (MOFs) composed of metal ions or metal ion clusters as "nodes" and organic ligands as "linkers" has been the subject of intense current research due to scientific interest in the creation of nanometer-sized spaces and their enormous potential in applications such as gas storage, separation, photonics, and heterogeneous catalysis.[1–15] In the area of porous materials the remarkable progress of PCPs as functional materials are mainly due to the fact that, compared to the other conventional solids such as zeolites and carbon-based materials, PCPs are synthesized under mild conditions and can easily be designed based on the modifications of organic components and variety of coordination geometries at the metal centers. Therefore, unlike zeolites and carbon-based materials, PCPs exhibit some characteristic features: highly regular channel structures, controllable pore size, guest responding flexible pores, and designable pore

*Metal-Organic Frameworks: Design and Application*, Edited by Leonard R. MacGillivray

surface functionality, which produce efficient pores and excellent porous functionalities. Since the extensive development of PCPs from early 1990s, much research effort has been devoted for the investigation of the high porosity, stability, control of pore shape and size, and framework flexibility. In contrast, the pore surface functionalizations of porous materials remain unexplored.

Pore surface modification using coordinatively unsaturated metal centers (UMCs) or open metal sites (OMSs) represents an excellent strategy, since highly selected Lewis-base guests can interact very strongly through coordination bonds with the OMSs. In addition, immobilizations of OMSs into the pore surface of PCPs are attractive area of research because such kind of framework provides porous functionalities like chromic sensor and excellent heterogeneous catalytic activity.[16–20] Despite their importance less attention has been devoted to OMSs, this is because of the difficulty of producing guest-accessible OMSs in pore surface, as organic-groups and/or counter anions tend to saturate the coordination geometry of the metal centers in a self-assembly process to give a framework where all the coordination sites are occupied. Immobilization of OMSs has been attempted to other porous solids like zeolites and carbon-based materials through ion exchange and impregnation but in these cases uniform distribution into the pore is not adequate and local environment around the OMSs are not well understood.[21,22] Taking the advantage of highly regular channel structures of PCPs if OMSs can be incorporated into the pore surface, a completely uniform arrangement could be realized. Thus, finding the methods to successfully incorporate the OMSs onto the pore surface has immense importance.

Another very good strategy used to functionalize the pore surface by introducing the organic functional groups to provide the guest-accessible functional organic sites (FOSs).[23–26] The advantage of FOS is that, since the pore walls of PCPs generally made of organic connector ligands, by modifying the organic ligands using varieties of organic functional groups, PCPs with regularly arranged and highly dense active sites could be obtained. However, in spite of their great importance for possible applications and availability of many functional groups to synthesis large number of various kind of PCPs, such systems are rare because of their synthetic difficulties, where either functional groups tend to coordinate with the metal centers (like highly basic functional groups) or in most cases self-assembly process produce non-porous framework. So, successful design of the PCPs containing FOSs in the pore surface is a challenging topic to make the next generation PCPs.

Mostly undeveloped but another good strategy to design the pore surface of PCP is post-synthetic pore surface modification.[27–29] Many PCPs are stable after removal of the guest molecules so the space inside the framework of the desolvated PCP is like reaction vessel and it is possible to do reaction into the channels to modify the pore surface. Free uncoordinated secondary functional groups of porous frameworks can be modified by reacting with other external species. This method is very important for modification of surface functionality because, organic ligands containing reactive functional groups are not stable under vigorous reaction condition (e.g., under solvothermal condition), or with such functional groups the synthesis of the framework is very difficult or even not possible. Using this method it is possible to generate many desired reactive group to modify the functionality of the pore surface that may enhance the efficiency of the pore surface which may be otherwise not possible.

In this chapter, three different ways to functionalize the pore surface of PCPs are described. In the first section, we have focused on the strategy to introduce OMSs onto the pore surface of PCPs. In the following section, PCPs where pore surface has been functionalized by the FOSs are discussed. Finally, in the last section, post-synthetic pore surface modifications of PCPs involving FOSs are illustrated.

## 5.2 PORE SURFACE WITH OMSs

### 5.2.1 Utilization of Metallo-Ligands (MLs)

Metallo-ligand (MLs) are metal-containing molecular complexes that contain two or more Lewis-base sites, pre-constructed from simple organic linkers combine with other metal ions to construct MOFs.[30] To introduce OMSs onto the channel surface of PCPs by the use of MLs there are several advantages like: (1) it is possible to make pre-designed heterometallic complexes, (2) by changing the metal ions in both the sites (MLs and connectors) framework functionality can be tuned, and (3) they can construct the framework through multi-coordination sites. By this two-step self-assembly approach first, an organic ligand is reacted with a metal ion (M1) to make a discrete mononuclear unit called metallo-ligand (ML). In the second step, these MLs are used as linkers and are connected by another metal ion (M2) used as node to construct PCPs (Scheme 5.1).

To construct PCPs with available OMSs in the surface using MLs as building block suitable choice of metal ions and organic ligands are very crucial. To provide enough coordination sites for both organic ligands and volatile solvent molecules in the MLs, metal ions (M1) with high-coordination number (6 or more) are desired. Low boiling solvents and metal ions with variable coordination numbers are preferred to synthesize MLs, since after the removal of the volatile molecules capping the OMSs, the resulting geometry of the metal centers must be stable. In this regard, Cu(II) metal ion displaying Jahn–Teller distortion is a good choice as it has versatile coordination chemistry and

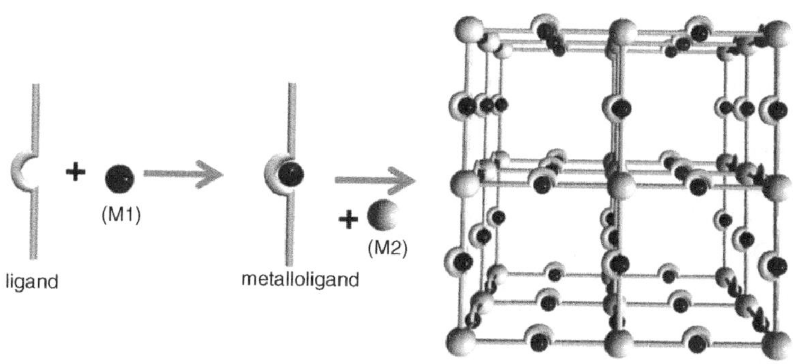


**SCHEME 5.1** Schematic representation of the stepwise formation of the PCP using ML.

due to the ligand field effect, the binding of the axial ligand is sufficiently weak to allow its release and the resulting square planar geometry is stable. On the other hand, to avoid polymeric compound by self-assembly process the organic ligands containing two coordination sites with different Lewis-basicity are desired. Macrocyclic groups or chelating ligands are used to form MLs and carboxylate, sulfonate, cynanide, and pyridine groups are used for further assembly of MLs with other metal ions to construct PCPs (Figure 5.1).[16–18,31–41]

Our group reported[17] one very good example of ML, $[Cu(salphdc)]^{2-}$ ($ML^{VII}$ in Figure 5.1, where $H_2$salphdc is *N,N'*-phenylenebis(salicylideneimine)dicarboxylic acid) and this ML has been used to construct a 3D coordination framework $\{Zn_3(OH)_2([ML^{VII})_2]$.2DMF. $ML^{VII}$ has a multi-coordination ability in addition to the above-mentioned advantages desired for a good ML. First the ML has been synthesized by the reaction of $Cu(OAc)_2.H_2O$ and $H_4$salphdc in a DMF/MeOH/$H_2O$ solution at room temperature, which provides the crystal of $[Cu(H_2salphdc)].H_2O$. This ML $[Cu(H_2salphdc)].H_2O$ was reacted with $Zn(NO_3)_2.6H_2O$ to afford the 3D coordination framework $\{Zn_3(OH)_2([ML^{VII})_2]$.2DMF. Here $\{Zn_3(OH)_2\}$s units are behaving as linker of the MLs, one Zn(II) is tetrahedrally coordinated to two hydroxide anions and two carboxylate oxygen atoms of $ML^{VII}$ and the other Zn(II) is octahedrally coordinated to two hydroxide anions and four carboxylate oxygen atoms of $ML^{VII}$ to form an infinite chain structure. These chains are further linked by $ML^{VII}$ to form a 3D porous structure. The 3D framework possesses one-dimensional large channels with a cross-section of approximately $14 \times 14$ Å$^2$ and a large free void space of about 53% considering the van der Waals radii of the constituting atoms (Figure 5.2). DMF and other disordered solvent molecules are present inside the channels and the framework is very stable after removal of the guest molecules from the channels as confirmed by the X-ray powder diffraction (XRPD) measurement. In the framework, the Zn(II) ions are coordinatively saturated with octahedral geometry (indicated as site B in Figure 5.2) but the Cu(II) ions from the ML has a square planar geometry and is coordinatively unsaturated (indicated as site A in Figure 5.2). So, the coordinatively unsaturated Cu(II) ions were fixed onto the pore surface of the framework with an intermetallic distance of 6.1 Å. Another important point is that the UMCs, the Cu(II) metal ions, were replaced by the other metal ions like Co(II) and Ni(II) and the structures were confirmed by the XRPD measurements. So these three compounds are very good examples of PCPs containing unsaturated ML, where OMSs are embedded in the pore surface and these OMSs environments are also changed by different UMCs.

Few other coordination polymers reported[16,35–37] from our group using another novel ML, $[Cu(2,4\text{-pydc})_2]^{2-}$ (where 2,4-pydc = 2,4-pyridinedicarboxylate). This ML (hereafter $ML^{I}$) has multi-coordination ability with two types of Lewis-base binding group, 4-carboxylate and 2-carboxylate moieties, and is suitable for the construction of both coordination and H-bonded frameworks. Reaction of $ML^{I}$ with Zn(II) in DMF provided the PCP, $\{Zn[ML^{I}](H_2O)_3(DMF)\}.DMF\}_n$. The $ML^{I}$ bridges with the Zn(II) ions to form a 1D chains. These chains are interconnected by the bridging $H_2O$ molecule between Zn(II) and Cu(II) to form a 2D networks with large grids. The bridging $H_2O$ molecules are supported by the oxygen atoms of the 4-carboxylate groups and also the chains are connected via hydrogen bonds between the

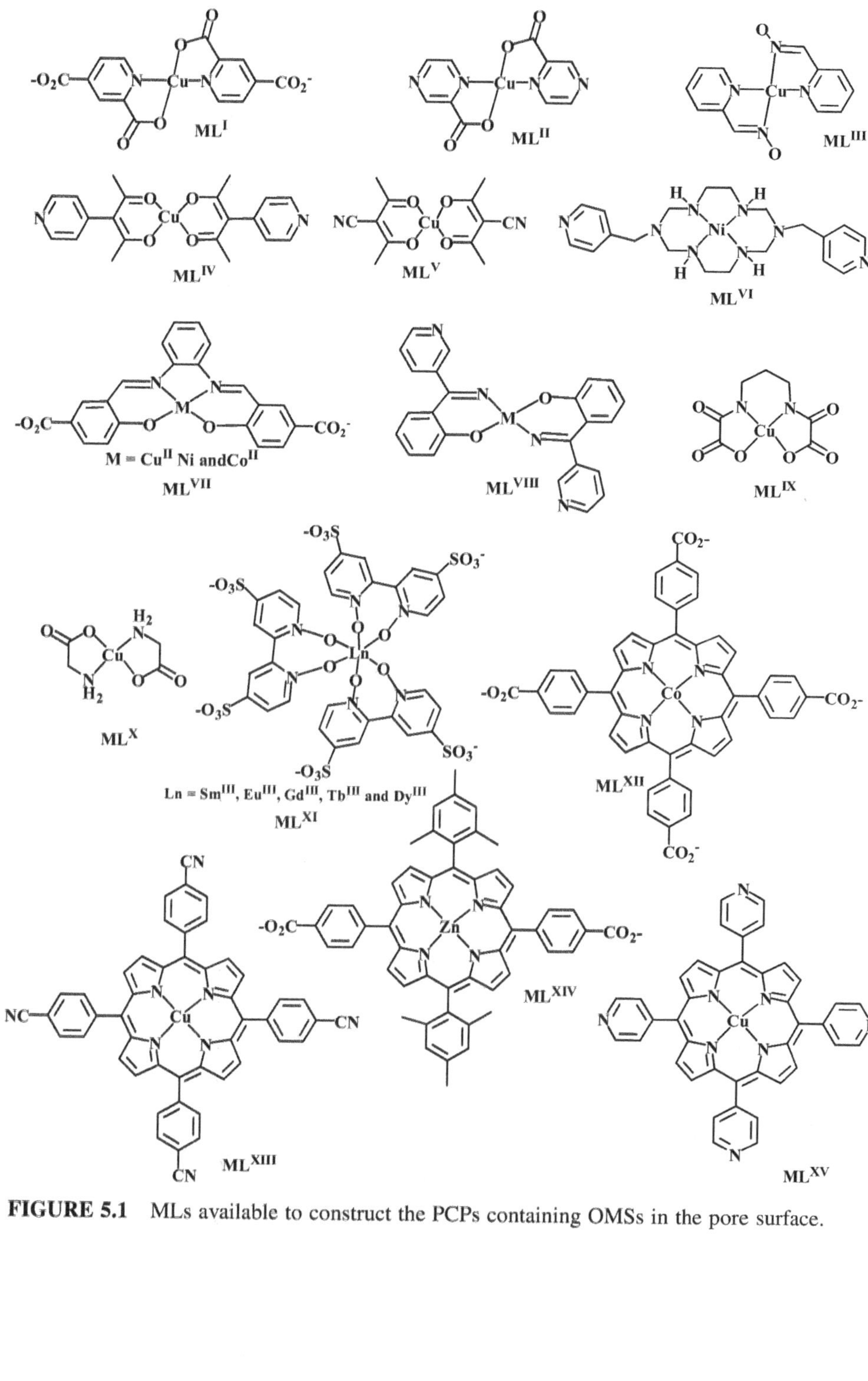


**FIGURE 5.1** MLs available to construct the PCPs containing OMSs in the pore surface.

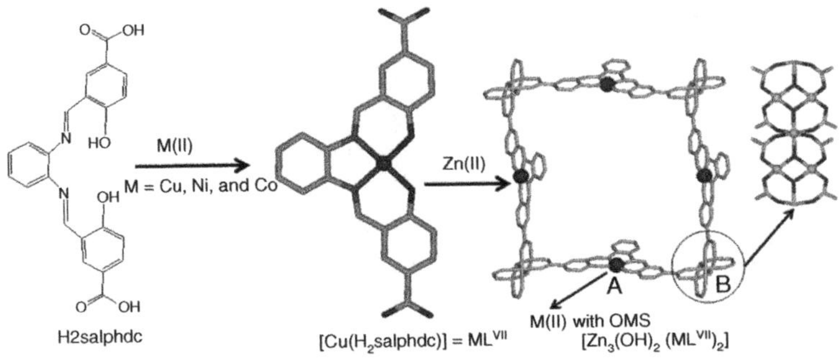


**FIGURE 5.2** Stepwise formation and structure of coordination polymer $\{Zn_3(OH)_2([ML^{VII})_2].2DMF$ using metalloligand $ML^{VII}$.

oxygen atoms of 2-carboxylate groups and the coordinated $H_2O$ molecule at the Zn(II) centers. Each 2D networks are connected via hydrogen bonds to form 1D rhombic channels with dimensions of ~15 × 5 Å$^2$ filled with DMF molecules. In the channels, Cu(II) ions of the ML facing the 1D channel space have axial coordination sites, available for the guest-coordination are now coordinated by the volatile DMF molecules. This same ML has also been used to make few other coordination polymers. Like, the reaction of $ML^{I}$ with Lewis-acid building blocks $[Fe(H_2O)_6]^{6+}$ and imidazole affords H-bonded assemblies with dimmer units of $[Cu_2(2,4\text{-pydc})_4]^{4-}$ and $[Cu_2(2,4\text{-pydc})_2(imidazole)_4]$ to form $[Fe(H_2O)_6][Cu(2,4\text{-pydc})_2]$ and $[Cu(2,4\text{-pydc})(imidazole)_2].2H_2O$, respectively. Each of these are one-dimensional ML arrays through supramolecular (π–π and hydrogen bonds) interactions, which are further linked by the hydrogen bonding interactions via $H_2O$ molecules.

Metal β-diketonate type ligand $[Cu(pyac)_2] = ML^{IV}$ where pyac = 3-(4-pyridyl) pentane-2,4-dionate) has been used as ML.[18] $ML^{IV}$ contains both Lewis-base site (N atoms) and Lewis-acid site ($Cu^{II}$), and is a good precursor for the construction of PCPs containing OMSs in the pore surface. Reaction of $ML^{IV}$ with $Cd(NO_3)_2$ and $CdCl_2$, acting as T-shaped and square-planar nodes respectively, produce 1D and 2D frameworks $\{[\{Cd(NO_3)_2\}_2(ML^{IV})_3].3MeOH.2H_2O\}_n$ and $\{[\{CdCl_2\}(ML^{IV})_2].6THF.0.5H_2O\}_n$. The 1D ladder frameworks $\{[\{Cd(NO_3)_2\}_2(ML^{IV})_3].3MeOH.2H_2O\}_n$ assemble without interpenetrations, stacked in ABCABC. . .packing to create interconnected channels of ~5.7 × 10.2 and 4.1 × 9.8 Å$^2$ along *a*- and *b*-axes respectively. The void space occupies about 60% of the crystal cell volume and filled with solvent molecules. On the other hand, the 2D square-grid layers of compound $\{[\{CdCl_2\}(ML^{IV})_2]. 6THF. O.5 H_2O\}_n$ are stacked in ABCDEF. . .packing, leading to a channel with dimensions of ~8.0 × 8.2 Å$^2$, with porosity is about 63%. Coordination of the solvent molecules to the Cu sites, exposed to the pore, in the crystal structure is the evident for the host–guest interaction ability for both the compounds. Similarly other metal β-diketonate type MLs (like $ML^{V}$, where L = 3-cyanoacetylacetone and M = $Cu^{II}$ have been used for the synthesis of coordination frameworks.[31]

A functional PCP, $\{[Co_{1.5}(ML^{XII})(pyridine)_3(H_2O)].\ 11\ pyridine\}_n$, based on the supramolecular assembly of carboxylate-substituted porphyrins ($ML^{XII}$) with cobalt ions was synthesized and characterized by the single-crystal X-ray structure.[40] The framework is stable after removal of the guest molecules, with void space of 74% of the total unit cell volume. The non-interpenetrated framework has oval shape channel with dimension of $9 \times 7$ Å$^2$ along *b* and *c* axes and another set of channels with dimensions of $14 \times 7$ Å$^2$ along *a* axis considering the van der Waals radii of atoms lining the pore. The surface of the pore containing OMSs at the axial sites of the Co(II) in the porphyrin ring is free to interact with the guest molecules. Fully desolvated compound shows the size, shape, and functional group selective sorption with a remarkable preference to the water and amines, which was confirmed from the measurement of their adsorption of various guests.

Macrocyclic porphyrin ligands with coordinating Lewis-base sites on peripheral groups can create PCPs with the surface containing OMSs at axial sites of the central metal ions. A robust porous Zn(II) metallo-porphyrin framework solid has been synthesized using $ML^{XIV}$ (Figure 5.1).[41] The three-dimensional interpenetrated framework of Zn(II) *trans*-biscarboxylate tetraarylporphyrins formed by the coordination of the carboxylate groups to the six edges of tetrahedral $Zn_4O^{6+}$ clusters, maintaining a charge-neutral framework. This cubic framework with channel dimensions of $4 \times 7$ Å$^2$ and a large free volume of 74% has a surface area (800 $m^2/g$) greater than a typical zeolite. The free volume of the framework are filled with disordered solvents, the framework remains intact after the removal of solvents and is robust to >400°C. The channel surface of the framework containing OMSs at the axial sites of the metal in the porphyrin rings are available for the interactions to the guest molecules, where $Zn_4O^{6+}$ clusters are behaving as coordinatively saturated linker of the MLs. Two other metallo-porphyrin based Cu(I) porous compounds with ligands containing cyanophenyl and pyridyl coordinating groups ($ML^{XIII}$ and $ML^{XV}$ in Figure 5.1) were synthesized and fully characterized.[7]

### 5.2.2 Removal of Volatile Solvents

Coordination sites of the metal ions in the framework structure are generally fully occupied by the coordination from the ligands. In some cases due to the lack of large number of coordination sites of the ligands or unfavorable coordination by the bulky ligand some of the coordination sites of the metal ions are left vacant. Unfilled coordination sites are then filled by other external agents, like counter anion of the metal salt, solvent molecules, or both together. Metal ions function as node in the frameworks if occupied by the volatile solvent molecules, such as water, acetone, alcohols, acetonitrile, and DMF, then it is possible to remove the solvent molecules by heating and/or vacuum, keeping the framework structure intact if the framework is rigid enough.[18,19,42–47] On the other hand, in some cases due to the flexible nature of the metal ion coordination geometry or the ligand flexibility, after removal of the solvent molecules size and shape of the framework structure changes.[48] In both the cases, removal of the solvent molecules create vacant site on the metal centers called as OMS or UMC.

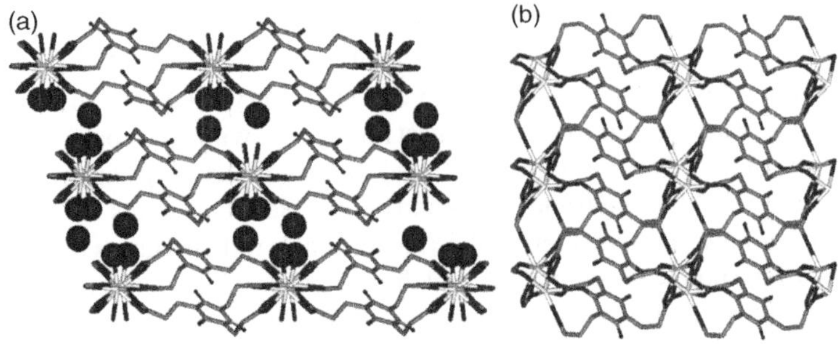


**FIGURE 5.3** Perspective view of the (a) 2D layers structure of $\{[Ce(tci)(H_2O)_2]\cdot 2H_2O\}_n$ and (b) 3D framework of $\{[Ce(tci)]\}_n$. Solvent water molecules are highlighted.

For example, recently we reported two isomorphous 2D compounds $\{[Ln(tci)(H_2O)_2]\cdot 2H_2O\}_n$ (where Ln = Ce(III) and Pr(III); $tciH_3$ = tris(2-carboxyethyl) isocyanurate) which contain both coordinated as well as free water molecules into the channels (Figure 5.3a).[48] After removal of all the water molecules, the vacant sites of the metal centers are coordinated by the oxygen atoms of the available secondary functional group. In addition, due to the ligand flexibility and variable coordination number of lanthanide metal ions the compound rearranges its bonding, which leads to the formation of 3D framework $\{[Ln(tci)]\}_n$ by interlayer bond formations (Figure 5.3b). Interestingly, this 3D compound is stable in presence of other solvents but when exposed to water vapor or just after keeping in the open air for long time, it transformed back to the original 2D layered structure. Both the transformations are directly visualized by the single-crystal-to-single-crystal (SCSC) transformations and also supported by the XRPD and solvent sorption measurements. This reversible transformation is due to the strong affinity of the water molecules to the vacant metal sites, which is coordinated by the other functional groups in the absence of water molecule. Removal of the volatile solvent molecules from the channel and its subsequent strong affinity toward the same solvent gives selective functional behavior. Like, the dehydrated compound can selectively absorb water molecules but reject other absorbates such as methanol, ethanol, acetone, and tetrahydrofuran.

Our group also reported, two isomorphous porous frameworks of Ln(II), $\{Ln_2(imidc)_2(H_2O)_3](H_2O)\}_n$, [Ln = Gd(III) and Er(III)] synthesized by hydrothermal method using multi-functional 4,5-imidazoledicarboxylic acid (imidc).[42] These two compounds have coordinated as well as free water molecules in the channels. Both the compounds are stable upto ~300°C and framework structure remains stable after removal of the water molecules. The XRPD patterns of the dehydrated compounds $\{Ln_2(imidc)_2]\}_n$ indicated that a certain structural change or distortion occurs after removal of the coordinated and guest water molecules, forming a porous framework with coordinatively unsaturated lanthanide ions. Both the compounds have zeolite-like network topology and exhibit selective absorption for $H_2O$ over MeOH and other guest molecules.

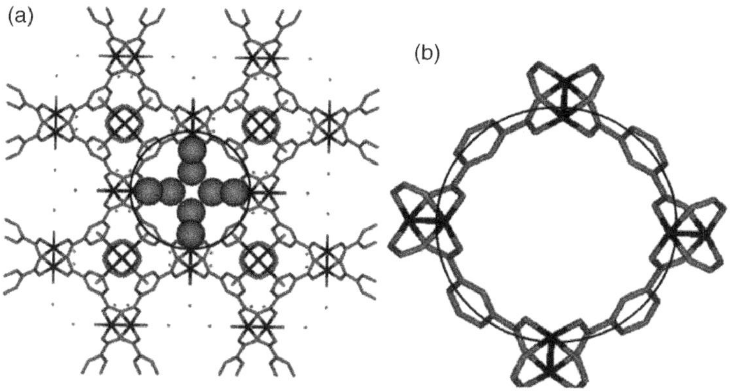


**FIGURE 5.4** Perspective view of the (a) framework of $[Cu_3(TMA)_2(H_2O)_3]_n$ and (b) one nanochannel with OMSs after removal of the guest molecules.

One Cu(II) nanoporous PCP $[Cu_3(TMA)_2(H_2O)_3]_n$ (where TMA = benzene-1,3,5-tricarboxylate) has been reported.[43] This compound has interconnected $[Cu_2(O_2CR)_4]$ units (where R is the aromatic ring), which create a three-dimensional system of channels with a pore size of 1 nm and has 40% solvent accessible pore volume (Figure 5.4). The compound has coordinated as well as free solvent molecules into the large channel and framework is stable after removal of all the guest molecules. The dehydrate framework generate a nanoporous material containing OMSs onto the channel surface, which is chemically functionalizable, like, aqua ligands can be replaced by pyridines. Perhaps the most interesting feature of this material is the presence of OMSs in the pore surface of the material that, after dehydration, is potentially available to act as a site where coordination to guest molecules can take place. Gas adsorption experiments have been carried out on the desolvated framework $[Cu_3(TMA)_2]_n$, which gives an adsorption capacity of 11.16 mmol $H_2$ per g (22.7 mg/g, 2.27 wt%) at 1 bar and 18 mmol per g (36.28 mg/g, 3.6 wt%) at 10 bar (both 77 K). Most interestingly, in the OMSs it can bind NO gas molecules, which has been proved by the infra red experiments. In addition to that, gravimetric adsorption measurements of NO at 196 K (1 bar) gives a large adsorption capacity of ~9 mmol/g, which is significantly greater than any other adsorption capacity reported on a porous solid, whereas, at 298 K the adsorption capacity at 1 bar is just over 3 mmol/g. The amount of NO recovered on exposure of the NO coordinated complex to water is enough to be biologically active and completely inhibiting platelet aggregation in platelet rich plasma as demonstrated by the chemiluminescence and platelet aggregometry experiments.[49] Similar to zeolites, crystalline PCPs with well-defined pores and surface-isolated Lewis-acid sites generated by the removal of the coordinated solvent molecules could also serve as size- or shape-selective heterogeneous catalysis. Recently, one sodalite-type compound $Mn_3[(Mn_4Cl)_3(BTT)_8(CH_3OH)_{10}]_2$ (where BTT = 1,3,5-benzene-tristetrazol-5-yl) with OMSs exposed to the channel surface of the framework catalyzes the cyanosilylation of aromatic aldehydes and ketones, as well as the Mukaiyama-aldol reaction in a size selective manner.[47]

### 5.2.3 Immobilization of Alkali Cations

Another novel approach for the generation of OMSs is by immobilization of alkali cations into the pore surface.[50–52] In addition to their various coordination modes, alkali cations are intriguing because of low polarizability, and unique affinity for the basic molecules ranging from strong base to a weak one. Alkali cations in zeolites are well-known to be catalytic sites for hydrocarbon transformation and also serve as guest recognition sites in biological systems. For the materials with high-performance pores, it is one of the challenges to find the synthetic routes to immobilize alkali cations onto the pore surface of the PCPs in a regular order and it is also very important that alkali cations should sit at the linker sites but not at connector sites of a framework, which can afford the guest-accessible pore surface.

For this purpose, design of heterometallic PCPs containing transition metal ions as connectors of a framework and alkali metal ions as affinity sites on the pore surface would be a good choice. Heterometallic microporous coordination polymer, $\{[CdNa(2\text{-stp})(dabco)_{0.5}(H_2O)].2H_2O\}_n$ (2-stp = 2-sulfonylterephthalate, dabco = 1,4-diazabicyclo[2,2,2]octane), was synthesized by the reaction of $Cd(NO_3)_2.4H_2O$ with 2-stpNa and dabco in $H_2O$/EtOH media under basic conditions using NaOH.[50] The compound possesses 1D channels along the *b* axis with a cross-section of approximately $4.9 \times 4.9\ Å^2$ and void volume of 20.6% without guest water molecules. After removal of all water molecules the compound is stable up to 300°C and maintains the framework, which has been confirmed by the XRPD measurement. To form the open framework the $-CO_2^-$ coordinates to the $Cd^{2+}$, whereas $-SO_3^-$ groups bind to $Na^+$ as active Lewis-acidic sites (Figure 5.5). After the removal of the coordinated water molecules the pentacoordinated sodium ions, densely immobilized at the 1D pore surface with their sixth site open for a guest and show strong adsorbate/adsorbent interaction for acetone and benzene.

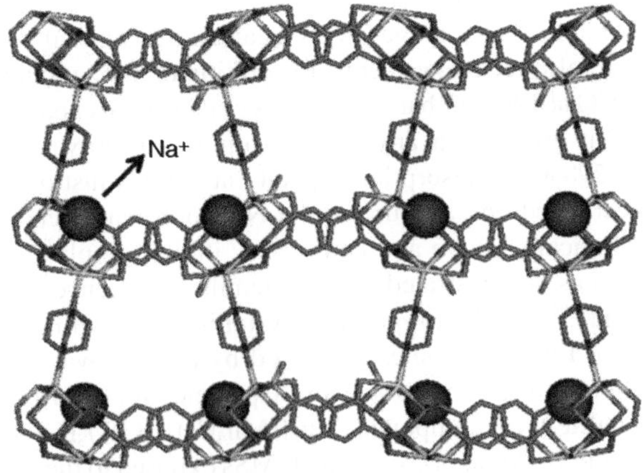


**FIGURE 5.5** Three-dimensional structure of $\{[CdNa(2\text{-stp})(dabco)_{0.5}(H_2O)].2H_2O\}_n$. $Na^+$ ions are highlighted to show the OMSs in the pore surface.

A 3D porous magnet $[KCo_7(OH)_3(ip)_6(H_2O)_4]\cdot 12H_2O$ (where $H_2ip$ is isophthalic acid) was prepared by solvothermal reaction of cobalt acetate, $H_2ip$, and KOH in $H_2O$–MeOH.[51] The framework contains 1D channels with large cages of diameter ~1.0 nm interconnected by the small aperture ($d \sim 4.5$ Å), with void space of 26.2%. After removal of the coordinated as well as free water molecules, the dehydrated compound $[KCo_7(OH)_3(ip)_6]$ is stable up to 300°C and afford reversible SCSC transformations. Dehydrated compound contains two kinds of coordinatively unsaturated OMSs, namely the Co(II) as well as the alkali cation $K^+$ which are regularly embedded in the pore surface of the 1D channels but no functional studies of the OMSs have been done for this compound.

A dinuclear Fe(III) unit was assembled to a 3D porous compound $\{K_6[Fe_2(\mu\text{-}O)(\mu\text{-}CO_3)(SO_3hda)_2].11H_2O.9MeOH\}_n$, that has 1D channels with cross-sectional diameter of approximately 9 Å.[52] In the framework, the dinuclear complexes of Fe (III) and $K^+$ ions form the wall of the honeycomb motif. This compound has large density of $K^+$ ions into the channel surface coordinated by volatile solvents $H_2O$ and MeOH. After removal of the solvent molecules, if the framework remains stable then this compound will provide a PCP with large 1D channel with high density of regularly implanted OMSs into the pore surface.

### 5.2.4 Use of Low-Coordinate Metal Ions

The monovalent coinage metals Ag(I) and Cu(I) can adopt diverse coordination environments such as liner, trigonal planar, tetrahedral, etc. and are attractive OMS candidates as soft Lewis acids. Ag(I) and Cu(I) retains large unoccupied space in the two- and three-coordination forms and is readily accessible for other soft Lewis-base molecules. Like, as the Ag(I) center has high affinity for unsaturated hydrocarbon molecules arises from Ag...$\pi$ interaction Ag(I)-exchanged zeolites were used in alkane–alkene separation studies. Whereas, coordinatively unsaturated Cu(I) is a fundamental metal ion in the biological systems and PCPs with guest accessible Cu(I) centers may potentially serve as solid-state metallo-enzyme mimics. In spite of their great potential, low coordination Ag(I)/Cu(I) immobilized PCPs are very rare because Ag(I)/Cu(I) tends to have energetically stable tetrahedral coordination geometry rather than linear or trigonal planar ones and Ag(I)/Cu(I) coordination complexes are very unstable and sensitive to heat, light, air, moisture, etc.[53–56]

Recently our group reported two isostructural binary metal bipyrazolates $\{[M_2(bpz)]\}_n$ ($M = Ag^I$, $Cu^I$), synthesized by the template-controlled reactions of 3,3′,5,5′-tetramethyl-4,4′-bipyrazole ($H_2bpz$) with $[Ag(NH_3)_2]^+$ or $[Cu(NH_3)_2]^+$.[53,54] Both the compound possesses guest-accessible coordinatively unsaturated bi-coordinated Ag(I)/Cu(I) trinuclear metal clusters with fourfold coordination networks (Figure 5.6). The 3D coordination networks contain $\{M_3(pz)_3\}$ trigonal subunits and a cavity is sandwiched between such trimers with an interplanar distance of 8.82 Å. These threefold symmetric cavities are interconnected through small apertures ($d \sim 3.4$ Å) to generate a 3D bicontinious channel system with void volume of 30.4%. These chemically stable bipyrazolates are thermally stable up to ~500°C. The guest-accessible OMSs in the pore surface of the PCPs with low coordinated

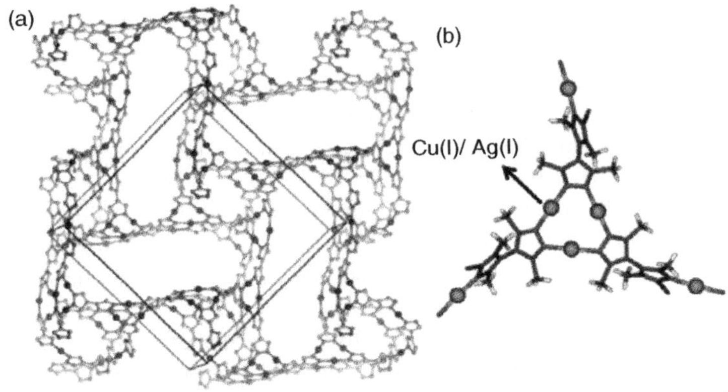


**FIGURE 5.6** (a) A single coordination network of $\{[M_2(bpz)]\}_n$ ($M = Ag^I/Cu^I$). (b) Coordination environment showing two-coordinate metal centers.

Ag(I)/Cu(I) centers have also been demonstrated to facilitate the accommodation of unsaturated hydrocarbons such as benzene, toluene, mesitylene, and acetylene via weak metal. . .π interactions.

Another nanoporous coordination polymer, $\{[Ag_6Cl(atz)_4]OH.6H_2O\}_n$, of bi-coordinated Ag(I) metal was synthesized by the slow evaporation of an ammonia solution of 3-amino-1,2,4-trizole (Hatz) and AgCl.[55] The compound possesses 1D columned channels of diameter 8.5 Å in the *c*-direction, with the guest accessible volume ~32.7%. The compound exhibits temperature-induced pore shape and size transformation and guest desorption–absorption triggered 3D net rearrangements between five- and sixfold interpenetrations.

## 5.3 PORE SURFACE WITH FUNCTIONAL ORGANIC SITES (FOS)

### 5.3.1 Pore Surface with Basic Functional Groups

PCPs with Lewis-acid structures, that is, coordinatively unsaturated sites have been widely explored, but the rational synthesis of PCPs with Lewis-basic functional groups are scarcely observed.[26,57–61] This is because of the difficulty of producing metal-free base groups on the pore surface, as basic groups tend to coordinate metal ions in a self-assembly process to give frameworks in which basic functional groups are completely blocked. The advantage of the PCPs containing basic functional group is that, there are a variety of organic functional groups that can serve as active base sites. Few reports on basic micropore surface reveal that significant sorption for selective binding or catalysis, exploration of their synthetic guidelines and functional uses are strongly required.

Two isomorphous 3D porous coordination frameworks $\{[Ln(tci).H_2O]\ 4H_2O\}_n$ ($Ln = Ce$ and Gd; $tciH_3 =$ tris(2-carboxyethyl) isocyanurate) were obtained from $Ln(NO_3)_3.xH_2O$ and a tripodal symmetrical ligand with a highly flexible arms

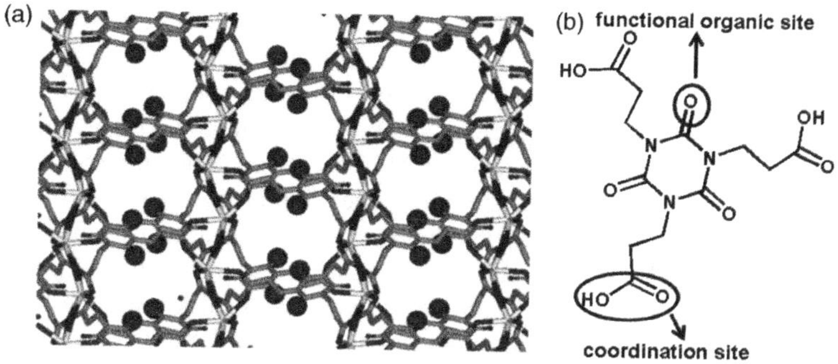


**FIGURE 5.7** (a) View of the 3D structure of $\{[Ln(tci).H_2O]\,4H_2O\}_n$. Basic functional groups in the pore surface are highlighted. (b) Functional site and coordination sites are marked in the ligand.

($-CH_2CH_2-$) and secondary basic functional groups ($-C{=}O$) by hydrothermal technique.[57] The framework contains 1D channels of diameter $2.85 \times 2.74$ Å$^2$ with void space of 21.1% of the total crystal volume occupied by four water molecules, which are H-bonded with the carbonyl groups of the ligand, coordinated water molecule, as well as with the carboxylate oxygen atoms (Figure 5.7). The SCSC structural changes have been observed after removal of the free water molecules and the cross-section of the 1D channels shrink to $2.52 \times 1.30$ Å$^2$ (void space of 13.2%). The framework shows size and affinity-based selective sorption via dynamic structural transformations. Two MeOH molecules entered per unit pore but even with the comparative size of the acetonitrile (dimension $3.80 \times 5.49$ Å$^2$) to methanol (dimension $3.80 \times 4.70$ Å$^2$), acetonitrile cannot enter into the small aperture of the channel. This is due to the fact that MeOH can make strong H-bonds with the hydrophilic pore surface which helps it to enter into the pore but since MeCN cannot make strong H-bonding it fails to enter into the pore. Other guest molecules like EtOH, THF, and $Me_2CO$ with very large size as compared to the channel aperture are not adsorbed; therefore, it is clear that selectivity arises from the very small size of the channel window comparative to the size of the adsorbates. So the free basic functional groups ($-C{=}O$) in the pore surface of the framework are playing here active roles for the selective sorption. PCPs containing other basic functional groups in the pore surface are also prepared. Like, $\{[Zn_3(\mu_3\text{-}OH)_3(2\text{-stp})(bpy)_{1.5}(H_2O)](EtOH)(2H_2O)\}_n$ and $\{[Zn_3(\mu_3\text{-}OH)_3(2\text{-stp})(dpe)_{1.5}](EtOH)(2H_2O)\}_n$ were prepared by the reaction of Zn $(NO_3)_2.4H_2O$ with 2-sulfonylterephthalate (2-stp) and 4,4′-bipyridine (bpy)/1,2-di(4-pyridyl)ethylene (dpe) and NaOH in $H_2O$–EtOH medium at room temperature.[58] These compounds contain free sulfonate groups on the pore surface, which provide Lewis-basic property for acid guest sorption.

A 3D PCP based on octahedral Cd(II) centers, $\{[Cd(4\text{-btapa})_2(NO_3)_2].6H_2O.2DMF\}_n$, was obtained from $Cd(NO_3)_2.4H_2O$ and a 3-connector ligand with amide groups (4-btapa = 1,3,5-benzene tricarboxylic acid tris[*N*-(4-pyridyl)amide]).[59] The amide groups, which acts as guest interaction sites, remain on the surfaces of channels with dimensions of $4.7 \times 7.3$ Å$^2$ and void space volume of 43.8%. The amide groups are ordered uniformly on the channel surfaces and the framework undergoes reversible dynamic structural transformation via an amorphous state in response to the removal and rebinding of guest molecules. Short-chain alcohols (methanol, ethanol, *n*-propanol, and *n*-butanol) exhibited guest inclusion with structural transformation, whereas long-chain alcohols (*n*-pentanol and *n*-hexanol) caused no structural change with guest adsorption. This selective inclusion system is based on not only the size and shape of the incoming guest but also its affinity for the amide group, as demonstrated by X-ray crystallographic analysis and adsorption measurements. The framework also selectively accommodated and activated guests in its channels because of the active amide groups. A well-known base catalysis Knoevenagel condensation reaction was catalyzed to demonstrate its selective heterogeneous base catalytic properties, which depends on the size of the reactants. The framework maintains its crystallinity after the reaction and is easily recycled. Catalytic behavior of the basic functional group into the pore surface of PCPs also observed in the nanochannel-promoted polymerization[61] and in the transesterification of ester.[26] A pillared-layered copper coordination polymer $[Cu_2(pzdc)_2(pyz)]_n$ (**CPL-1**; pydc = pyrazine-2,3-dicarboxylate and pyz = pyrazine) with a pore size of 4 × 6 Å has basic surface oxygen atoms that act as specific adsorption sites for acetylene molecules, as a consequence of a double hydrogen-bonding interaction. By the introduction of more acidic monosubstituted acetylenes (like methyl propiolate where acetylenes bearing an electron-withdrawing substitution group) in the channels produced reactive acetylide species by C–H bond dissociation, which subsequently initiated anionic polymerization of the substituted acetylenes. Moreover, the narrow nanochannel structure directed the selectivity toward a polymerization with *trans* addition because of the prohibitive steric demand for the formation of trisubstituted benzenes and *cis* polymers (Figure 5.8).[61]

Acetylene ($C_2H_2$) is well-known as a highly reactive molecule: it cannot be compressed above 0.2 MPa or it explodes without oxygen, even at room temperature. In addition to this, acetylene is one of the key molecules as a starting material for many chemical products and electric materials in the petrochemical and electronic industries. To obtain highly pure $C_2H_2$ for the preparation of these materials, it is important to separate $C_2H_2$ from the mixture gas containing carbon dioxide ($CO_2$, which is commonly present in many industrial processes) impurities without a large

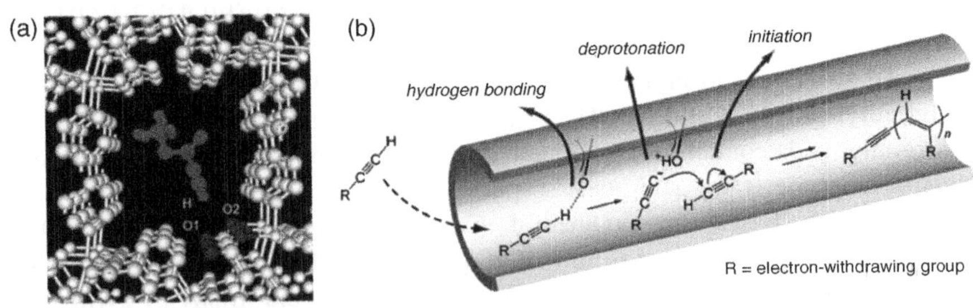


**FIGURE 5.8** (a) Optimized structure of MP incorporated in a nanochannel. (b) Mechanism of the nanochannel-promoted polymerization of acidic acetylenes in porous coordination polymers.

expenditure of energy. Zeolites and activated carbons are known to be materials capable of accommodating large amounts of both gases. No relevant difference in capacity has been observed, because these molecules are similar to one another in equilibrium sorption parameters, related physico-chemical properties, and molecular size and shape. It thus appears that potential, safe materials for $C_2H_2$ separation and storage should be investigated. Our group reported high levels of selective sorption of acetylene molecules as compared to a very similar molecule, carbon dioxide, onto the basic organic group functionalized surface of CPL-1. *In situ* Raman spectra and maximum entropy method/Rietveld analysis using synchrotron XRPD study revealed that the acetylene molecules are held at a periodic distance from one another by hydrogen bonding between two non-coordinated oxygen atoms in the nanoscale pore wall of the PCP and the two hydrogen atoms of the acetylene molecule. This permits the stable storage of acetylene at a density 200 times the safe compression limit of free acetylene at room temperature.[60]

### 5.3.2 Redox Active Pore Surface

Charge transfer (CT) interactions play a significant role in various functional materials. By incorporating a redox active module into the pore surface of the framework it can act as an interaction site to react with certain chemicals to produce reduced frameworks and oxidized substrate and vice versa, or it can provide selective accommodation or sensing properties based on the combination of CT and van der Waals interactions.[27,62–67] In particular, redox active PCPs are useful but only a few examples are known, because they can oxidize or reduce certain substrates and include them in the PCPs.

7,7,8,8-Tetracyano-*p*-quinodimethane (TCNQ) is a well-known multi-redox active ligand that can act as a good acceptor and a weak or a strong donor when its valence is 0, −1, or −2, respectively. A redox active 3D coordination framework $\{[Zn(TCNQ)bpy].6MeOH\}_n$ was synthesized by reacting $Zn(NO_3)_2.6H_2O$ with LiTCNQ and bpy (4,4′-bipyridine) in MeOH.[62] The 3D framework possesses 2D channels with cross-sections of approximately $3.4 \times 5.9$ Å$^2$ and a void space of about 50.7%. The charge number of the TCNQ is estimated to be −2 in this complex, so a dense array of strong donor sites ($TCNQ^{2-}$) is formed on the pore surface of the open framework, providing a highly electron-rich surface for the guest molecules. The framework shows the ability to accommodate various guest molecules by guest exchange method. Like, just by immersing into benzene, MeOH can be exchanged with benzene within 10 s with an apparent color change from yellow to red observed by the SCSC transformation. Other guests (toluene, ethylbenzene, anisole, benzonitrile, and nitrobenzene) also provide reversible changes in color similar to that for benzene, except for anisole and nitrobenzene, where the occlusion of anisole and nitrobenzene formed light-yellow and dark-brown crystals, respectively. The visible spectra showed absorption bands in the order which are associated with the electron-accepting characteristics of the guest molecule. In particular, the band associated with nitrobenzene was quite broad, extending out to 800 nm (Figure 5.9). This is ascribed to the CT interaction between TCNQ in the pore surface and the aromatic molecule.

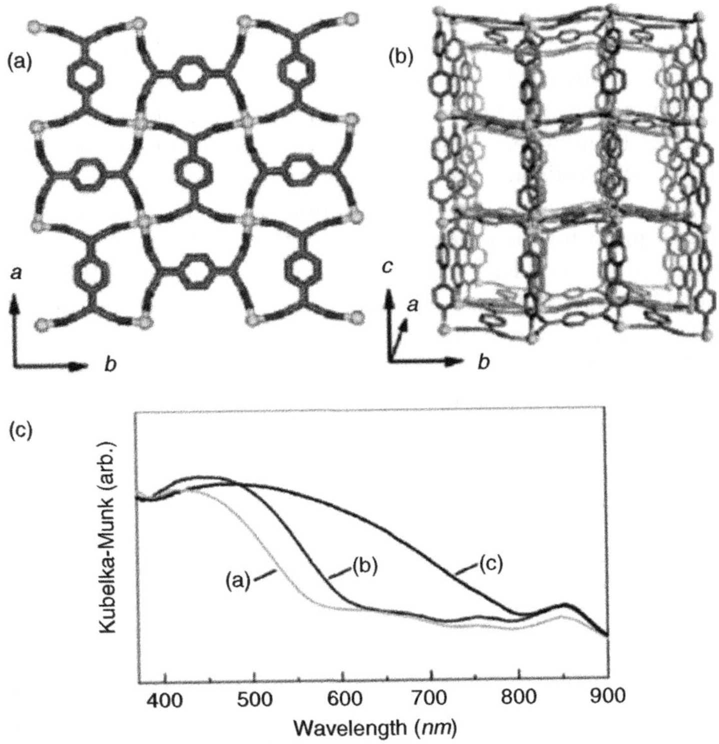


**FIGURE 5.9** (a) Top view and (b) side view of the framework [Zn(TCNQ)bpy].6MeOH$\}_n$. (c) Diffuse reflectance UV–Vis spectra of (a) anisole, (b) benzene, and (c) nitrobenzene exchanged compounds.

Few other PCPs containing redox active pore surface with functional properties have been reported. Like, a 3D porous framework of $\{[Zn_2(adc)_2(dabco)]\}_n$ (adc = 9,10-anthracenedicarboxylate, dabco = 1,4-diazabicyclo-[2.2.2]octane) was synthesized and its photoluminescent properties originated from the host–guest CT interaction has been demonstrated.[63] The confinement of the electron donor guests (*N*, *N*-dimethylaniline, *N*-methylaniline, *N,N*-dimethyl-*p*-toluidine) affords an efficient, photo-induced CT with the anthracene moieties in the pore surface of the PCP. To show that the chemical reduction of the pore surface as a unique approach to enhancing sorption, a redox active ligand diPyNI (*N,N′*-di-(4-pyridyl)-1,4,5,8-naphthalenetetracarboxydiimide) which is reversibly reducible at −0.79 and −1.30 V versus the ferrocene couple has been used to synthesize a 3D porous framework $\{Zn_2(NDC)_2(diPyNI)\}_n$ (NDC = 2,6-naphthalenedicarboxylate). By doping with alkali metal cation ($Li^+$) into the framework, the ligand redox activity has been exploited and that substantially enhance both nitrogen and hydrogen gas uptake.[27] Three porous frameworks have been synthesized from $Ni^{II}$ macrocyclic complex $\{[Ni(cyclam)]^{2+}\}$(where cyclam = 1,4,8,11-tetraazacyclotetradecane) and carboxylate ligands 4,4′-biphenyldicarboxylate (bpdc), 1,1′-biphenyl-2,2′,6,6′-tetracarboxylate and methanetetrabenzoate (MTB). These PCPs produced small (<5 nm) silver, gold, and palladium nanoparticles when they were immersed in $Ag^{I}$ and $Au^{III}$ and $Pd^{II}$solutions, respectively, owing to the redox reactions between the $Ni^{II}$ macrocyclic species incorporated in the coordination polymers and the metal ions.[64–67]

### 5.3.3 Chiral Pore Surface

Homochiral PCPs, which contain regular asymmetric centers incorporated into the pore surface of the framework, are capable of enantioselective separation and/or catalysis.[20,26,68–72] For the synthesis of homochiral porous frameworks, three strategies have been used: (1) use of a rigid homochiral organic ligand as a spacer to link adjacent metal centers, (2) use of a homochiral ligand as an auxiliary pendant which does not directly participate in the formation of a framework backbone, but forces the framework to adopt a specific chiral topology, and (3) use of a readily available homochiral organic ligand and metal ion to form homochiral secondary building units, which in turn, are linked together by rigid spacers to build a framework, in a one-pot reaction. With a judicious choice of homochiral organic molecule, metal ion, and rigid connector with more than one coordination site, these above approaches can provide many PCPs with chiral pore surface. In spite of enormous importance, the synthesis of robust homochiral PCPs with potential for application is still challenging, this is due to the facts that most of them are not robust enough to show permanent porosity, nor porous enough to be useful for selective sorption or catalytic transformation of organic molecules.

A paddlewheel-stabilized twofold interpenetrating PCP, $\{Zn_2(bpdc)_2L.10DMF.8H_2O\}_n$ was obtained by sealed-vial solvothermal synthesis of $bpdcH_2$, zinc nitrate, and a chiral ligand (R,R)-(2)-1,2-cyclohexanediamino-*N*,*N*′-bis(3-*tert*-butyl-5-(4-pyridyl)salicylidene)$Mn^{III}Cl$, (CL1) in DMF solution.[20] The framework possesses rhombic 1D channels in the *c* and *a* directions with dimensions of $6.2 \times 15.7$ and $6.2 \times 6.2$ Å$^2$, respectively with large solvent accessible void space of 57%. The compound featuring chiral pore surface is highly effective as an asymmetric catalyst for olefin epoxidation with enantiomeric excess, examined with 2,2-dimethyl-2H-chromene as a substrate and 2-(*tert*-butylsulfonyl) iodosylbenzene as an oxidant (Figure 5.10). Such enantioselective heterogeneous catalytic behaviors were also observed in other homochiral frameworks with different kind of chiral centers in the pore surface. The enantiopure chiral ligand CL2 (Figure 5.10a) used to produce a 2D homochiral porous framework $\{[Zn3(\mu_3\text{-}O)(1\text{-}H)_6].2H_3O.12H2O\}_n$.[26] The framework possesses large chiral 1D equilateral triangle channels of cross-section ~13.4 Å with a void space volume of ~47%. The compound shows the enantioselective inclusion of metal complexes $[Ru(2,2'\text{-}bpy)_3]Cl_2$, where (bpy = bipyridine), in its pores and catalyzes a transesterification reaction in an enantioselective manner. Lewis acidic (BINOLate)$Ti(O^{i}Pr)_2$ compounds are active catalysts for the addition of $ZnEt_2$ to aromatic aldehydes to afford chiral secondary alcohols, this idea has been used for the heterogeneous asymmetric catalysis of a PCP containing chiral dihydroxy groups. Axially chiral bridging ligand (*R*)-6,6′-dichloro-2,2′-dihydroxy-1,1′-binaphthyl-4,4′-

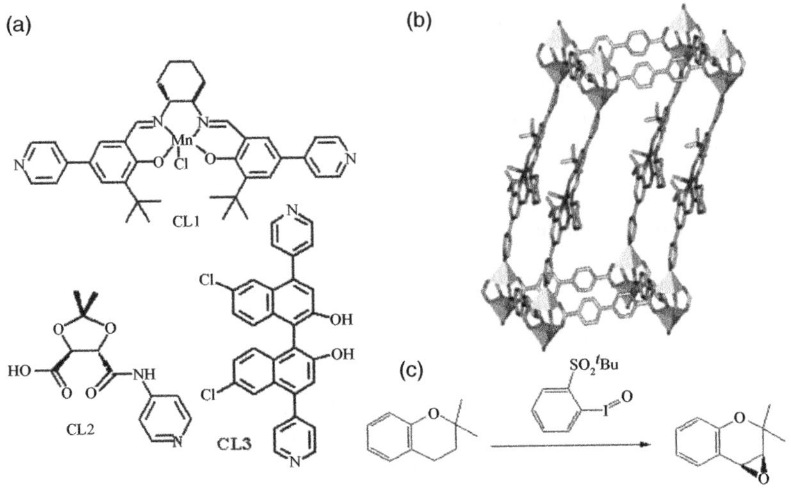


**FIGURE 5.10** View of the (a) structure of the chiral ligands, (b) and (c) single network unit of $\{Zn_2(bpdc)_2(CL1).10DMF.8H_2O\}_n$, and schematic representation of its catalytic epoxidation reaction.

bipyridine (CL3 in Figure 5.10a) which contains the bipyridyl primary functionality and orthogonal chiral 2,2′-dihydroxy secondary functionality was used to construct homochiral porous framework $\{[Cd_3Cl_6(CL3)_3].4DMF.6MeOH.3H_2O\}_n$.[69] The non-interpenetrating 3D network possesses large chiral channels of ~1.6 × 1.8 nm cross-section. Treatment of the as-synthesized compound containing free 2,2′-dihydroxy secondary functional group with excess $Ti(O^iPr)_4$ led to an highly active and enantioselective catalyst for the synthesis of chiral secondary alcohols by the addition of $ZnEt_2$ to aromatic aldehydes.

### 5.3.4 Hydrophobic Pore Surface

Another approach by which we can design the pore surface of the PCPs is by introducing hydrophobic group into the channel to make hydrophobic pore surface. Although most of the PCPs are made of organic ligands with aromatic rings which are hydrophobic in nature, by adding some hydrophobic groups with the aromatic rings it is possible to design the framework with absolute hydrophobic pore.[73–75]

Mainly due to the difficulties in the synthesis and proper design very few porous frameworks containing absolute hydrophobic channel with some functional behaviors are reported. Recently, a porous metal azolate framework $[Cu(etz)]_\infty$ (Hetz = 3,5-diethyl-1,2,4-triazole) has been reported.[73] This framework possesses an NbO type cuprous triazolate scaffold and a CsCl type hydrophobic channel system, in which the large cavities are interconnected by small apertures with pendant ethyl groups. Due to the kinetically controlled process, flexibility arises from the ethyl group in the pore. The framework can adsorb large amounts of organic molecules like MeOH, EtOH, MeCN, benzene, as well as gas molecules like $N_2$, but due to the hydrophobic pore

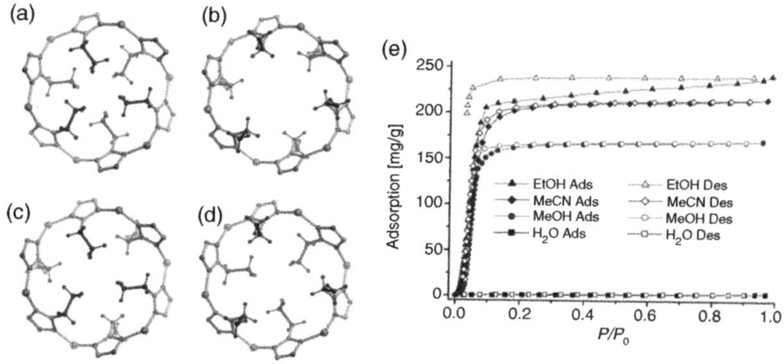


**FIGURE 5.11** (a–d) Perspective view of the four different states of the pore containing hrdrophobic ethyl groups derived from the crystal structure of $[Cu(etz)]_\infty$, and (e) $H_2O$, MeOH, EtOH, and MeCN sorption isotherms.

surface it excludes comparatively smaller size $H_2O$ molecules (Figure 5.11). A hydrophobic and solvent-free channel frameworks $\{Zn(tbip)\}_n$ (tbip = 5-*tert*-butyl isophthalate) was synthesized by hydrothermal technique using water as a solvent and using $H_2$tbip as the ligand, which contain bulky aliphatic groups.[74] This compound can also adsorp gas molecules like Ar, $N_2$, $H_2$ and solvents molecules like MeOH and dimethyl ether but rejected $H_2O$ due the complete hydrophobic nature of the pore surface. Even without addition of extra hydrophobic groups in the framework, flexible compounds sometime show characteristic properties of hydrophobic pore because adsorption properties of flexible PCPs are heavily dependent on both the structural transition mechanism and pore functionality. Like an interdigitated PCP, $\{[Zn_2(ip)_2(bpy)_2].DMF\}_n$ was prepared from $Zn(NO_3)_2.6H_2O$, isophthalic acid (Hip), and 4,4′-bipyridine (bpy) in DMF solution.[75] The guest free compound shows selective absorption of MeOH over $H_2O$ at the low pressure region, this arises from the hydrophobicity-dependent selectivity with the structural transformation.

### 5.3.5 π-Lewis Acidic/Basic Pore Surface

The investigation and discovery of PCPs with π-Lewis acidic/basic pore surface is significant to the design of a new type of interaction site that makes the surface functionality of the pores useful.[76–78] Like for the highly selective separation of molecules having similar physical properties we need to design the high recognition capability system, similar to a biological system, which requires the flexible structure and the specific binding sites such as cation–π, H–π, and H-bonding-type interaction to create suitable space for the target molecules. Inclusion of the aromatic π-moieties in the pore surface could be a key component for the effective shape-selective and site-selective confinement of guest molecules due to their large surface area and feasibly modifiable functionality, which is responsible for their interaction with guests. In spite of the importance of these interactions experimental studies within porous frameworks are still rare.

To construct PCPs with π-Lewis acidic pore surface, hexaazatriphenylene (hat) derivatives as building units are useful as they have electron-deficient heterocyclic cores. A 3D coordination polymer, $\{[Cu^{I}_3(CN)_3\{hat(CN)_3(OEt)_3\}]\cdot 3THF\}_n$, was prepared by the slow intermixing of a solution of $[Cu^{I}_2(m\text{-}\mu^2,\mu^2\text{-benzoquinone})(OAc)_2]$ in ethanol and a solution of $hat(CN)_6$ in THF.[76] There are two kind of pores in the framework: one is a large 1D channel with a cross-section of $10 \times 10$ Å$^2$ and the other pore is just a pocket located along the channel, in which the hat units form a floor and a ceiling and whose portals open to three adjacent channels. The framework is not stable after removal of the guest THF molecules but can recover when soaked with THF, 1,4-dioxane, *N,N*-dimethylformamide (DMF) or in an ethanol solution of pyrazine. In contrast, when soaked in hydrocarbons (hexane, cyclohexane, and benzene), linear ethers and alcohols (diethylether, methanol, and ethanol), and a large-molecule compound (4,4′-bipyridyl) failed to recover the framework. This results show that the structural recovery occurs only for guest molecules with at least one electronegative atom, X, and which are the right size for the cavity, thus indicating that the origin of the guest-responsive phase transformation is the guest-selective inclusion in the cavity promoted by X–π(hat) interactions. Therefore, the neutral organic guest molecules can be confined within the cavities of a PCP through the interaction between electronegative atoms and π-acidic pore surface (Figure 5.12).

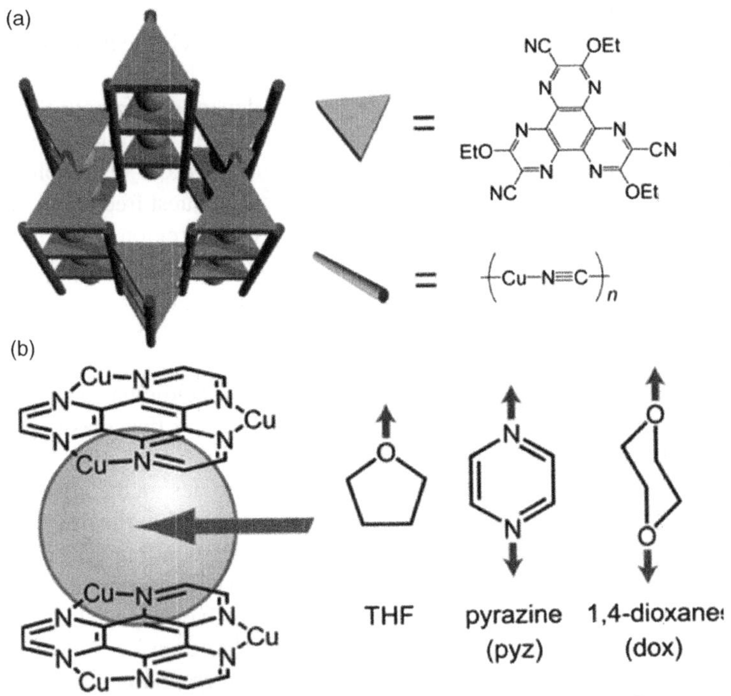


**FIGURE 5.12** Schematic representation of the (a) framework $\{[Cu^{I}_3(CN)_3\{hat(CN)_3(OEt)_3\}]\cdot 3THF\}_n$ and (b) inclusion of solvent guest molecules into pores comprising π planes.

Another electron deficient ligand 2,4,6-tris(4-pyridyl)triazine (tpt) has been used to construct 3D coordination network $\{(ZnI_2)_3(tpt)_2.5.5(nitrobenzene)\}_n$. The guest nitrobenzene molecules were replaced by the electron-rich guest molecules like triphenylene, anthracene, and perylene and crystallographic analysis showed efficient stacking of the guest and ligand, suggested donor–acceptor interactions between the electron-deficient π-Lewis acidic pore surface and electron-rich guests. This was also supported by a significant color change during the formation of the inclusion complex.[77] On the other hand, to construct PCP with π-Lewis basic pore surface electron-rich ligands like TCNQ have been used. A 3D framework $\{[Zn(\mu_4\text{-TCNQ-TCNQ})bpy]\cdot 1.5benzene\}_n$ with highly selective capacity to separate benzene from cyclohexane was synthesized by reacting $Zn(NO_3)_2.6H_2O$ with LiTCNQ and bpy in a MeOH/benzene mixture. The guest benzene molecule is accommodated strongly in the cavity with the size effect and H–π interaction with the electron-rich π-Lewis basic pore surface of the host framework.[78]

## 5.4 POST-SYNTHETIC PORE SURFACE MODIFICATIONS

Most of the robust and porous PCPs are synthesized by solvothermal complexation of organic ligands and metal ions in sealed vessels at higher temperature and pressure, where organic ligands with limited functional groups which are thermally stable and non-coordinating can be utilized. On the other hand, active metal centers are very effective for the efficient performance for catalytic activity and $H_2$ gas storage but it is very difficult to construct PCPs with desired active metal centers. Therefore, clearly there is a need for post-synthetic further improvement of the pore surface active sites: in the metal centers as well as the organic functional groups. One very simple and most common way to modify the metal centers in the pore surface of the PCPs after the synthesis of the parent compound, that is post-synthetic pore surface modification, is by the removal of the volatile solvent molecules from the metal centers to produce OMSs as active functional sites. We have already discussed about this in Section . Here we will discuss about other different methods of post-synthetic pore surface modifications.[26–29,79–82]

Recently reported a straightforward method for functionalizing the benzene ring in the pore surface of the compound $\{Zn_4O(BDC)_3\}_n$ (BDC = 1,4-benzene dicarboxylate) with photoactive $Cr(CO)_3$ group under mild condition compatible with the framework stability.[79] The resulting compound $\{Zn_4O[(BDC)Cr(CO)_3]_3\}_n$ was further modified by the photolysis under a flow of $N_2$ $\{Zn_4O[(BDC)Cr(CO)_2(H_2)]_3\}_n$ (Figure 5.13a). In an another report, a similar compound with identical topology but one extra $-NH_2$ group in the benzene ring has also been modified. The amino group of the benzene ring does not coordinate to the metal centers and are potentially available for undergoing a variety of organic transformations. When treated with acetic anhydride the amino groups have been modified by the acetyl groups maintaining the framework intact (Figure 5.13b).[80] Recently this approach has been extended to the stepwise modifications by two different functional groups.[81] Direct covalent modification of the organic functional groups in the pore surface of the PCP has also been

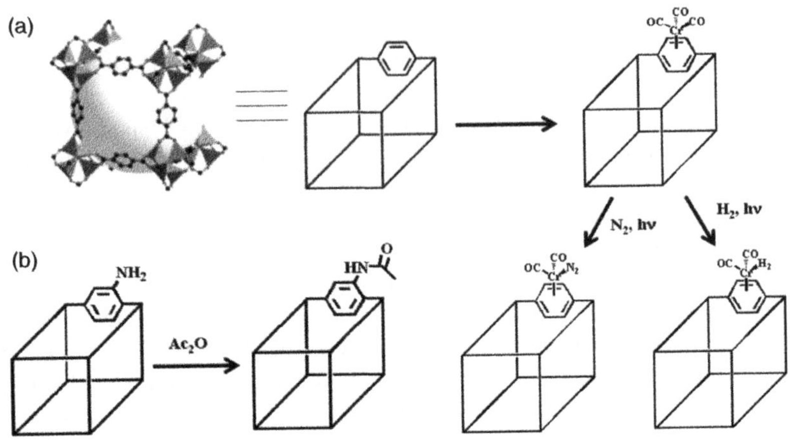


**FIGURE 5.13** Scheme for the post-synthetic pore surface modifications. (a) Reaction of $\{Zn_4O(BDC)_3\}_n$ with $Cr(CO)_6$ to generate $\{Zn_4O[(BDC)\ Cr(CO)_3]_3\}_n$ followed by photolysis under $N_2$ or $H_2$ to afford $\{Zn_4O[(BDC)\ Cr(CO)_2(N_2)]_3\}_n$ and $\{Zn_4O[(BDC)\ Cr(CO)_2(H_2)]_3\}_n$. (b) Acetylation reaction of the similar compound.

performed, where uncoordinated pyridyl groups exposed to the pore surface has been modified by the *N*-alkylation with iodomethane and iodohexane.[26] Chemical reduction of the redox active ligand diPyNI in the pore surface of PCP has also been used to improve the performance of the framework as discussed previously in Section [27] One sodalite-type compound $Mn_3[(Mn_4Cl)_3(BTT)_8(CH_3OH)_{10}]_2$ (also discussed in Section 2.2.) contain guest $Mn^{2+}$ ions in the pore surface of the framework, this metal ion has been exchanged with other monovalent and divalent cations to improve the performance of the pore surface for the $H_2$ adsorption study.[28] To generate active catalytic sites in the pore surface, as-synthesized homochiral Cd(II) PCP containing free 2,2′-dihydroxy secondary functional group treated with excess $Ti(O^iPr)_4$ which led to an highly active and enantioselective catalyst with Lewis acidic (BINOlate)Ti $(O^iPr)_2$ site for the synthesis of chiral secondary alcohols by the addition of $ZnEt_2$ to aromatic aldehydes.[29] Two step pore surface modification has also been performed in the PCPs by combining two different approaches: in the first step volatile solvent molecules have been removed to generate UMCs and then coordinating organic functional groups have been attached as grafting agent to produce the organic functional group-grafted framework. Like, MIL-101 has been modified with ethylenediamine and diethylenetriamine as grafting agents, to produce the amine-grafted MIL-101, exhibiting the remarkably high activities in the Knoevenagel condensation.[82]

## 5.5 SUMMARY AND PERSPECTIVES

This chapter summarizes the latest developments in the area of pore surface engineering of porous metal-organic coordination frameworks in terms of design, synthesis, characterization, and their functional behaviors like selective sorption and catalytic processes. Since the discovery of the PCPs much effort has been made to develop the rigid framework with large surface area by the structural modification and their gas/solvent sorption properties have been extensively studied. The functionalization of the pore surface of the PCPs with high ordering and density of guest-accessible sites can effectively enhance the sorption properties along with many other properties like selective catalytic activity by the precise confinement of desired guest. PCPs with pore surface of diverse functional sites can be designed via a modular approach. The facile tunabilty of such modular building blocks should allow precise engineering of shape, size, and functionalities of the pore surface and the efficient activity of the synthesized compounds may contribute to the next generation of functional porous materials.

## ACKNOWLEDGMENT

Dr. S. K. Ghosh is grateful to JSPS for a post-doctoral fellowship.

## REFERENCES

1. Kitagawa, S.; Kitaura, R.; Noro, I. S. Functional porous coordination polymers. *Angew. Chem. Int. Ed.* **2004**, *43*, 2334–2375.
2. Kitagawa, S.; Matsuda, R. Chemistry of coordination space of porous coordination polymers. *Coord. Chem. Rev.* **2007**, *251*, 2490–2509.
3. Yaghi, O. M.; O'Keeffe, M.; Ockwig, N. W.; Chae, H.; Eddaoudi, M.; Kim, J. Reticular synthesis and the design of new materials. *Nature* **2003**, *423*, 705–714.
4. Férey, G. Hybrid porous solids: past, present, future. *Chem. Soc. Rev.* **2008**, *37*, 191–214.
5. Papaefstathiou, G. S.; MacGillivary, L. R. Inverted metal–organic frameworks: solid-state hosts with modular functionality. *Coord. Chem. Rev.* **2003**, *246*, 169–184.
6. Evans, O. R.; Lin, W. Crystal engineering of NLO materials based on metal-organic coordination networks. *Acc. Chem. Res.* **2002**, *35*, 511–512.
7. Abrahams, B. F.; Hoskins, B. F.; Michall, D. M.; Robson, R. Assembly of porphyrin building blocks into network structures with large channels. *Nature* **1994**, *369*, 727–729.
8. Rao, C. N. R.; Natarajan, S.; Vaidhyanaathan, R. Metal carboxylates with open architectures. *Angew. Chem. Int. Ed.* **2004**, *43*, 1466–1496.
9. Kepert, C. J. Advanced functional properties in nanoporous coordination framework materials. *Chem. Commun.* **2006**, 695–700.
10. Bradshaw, D.; Claridhe, J. B.; Cussen, E. J.; Prior, T. J.; Rosseinsky, M. J. Design, chirality, and flexibility in nanoporous molecule-based materials. *Acc. Chem. Res.* **2005**, *38*, 273–282.
11. Liu, Y.; Eubank, J. F.; Cairns, A. J.; Eckert, J.; Kravtsov, V. C.; Luebke, R.; Eddaoudi, M. Assembly of metal–organic frameworks (MOFs) based on indium-trimer building blocks: a porous MOF with soc topology and high hydrogen storage. *Angew. Chem. Int. Ed.* **2007**, *46*, 3278–3283.

12. Ma, B. Q.; Mulfort, K. L.; Hupp, J. T. Microporous pillared paddle-wheel frameworks based on mixed-ligand coordination of zinc ions. *Inorg. Chem.* **2005**, *44*, 4912–4914.
13. Dinĉa, M.; Yu, A. F.; Long, J. R. Microporous metal-organic frameworks incorporating 1,4-benzeneditetrazolate: syntheses, structures, and hydrogen storage properties. *J. Am. Chem. Soc.* **2006**, *128*, 8904–8913.
14. Lin, X.; Blake, A. J.; Wilson, C.; Sun, X. Z.; Champness, N. R.; George, M. W.; Hubberstey, P.; Mokaya, R.; Schröder, M. High $H_2$ adsorption by coordination-framework materials. *J. Am. Chem. Soc.* **2006**, *128*, 10745–10753.
15. Lee, E. Y.; Suh, M. P. A robust porous material constructed of linear coordination polymer chains: reversible single-crystal to single-crystal transformations upon dehydration and rehydration. *Angew. Chem. Int. Ed.* **2004**, *43*, 2798–2801.
16. Noro, S.; Kitagawa, S.; Yamashita, M.; Wada, T. New microporous coordination polymer affording guest-coordination sites at channel walls. *Chem. Commun.* **2002**, 222–223.
17. Kitaura, R.; Onoyama, G.; Sakamoto, H.; Matsuda, R.; Noro, S.; Kitagawa, S. Immobilization of a metallo schiff base into a microporous coordination polymer. *Angew. Chem., Int. Ed.* **2004**, *43*, 2684–2687.
18. Chen, B.; Fronczek, F. R.; Maverick, A. W. Porous Cu-Cd mixed-metal-organic frameworks constructed from $Cu(Pyac)_2$ {Bis[3-(4-pyridyl)pentane-2,4-dionato]copper(II)}. *Inorg. Chem.* **2004**, *43*, 8209–8211.
19. Chen, B.; Ockwig, N. W.; Millward, A. R.; Contreras, D. S.; Yaghi, O. M. High $H_2$ adsorption in a microporous metal–organic framework with open metal sites. *Angew. Chem. Int. Ed.* **2005**, *44*, 4745–4749.
20. Cho, S. H.; Ma, B.; Nguyen, S. T.; Hupp, J. T.; Albrecht-Schmitt, T. E. A metal–organic framework material that functions as an enantioselective catalyst for olefin epoxidation. *Chem. Commun.* **2006**, 2563–2565.
21. Sharma, A. C.; Borovik, A. S. Design, synthesis, and characterization of templated metal sites in porous organic hosts: application to reversible dioxygen binding. *J. Am. Chem. Soc.* **2000**, *122*, 8946–8955.
22. Ozin, G. A.; Gil, C. Intrazeollte organometallics and coordination complexes: internal versus external confinement of metal guests. *Chem. Rev.* **1989**, *89*, 1749–1764.
23. Kitaura, R.; Fujimoto, K.; Noro, S.; Kondo, M.; Kitagawa, S. A pillared-layer coordination polymer network displaying hysteretic sorption: [Cu2(pzdc)2(dpyg)]n (pzdc_ Pyrazine-2,3-dicarboxylate; dpyg_1,2-Di(4-pyridyl)-glycol). *Angew. Chem. Int. Ed.* **2002**, *41*, 133–135.
24. Shin, D. M.; Lee, I. S.; Chung, Y. K. Self-assemblies of new rigid angular ligands and metal centers toward the rational construction and modification of novel coordination polymer networks. *Inorg. Chem.* **2003**, *42*, 8838–8846.
25. Custelcean, R.; Gorbunova, M. G. A metal-organic framework functionalized with free carboxylic acid sites and its selective binding of a $Cl(H_2O)_4^-$ cluster. *J. Am. Chem. Soc.* **2005**, *127*, 16362–16363.
26. Seo, J. S.; Whang, D.; Lee, H.; Jun, S. I.; Oh, J.; Jeon, Y. J.; Kim, K. A homochiral metal-organic porous material for enantioselective separation and catalysis. *Nature* **2000**, *404*, 982–986.
27. Mulfort, K. L.; Hupp, J. T. Chemical reduction of metal-organic framework materials as a method to enhance gas uptake and binding. *J. Am. Chem. Soc.* **2007**, *129*, 9604–9605.

28. Dinĉa, M.; Long, J. R. High-enthalpy hydrogen adsorption in cation-exchanged variants of the microporous metal-organic framework $Mn_3[(Mn_4Cl)_3(BTT)_8(CH_3OH)_{10}]_2$. *J. Am. Chem. Soc.* **2007**, *129*, 11172–11176.

29. Wu, C. -D.; Hu, A.; Zhang, L.; Lin, W. A homochiral porous metal-organic framework for highly enantioselective heterogeneous asymmetric catalysis. *J. Am. Chem. Soc.* **2005**, *127*, 8940–8941.

30. Kitagawa, S.; Noro, S.; Nakamura, T. Pore surface engineering of microporous coordination polymers. *Chem. Commun.* **2006**, 701–707.

31. Burrows, A. D.; Casser, K.; Mahon, M. F.; Warren, J. E. The stepwise formation of mixed-metal coordination networks using complexes of 3-cyanoacetylacetonate. *Dalton Trans.* **2007**, 2499–2509.

32. Bi, W.; Mercier, N.; Louvain, N.; Latroche, M. Lead halide layers linked by *trans*-$Cu(Gly)_2$ (Gly = $^{-}O_2C\text{-}CH_2\text{-}NH_2$) lillars in heterometallic glycinate based organic-inorganic hybrids. *Eur. J. Inorg. Chem.* **2006**, 4225–4228.

33. Chandler, B. D.; Cramb, D. T.; Shimizu, G. K. H. Microporous metal-organic frameworks formed in a stepwise manner from luminescent building blocks. *J. Am. Chem. Soc.* **2006**, *128*, 10403–10412.

34. Doble, D. M. J.; Benison, C. H.; Blake, A. J.; Fenske, D.; Jackson, M. S.; Kay, R. D.; Li, W.-S.; Schröder, M. Template assembly of metal aggregates by imino-carboxylate ligands. *Angew. Chem. Int. Ed.* **1999**, *38*, 1915–1918.

35. Noro, S.; Kitagawa, S.; Wada, T. Hydrogen-bonding assemblies constructed from metalloligand building blocks and $H_2O$. *Inorg. Chim. Acta.* **2005**, *358*, 423–428.

36. Noro, S.; Kitagawa, S.; Yamashita, M.; Wada, T. Novel 2-dimensional coordination polymer constructed from a multi-functional metalloligand. *CrystEngComm* **2002**, *4*, 162–164.

37. Noro, S.; Miyasaka, H.; Kitagawa, S.; Wada, T.; Okubo, T.; Yamashita, M.; Mitani, T. Framework control by a metalloligand having multicoordination ability: new synthetic approach for crystal structures and magnetic properties. *Inorg. Chem.* **2005**, *44*, 133–146.

38. Vreshch, V. D.; Chernega, A. N.; Howard, J. A. K.; Sieler, J.; Domasevitch, K. V. Two-step construction of molecular and polymeric mixed-metal Cu(Co)/Be complexes employing functionality of a pyridyl substituted acetylacetonate. *Dalton Trans.* **2003**, 1707–1711.

39. Abrahams, B. F.; Hoskins, B. F.; Michail, D. M.; Robson, R. Assembly of porphyrin building blocks into network structures with large channels. *Nature* **1994**, *369*, 727–729.

40. Kosal, M. E.; Chou, J.-H.; Wilson, S. R.; Suslick, K. S. A functional zeolite analogue assembled from metalloporphyrins. *Nat. Mater.* **2002**, *1*, 118–121.

41. Smithenry, W.; Wilson, S. R.; Suslick, K. S. A robust microporous zinc porphyrin framework solid D. *Inorg. Chem.* **2003**, *42*, 7719–7721.

42. Maji, T. K.; Mostafa, G.; Chang, H.; Kitagawa, S. Porous lanthanide–organic framework with zeolite-like topology. *Chem. Commun.* **2005**, 2436–2438.

43. Chui, S. S. Y.; Lo, S. M. F.; Charmant, J. P. H.; Orpen, A. G.; Williams, I. D. A chemically functionalizable nanoporous material $[Cu_3(TMA)_2(H_2O)_3]_n$. *Science* **1999**, *283*, 1148–1150.

44. Suh, M. P.; Cheon, Y. E.; Lee, E. Y. Reversible transformation of $Zn^{II}$ coordination geometry in a single crystal of porous metal-organic framework $[Zn_3\,(ntb)_2\,(EtOH)_2]\cdot$4EtOH. *Chem. Eur. J.* **2007**, *13*, 4208–4215.

45. Jia, J.; Lin, X.; Blake, A. J.; Champness, N. R.; Hubberstey, P.; Shao, L.; Walker, G.; Wilson, C.; Schröder, M. Triggered ligand release coupled to framework rearrangement: generating crystalline porous coordination materials. *Inorg. Chem.* **2006**, *45*, 8838–8840.
46. Chen, B.; Eddaoudi, M.; Reineke, T. M.; Kampf, J. W.; O'Keeffe, M.; Yaghi, O. M. $Cu_2(ATC).6H_2O$: design of open metal sites in porous metal-organic crystals (ATC: 1,3,5,7-adamantane tetracarboxylate). *J. Am. Chem. Soc.* **2000**, *122*, 11559–11560.
47. Horike, S.; Dinĉa, M.; Tamaki, K.; Long, J. R. Size-selective Lewis acid catalysis in a microporous metal-organic framework with exposed $Mn^{2+}$ coordination sites. *J. Am. Chem. Soc.* **2008**, *130*, 5854–5855.
48. Ghosh, S. K.; Zhang, J. P.; Kitagawa, S. Reversible topochemical transformation of a soft crystal of a coordination polymer. *Angew. Chem., Int. Ed.* **2007**, *46*, 7965–7968.
49. Xiao, B.; Wheatley, P. S.; Zhao, X.; Fletcher, A. J.; Fox, S.; Rossi, A. G.; Megson, I. L.; Bordiga, S.; Regli, L.; Thomas, K. M.; Morris, R. E. High-capacity hydrogen and nitric oxide adsorption and storage in a metal-organic framework. *J. Am. Chem. Soc.* **2007**, *129*, 1203–1209.
50. Horike, S.; Matsuda, R.; Tanaka, D.; Mizuno, M.; Endo, K.; Kitagawa, S. Immobilization of sodium ions on the pore surface of a porous coordination polymer. *J. Am. Chem. Soc.* **2006**, *128*, 4222–4223.
51. Cheng, X.-N.; Zhang, W.-X.; Lin, Y.-Y.; Zheng, Y.-Z.; Chen, X.-M. A dynamic porous magnet exhibiting reversible guest-induced magnetic behavior modulation. *Adv. Mater.* **2007**, *19*, 1494–1498.
52. Schmitt, W.; Hill, J. P.; Juanico, M. P.; Caneschi, A.; Costantino, F.; Anson, C. E.; Powell, A. K. Supramolecular coordination assemblies of dinuclear $Fe^{III}$ complexes. *Angew. Chem. Int. Ed.* **2005**, *44*, 4187–4192.
53. Zhang, J.-P.; Horike, S.; Kitagawa, S. A flexible porous coordination polymer functionalized by unsaturated metal clusters. *Angew. Chem. Int. Ed.* **2007**, *46*, 889–892.
54. Zhang, J.-P.; Kitagawa, S. Supramolecular isomerism, framework flexibility, unsaturated metal center, and porous property of Ag(I)/Cu(I) 3,3′, 5,5′-tetrametyl-4,4′-bipyrazolate. *J. Am. Chem. Soc.* **2008**, *130*, 907–917.
55. Zhang, J.-P.; Lin, Y.-Y.; Huang, X.-C.; Chen, X.-M. Copper(I) 1,2,4-triazolates and related complexes: studies of the solvothermal ligand reactions, network topologies, and photoluminescence properties. *J. Am. Chem. Soc.* **2005**, *127*, 5495–5506.
56. Zhang, J.-P.; Chen, X.-M. Crystal engineering of binary metal imidazolate and triazolate frameworks. *Chem. Commun.* **2006**, 1689–1699.
57. Ghosh, S. K.; Zhang, J.-P.; Kitagawa, S. A dynamic, isocyanurate-functionalized porous coordination polymer. *Angew. Chem., Int. Ed.* **2008**, *47*, 3403–3406.
58. Horike, S.; Bureekaew, S.; Kitagawa, S. Coordination pillared-layer type compounds having pore surface functionalization by anionic sulfonate groups. *Chem. Commun.* **2008**, 471–473.
59. Hasegawa, S.; Horike, S.; Matsuda, R.; Furukawa, S.; Mochizuki, K.; Kinoshita, Y.; Kitagawa, S. Three-dimensional porous coordination polymer functionalized with amide groups based on tridentate ligand: selective sorption and catalysis. *J. Am. Chem. Soc.* **2007**, *129*, 2607–2614.
60. Matsuda, R.; Kitaura1, R.; Kitagawa, S.; Kubota, Y.; Belosludov, R. V.; Kobayashi, T. C.; Sakamoto, H.; Chiba, T.; Takata, M.; Kawazoe, Y.; Mita, Y. Highly controlled acetylene accommodation in a metal–organic microporous material. *Nature* **2005**, *436*, 238–241.

61. Uemura, T.; Kitaura, R.; Ohta, Y.; Nagaoka, M.; Kitagawa, S. Nanochannel-promoted polymerization of substituted acetylenes in porous coordination polymers. *Angew. Chem. Int. Ed.* **2006**, *45*, 4112–4116.

62. Shimomura, S.; Matsuda, R.; Tsujino, T.; Kawamura, T.; Kitagawa, S. TCNQ dianion-based coordination polymer whose open framework shows charge-transfer type guest inclusion. *J. Am. Chem. Soc.* **2006**, *128*, 16416–16417.

63. Tanaka, D.; Horike, S.; Kitagawa, S.; Ohba, M.; Hasegawa, M.; Ozawac, Y.; Toriumi, K. Anthracene array-type porous coordination polymer with host–guest charge transfer interactions in excited states. *Chem. Commun.* **2007**, 3142–3144.

64. Choi, H. J.; Suh, M. P. Dynamic and redox active pillared bilayer open framework: single-crystal-to-single-crystal transformations upon guest removal, guest exchange, and framework oxidation. *J. Am. Chem. Soc.* **2004**, *126*, 15844–15851.

65. Moon, H. R.; Kim, J. H.; Suh, M. P. Redox-active porous metal–organic framework producing silver nanoparticles from $Ag^{I}$ ions at room temperature. *Angew. Chem. Int. Ed.* **2005**, *44*, 1261–1265.

66. Suh, M. P.; Moon, H. R.; Lee, E. Y.; Jang, S. Y. A redox-active two-dimensional coordination polymer: preparation of silver and gold nanoparticles and crystal dynamics on guest removal. *J. Am. Chem. Soc.* **2006**, *128*, 4710–4718.

67. Cheon, Y. E.; Suh, M. P. Multifunctional fourfold interpenetrating diamondoid network: gas separation and fabrication of palladium nanoparticles. *Chem. Eur. J.* **2008**, *14*, 3961–3967.

68. Kesanli, B.; Lin, W. Chiral porous coordination networks: rational design and applications in enantioselective processes. *Coord. Chem. Rev.* **2003**, *246*, 305–326.

69. Wu, C.-D.; Hu, A.; Zhang, L.; Lin, W. A homochiral porous metal-organic framework for highly enantioselective heterogeneous asymmetric catalysis. *J. Am. Chem. Soc.* **2005**, *127*, 8940–8941.

70. Vaidhyanathan, R.; Bradshaw, D.; Rebilly, J.-N.; Barrio, J. P.; Gould, J. A.; Berry, N. G.; Rosseinsky, M. J. A family of nanoporous materials based on an amino acid backbone. *Angew. Chem., Int. Ed.* **2006**, *45*, 6495–6499.

71. Dybtsev, D. N.; Yutkin, M. P.; Peresypkina, E. V.; Virovets, A. V.; Serre, C.; Férey, G.; Fedin, V. P. Isoreticular homochiral porous metal-organic structures with tunable pore sizes. *Inorg. Chem.* **2007**, *46*, 6843–6845.

72. Zhang, J.; Bu, X. Chiralization of diamond nets: stretchable helices and chiral and achiral nets with nearly identical unit cells. *Angew. Chem., Int. Ed.* **2007**, *46*, 6115–6118.

73. Zhang, J. -P.; Chen, X. -M. Exceptional framework flexibility and sorption behavior of a multifunctional porous cuprous triazolate framework. *J. Am. Chem. Soc.* **2008**, *130*, 6010–6017.

74. Pan, L.; Parker, B.; Huang, X.; Olson, D. H.; Lee, J. Y.; Li, J. Zn(tbip) (H2tbip= 5-*tert*-butyl isophthalic acid): a highly stable guest-free microporous metal-organic framework with unique gas separation capability. *J. Am. Chem. Soc.* **2006**, *128*, 4180–4181.

75. Horike, S.; Tanaka, D.; Nakagawa, K.; Kitagawa, S. Selective guest sorption in an interdigitated porous framework with hydrophobic pore surfaces. *Chem. Commun.* **2007**, 3395–3397.

76. Tanaka, D.; Masaoka, S.; Horike, S.; Furukawa, S.; Mizuno, M.; Endo, K.; Kitagawa, S. Porous coordination polymer with π Lewis acidic pore surfaces, $\{[Cu_3(CN)_3\{hat(CN)_3(OEt)_3\}]\cdot 3THF\}_n$. *Angew. Chem. Int. Ed.* **2006**, *45*, 4628–4631.

77. Ohmori, O.; Kawano, M.; Fujita, M. Crystal-to-crystal guest exchange of large organic molecules within a 3D coordination network. *J. Am. Chem. Soc.* **2004**, *126*, 16292–16293.

78. Shimomura, S.; Horike, S.; Matsuda, R.; Kitagawa, S. Guest-specific function of a flexible undulating channel in a 7,7,8,8-tetracyano-p-quinodimethane dimer-based porous coordination polymer. *J. Am. Chem. Soc.* **2007**, *129*, 10990–10991.

79. Kaye, S. S.; Long, J. R. Matrix isolation chemistry in a porous metal-organic framework: photochemical substitutions of $N_2$ and $H_2$ in $Zn_4O[(\eta^6$-1,4-benzenedicarboxylate)$Cr(CO)_3]_3$. *J. Am. Chem. Soc.* **2008**, *130*, 806–807.

80. Wang, Z.; Cohen, S. M. Post-synthetic covalent modification of a neutral metal-organic framework. *J. Am. Chem. Soc.* **2007**, *129*, 12368–12369.

81. Wang, Z.; Cohen, S. M. Tandem modification of metal–organic frameworks by a post-synthetic approach. *Angew. Chem., Int. Ed.* **2008**, *47*, 4699–4702.

82. Hwang, Y. K.; Hong, D. Y.; Chang, J.-S.; Jhung, S. H.; Seo, Y.-K.; Kim, J.; Vimont, A.; Daturi, M.; Serre, C.; Férey, G. Amine grafting on coordinatively unsaturated metal centers of MOFs: consequences for catalysis and metal encapsulation. *Angew. Chem., Int. Ed.* **2008**, *47*, 4144–4148.

# 6

# RATIONAL DESIGN OF NON-CENTROSYMMETRIC METAL-ORGANIC FRAMEWORKS FOR SECOND-ORDER NONLINEAR OPTICS

WENBIN LIN AND SHUTING WU

*Department of Chemistry, University of North Carolina at Chapel Hill, Chapel Hill, NC 27599, USA*

## 6.1 INTRODUCTION

The properties of solid-state materials are dependent not only upon their building blocks but also on the arrangement of these constituents within the crystal lattice. For example, many solid-state properties including pyroelectricity, piezoelectricity, ferroelectricity, and second-order nonlinear optics (NLO) only exist in non-centrosymmetric or polar bulk materials.[1] Despite the practical importance of these properties in many technological applications, the discovery of non-centrosymmetric solids still relies on Edisonian approaches, in part due to the lack of successful strategies for modular synthesis of solid-state materials.

Metal-organic frameworks (MOFs) are a class of hybrid materials that are built from metal connecting points and multidentate organic bridging ligands.[2] They are also called coordination polymers or metal-organic coordination networks. Unlike polymeric inorganic materials such as metal oxides, MOFs are typically synthesized under mild conditions and therefore allow for fine-tuning of their properties by systematic variation of the metals with similar coordination modes and/or bridging

*Metal-Organic Frameworks: Design and Application,* Edited by Leonard R. MacGillivray

ligands of the same functionalities but different lengths. They provide a platform for the modular synthesis of solid-state materials with a range of interesting properties. In this chapter, we illustrate how MOFs can be rationally designed to crystallize in non-centrosymmetric spaces groups and their potential in second-order nonlinear optical applications.

## 6.2 DESIGN STRATEGIES FOR NON-CENTROSYMMETRIC METAL-ORGANIC FRAMEWORKS

The easiest way to constructing non-centrosymmetric MOFs is to use chiral bridging ligands since the resulting MOFs must crystallize in a chiral space group which is necessarily non-centrosymmetric.[2,3] Although such MOFs will crystallize in non-centrosymmetric space groups, they cannot necessarily ensure non-centrosymmetric arrangement of all the molecular dipoles of the building blocks. Non-centrosymmetric alignment of the molecular dipoles is key to many of the solid properties (including second-order NLO and ferroelectricity) that are important for technological applications. Alternative approaches are needed for the construction of non-centrosymmetric MOFs by controlling spatial arrangements of molecular building blocks in the solid state. In a typical molecular solid, the molecular dipoles are the dominant interactions which tend to adopt centrosymmetric arrangements as a result of the thermodynamic preference. Predictably synthesizing non-centrosymmetric solids still presents a great challenge to solid-state chemists and materials scientists.

Recent emergence of MOFs offers a potential solution to the rational design (crystal engineering) of non-centrosymmetric solids.[4–6] MOFs with desired topologies can be assembled from their constituent building blocks via strong and highly directional metal-ligand coordination bonds. Because a single strong interaction dominates in MOFs, other weak intermolecular interactions play a less important role in the crystal packing of these materials. Coordination bonds can thus be utilized to counteract deleterious centrosymmetric intermolecular interactions which are dominant in the solid state. Non-centrosymmetric MOFs with desired topologies can be rationally designed by taking advantage of well-defined metal coordination geometries in combination with carefully chosen rigid bridging ligands. The modular nature of such an approach potentially allows precise engineering of a multitude of properties via systematic tuning of the molecular modules. Optimization of NLO-active materials becomes possible via crystal engineering of non-centrosymmetric MOFs.

## 6.3 NON-CENTROSYMMETRIC METAL-ORGANIC FRAMEWORKS FOR SECOND-ORDER NONLINEAR OPTICAL APPLICATIONS

Second-order NLO materials are a class of substances that have the unique capability to double the frequency of an intense beam of laser light (i.e., second harmonic generation, SHG). They are key to the future photonics technology in frequency conversion and electro-optics. SHG describes the capacity of a material to double the frequency of incident light, and is a direct result of bulk second-order NLO susceptibility ($\chi^2$) which is a third rank tensor and will vanish in a centrosymmetric environment.[7] In addition to the non-centrosymmetric requirement, ideal second-order NLO materials must have other attributes including transparency in the relevant wavelengths, ability to withstand laser irradiation, and chemical and mechanical stability.

Given the importance of second-order NLO materials, there has been an intense research effort devoted to this area over the past few decades. Molecule-based NLO materials have received particular interest because their optical properties can be readily fine-tuned through subtle changes in the molecular structure using the tools of synthetic organic chemistry.[1] Guided by sophisticated theoretical models, many NLO chromophores with extremely high first hyperpolarizability ($\beta$) have been synthesized.[1] However, organic NLO materials have not found practical applications, primarily due to the inability to reliably induce stable non-centrosymmetric arrangements of molecular chromophores in these systems; inorganic materials such as lithium niobate and potassium dihydrogen phosphate (KDP) still dominate in frequency conversion, optical signal processing, and electro-optic devices. Since efficient NLO chromophores typically contain a good electron donor and acceptor connected through a conjugated bridge, they are electronically asymmetric and highly dipolar and tend to adopt centrosymmetric arrangements in the solid state due to the dominance of centrosymmetric dipole–dipole repulsions. Non-centrosymmetric MOFs present a unique approach to aligning strong molecular dipoles in the solid state by counteracting the deleterious centrosymmetric dipole–dipole interactions with strong and directional metal-ligand coordination bonds. This chapter illustrates the rational synthesis of non-centrosymmetric MOFs based on several different network topologies, namely, 3D diamondoid networks (DNs), 2D polymeric grids, octupolar frameworks, and 1D polymeric and related helical structures.

### 6.3.1 3D Diamondoid Networks

The most obvious choice of 3D network topology for crystal engineering of non-centrosymmetric solids is the DN. A DN can be readily obtained by simply linking a 4-connected nodes with an exo-bidentate linear spacer (i.e., a linear bridging ligand),[8,9] and indeed, numerous MOFs with the diamondoid topology have been synthesized over the past two decades. DNs are not predisposed to pack in centrosymmetric space groups owing to the lack of inversion centers on the 4-connected nodes (connecting points). Many DNs are centrosymmetric because of the presence of inversion centers on the symmetrical bridging ligands. Lin et al. first hypothesized that a non-centrosymmetric DN will arise if unsymmetrical bridging ligands are used to connect tetrahedral metal centers, as a result of the absence of an inversion center on both the 4- and 2-connected nodes.[10]

Because typical bridging ligands are long with a large aspect ratio, MOFs with the diamondoid topology have a high propensity to interpenetrate in order to fill the large void space. Despite the inherent non-centrosymmetric nature of individual DNs, an even number-fold interpenetration could potentially lead to inversion centers relating

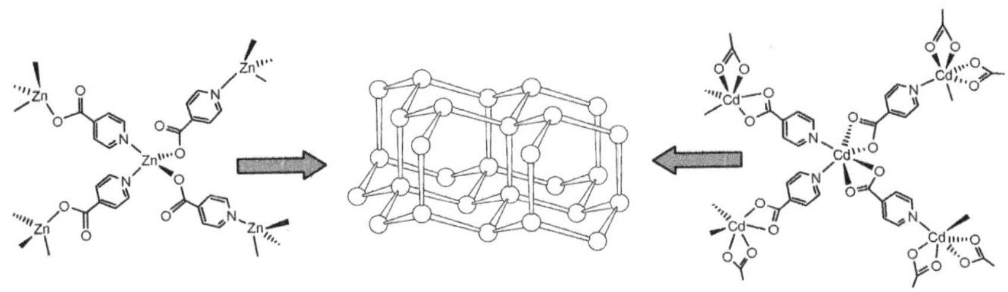

**SCHEME 6.1** Linking of *p*-pyridinecarboxylate ligands by $Zn^{2+}$ and $Cd^{2+}$ ions to lead to 3D MOFs with the diamondoid topology.

pairs of mutually independent diamondoid nets. On the other hand, an odd number-fold interpenetrated DN synthesized from unsymmetrical bridging ligands will be necessarily non-centrosymmetric.

Lin et al. showed that unsymmetrical rigid linear *p*-pyridinecarboxylate bridging ligands (spacers) could be used to link tetrahedral (or *pseudo*-tetrahedral) metal centers (nodes) to form non-centrosymmetric DNs. In order to synthesize optically transparent MOFs, Lin et al. focused on the $d^{10}$ metal ions $Zn^{2+}$ and $Cd^{2+}$ as connecting points to avoid unwanted optical losses from d $\rightarrow$ d transitions in the visible region. When coordinated to both pyridyl and carboxylate groups in monodentate fashion of the *p*-pyridinecarboxylate ligands, the tetrahedral $Zn^{2+}$ ions will link with each other to form 3D MOFs with a high propensity of adopting a DN structure (Scheme 6.1). The six-coordinating $Cd^{2+}$ centers shown in Scheme 6.1 will have a *pseudo*-tetrahedral extension and also have the tendency to lead to the formation of a DN.

Rigid conjugate bridging ligands were chosen in order to minimize potential packing complexity due to the ligand conformational flexibility. The use of an unsymmetrical linking group such as 4-pyridylcarboxylate also introduces electronic asymmetry (with the necessary push–pull effect), while the rigidity of the bridging ligands is synergistic with good conjugation between the electron donors and acceptors. Both of these factors are key to strong second-order NLO effects in the MOFs with the diamondoid topology.

As shown in Scheme 6.2, a family of Zn and Cd pyridinecarboxylates was synthesized under solvothermal conditions.[10–12] The elevated temperatures used for solvothermal syntheses allowed the formation of pyridinecarboxylate ligands from a wide variety of cyano, ester, and carboxylic acid precursors (Scheme 6.2, $\mathbf{L_1}$–$\mathbf{L_5}$). Bis(isonicotinato)zinc, **1**, was for example synthesized from a solvothermal reaction between $Zn(ClO_4)_2{\cdot}6H_2O$ and 4-cyanopyridine ($\mathbf{L_1}$) at 130°C. As shown in Figure 6.1, compound **1** adopts a threefold interpenetrated DN structure.[10] Each $Zn^{II}$ center in **1** is four coordinating by binding to two pyridine groups and two monodentate carboxylate groups of four different isonicotinate ligands. The $Zn^{II}$ centers thus have a tetrahedral geometry and extend three-dimensionally to form a DN (Figure 6.1a). The large Zn–Zn separations of 8.74 and 8.81 Å in **1** lead to a large void which is filled via the formation of a threefold interpenetrated DN structure (Figure 6.1b). As a result of

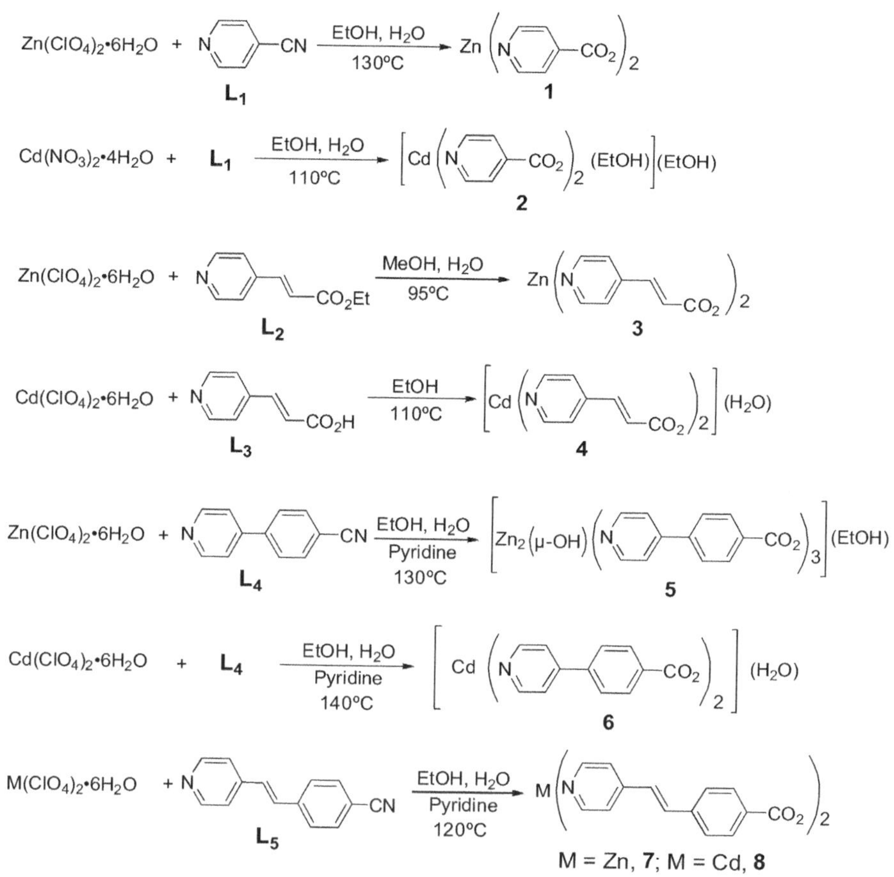


**SCHEME 6.2** Synthesis of diamondoid networks based on the *p*-pyridinecarboxylate bridging ligands.

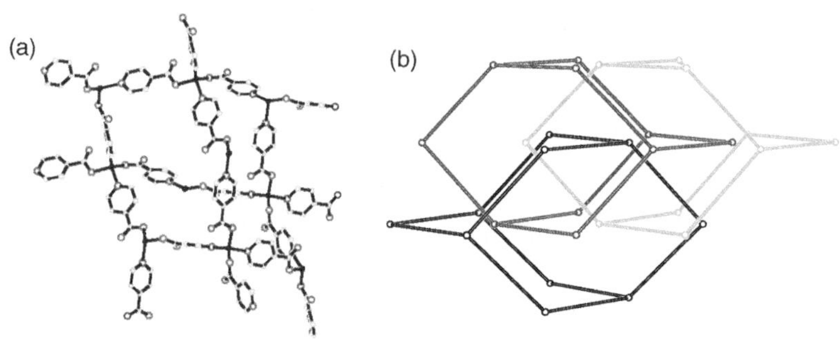


**FIGURE 6.1** (a) A view of the diamondoid network of **1**. (b) Diagram showing the threefold interpenetration in **1**.

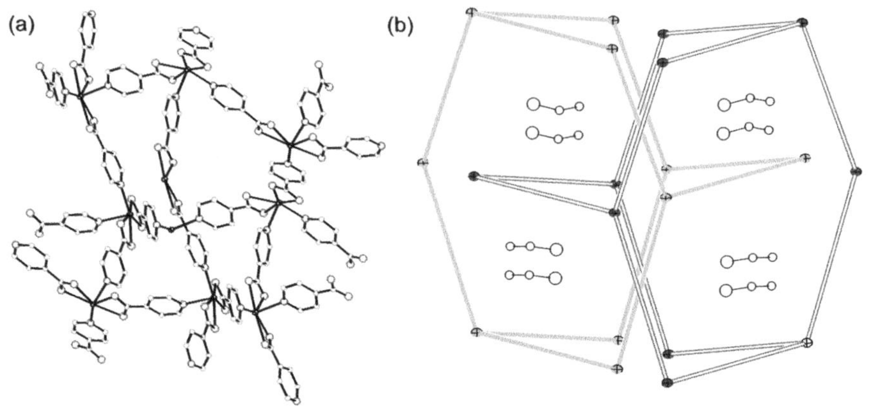


**FIGURE 6.2** (a) A view of the diamondoid network of **2**. The coordinating ethanol molecules were omitted for clarity. (b) Diagram showing the twofold interpenetration in **2** and the inclusion of ethanol molecules.

the unsymmetrical nature of isonicotinate groups and threefold interpenetration, **1** crystallizes in the chiral space group $P2_12_12_1$.

When $Cd^{2+}$ ions were used as the metal connecting points, a twofold interpenetrated DN [Cd(isonicotinato)$_2$(EtOH)][EtOH], **2**, was obtained under similar reaction conditions.[12] **2** adopts a twofold interpenetrated DN, presumably owing to the tendency of $Cd^{2+}$ ions to adopt a higher coordination number than $Zn^{2+}$ ions (Figure 6.2a). The coordination of an ethanol molecule prohibits the formation a threefold interpenetrated DN. The void space remaining after twofold interpenetration is effectively filled by included ethanol guest molecules (Figure 6.2b). **2** contains a center of inversion relating the two independent networks, and crystallizes in the centrosymmetric space group *Pbca*.

As the *p*-pyridinecarboxylate bridging ligands become longer, the void space within a single DN becomes larger. As a result, more interpenetrating nets are needed in order to effectively fill the space. For example, Zn(4-pyridylacrylate)$_2$, **3**, and [Cd(4-pyridylacrylate)$_2$]($H_2O$), **4**, both adopt a fivefold interpenetrated DN and crystallize in non-centrosymmetric space group *Cc* (Figure 6.3).[11] The longer 4-(4-pyridyl)benzoate and 4-[2-(4-pyridyl)ethenyl]benzoate ligands afforded Zn and Cd *p*-pyridinecarboxylate DNs with seven- and eightfold interpenetration, respectively (Scheme 6.2 and Figure 6.4).[13] It is interesting to note that both **7** and **8** crystallize in the chiral space group *C*2 in spite of an eightfold interpenetration. These results show that an even number-fold interpenetration can, but may not lead to the formation of centrosymmetric solids. The same network topologies observed for **3/4** and **7/8** pairs indicate that the formation of diamondoid solids depends mostly on the nature of the linking group, but is not sensitive to the local metal coordination environments.

Compound **5** does not follow the same trend as the rest of the compounds shown in Scheme 6.2 due to the formation of hydroxo-bridged dizinc secondary building units (SBUs) (Figure 6.5).[11] The dizinc SBU in **5** adopts a *pseudo*-tetrahedral arrangement

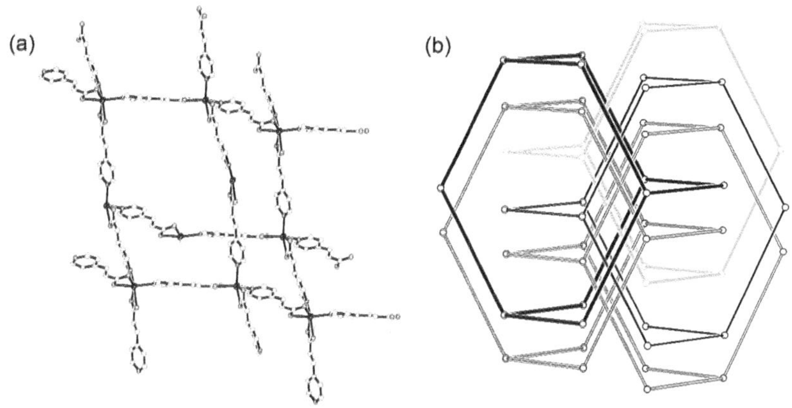


**FIGURE 6.3** (a) A view of the diamondoid network of **3**. (b) Diagram showing the fivefold interpenetration in **3** and **4**.

of 4-(4-pyridyl)benzoate linking groups. Each dizinc SBU is connected to four adjacent dizinc SBUs to form a DN via two distinct linking groups. Two of the four linking groups are the expected unsymmetrical 4-(4-pyridylbenzoate) bridging ligands, while the other two are symmetrical bis[4-(4-pyridyl)benzoate] double bridges which are related by an inversion center. As a result, the individual DNs in **5** are inherently centrosymmetric. **5** thus crystallizes in the centrosymmetric space group $P2_1/n$ in spite of fivefold interpenetration. The formation of symmetrical bis[4-(4-pyridyl)benzoate] bridges in **5** is akin to using symmetrical 4,4-bipyridine or 1,4-benzenedicarboxylate as bridging groups. This result emphasizes the importance of unsymmetrical linking groups in the synthesis of non-centrosymmetric MOFs.

It is clear from the above discussion that metal *p*-pyridinecarboxylate frameworks have a high propensity to form diamondoid solids. Moreover, the combination of unsymmetrical bridging ligands and metal centers with appropriate geometry ensures the formation of non-centrosymmetric solids. While an even number-fold interpenetration can (but may not) introduce unwanted inversion centers and lead to bulk

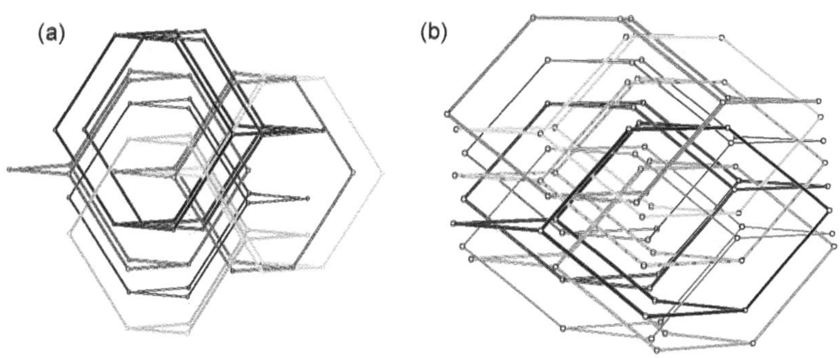


**FIGURE 6.4** (a) Diagram showing the sevenfold interpenetration in **6**. (b) Diagram showing the eightfold interpenetration in **7** and **8**.

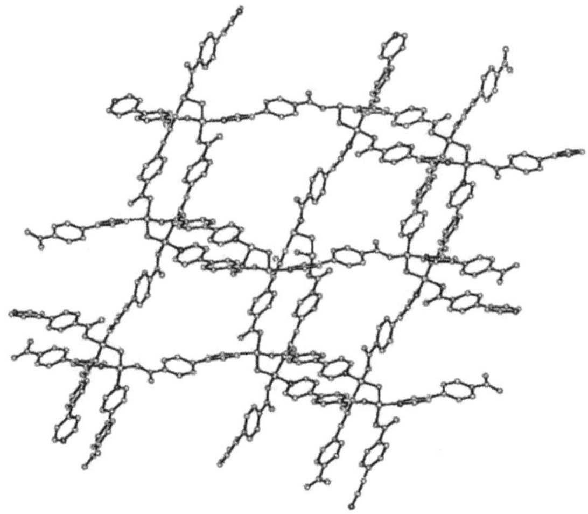

**FIGURE 6.5** Diamondoid structure of **5**. The black, light gray, dark gray, and gray circles represent Zn, O, N, and C atoms, respectively.

centrosymmetry, an odd number-fold DN with unsymmetrical bridging ligands will necessarily crystallize in a non-centrosymmetric space group. Furthermore, the degree of interpenetration is entirely dependent upon the length of the bridging ligand (Figure 6.6). DNs based on metal *p*-pyridinecarboxylate frameworks present the most robust topology for crystal engineering of non-centrosymmetric solids.

Non-centrosymmetric solids can also be rationally synthesized based on DNs built from other bridging ligands. Using a very similar *in situ* ligand formation strategy, Xiong et al. successfully synthesized several Zn tetrazole MOFs **9–11** with the diamondoid topology by solvothermal reactions of nitriles, sodium azide, and $Zn^{2+}$ salts (Scheme 6.3).[14] The *in situ* synthesized tetrazaole ligands bind to four-coordinate $Zn^{2+}$ centers via two nitrogen atoms as shown in Scheme 6.3. As a result of

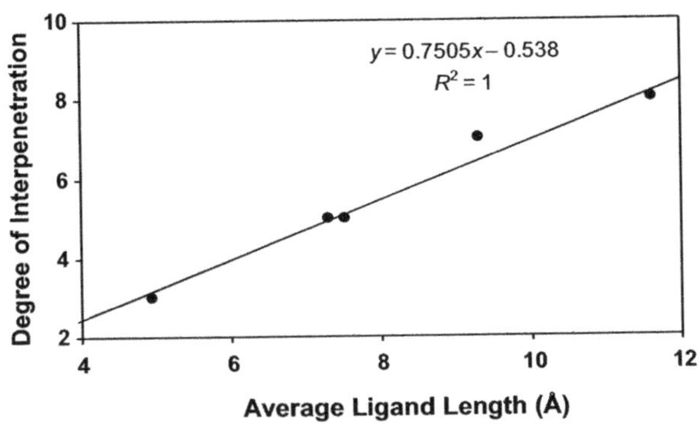


**FIGURE 6.6** Dependence of degree of interpenetration of diamondoid networks on the length of bridging *p*-pyridinecarboxylate ligands.

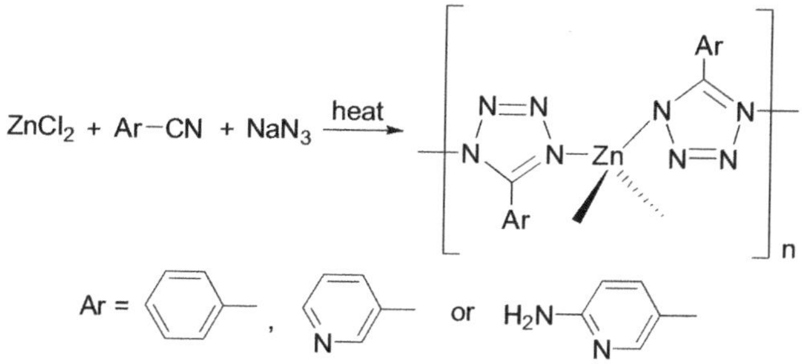


**SCHEME 6.3** Synthesis of Zn diamondoid networks 9–11 based on the *in situ* generated tetrazole bridging ligands.

the pseudo-tetrahedral extension of the $Zn^{2+}$ ions and linear nature of the tetrazole ligands, **9–11** adopt DN structures (Figure 6.7). Because of shorter Zn···Zn separations in these DNs and the presence of bulky aryl groups on the tetrazole ligands, the DNs in **9–11** are non-interpenetrated. Unlike the *p*-pyridinecarboxylate series, the length of the tetrazole ligands cannot be readily tuned, which limits the potential of the Zn-tetrazole MOFs.

The NLO properties of non-centrosymmetric DNs were evaluated using the Kurtz powder techniques.[15] Measurements were carried out on ground and sieved samples with a fundamental wavelength of 1064 nm. Consistent with their non-centrosymmetric structures, **1**, **3**, **4**, and **6–11** are SHG-active (Table 6.1). The powder SHG capabilities of **6–8** and **11** are very high and approach that of technologically important lithium niobate. The increased ligand length results in enhanced SHG intensities. This correlation between SHG intensities and ligand lengths suggests that the electronic asymmetry in the *p*-pyridinecarboxylate ligands constitutes the molecular origin of second-order nonlinearities of these solids. These results show that efficient NLO materials can be synthesized when all the constituents are efficiently aligned to function cooperatively in the solids, and point to the great potential of crystal

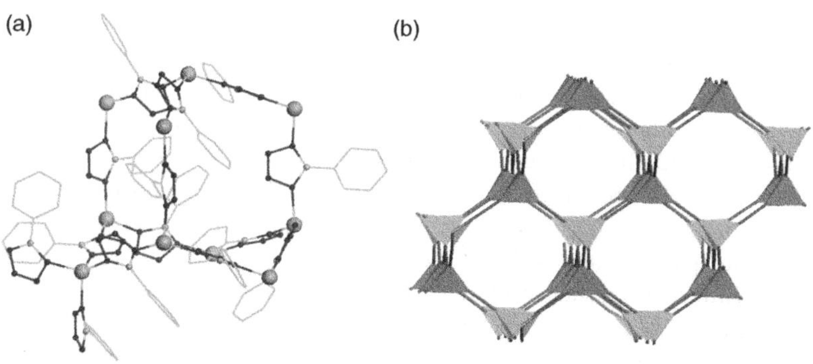


**FIGURE 6.7** (a) A view of the diamondoid network of **9**. (b) Diagram showing the diamondoid network of **9** with tetrahedral representing the Zn coordination environments.

**TABLE 6.1 Second-Order NLO Properties of Non-Centrosymmetric Diamondoid Networks**

| Compound | Degree of Interpenetration | Space Group | $I^{2\omega}/I^{2\omega}$(quartz) |
|---|---|---|---|
| **1** | Threefold | $P2_12_12_1$ | 1.5 |
| **3** | Fivefold | *Cc* | 126 |
| **4** | Fivefold | *Cc* | 18 |
| **6** | Sevenfold | *Ia* | 310 |
| **7** | Eightfold | *C*2 | 400 |
| **8** | Eightfold | *C*2 | 345 |
| **9** | None | *I*-42*d* | 100 |
| **10** | None | *I*-42*d* | 40 |
| **11** | None | *Fdd*2 | 500 |
| **$LiNbO_3$** | N/A | *R*3*c* | 600 |

engineering of non-centrosymmetric solids based on 3D diamondoid coordination networks for second-order NLO applications.

### 6.3.2 2D Grid Structures

Lin et al. also demonstrated the synthesis of non-centrosymmetric solids based on 2D MOFs with grid structures.[16] In contrast to the linear *p*-pyridinecarboxylate bridging ligands which have a high propensity to link tetrahedral (or *pseudo*-tetrahedral) metal centers to form 3D DNs, the bent configuration of an *m*-pyridinecarboxylate ligand can allow the metal centers to adopt infinite 2D grid structures. The metal centers in either a *cis*-octahedral (with chelating carboxylate groups) or a tetrahedral (with monodentate carboxylate groups) environment can have at most $C_{2v}$ symmetry and thus cannot possess an inversion center. Non-centrosymmetric 2D grids can result when these metal centers are linked with unsymmetrical bifunctional bridging ligands. Under favorable conditions, these 2D grids can pack into a non-centrosymmetric solid. Unsymmetrical *m*-pyridinecarboxylates also possess intrinsic electronic asymmetry that is essential for desired second-order NLO properties.

As shown in Scheme 6.4, a series of 2D MOFs were synthesized from solvothermal reactions between *m*-pyridinecarboxylate precursors **$L_6$–$L_9$** and transparent $d^{10}$ metal centers.[16,17] The Zn centers in Zn(nicotinate)$_2$, **12**, adopt a distorted octahedral geometry by coordinating to two carboxylate groups and two pyridyl nitrogen atoms of four different nicotinate groups in a *cis*-configuration. The Zn centers thus exhibit the *C*2 symmetry and are linked by the bent nicotinate ligands to form a chiral infinite grid structure. Individual grids stack via interdigitation of 3-pyridyl groups to form a chiral solid in space group $P4_32_12$ (Figure 6.8).

As *m*-pyridinecarboxylate bridging ligands grow longer, the open space within each 2D grid will become much larger.[17] Individual grids can then interweave in order to reduce the void space. Such a behavior is clearly illustrated by the structure of Cd{3-[2-(4-pyridyl)ethenyl]benzoate}$_2$, **13**, which adopts an infinite 2D grid structure (Figure 6.9). The very void space in **13** that is created by the long bridging ligand is

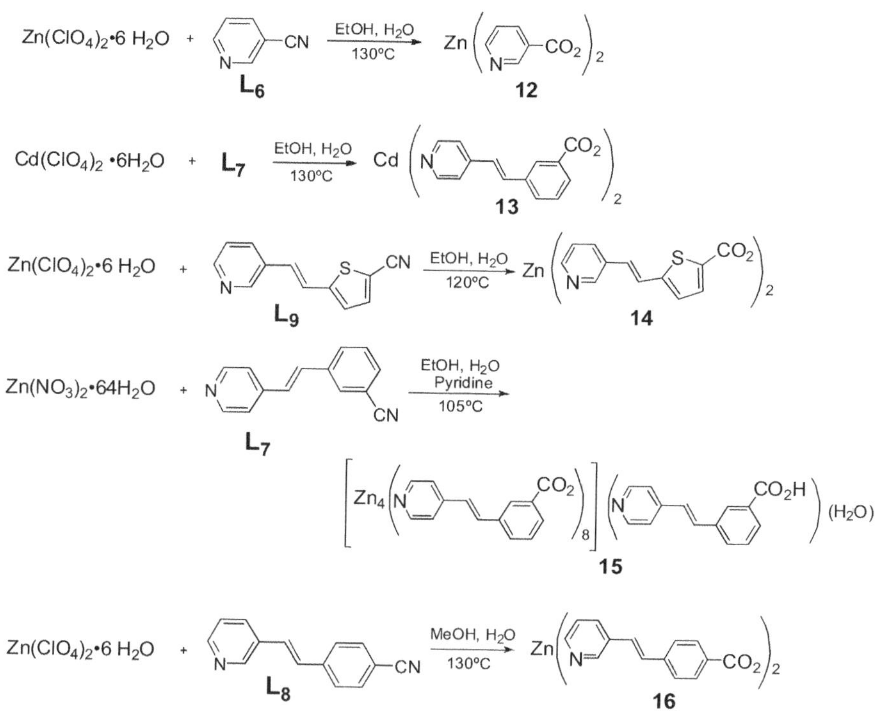


**SCHEME 6.4** Synthesis of 2D MOFs based on $d^{10}$ metal connecting points and bent *m*-pyridinecarboxylate bridging ligands.

effectively reduced by threefold interweaving. **13** crystallizes in the non-centrosymmetric *Fdd2* space group because of an odd number-fold interweaving. On the other hand, an even number-fold interweaving can bring an inversion center to lead to a centrosymmetric solid.

The deleterious consequence of an even number-fold interweaving is illustrated by the synthesis of Zn{5-[2-(3-pyridyl)ethenyl]thiophene-2-carboxylate}$_2$, **14**.[17] The Zn center coordinates to two pyridyl nitrogen atoms and two carboxylate oxygen atoms of

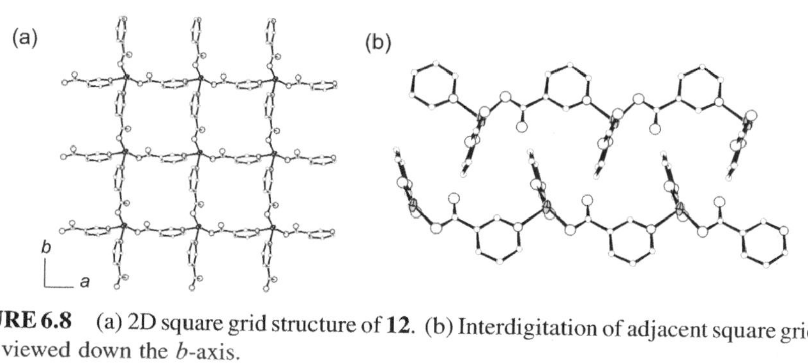


**FIGURE 6.8** (a) 2D square grid structure of **12**. (b) Interdigitation of adjacent square grids in **12** as viewed down the *b*-axis.

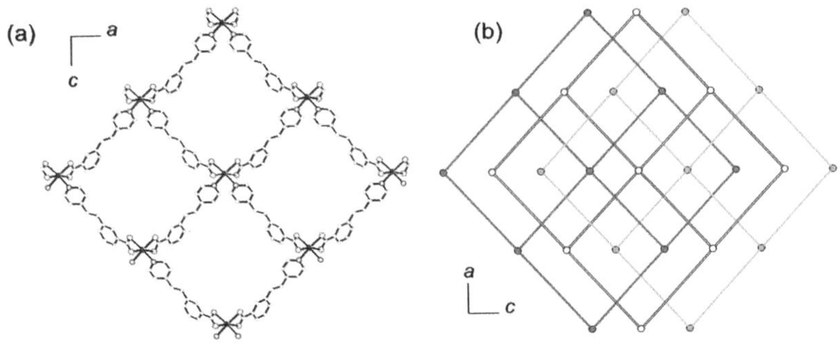


**FIGURE 6.9** (a) A view of the 2D rhombohedral grid structure of **13** showing the presence of large void space. (b) A schematic showing the interweaving of three independent rhombohedral grids in the *ac* plane in **13**.

four different 3-[2-(3-pyridyl)ethenyl]-thiophenecarboxylate ligands to result in an infinite 2D rhombohedral grid with a large Zn–Zn separation of 13.33 Å (Figure 6.10). There is thus a large void space inside each rhombohedral grid in **14**, resulting in a twofold interweaved structure. The two interweaved rhombohedral grids are structurally identical and are related to each other through an inversion center. As a result, **14** crystallizes in the centrosymmetric space group *Pcca*.

An even number-fold interweaving can but does not necessarily lead to a centrosymmetric solid. $Zn_4$\{3-[2-(4-pyridyl)ethenyl]benzoato\}$_8$·\{3-[2-(4-pyridyl)ethenyl]-benzoic acid\}·($H_2O$), **15**, is built from four independent 2D grids that stack in an *ABC*-type sequence (Figure 6.11).[17] Two of the grids exhibit twofold interweaving (*A* sequence) and the other two grids interdigitate through π–π stacking (*B* and *C* sequences). The void space in **15** is thus filled through twofold interweaving and inclusion of a free 3-[2-(4-pyridyl)ethenyl]benzoic acid and water guest molecules. Despite the presence of twofold interweaving and inclusion of guest molecules, **15** crystallizes in the non-centrosymmetric space group *Cc*.

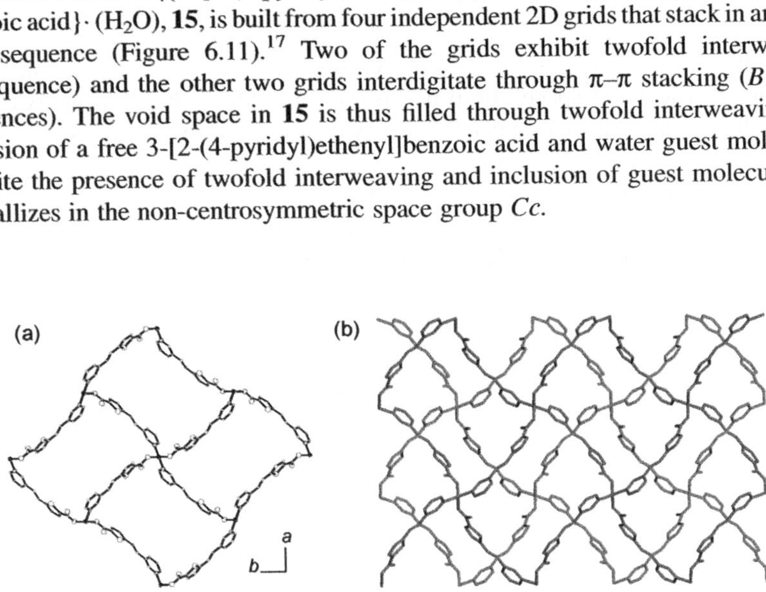


**FIGURE 6.10** (a) 2D square grid structure of **14**. (b) Twofold interweaving of the 2D grids in **14**. The two networks are related by an inversion center.

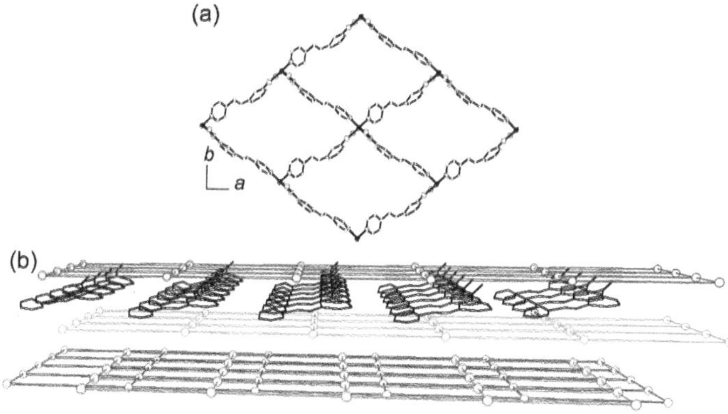


**FIGURE 6.11** (a) A 2D rhombohedral grid in **15**. (b) Stacking of four independent 2D rhombohedral grids along the *c*-axis.

In addition to the potential introduction of centrosymmerty via an even number-fold interweaving, an inversion center can be introduced when the 2D networks stack along the third dimension to form a bulk solid. This is because that stacking along the third dimension is still dictated by weaker non-covalent interactions which can force adjacent layers of 2D grids to be related via inversion centers. It is thus necessary to uncover novel packing arrangements that exhibit better interlayer registry and inhibit the occurrence of centrosymmetry.

The pleated sheet topology provides an interesting solution to such a problem. In Zn {4-[2-(3-pyridyl)ethenyl]benzoato}$_2$·($H_2O$), **16**, the Zn centers coordinating to two carboxylate groups in a monodentate fashion and to two pyridyl nitrogen atoms to lead to a highly distorted rhombohedral 2D grid structure.[17] Adjacent rows of Zn centers within each 2D grid are offset by 2.93 Å along the *c*-axis to result in a pleated topology (Figure 6.12). The puckered nature of 2D grids renders better interlayer registry, and adjacent 2D grids are related only by translational symmetry. The Zn centers in **16** thus have the same chirality to lead to a chiral solid that crystallizes in space group $P2_12_12$. The pleated topology of **16** provides an effective means for controlling the stacking of 2D grids, and represents an interesting structural motif for the synthesis of non-centrosymmetric 2D MOFs.

Non-centrosymmetric 2D MOFs based on grid structures can also be prepared using other bridging ligands. For example, Xiong et al. prepared a Zn-tetrazole 2D MOF **17** by a generating the tetrazole bridging ligand *in situ* under solvothermal conditions.[14] As shown in Figure 6.13, the Zn centers adopt distorted tetrahedral coordination and are linked via the pyridyl nitrogen atom and one of the nitrogen atoms of the tetrazole groups to form a 2D network that crystallizes in the non-centrosymmetric space group *Fdd*2.

It is evidently more challenging to engineer non-centrosymmetric solids based on 2D grids than 3D DNs. Not only the degree of interweaving but also the stacking of 2D grids along the third dimension must be carefully controlled to ensure non-centro-symmetry of the bulk solid. Compounds **12**, **13**, and **15–17** are all SHG-active, in

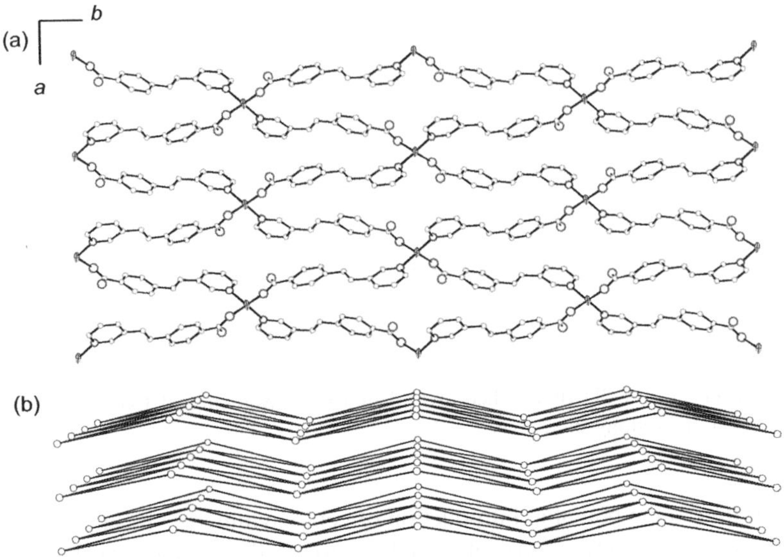


**FIGURE 6.12** (a) 2D rhombohedral grid of **16**. (b) Stacking of adjacent 2D pleated sheets in **16**.

agreement with their non-centrosymmetric structures (Table 6.2). Compounds **13** and **17** have powder SHG efficiency higher than that of technologically important $LiNbO_3$. All these 2D coordination networks are thermally robust and optically transparent, and they are therefore attractive candidates for applications in second-order nonlinear optics.

### 6.3.3 Octupolar MOFs

Octupolar chromophores refer to donor-acceptor-substituted molecules with threefold rotational symmetry and have been shown to have an improved transparency/optical

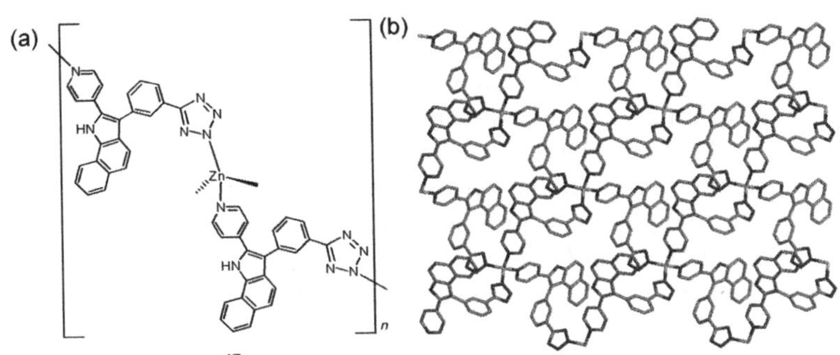


17

**FIGURE 6.13** (a) Chemical structure of **17**. (b) A view of the 2D grid structure of **17**.

**TABLE 6.2 Second-Order NLO Properties of Non-Centrosymmetric 2D Networks**

| Compound | Space Group | $I^{2\omega}/I^{2\omega}$(quartz) |
|---|---|---|
| **12** | $P4_32_12$ | 2 |
| **13** | $Fdd2$ | 800 |
| **15** | $Cc$ | 400 |
| **16** | $P4_32_12$ | Sample decomp. |
| **17** | $Fdd2$ | 5000 |
| **$LiNbO_3$** | $R3c$ | 600 |

nonlinearity trade-off as compared to traditional dipolar chromophores as a result of the presence of four significant components of molecular hyperpolarizability (β).[18] Lin et al. synthesized the first octupolar MOF with SHG activity.[19] A hydro(solvo) thermal reaction between $Cd(ClO_4)_2 \cdot 6H_2O$ and ethyl 4-[2-(4-pyridyl)ethenyl]benzoate under basic conditions resulted in a novel octupolar 2D MOF $[Cd_3(\mu_3\text{-}OH)$(pyridine)$_6$(4-[2-(4-pyridyl)ethenyl]benzoate)$_3](ClO_4)_2$, **18**, which adopts an infinite 2D coordination network based on an unprecedented "basic" tricadmium carboxylate core (Figure 6.14). The tricadmium carboxylate clusters are connected by unsymmetrical pyrdinecarboxylate linking groups to result in a chiral infinite 2D layer. The 2D layers in **18** stack along the *c*-axis in an interdigitated fashion to result in a chiral bulk solid (space group *R*32). Kurtz powder SHG measurements indicated that **18** has a powder SHG intensity of 130 versus α-quartz, consistent with its chiral nature.

Cui et al. recently synthesized octupolar 3D anionic MOFs that showed very high SHG activity.[20] As shown in Figure 6.15, complex $[(H_2NMe_2)_2Cd_3(C_2O_4)_4] \cdot MeOH \cdot 2H_2O$, **19**, crystallized in the non-centrosymmetric space group *I*-43*d*. There are two polyhedral cages, $3^55^3$ and $3^45^4$, in the open framework. The framework shows high ion-exchange capacities for cations $NH_4^+$, $Na^+$, and $K^+$. Interestingly, the SHG property of **19** is strongly influenced by nature of the cations.

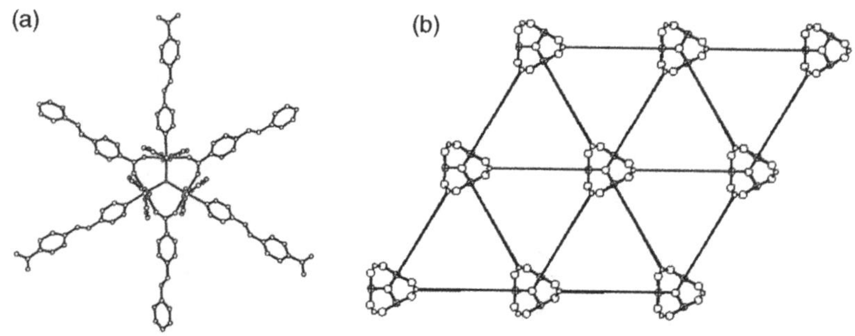


**FIGURE 6.14** (a) A view of the octupolar $[Cd_3(\mu_3\text{-}OH)$(4-[2-(4-pyridyl)ethenyl]benzoate)$_6$(py)$_6]^{2+}$ building block in **18**. (b) A schematic showing the 2D sheet in **18** down the *c*-axis.

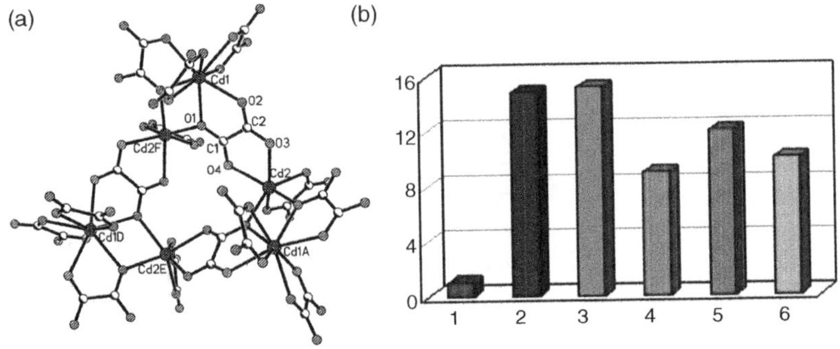


**FIGURE 6.15** (a) A view of the octupolar 3D anionic open framework of **19**. (b) A comparison of the SHG intensities of KDP, **19**, the $NH_4^+$, $Na^+$, $K^+$ exchanged samples of **19**, and **18** (from left to right; the intensity of KDP is set as 1).

### 6.3.4 1D Coordination Polymers and Related Helical Structures

Although it is possible to design non-centrosymmetric 1D chains, the lack of control in two other dimensions makes this approach much more challenging.[21] Teo et al. first demonstrated the synthesis of 1D NLO-active materials based on polymeric cadmium chalcogenocyanates (Scheme 6.5).[22,23] Anionic $[Cd(XCN)_3^-]_\infty$ (X = S or Se) chains can adopt non-centrosymmetric structures by taking advantage of strong *trans*-effects to force the coordinating chalcogen atoms to lie in positions *trans* to the coordinating nitrogen atoms (Scheme 6.5a). The packing of the individual non-centrosymmetric $[Cd(XCN)_3^-]_\infty$ chains however strongly depends upon the symmetry of the cations. For instance, the cation in compound **21** is dimeric and contains an inversion center. As a result, **21** crystallizes in the centrosymmetric space group $P2_1/n$. On the other hand, the monomeric cation [18-crown-6]$K^+$ in **20** lacks an inversion center, leading to the formation of a non-centrosymmetric solid. The same cation dependence was observed in compounds **22–25**. The tetraalkylammonium cations in **22–25** do not possess inversion centers and lead to non-centrosymmetric solids. Compounds **20**, **22**, and **23** have SHG efficiencies 15–30 times of α-quartz.

(a)

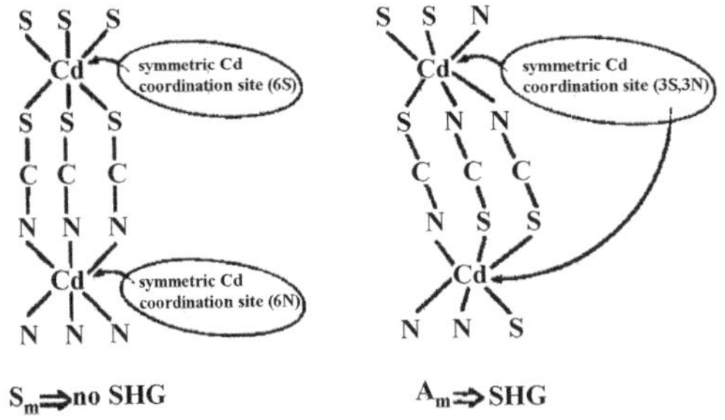


$S_m$ ⇒ no SHG    $A_m$ ⇒ SHG

(b)

$[(18\text{-crown-6})K][Cd(SCN)_3]$
**20**

$[(18\text{-crown-6})_2Na_2(H_2O)_2]_{0.5}[Cd(SCN)_3]$
**21**

$[R_4N][Cd(XCN)_3]$

R=Et, X=S (**22**), Se (**23**);
R=Me, X=S (**24**), Se (**25**)

**SCHEME 6.5** (a) Schematic representation of centrosymmetric and non-centrosymmetric $[Cd(SCN)_3^-]_\infty$ chains. (b) Chemical formulae of compounds **20–25**.

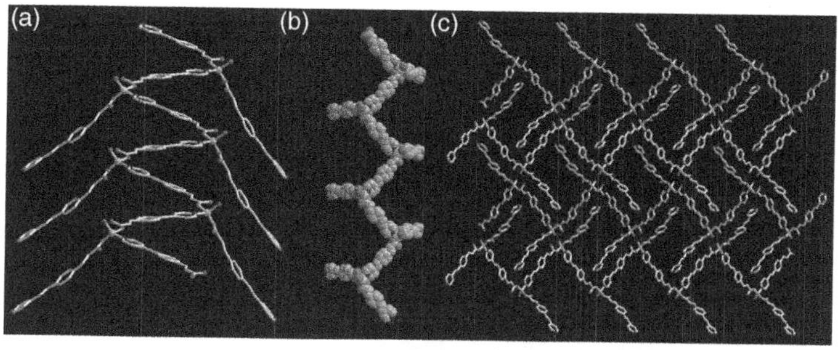


**FIGURE 6.16** (a) A view of caterpillar-shaped 1D chain in **26**. (b) A space-filling model of the 1D chain. (c) A view of **26** down the *c*-axis. Interdigitation via edge-to-face π–π interactions is clearly visible.

However, **24** and **25** display negligible SHG efficiencies because adjacent $[Cd(XCN)_3^-]_\infty$ chains in **24** and **25** adopt an anti-parallel arrangement to result in effective cancellation of dipole moments of individual chains which in turn nullifies any SHG effect.

Lin et al. synthesized a novel non-centrosymmetric 1D coordination network with an interlocked structure Zn(4-[2-(4-pyridyl)vinyl]cinnamate)$_2$($H_2O$), **26**.[24] Each Zn center is connected to two neighboring Zn centers through the bridging ligands to form a 1D zig-zag chain. The 4-[2-4-pyridyl)vinyl]cinnamate bridging ligands with non-coordinating pyridyl groups are dangling from the zig-zag chain to give a caterpillar-shaped structure (Figure 6.16). These caterpillar-shaped 1D chains are running along mutually perpendicular [1/4 0 7/4] and [−1/4 0 7/4] directions. The parallel-running chains are held together by strong face-to-face π–π stacking interactions. These 1D chains further form a highly interdigitated structure via edge-to-face π–π stacking interactions of the dangling bridging ligands (Figure 6.16c). These interactions self-assemble the caterpillar-shaped 1D chains into an open framework structure with open channels of ~4.6 × 11.8 Å that are running along the *c*-axis and occupied by included acetonitrile and water molecules. As shown in Figure 6.16c, all the dipoles have cancelled each other along both the *a* and *b* axes, and the polar axis of **26** thus lies along the *c*-axis. The non-centrosymmetric structure of **26** is a direct consequence of extensive π–π interactions. The face-to-face π–π interactions ensure the polarity of all the parallel 1D chains along both [1/4 0 7/4] and [−1/4 0 7/4] directions, while the edge-to-face π–π interactions between the dangling bridging ligands define the polarity of **26** along the *c*-axis. **26** exhibits a powder SHG intensity of 75 versus α-quartz.

Poeppelmeier et al. has synthesized non-centrosymmetric solids based on transition metal oxyfluoride anions $[MO_2F_{6-x}]^{2-}$ (M = V, Nb, Ta, $x = 1$; M = Mo, W, $x = 2$).[25,26] Just like the niobate anion in $LiNbO_3$, $[MO_2F_{6-x}]^{2-}$ are intrinsically non-centrosymmetric due to a distortion of the metal from the center of the octahedron. The $[MO_2F_{6-x}]^{2-}$ anions however often crystallize in centrosymmetric space groups owing to orientational disorder of the oxyfluoride anions. Co-crystallization with two different cations has led to the ordering of the

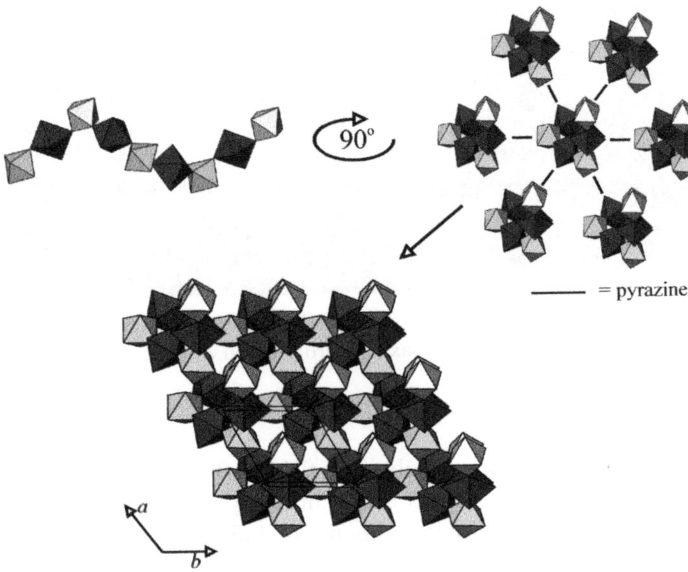


**FIGURE 6.17** Polyhedral presentation of a helical chain of **27** and the linking of neighboring helices to form a chiral solid.

$[MO_2F_{6-x}]^{2-}$ anions, which effectively reduces the local symmetry about the $[MO_2F_{6-x}]^{2-}$ anion and the number of possible orientations, thereby promoting the formation of non-centrosymmetric 1D chains. As discussed above, packing these anionic 1D chains into a non-centrosymmetric solid still remains a challenge. Poeppelmeier et al. addressed this challenge by synthesizing a helical metal-oxyfluoride Zn(pyrazine)$(H_2O)_2MoO_2F_4$, **27** (Figure 6.17). Widely found in nature, helical structural motifs exhibit axial chirality, and solids based on helical structural motifs can crystallize in non-centrosymmetric space groups. The $[MoO_2F_4]^{2-}$ anions in **27** are orientationally ordered and coordinated *cis* to $[Zn(pyrazine)_2(H_2O)_2]^{2+}$ cations in an alternating fashion to result in $3_1$ helical chains running along the *c*-axis. All the helices in **27** have the same handedness and are linked by pyrazine bridges to result in a chiral 3D solid.

The helical chains have also been used by Lin et al. to form non-centrosymmetric solids. In Zn{Z-[4-2-(3-pyridyl)ethenyl]benzoate}$_2$·0.5$H_2O$, **28**, the skewed conformation of the bridging ligand induces the formation of sixfold helices.[27] Just as in **27**, neighboring helices of **28** have the same handedness and are covalently bridged by pyridylethenylbenzoate ligands to result in a chiral 3D MOF (Figure 6.18). Owing to non-conjugate nature of the Z-conformation of the bridging ligand, **28** exhibits a modest powder SHG efficiency of six times of α-quartz. Although **27** and **28** are 3D MOFs, they are both built from helical chains. Non-centrosymmetric nature of **27** and **28** reinforces the recurring theme of this chapter: the probability of obtaining a non-centrosymmetric solid significantly increases when it adopts a 3D framework structure.

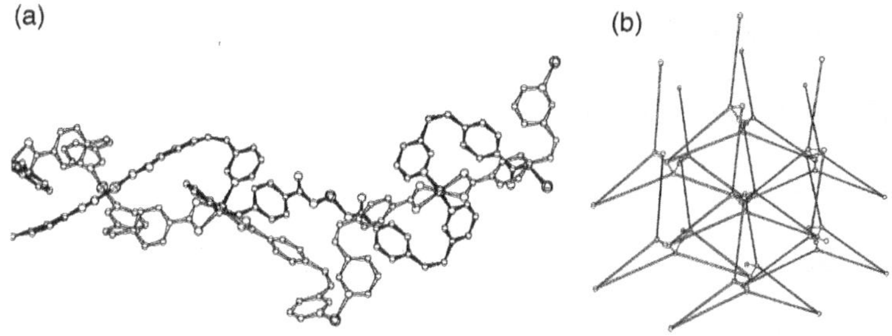


**FIGURE 6.18** (a) A view of the cross-link between a primary and a secondary helix in **28**. (b) A perspective view of **28** slightly away from the *c* axis showing the 3D network resulting from the crosslinking of primary and secondary helices. Only Zn atoms are shown.

## 6.4 CONCLUSIONS AND OUTLOOK

We have surveyed the successful strategies toward a new class of second-order NLO materials based on MOFs in this chapter. Crystal engineering of non-centrosymmetric solids becomes a reality by exploiting the strong and highly directional nature of metal-ligand coordination bonds. Dominant metal-ligand coordination can be used to direct the assembly of MOFs with desired topologies. 3D DNs provide the most robust topology for the crystal engineering of non-centrosymmetric solids and reduce such a daunting task to a simple choice of bridging ligands with appropriate length. The synthesis of non-centrosymmetric coordination networks with lower dimensionality is less predictable. For example, centrosymmetry in 2D grids can arise from an even number-fold interweaving as well as from the stacking of non-centrosymmetric sheets through centrosymmetric symmetry operations. Well-chosen metal centers and bridging ligands can lead to pleated sheets with an odd number-fold interweaving which will necessarily be non-centrosymmetric. The exploitation of octupolar MOFs for NLO application was also illustrated. Octupolar MOFs can have an improved transparency/optical nonlinearity trade-off as a result of the presence of four significant components of molecular hyperpolarizability. Finally, the synthesis of non-centrosymmetric solids based on 1D and related helical networks was also explored, albeit with much less predictability. Non-centrosymmetric MOFs can have a favorable combination of thermal and chemical stability, solubility characteristics, transparency, and second-order optical nonlinearity, and are thus potential candidates for applications in electro-optic devices.

The research outlined in this chapter serves as one of nice examples in which functional solid-state materials can be rationally designed. Although non-centrosymmetric and polar MOFs can also be explored for ferroelectric and piezoelectric applications, inferior mechanical properties of MOFs present a serious disadvantage in these applications. Further work is needed to extend successful crystal engineering strategies to the synthesis of new materials for other applications such as heterogeneous asymmetric catalysis, gas storage, and molecular separations.[3,28] A new generation of functional materials can be expected via modular synthesis of MOFs.

## ACKNOWLEDGMENTS

We thank National Science Foundation for supporting the research program in the Lin group.

## REFERENCES

1. Marks, T. J.; Ratner, M. A. Design, synthesis, and properties of molecule-based assemblies with large 2nd-order optical nonlinearities. *Angew. Chem. Int. Ed. Engl.* **1995**, *34*(2), 155–173.
2. Lin, W. Metal-organic frameworks for asymmetric catalysis and chiral separations. *MRS Bull.* **2007**, *32*(7), 544–548.
3. Lin, W. Homochiral porous metal-organic frameworks: why and how? *J Solid State Chem.* **2005**, *178*(8), 2486–2490.
4. Evans, O. R.; Lin, W. Crystal engineering of NLO materials based on metal-organic coordination networks. *Acc. Chem. Res.* **2002**, *35*(7), 511–522.
5. Yaghi, O. M.; O'Keeffe, M.; Ockwig, N. W.; Chae, H. K.; Eddaoudi, M.; Kim, J. Reticular synthesis and the design of new materials. *Nature* **2003**, *423*(6941), 705–714.
6. Ayyappan, P.; Evans, O. R.; Cui, Y.; Wheeler, K. A.; Lin, W. B. Nonlinear optically active polymeric coordination networks based on metal m-pyridylphosphonates. *Inorg. Chem.* **2002**, *41*(20),4978–4980.
7. Lebeau, B.; Brasselet, S.; Zyss, J.; Sanchez, C. Design, characterization, and processing of hybrid organic-inorganic coatings with very high second-order optical nonlinearities. *Chem. Mater.* **1997**, *9*(4), 1012–1020.
8. Hoskins, B. F.; Robson, R. Infinite polymeric frameworks consisting of 3 dimensionally linked rod-like segments. *J. Am. Chem. Soc.* **1989**, *111*(15), 5962–5964.
9. Hoskins, B. F.; Robson, R. Design and construction of a new class of scaffolding-like materials comprising infinite polymeric frameworks of 3-D-linked molecular rods—a reappraisal of the $Zn(Cn)_2$ and $Cd(Cn)_2$ structures and the synthesis and structure of the diamond-related frameworks [N(Ch3)4][Cuiznii(Cn)4] and Cui[4,4′,4″,4‴-tetracyanotetraphenylmethane]$Bf_4$. $XC_6H_5NO_2$. *J. Am. Chem. Soc.* **1990**, *112*(4), 1546–1554.
10. Evans, O. R.; Xiong, R. G.; Wang, Z. Y.; Wong, G. K.; Lin, W. B. Crystal engineering of acentric diamondoid metal-organic coordination networks. *Angew. Chem. Int. Ed.* **1999**, *38*(4), 536–538.
11. Evans, O. R.; Lin, W. B. Crystal engineering of nonlinear optical materials based on interpenetrated diamondoid coordination networks. *Chem. Mater.* **2001**, *13*(8), 2705–2712.
12. Evans, O. R.; Wang, Z. Y.; Xiong, R. G.; Foxman, B. M.; Lin, W. B. Nanoporous, interpenetrated metal-organic diamondoid networks. *Inorg. Chem.* **1999**, *38*(12), 2969–2973.
13. Lin, W. B.; Ma, L.; Evans, O. R. NLO-active zinc(II) and cadmium(II) coordination networks with 8-fold diamondoid structures. *Chem. Commun.* **2000**, (22), 2263–2264.
14. Zhao, H.; Qu, Z. R.; Ye, H. Y.; Xiong, R. G. In situ hydrothermal synthesis of tetrazole coordination polymers with interesting physical properties. *Chem. Soc. Rev.* **2008**, *37*, 84–100.

15. Kurtz, S. K.; Perry, T. T. A powder technique for evaluation of nonlinear optical materials. *J. Appl. Phys.* **1968**, *39*(8), 3798–3813.

16. Lin, W. B.; Evans, O. R.; Xiong, R. G.; Wang, Z. Y. Supramolecular engineering of chiral and acentric 2-D networks. Synthesis, structures, and second-order nonlinear optical properties of bis(nicotinato)zinc and bis{3-[2-(4-pyridyl)ethenyl]benzoato}cadmium. *J. Am. Chem. Soc.* **1998**, *120*(50), 13272–13273.

17. Evans, O. R.; Lin, W. B. Rational design of nonlinear optical materials based on 2-D coordination networks. *Chem. Mater.* **2001**, *13*(9), 3009–3017.

18. Di Bella, S.; Fragala, I.; Ledoux, I.; Zyss, J. Dipolar donor-acceptor-substituted Schiff base complexes with large off-diagonal second-order nonlinear optical tensor components. *Chem. Eur. J.* **2001**, *7*(17), 3738–3743.

19. Lin, W. B.; Wang, Z. Y.; Ma, L. A novel octupolar metal-organic NLO material based on a chiral 2-D coordination network. *J. Am. Chem. Soc.* **1999**, *121*(48), 11249–11250.

20. Liu, Y.; Li, G.; Li, X.; Cui, Y. Cation-dependent nonlinear optical behavior in an octupolar 3D anionic metal-organic open framework. *Angew. Chem. Int. Ed. Engl.* **2007**, *46*(33), 6301–6304.

21. Chen, C. T.; Suslick, K. S. One-dimensional coordination polymers—applications to material science. *Coord. Chem. Rev.* **1993**, *128*(1–2), 293–322.

22. Zhang, H.; Wang, X. M.; Teo, B. K. Molecular design and crystal engineering of a new series of inorganic polymers separated by organic spacers: Structures of [(18$C_6$)K]-[Cd(SCN)(3)] and [(18$C_6$)(2)Na-2($H_2O$)(2)](1/2)[Cd(SCN)(3)]. *J. Am. Chem. Soc.* **1996**, *118*(47), 11813–11821.

23. Zhang, H.; Wang, X. M.; Zhang, K. C.; Teo, B. K. Molecular and crystal engineering of a new class of inorganic cadmium-thiocyanate polymers with host-guest complexes as organic spacers, controllers, and templates. *Coord. Chem. Rev.* **1999**, *183* 157–195.

24. Ayyappan, P.; Sirokman, G.; Evans, O. R.; Warren, T. H.; Lin, W. B. Non-linear optically active zinc and cadmium p-pyridinecarboxylate coordination networks. *Inorg. Chim. Acta* **2004**, *357*(13), 3999–4004.

25. Maggard, P. A.; Stern, C. L.; Poeppelmeier, K. R. Understanding the role of helical chains in the formation of noncentrosymmetric solids. *J. Am. Chem. Soc.* **2001**, *123*(31), 7742–7743.

26. Halasyamani, P. S.; Poeppelmeier, K. R. Noncentrosymmetric oxides. *Chem. Mater.* *10*(10),**1998**, 2753–2769.

27. Evans, O. R.; Wang, Z. Y.; Lin, W. B. An unprecedented 3D coordination network composed of two intersecting helices. *Chem. Commun.* **1999**, (18), 1903–1904.

28. Wu, C.-D.; Hu, A.; Zhang, L.; Lin, W. A homochiral porous metal-organic framework for highly enantioselective heterogeneous asymmetric catalysis. *J. Am. Chem. Soc.* **2005**, *127*(25), 8940–8941.

# 7

# SELECTIVE SORPTION OF GASES AND VAPORS IN METAL-ORGANIC FRAMEWORKS

HYUNUK KIM[1], HYUNGPHIL CHUN[2], AND KIMOON KIM[1]

[1]*National Creative Research Initiative Center for Smart Supramolecules, Department of Chemistry and Division of Advanced Materials Science, Pohang University of Science and Technology, Pohang 790-784, Republic of Korea*
[2]*Department of Applied Chemistry, College of Science and Technology, Hanyang University, 1271 Sadong, Ansan 426-791, Republic of Korea*

## 7.1 INTRODUCTION

In principle, no two gases can have identical sorption properties on a given solid surface. This is as evident as the differences in the boiling points of common gases. Besides, simple diatomic molecules are more or less similar in terms of the strength of non-bonded interactions with respect to light atoms of a solid surface. Therefore, any porous material would adsorb, for example, more nitrogen than hydrogen at 77 K or higher temperatures, and more carbon dioxide than methane at 196 K or higher temperatures. This suggests that, at suitably adjusted temperature and pressure, some degree of gas separation can be expected to occur when porous metal-organic frameworks (MOFs) are used as the sorption medium. There are, however, reports of highly selective sorption for light gases, and even an almost complete exclusion for most gases but one or two is observed. The ability of MOFs to selectively adsorb chemical vapors is even more impressive, which is probably due to the stronger interactions between adsorbates and adsorbents. These interesting phenomena should have their origins related to either physical or chemical environments of pores

*Metal-Organic Frameworks: Design and Application,* Edited by Leonard R. MacGillivray

(or maybe both contributing unequally). This article intends to have a close look at these matters, and wishes to figure out what the current status for academic research and future prospects for practical applications are. In analyzing crystal structures, it is often informative to visualize the Connolly surface[1] of frameworks because it clearly reveals the size and shape of accessible voids. In doing so, we set the probe radius to 1.4 Å in accordance with the kinetic diameter of hydrogen.

Currently, there are plenty of chances for porous MOFs to find applications in industry. The production of nitrogen and oxygen from air, ammonia synthesis, petrochemicals separations, and hydrogen purification during methane reforming are representative examples. However, it may be wise to think the physisorption in MOFs as a complement for existing distillation methods rather than a complete replacement of the current separation and purification technologies.[2] A few recent papers dealing with the practical aspects of gas separation will be mentioned in a later section.

## 7.2 SELECTIVE GAS SORPTION BY MOFs

### 7.2.1 Size-Selective Gas Sorption

***7.2.1.1 Preferential Sorption Due to Inherent Size Limitation of Pores*** The first case of selective sorption of a specific gas by MOFs has been reported for a lanthanide–organic framework, $[Er_2(pda)_3(OH_2)]$ (pda = 1,4-phenylenediacetate).[3] The dehydrated form $[Er_2(pda)_3]$ obtained by a thermal activation preferentially adsorbs a small amount of $CO_2$ (~24 mg/g at 1 bar and 273 K), while it shows non-porous behavior with respect to $N_2$ and Ar at 77 K. No significant change is noted in the positions of the major peaks in powder diffractions before and after the thermal activation process, meaning that the X-ray single-crystal structure can be safely used for the visualization of the pore environment responsible for the selective sorption. As shown in Figure 7.1, $[Er_2(pda)_3]$ has one-dimensional (1D) channels developed along the crystallographic *c*-axis. Both the aromatic and the polar carboxylate moieties are exposed to the inner surface of the channels that are corrugated but without a notable bottle-neck structure. The authors claim that the inherent selectivity of the framework towards $CO_2$ is due to the combination of size-exclusion and host–guest effects, such as dipole-induced dipole and quadrupole moment of $CO_2$ interacting with the electric field gradient of the framework. The analysis of the Connolly surface of $[Er_2(pda)_3]$ reveals that the corrugated 1D channels have the free passage smaller than 3 Å. Therefore, considering the kinetic diameters of $CO_2$, Ar, and $N_2$ which are 3.30, 3.40, and 3.64 Å (Table 7.1), respectively, the framework is likely to reject most gases unless there are some favorable interactions. Although it is unclear weather lanthanide-carboxylate structures in general exert particularly favorable influences towards $CO_2$ molecules, the measured isosteric heat of adsorption of $CO_2$ at 273 K (30 kJ/mol) seemingly supports the claim by the authors, as the value is somewhat higher than that determined for $CO_2$ in silicalite-I (20 kJ/mol). It is also possible that once adsorbed, the $CO_2$ molecules may be occluded in the pores due to the small free passages. The presence of the large hysteresis observed in the $CO_2$ sorption isotherms supports this possibility.

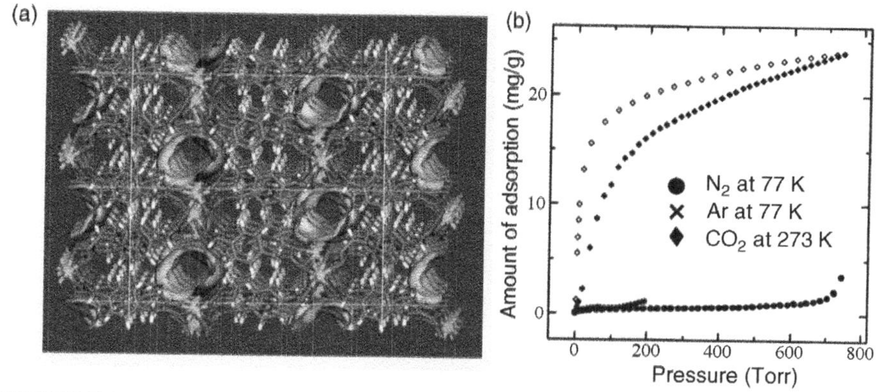


**FIGURE 7.1** (a) The Connolly surface of [$Er_2(pda)_3$] with the pore walls. (b) Gas sorption isotherms for $N_2$, Ar (77 K), and $CO_2$ (273 K).

Another porous MOF that shows size-selective sorption due to small pore openings within the rigid framework is microporous metal formates [$M(HCOO)_2$] that contain 1D zigzag channels. $Mn(HCOO)_2$, the first of the series, does not adsorb $N_2$ and Ar, but uptakes a significant amount of $H_2$ at 78 K.[4] The sorption isotherm of $H_2$ is that of typical microporous materials. The selective sorption of $H_2$ by $Mn(HCOO)_2$ has been attributed to the bottle-neck structure of the 1D channels in which the widths of narrow

**TABLE 7.1 Kinetic Diameter (in Å) for Some Molecules**

| | | | | | |
|---|---|---|---|---|---|
| He[a] | 2.6 | HBr[a] | 3.5 | $C_3H_6$[a] | 4.5 |
| $NH_3$[a] | 2.6 | $CS_2$[a] | 3.6 | Acetonitrile[b] | 4.28 |
| $H_2O$[a] | 2.65 | $H_2S$[a] | 3.6 | 1-Propanol[b] | 4.69 |
| Ne[a] | 2.75 | Kr[a] | 3.6 | Acetone[b] | 4.69 |
| $H_2$[a] | 2.89 | $SO_2$[a] | 3.6 | 2-Propanol[b] | 4.7 |
| NO[a] | 3.17 | $N_2$[a] | 3.64 | THF[b] | 4.86 |
| $Cl_2$[a] | 3.2 | CO[a] | 3.76 | Pyridine[b] | 4.96 |
| HCl[a] | 3.2 | $CH_4$[a] | 3.8 | *iso*-$C_4H_{10}$[a] | 5.0 |
| $C_2H_2$[a] | 3.3 | Methanol[b] | 3.8 | 1-Butanol[b] | 5.05 |
| $CO_2$[a] | 3.3 | $C_2H_4$[a] | 3.9 | Benzene[a] | 5.85 |
| $N_2O$[a] | 3.3 | Xe[a] | 3.96 | *p*-Xylene[c] | 5.85 |
| Ar[a] | 3.4 | Tetrahydrothiophene[b] | 4.19 | Cyclohexane[a] | 6.0 |
| $O_2$[a] | 3.46 | Ethanol[b] | 4.3 | *o*-Xylene[d] | 6.4 |
| $Br_2$[a] | 3.5 | Dimethyl ether[b] | 4.39 | *m*-Xylene[c] | 6.8 |

[a] D. W. Breck, *Zeolite Molecular Sieves*, John Wiley & Sons, New York, 1973; p. 636.
[b] M. E. van Leeuwen, Derivation of Stockmayer potential parameters for polar fluids. *Fluid Phase Equlib.* **1994**, 99, 1–18.
[c] Y. Yang, J. -W. Qian, L. -J. Xuan, A. -F. An, L. Zhang, C. -J. Gao, Preparation and pervaporation of a palygorskite/polyacrylamide inorganic-organic hybrid membrane for separating m-/p-xylene isomers. *Desalination* **2006**, *193*, 193–201.
[d] F. Letellier, J. Blanchard, K. Fajerwerg, C. Louis, M. Breysse, D. Guillaume, D. Uzio, Search for confinement effects in mesoporous supports: hydrogenation of o-xylene on Pt°/MCM-41. *Catal. Lett.* **2006**, *110*, 115–124.

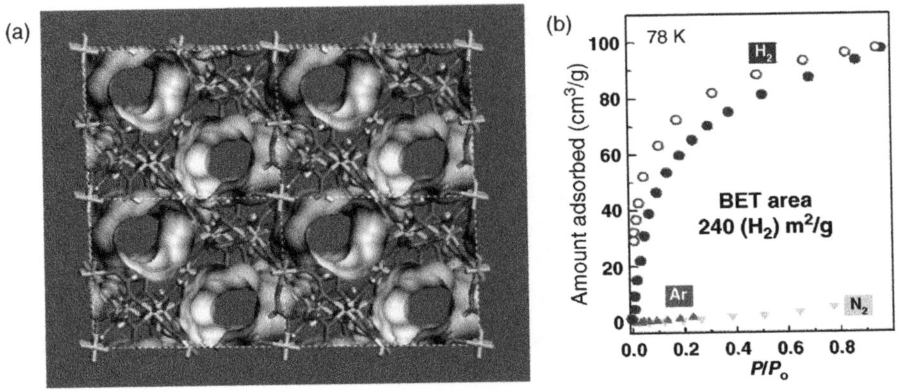


**FIGURE 7.2** (a) The Connolly surface of [$Mn(HCOO)_2$] with the pore walls. (b) Gas sorption isotherms of $N_2$, Ar, and $H_2$ at 78 K for [$Mn(HCOO)_2$].

passages are in the range 2.5–3.0 Å (Figure 7.2). Note that the kinetic diameters of $H_2$, Ar, and $N_2$ are 2.8, 3.4, and 3.64 Å, respectively (Table 7.1).

It is also interesting to note that $M(HCOO)_2$ (M = Mg, Mn) show a significant storage capacity for $C_2H_2$ with a remarkable selectivity over $CO_2$ and $CH_4$.[5] When the temperature increases from 195 to 298 K, the amount of adsorbed acetylene decreases from 68 to 51 $cm^3/g$ for $Mn(HCOO)_2$ (from 73 to 66 $cm^3/g$ for $Mg(HCOO)_2$). However, the decrease in the sorption of $CO_2$ by $Mn(HCOO)_2$ on going from 195 to 298 K is more pronounced (from 105 to 38 $cm^3/g$), giving rise to the selectivity. Mg $(HCOO)_2$ is expected to show even an improved selectivity at ambient temperatures (Figure 7.3).

The low sorption for $CH_4$ at 298 K by both formates is not surprising since the boiling point of $CH_4$ is low at 111.6 K. However, the selective sorption of $C_2H_2$ over $CO_2$ is somewhat remarkable because the two gases are similar in molecular size (3 Å × 5 Å) and sublimation point (189.2 vs. 194.7 K). The crystal structures of the acetylene-adsorbed $M(HCOO)_2$ reveal no particular interaction but van der Waals contacts with the framework atoms. Therefore, such discrimination is presumably due

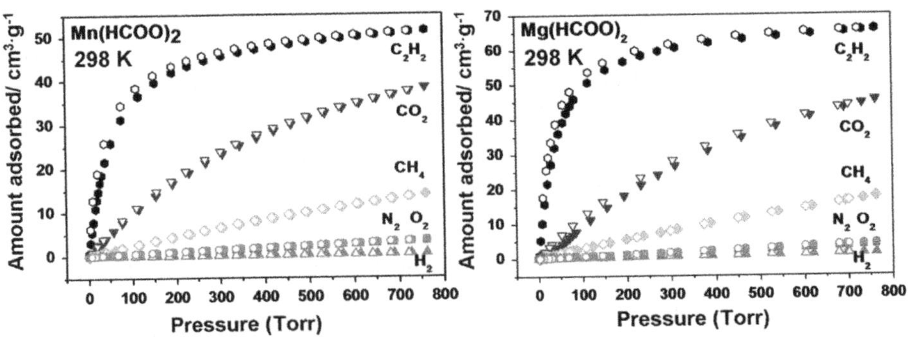


**FIGURE 7.3** The gas sorption isotherms for $M(HCOO)_2$ (M = Mg, Mn) at 298 K.

to the slightly larger size of $C_2H_2$ molecule compared to $CO_2$, which provides the former with more effective van der Waals interactions with the framework walls. The fact that $Mn(HCOO)_2$ adsorbs 139 mg MeOH per gram material at 298 K[6] also rules out the size-exclusion as the main reason for the selectivity. Rather, it can be concluded that the selective sorption of metal formates may originate from the size-exclusion for non-polar, small gas molecules, or from the combination of small pores and non-bonded interactions between the adsorbate and the adsorbent for large polar guests.

An independent research has discovered that $Mg(HCOO)_2$ adsorbs both $H_2$ and $N_2$ with significant uptakes at 77 K and 1 atm (70 and 88 $cm^3/g$ for $H_2$ and $N_2$, respectively),[7] and that $Co(HCOO)_2$ adsorbs 108 $cm^3/g$ of $N_2$ at 77 K.[8] This contradicting behavior by a series of isostructural frameworks is puzzling indeed, and would require more detailed and careful measurements under controlled environments.

***7.2.1.2 Size-Selectivity Imposed by Framework Interpenetration*** When the organic linkers of MOFs are large enough compared to the size of secondary building units (SBUs), the resulting frameworks often form interpenetrated networks. One of such frameworks having a doubly interpenetrated α-Po nets based on paddlewheel SBUs, $[Cu_2(fum)_2(bpe)]$ (fum = fumarate, bpe = *trans*-bis-(4-pyridyl)ethylene), shows the selective sorption of $H_2$ over $N_2$, Ar, and CO at 77 K.[9] As shown in Figure 7.4, the interpenetration leaves only small pocket-like pores, and the selectivity is believed to originate from size-exclusion.

Similarly, $[Zn_2(azdc)_2(bpe)]$ (azdc = 4,4′-azobenzenedicarboxylate) has paddle-wheel SBUs, and the long linkers allow a triple interpenetration.[10] Selective sorption of $H_2$ over $N_2$ and CO at 77 K, and of $CO_2$ over $CH_4$ at 195 K is observed, which is believed to be a result of size-exclusion.

Another paddlewheel-based primitive cubic net, $[Co_2(ndc)_2(bpy)]$, is known to adsorb more $H_2$ than $N_2$ at 77 K when activated at 150 °C.[11] When activated at 200 °C, the two-fold interpenetrating nets of $[Co_2(ndc)_2(bpy)]$ slip off, resulting in the decrease of the pore sizes (Figure 7.5). The shift in the X-ray diffraction peaks to higher angles and almost non-porous behavior towards $N_2$ sorption after activation at

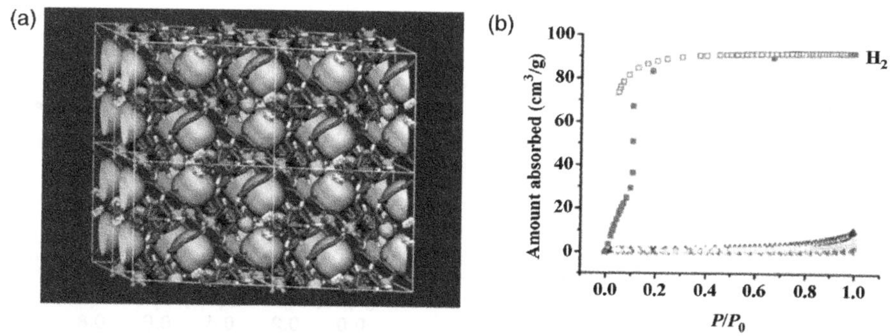


**FIGURE 7.4** (a) The Connolly surface of $[Cu_2(fum)_2(bpe)]$ with the inner walls of pocket-like pores. (b) The gas sorption isotherms for H2 (squares) over N2 (triangles), Ar (inverted triangles), and CO (diamonds) at 77K. Solid symbol = adsorption, open symbol = desorption.

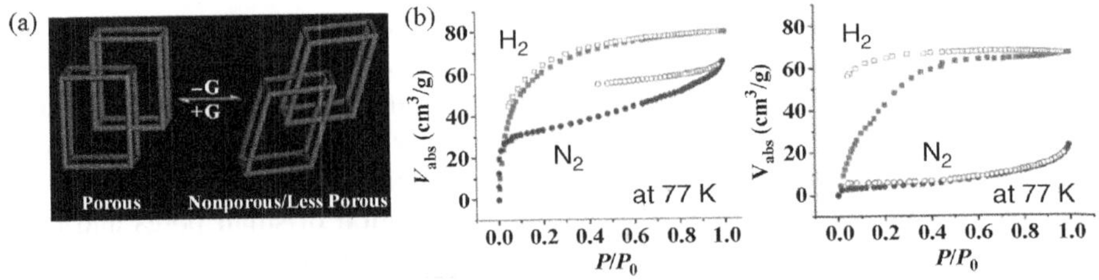


**FIGURE 7.5** (a) Schematic illustration of the structural transformation. (b) Gas sorption isotherms for $[Co_2(ndc)_2(bpy)]$ after activation at 150 °C (left) and 200 °C (right).

200 °C are presented as the evidences. The sorption of smaller $H_2$ molecules is not affected by the thermal activation process.

***7.2.1.3 Selective Sorption Depending on the Degree of Hydration*** The as-synthesized form of $[Zn_4(O)(anthdc)_3(OH_2)_3]$ (PCN-13, anthdc = 9,10-anthracene-dicarboxylate) includes two DMF guests per formula unit. The guest can be exchanged with MeOH which is then easily removed under vacuum at room temperature.[12] This "desolvated" framework shows the selective sorption of $H_2$ and $O_2$ over $N_2$ and CO at 77 K (Figure 7.6a). This result suggests that the pore opening of "desolvated" PCN-13 should be between 3.46 and 3.64 Å in diameter, which is consistent with the aperture size (3.5 Å × 3.5 Å) observed by X-ray crystallography (Figure 7.7a). Further heating (150 °C) of the desolvated material under vacuum to remove the aqua ligands gives "dehydrated" solid with enlarged pores (Figure 7.7b). As expected, this dehydrated form does not discriminate gases with different kinetic diameters; a significant sorption of $N_2$ and CO is observed (Figure 7.6b). Also, the $H_2$ uptake increases by 35% compared to the desolvated form, which was attributed to the formation of coordinatively unsaturated Zn centers.

A 3D sodalite network of $[Cu(2\text{-pymo})_2]$ (2-pymo = 2-pyrimidinolate) can discriminate $H_2$ from $N_2$ at 77 K when activated at 105 °C, though the $H_2$ uptake is as small

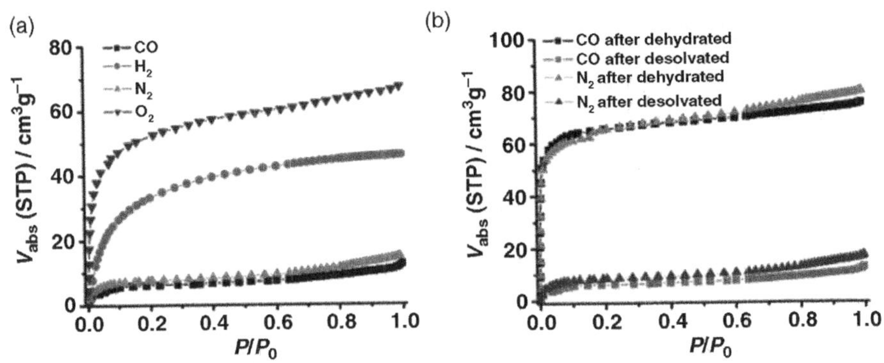


**FIGURE 7.6** Gas sorption isotherms of "desolvated" (a) and "dehydrated" (b) PCN-13 at 77 K.

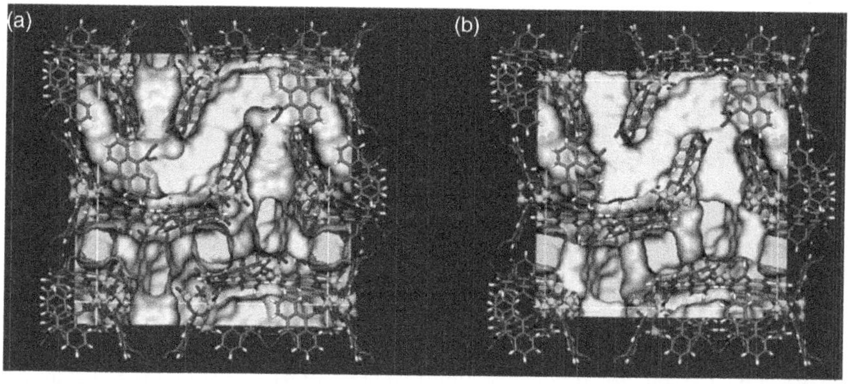


**FIGURE 7.7** The Connolly surface of "desolvated" (a) and "dehydrated" (b) PCN-13 with the inner pore walls.

as 0.33 wt%.[13] When fully evacuated at 120 °C, the $H_2$ uptake increases to 0.86 wt% at the same condition; however, the sorption of $N_2$ also increases to a level higher than that of $H_2$, thereby loosing the selective sorption property. This is probably caused by the complete removal of guest water molecules from the framework in a manner similar to the case of PCN-13. Accompanying the dehydration is a large shrinkage of the unit cell volume from 3852 to 3422 Å$^3$, and void volume from 32.6 to 27.8%.

### 7.2.2 Selective Sorption in Flexible and/or Dynamic Frameworks

***7.2.2.1 Selectivity Arising from the Flexibility of Frameworks*** A six-connected net based on pinwheel-like trinuclear SBUs, $Mg_3(ndc)_3(def)_4$, is desolvated to give $Mg_3(ndc)_3$ which shows the selective sorption of $H_2$ and $O_2$ over $N_2$ and CO at 77 K (Figure 7.8).[14] The framework is believed to be highly flexible, and the pore openings

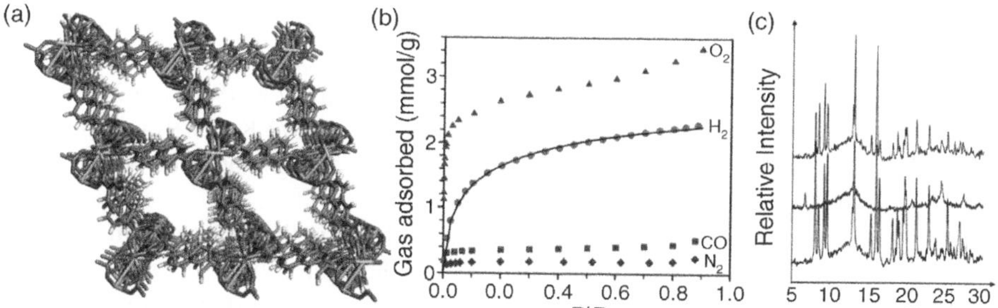


**FIGURE 7.8** (a) The crystal structure of $[Mg_3(ndc)_3(def)_4]$ after removing the coordinated def molecules. (b) Gas sorption isotherms for the uptake of $O_2$ and $H_2$ over $N_2$ and CO in $Mg_3(ndc)_3$ at 77 K. The solid line represents a Langmuir–Freundlich fit to the $H_2$ isotherm. (c) Powder XRD patterns for $[Mg_3(ndc)_3(def)_4]$: as-synthesized (bottom), evacuated (middle), and resolvated with def (top).

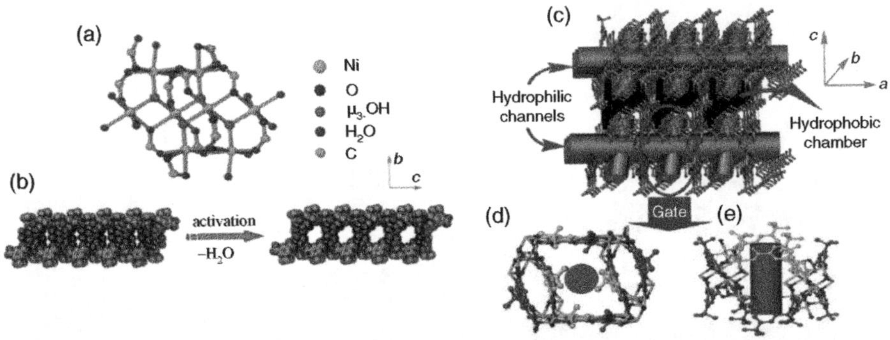


**FIGURE 7.9** Crystal structure of $[Ni_8(tbip)_6(\mu_3\text{-}OH)_4(OH_2)_8]\cdot 8H_2O$. (a) Structure of the octanickel cluster. (b) Structures of solvated and desolvated trilayers. The desolvated trilayer displays hydrophilic channels along the *a*-axis. (c) Two trilayers pack along the *c*-axis to form hydrophobic chambers. Top and side views of tbip pairs.

constrict upon the removal of coordinated def molecules giving rise to the observed size-selectivity. The pore openings in the active form is believed to be approximately 3.5 Å based on the kinetic diameters of the gas molecules. The material recovers its original phase when resolvated with def solvents, as shown in the powder diffraction patterns.

***7.2.2.2 Temperature-Dependent Selectivity (Dynamic Framework)*** $[Ni_8(tbip)_6(\mu_3\text{-}OH)_4(OH_2)_8]\cdot 8H_2O$ (tbip = 5-*tert*-butyl isophthalate) (MAMS-1) is a bilayer-type 2D framework in which $[Ni_8(\mu_3\text{-}OH)_4]$ moieties are sandwiched between layers of the *t*-butyl groups of tbip ligands (Figure 7.9).[15] Therefore, the predominant interactions between two adjacent sheets are mostly van der Waals in nature. The TGA and gas sorption measurements show that the material is non-porous when only the guest water molecules are removed by evacuation at temperatures below 120 °C. When heated to 200 °C under vacuum, however, the material is fully dehydrated to give $[Ni_8(tbip)_6(\mu_3\text{-}OH)_4]$ which shows a selective sorption of $H_2$ (73 $cm^3$/g) over $O_2$, $N_2$, and CO at 77 K. Remarkably, it was found that the selectivity depends on the measurement temperatures. That is, $O_2$ is adsorbed at 87 K (94.6 $cm^3$/g), $N_2$ at 113 K (54.2 $cm^3$/g), $CH_4$ at 143 K (71.4 $cm^3$/g), $C_2H_4$ at 175 K (77.9 $cm^3$/g), $C_3H_6$ at 241 K (84.6 $cm^3$/g), and *iso*-$C_4H_{10}$ at 295 K (98.1 $cm^3$/g), but not below these temperatures (Figure 7.10).

The framework structure of the as-synthesized form is thought to be maintained intact after activation at 200 °C based on XRPD. Therefore, the temperature-dependent size-selectivity is explained in terms of dynamic gating effects. A pair of tbip ligands acts as a gate at the interface between hydrophobic storage chambers and hydrophilic channels. The authors explain that the opening of the gate is controlled by the amplitude of the thermal vibration. The plot of molecular sizes of adsorbed guests versus the adsorption temperatures shows a linear relationship, which indirectly supports the conjecture. The hypothesis may be further verified by pressure- and

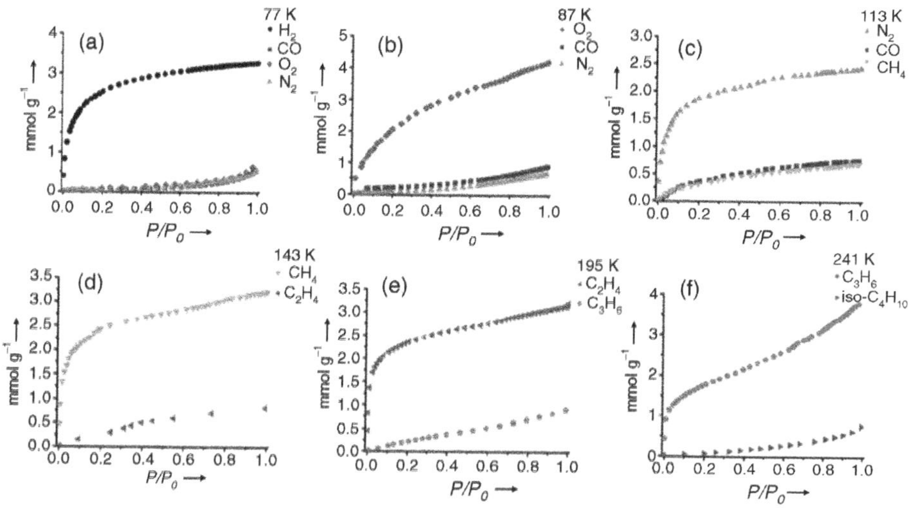


**FIGURE 7.10** Gas sorption isotherms measured at different temperatures: (a) 77 K, (b) 87 K, (c) 113 K, (d) 143 K, (e) 195 K, and (f) 241 K.

temperature-controlled measurements of diffraction patterns, if necessary. The authors exclude the possibility of simple temperature-dependent expansion of the framework based on the consistent unit cell parameters in the range 110–270 K; however, it is unlikely that the unit cell measurements had been carried out for the evacuated crystals under vacuum. Nevertheless, the temperature-dependent size-selectivity of the MOF-based "mesh-adjustable" molecular sieve is an excellent demonstration of potential applications in gas separations by temperature-swing adsorption techniques.

The analysis of the Connolly surface for the completely dehydrated structure (Figure 7.11) reveals that most of the solvent-accessible voids exist in the space between the octanickel clusters, which is described as the "hydrophilic channels" in the paper. The channels running along the *a*-axis have the width in the range 5.2–11.1 Å and the free passage of 3.1–5.9 Å. The free opening of these channels is critically affected by coordinated water molecules, if any. Meanwhile, the interlayer voids

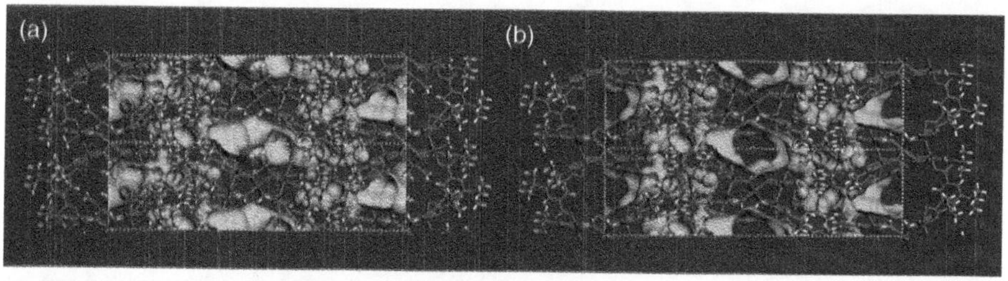


**FIGURE 7.11** The Connolly surface of desolvated (a) and dehydrated $[Ni_8(tbip)_6(\mu_3\text{-}OH)_4(OH_2)_8]$ (b) with the inner walls of the pores.

surrounded by the *t*-butyl groups are found pretty much isolated with very narrow openings. Therefore, the presentation of the structure as a completely evacuated and rigid framework cannot explain the observed gas sorption behavior, and some kinds of dynamic processes involving not only the *t*-butyl groups but also the $Ni_8$ cluster units should be necessary.

### 7.2.3 Selectivity Apparently Independent of Size-Exclusion

$[Co_3(2,4\text{-pdc})_2(\mu_3\text{-OH})_2]\cdot 9H_2O$ (2,4-pdc = pyridine-2,4-dicarboxylate) is a 3D framework having straight 1D channels.[16,17] It allows the full evacuation of guest water molecules without collapsing the framework as demonstrated by the successful single-crystal X-ray structure determination of an evacuated sample at 102 °C (Figure 7.12). The 1D channels running along the *c*-axis are wide enough for the passage of small guest molecules; the free opening is larger than 5 Å; and the corrugated channel has a width in the range 7.4–9.3 Å. Therefore, it is very surprising to learn that the material activated at 573 K shows the selective sorption of $H_2$ over $N_2$ at 77 K, $O_2$ over $N_2$ and Ar at 87 K, and $CO_2$ over $CH_4$ at 196 K (Figure 7.13). The authors took advantage of these selective sorption properties to demonstrate a potential application in the gas chromatographic separation of light gases.

As shown in Figure 7.12, the 1D channels in $[Co_3(2,4\text{-pdc})_2(\mu_3\text{-OH})_2]$ (CUK-1) are not bottle-necked nor much corrugated. Therefore, the prime question to ask is whether the single-crystal structure truly represents the active form of the bulk phase used for the gas sorption studies which is prepared by slow heating from room temperature to 573 K under vacuum for just 1 h. The FT-IR studies of the as-synthesized material reveals that it does not reach to a complete dehydration when heated to 423 K for 1 h, explaining why $O_2$ is not adsorbed in the materials activated at temperatures below 423 K. Also, the sorption of $O_2$ at 87 K for the sample activated at 573 K is found irreversible and the temperature has to be raised for a complete desorption. Combined, these results suggest that the activation processes have to be chosen conservatively to ensure a complete evacuation and that a dynamic change occurs in the physical environment of the channels as a function of temperature and

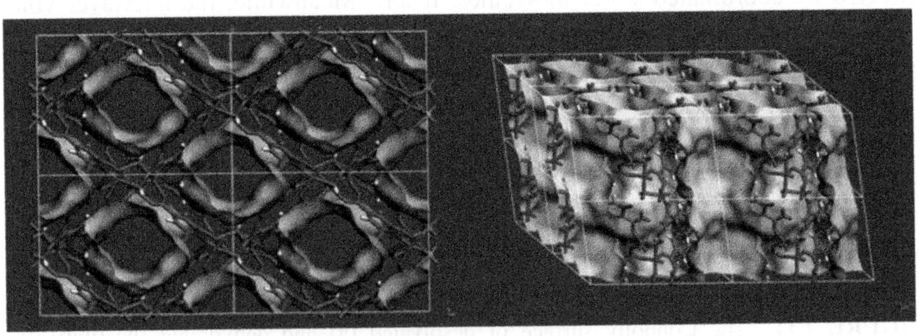

**FIGURE 7.12** The Connolly surface of dehydrated CUK-1 with the inner walls.

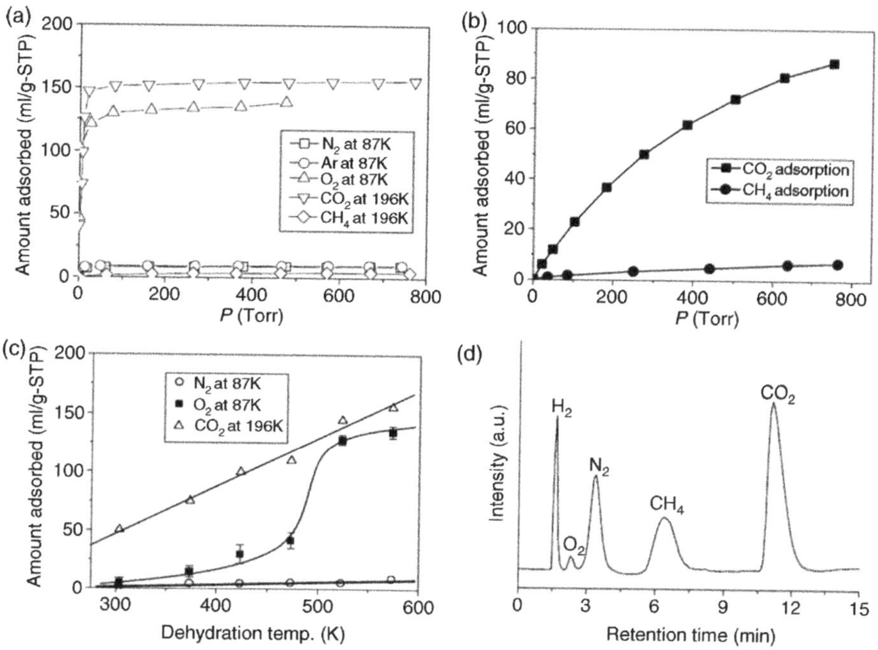


**FIGURE 7.13** (a) Sorption uptake of selected gases at low temperature and (b) reversible sorption uptakes of $CO_2$ and $CH_4$ at 298 K in dehydrated CUK-1. (c) Sorption uptakes of $N_2$, $O_2$, and $CO_2$ measured at 87 K ($N_2$ and $O_2$) and 196 K ($CO_2$) for CUK-1 according to activation temperature. Error bars denote the standard deviation of three measurements. (4) A gas chromatogram of a light gas mixture separated on a CUK-1 column. The composition of the gas mixture in a gas sampling loop (0.25 mL) is $H_2/O_2/N_2/CH_4 = 0.6{:}2{:}28{:}10{:}27$ (mol%).

pressure. Unfortunately, powder diffraction patterns are not available for partially or completely dehydrated samples, or more desirably, at various temperatures.

Another example for the selectivity not originating from size-exclusion is known.[18] $[Cd_3(aptz)(\mu_3\text{-}OH)_2(OH_2)_2]\cdot DMF$ (aptz = 4-aminophenyl-1H-tetrazole) forms 2D sheets that are interdigitated by hydrogen bonds to give well-defined 1D channels. The single-crystal structure determined after a complete desolvation, including the removal of the aqua ligands, shows that there is no significant change in the framework. The effective pore opening of the rather straight channel is at least 5 Å, as shown in Figure 7.14, and therefore, no selectivity is expected in the sorption of light gases. However, the material is found non-porous in terms of $N_2$ sorption while $H_2$ (77 K) and $CO_2$ (195 K) show type I sorption isotherms, albeit with significant hysteresis. A very slow equilibration is noted in the measurements of $H_2$ and $CO_2$ sorption; however, it still does not explain why the material rejects $N_2$. One possible explanation is again a dynamic structural change between the interdigitated 2D sheets which is unnoticed in the single-crystal diffractions. Or, the observed selectivity may be a result of complicated interactions between the gas molecules and the framework surface. The presence of multiple sites for specific interactions, changes in polarizabilities or

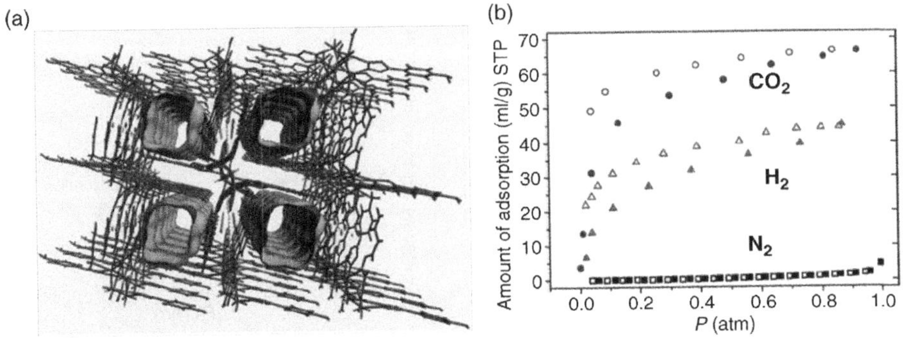


**FIGURE 7.14** (a) The top view of $[Cd_3(aptz)(\mu_3\text{-}OH)_2(OH_2)_2]$ with 2 × 2 arrays of the 1D channels. (b) Gas sorption isotherms of $N_2$ (77 K), $H_2$ (77 K), and $CO_2$ (195 K) (solid symbol = adsorption, open symbol = desorption).

poliarities, and other secondary interactions such as dipole–dipole and donor–acceptor affinity have been suggested for the selectivity not based on size-exclusion.

### 7.2.4 Selective Sorption at Higher Temperatures and Pressures

The sorption of $C_2H_2$ in a pillared layer network $[Cu_2(pzdc)_2(pyz)]$ (pzdc = pyrazine-2,3-dicarboxylate; pyz = pyrazine) is quantitative and constant in the temperature range 270–310 K while the sorption of $CO_2$ is much less effective as the temperature becomes higher (Figure 7.15).[19] Note that the sizes of $CO_2$ and $C_2H_2$ are similar

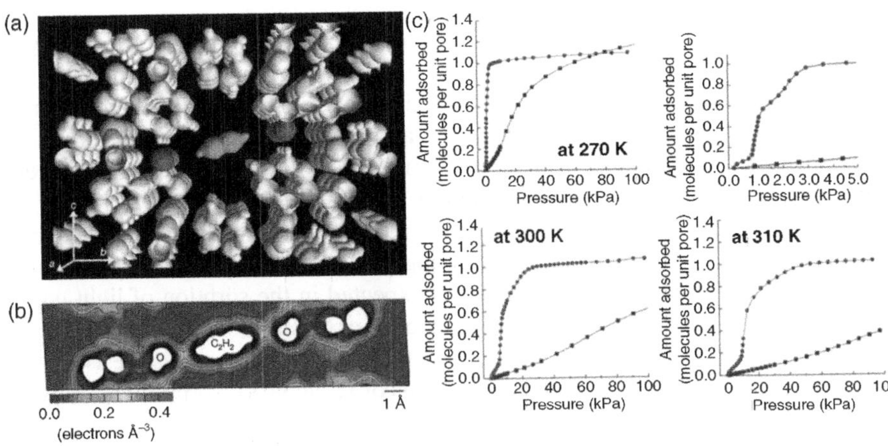


**FIGURE 7.15** (a) MEM electron densities of $C_2H_2$ adsorbed $[Cu_2(pzdc)_2(pyz)]$ at 170 K, (b) MEM electron density distribution views on the 2D section defined by the molecular axis of the $C_2H_2$ molecule, and (c) adsorption isotherms for $C_2H_2$ (circles) and $CO_2$ (squares) on $[Cu_2(pzdc)_2(pyz)]$. (The inset shows the lower $P/P_0$ region at 270 K.)

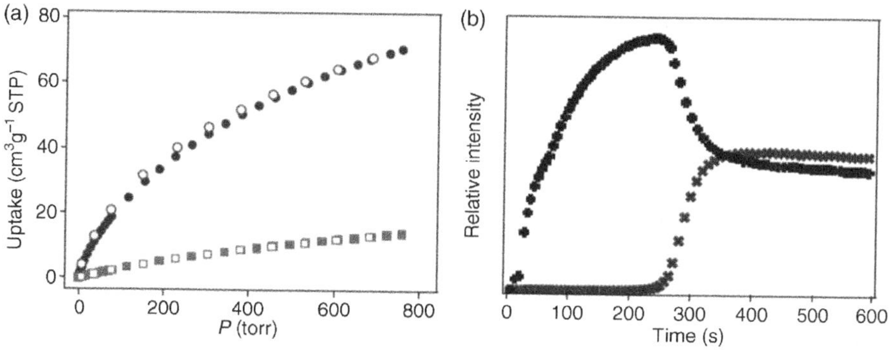


**FIGURE 7.16** (a) Gas adsorption isotherms of $CO_2$ (circles) and $CH_4$ (squares) on ZIF-20 at 273 K (solid symbol = adsorption, open symbol = desorption). (b) Separation of $CO_2$ (× marks) and $CH_4$ (+ marks) on ZIF-20 using a $CO_2/CH_4$ gas mixture. The relative intensities of each gas passing through the ZIF-20-packed column were obtained using a mass spectrometer to detect ion peaks at $m/z = 44$ ($CO_2$) and 16 ($CH_4$).

(3 Å × 5 Å), and the 1D channels of $[Cu_2(pzdc)_2(pyz)]$ have the cross-section of 4 Å × 6 Å. The MEM/Rietveld analysis using in situ synchrotron XRPD patterns reveals that adsorbed $C_2H_2$ molecules are confined within the pores by non-bonded contacts, such as van der Waals interactions, hydrogen bonding, and electron delocalization.

$[Zn(pur)_2]\cdot(DMF)_{0.75}\cdot(H_2O)_{1.5}$ (pur = purinate) (ZIF-20) is a 3D framework having the topology of zeolite A (LTA) which is a tiling of cubes, truncated octahedral, and truncated cuboctahedra.[20] At cryogenic temperature, it does adsorb a significant amount of $H_2$, Ar, and $N_2$. At 273 K, however, there is a marked preference for $CO_2$ over $CH_4$, and the separation ability is demonstrated using a mixture of $CO_2/CH_4$ at 1.4 bar in breakthrough experiments (Figure 7.16). The selective sorption is attributed partly to the small pore apertures (2.8 Å) and partly to the polar pore walls originating from uncoordinated nitrogen atoms of the ligands.

Similarly, materials that do not discriminate light gases at 77 K but showing preferential sorption of $CO_2$ over $CH_4$ at 195 K or higher temperatures have been reported.[21,22] These results, however, should be taken with caution because a certain degree of selectivity for $CO_2$ over $CH_4$ is always expected in physisorption processes considering the differences in the phase diagram of the two gases.

$[Zn_4(O)(btb)_2]$ (btb = benzenetribenzoate), better known as MOF-177, that uptakes a large amount of $N_2$ at 77 K shows a preferential sorption of $O_2$ over $N_2$ (1.8:1) at 1 atm and 298 K (Figure 7.17).[23] The concave isotherm for $N_2$ is compared to that of $O_2$ which is linear, and attributed to the lack of electric charges on the surfaces of $[Zn_4(O)(btb)_2]$ because the quadrupole moment of $N_2$ interacts strongly with charges.

$[Cu(dhbz)_2(bpy)]\cdot H_2O$ (dhbz = 2,5-dihydroxybenzoate) has 1D channels with a cross-section of 3.6 Å × 4.2 Å (Figure 7.18a).[24] The water-filled channels are formed by 2D sheets interdigitated with π–π stacking interactions. When dehydrated, only a very low sorption of $N_2$ is observed at 77 K. Before the dehydration, the channels

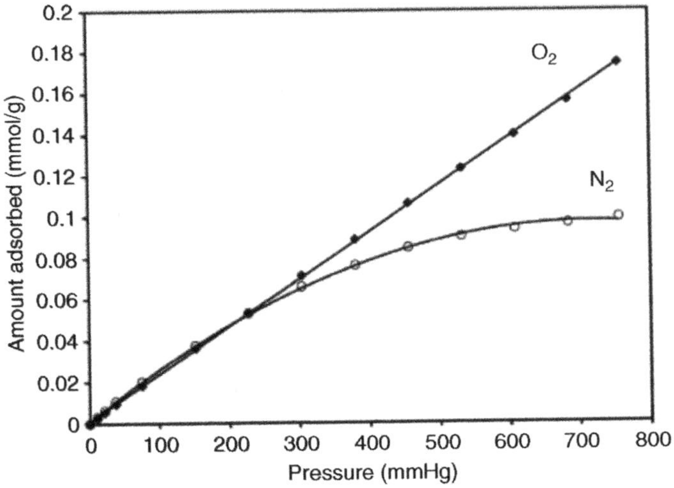


**FIGURE 7.17** Gas adsorption isotherms of $N_2$ and $O_2$ on MOF-177 at 298 K.

(Figure 7.18a) appear to be large enough for the adsorption of $N_2$ (kinetic diameter 3.64 Å). However, the pore size of the channels shrinks upon the dehydration as evidenced by XRPD, which may explain the low sorption of $N_2$. At high pressures and 298 K, however, an abrupt increase in the sorption of $N_2$ is observed at around 50 atm. A large hysteresis is observed in the sorption/desorption isotherms, and the breakthrough points from each traces are defined as the gate-opening and gate-closing pressures, 50 and 30 atm for $N_2$, respectively. The high-pressure sorptions of $O_2$ and $CH_4$ at 298 K also reveal similar patterns but with different gate-opening pressures (Figure 7.18). Therefore, the flexible and dynamic nature of the 2D coordination polymer should be responsible for these selectivities at high pressures.

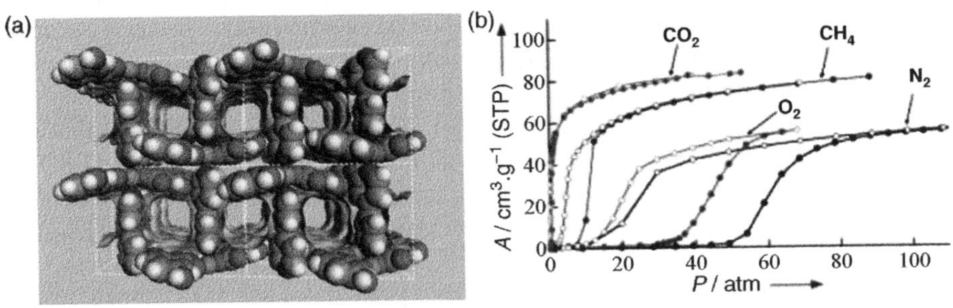


**FIGURE 7.18** (a) The Connolly surface of [Cu(dhbz)$_2$(bpy)] with the pore walls. (b) Adsorption (filled circles) and desorption (open circles) isotherms of $N_2$, $CH_4$, $CO_2$, and $O_2$ at 298 K.

## 7.3 SELECTIVE VAPOR SORPTION BY MOFs

### 7.3.1 Selective Sorption of Organic Vapors over $N_2$ or Other Light Gases

A functionalized porous network [Cd(4-btapa)$_2$(NO$_3$)$_2$]·6H$_2$O·2DMF is synthesized from Cd(NO$_3$)$_2$·4H$_2$O and a tridentate amide ligand (4-btapa = 1,3,5-benzene tricarboxylic acid tris[N-(4-pyridyl)amide]) in which attractive interactions between the framework and the alcoholic guests are expected.[25] The amide groups are exposed to the surfaces of channels having a dimension of 4.7 Å × 7.3 Å (Figure 7.19a). Upon removal of solvent molecules it looses its crystallinity as evidenced by XRPD. The activated framework shows no $N_2$ adsorption at 77 K indicating that the framework does not maintain the void channels. However, the guest-free framework can adsorb short-chain alcohols (methanol, ethanol, *n*-propanol, and *n*-butanol) (Figure 7.19b) but not long-chain alcohols such as *n*-pentanol and *n*-hexanol. The sorption of short-chain alcohols induces the amorphous-to-crystalline transformation through hydrogen bonding between the framework and the guests. Although *n*-butanol, *n*-pentane, and *n*-pentene are similar in size and shape, the porous framework cannot include *n*-pentane and *n*-pentene due to the lack of ability to form hydrogen bonding with the amide groups exposed to channels.

In the systematic investigations of pillared layer networks, Kitagawa and coworkers have synthesized [Cd(pzdc)(azpy)]·2H$_2$O and [Cd(pzdc)(bpee)]·1.5H$_2$O where pzdc is pyrazine-2,3-dicarboxylate, and azpy and bpee are py-N=N-py and py-CH=CH-py, respectively (py = pyridine).[26] The frameworks are assembled from Cd(pzdc) layers pillared by azpy and bpee ligands. Sorption properties of [Cd(pzdc)(azpy)] and [Cd(pzdc)(bpee)] are different even though the organic ligands used are similar. The adsorption isotherms of $N_2$ (kinetic diameter; 3.64 Å$^2$) at 77 K for both porous materials reveal that $N_2$ molecules are unable to diffuse into the channels at low

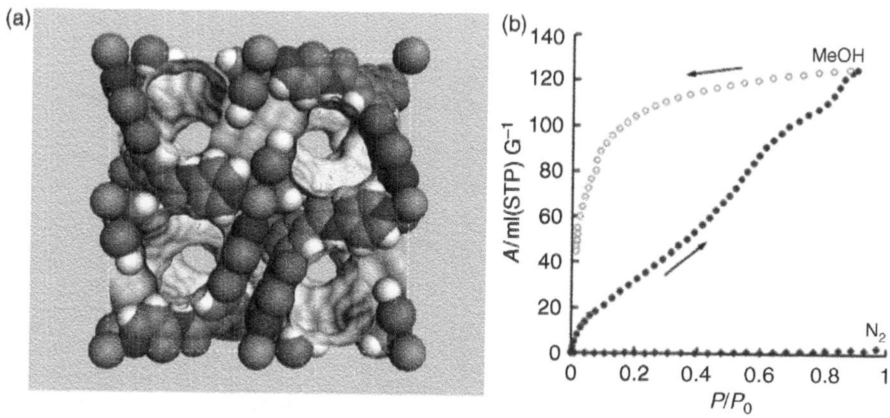


**FIGURE 7.19** (a) The Connolly surface of [Cd(4-btapa)$_2$(NO$_3$)$_2$] with the pore walls. (b) Gas adsorption–desorption isotherms of [Cd(4-btapa)$_2$(NO$_3$)$_2$] for $N_2$ and MeOH. $P_0$ is the saturated vapor pressure, 101.3 kPa, of $N_2$ (77 K), and 16.94 kPa, of methanol (298 K).

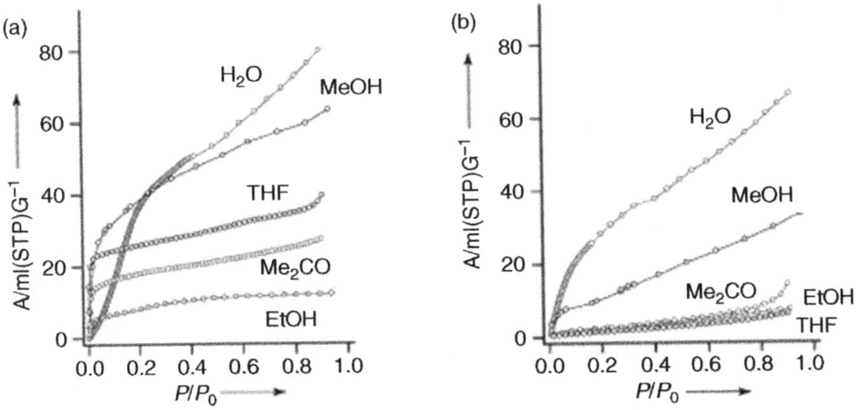


**FIGURE 7.20** Vapor sorption of [Cd(pzdc)(azpy)] (a) and [Cd(pzdc)(bpee)] (b) at 298 K, $P_0$ is the saturated vapor pressure at 298 K; 3.17 kPa ($H_2O$), 16.94 kPa (MeOH), 7.87 kPa (EtOH), 23.45 kPa (THF), and 30.59 kPa ($Me_2CO$).

temperature. On the other hand, $H_2O$ (2.65 Å$^2$), MeOH (3.8 Å$^2$), EtOH (4.3 Å$^2$), THF (4.86 Å$^2$), and $Me_2CO$ (4.69 Å$^2$) are able to diffuse into [Cd(pzdc)(azpy)] at 298 K, irrespective of whether or not it is similar in size to $N_2$ (Figure 7.20a), indicating that the size is not a crucial factor for adsorption in this case. Interestingly, however, the adsorption properties of [Cd(pzdc)(bpee)] show the selectivity of vapors; $H_2O$ and MeOH are adsorbed, whereas EtOH, THF, and $Me_2CO$ molecules are not (Figure 7.20b). This selectivity seems to arise from the size of the channel windows (3.5 Å × 4.5 Å) in [Cd(pzdc)(bpee)].

A different pillared layer framework [$Cu_2$(pzdc)$_2$(dpyg)] (dpyg = 1,2-dipyridylglycol) (Figure 7.21a) has also been studied by Kitagawa and coworkers, in which a functional pillar has been used to modulate the sorption properties for $CH_4$, MeOH,

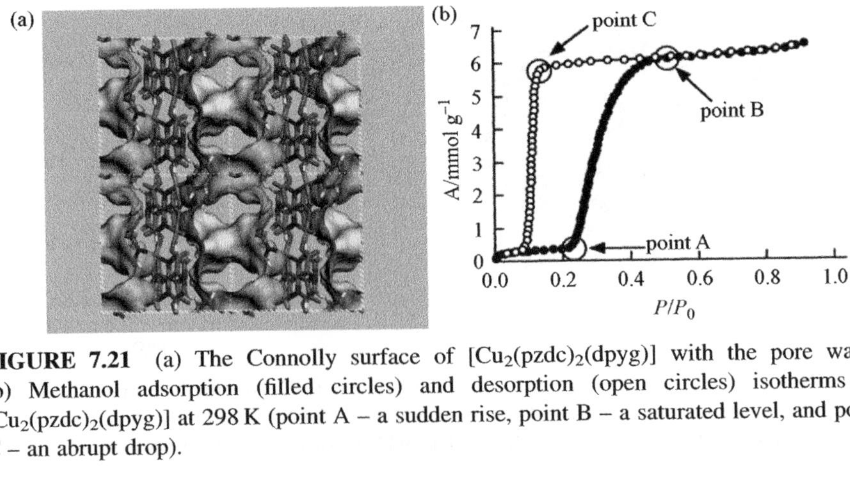


**FIGURE 7.21** (a) The Connolly surface of [$Cu_2$(pzdc)$_2$(dpyg)] with the pore walls. (b) Methanol adsorption (filled circles) and desorption (open circles) isotherms of [$Cu_2$(pzdc)$_2$(dpyg)] at 298 K (point A – a sudden rise, point B – a saturated level, and point C – an abrupt drop).

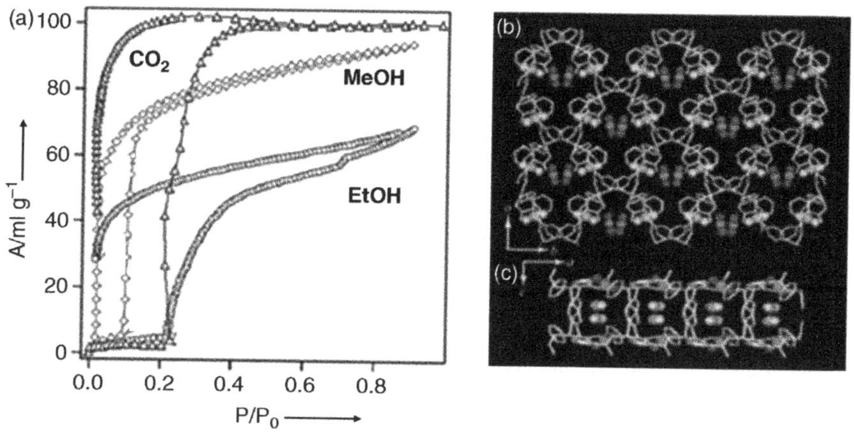


**FIGURE 7.22** (a) Sorption isotherms of $CO_2$ (195 K), MeOH, and EtOH (298 K) for [Cu (pyrdc)(bpp)]. Honeycomb-like 2D layer with $CO_2$ molecules along the *c*-axis (b) and *b*-axis (c).

and $H_2O$.[27] No adsorption of $CH_4$ indicates that channels (window size 3.3 Å) are not large enough to accommodate $CH_4$ molecules (kinetic diameter 3.8 Å). On the other hand, MeOH, which is similar in size to $CH_4$, is adsorbed with a large hysteresis between adsorption and desorption (Figure 7.21b). Therefore, the authors explained that the sorption affinity to MeOH relates to the hydrogen bonding interactions between the MeOH and the OH groups of the dpyg ligands. A similar adsorption isotherm is also observed for $H_2O$ vapor.

A similar behavior is observed in the 2D bilayer open framework [Cu(pyrdc)(bpp)]·$5H_2O$ (pyrdc = pyridine-2,3-dicarboxylate; bpp = 1,3-bis(4-pyridyl)-propane) that has a dynamic property with bond breaking and bond formation triggered by dehydration–rehydration.[28] This process occurs without the loss of its structural integrity. The dehydrated framework exhibits the adsorption of polar guest molecules, but not $N_2$ and $O_2$ (Figure 7.22a). The adsorption–desorption profiles with the large hysteresis for $CO_2$, MeOH, and EtOH relate to the framework transformation from a non-porous to porous phase in the solid state. The crystal structure with $CO_2$ inclusion (Figure 7.22b) shows the strong hydrogen bonding interactions between the guest and the channel walls. Therefore, the selective gate-opening for different guest molecules can be attributed to the hydrogen bonding interactions between the guests and the apohost framework.

### 7.3.2 Size-Selectivity Among Different Vapors

A zeotype ionic crystal composed of polyoxometalates and macrocations, $Cs_5[Cr_3O(OOCH)_6(H_2O)_3][\alpha\text{-}CoW_{12}O_{40}]\cdot 7.5H_2O$, has been reported to be effective in the separation of water from an ethanol/water azeotropic mixture.[29] The schematic illustration of the synthesis and the crystal structure (*ab* plane) of $Cs_5[Cr_3O(OOCH)_6(H_2O)_3][\alpha\text{-}CoW_{12}O_{40}]\cdot 7.5H_2O$ are shown in Figure 7.23.

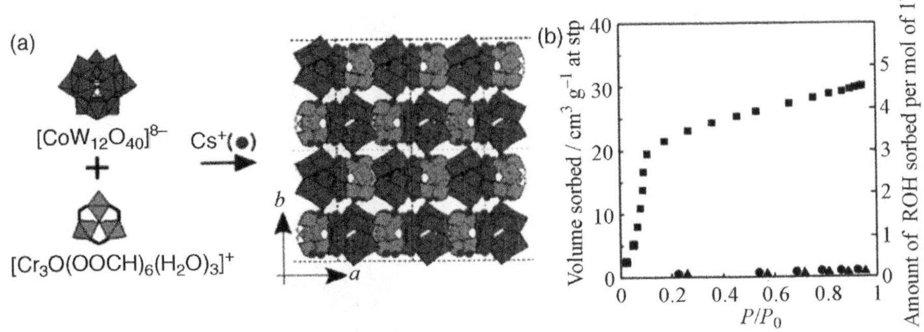


**FIGURE 7.23** (a) Synthesis and crystal structure (*ab* plane) of $Cs_5[Cr_3O(OOCH)_6(H_2O)_3]$ $[\alpha\text{-}CoW_{12}O_{40}]\cdot 7.5H_2O$,water (■, $P_0 = 3.06$ kPa), methanol (●, $P_0 = 15.6$ kPa), and ethanol (▲, $P_0 = 6.67$ kPa) sorption isotherms of the evacuated material at 298 K.

Figure 7.23b shows the sorption isotherms of the guest-free material for water, methanol, and ethanol at 298 K, with striking differences in the uptake behavior. The amount of water uptake sharply increases in the low relative pressure region. On the other hand, the amounts of methanol and ethanol sorption are smaller than those of the surface adsorption. These results are understood in terms of size-selective sorption. That is, activated $Cs_5[Cr_3O(OOCH)_6(H_2O)_3][\alpha\text{-}CoW_{12}O_{40}]$ can take water molecules (2.65 Å$^2$) into the winding hydrophilic channels but exclude larger methanol (3.8 Å$^2$) and ethanol (4.3 Å$^2$) through the small windows (2.7 Å).

$[Zn_2(ip)_2(bpy)_2]\cdot DMF$ (ip = isophthalte, bpy = 4,4′-bipyridyl) (CID-1) reported by Kitagawa and coworkers is characterized as an interdigitated 1D double-chain structure.[30] The guest-free cavities surrounded by organic ligands have hydrophobic environments with a cross-section of 5 Å × 6 Å. While this material shows no sorption of $N_2$ and $O_2$, the vapors of EtOH and MeOH are adsorbed quantitatively; two methanol or one ethanol per formula unit (Figure 7.24). The sorption of benzene in the pores of $[Zn_2(ip)_2(bpy)_2]$ is negligible mostly due to size-exclusion.

Also interesting is that water vapors are not adsorbed in the pressure range from 0 to 14 mmHg, and a sudden increase is observed above the pressure. The absence of water sorption at low pressures indicates that the framework has a hydrophobic surface resulting in weak interactions with the water molecules.

A similar sorption behavior is known for a pillared layer framework, $[Cu_2(pzdc)_2(bpy)]\cdot H_2O$ (CPL-2), that has a rectangular shape with 5.6 Å × 7.2 Å.[31] Guest-free CPL-2 exhibits different sorption behavior to benzene and water at low pressure regions, reflecting the difference of affinity between the host framework and the guest molecules (Figure 7.25). The adsorption isotherm of CPL-2 for benzene shows a typical type I behavior with one molecule per unit pore (Figure 7.25a). The slow adsorption of water at low pressure, however, reflects low affinity with the framework. The authors put these results to conclude that CPL-2 possesses two types of pores with hydrophobic and hydrophilic environments.

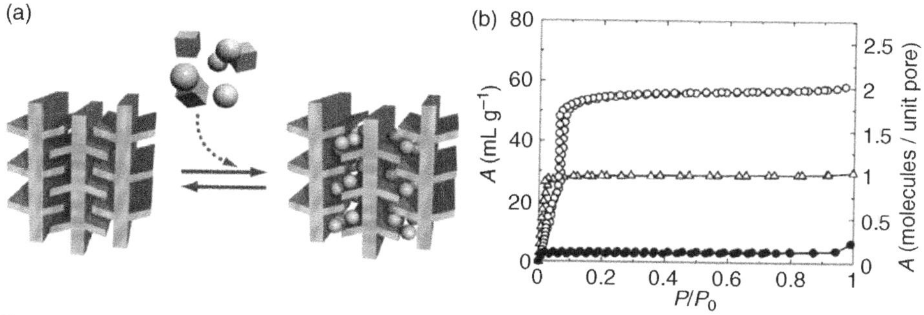


**FIGURE 7.24** (a) Representation of selective guest sorption of a coordination polymer with interdigitated structure. (b) Adsorption and desorption isotherms of methanol (open circles), ethanol (open triangles), and benzene (closed circles) for activated CID-1 at 298 K.

### 7.3.3 Selective Sorption of Organic Vapors over Water

A microporous material [Zn(tbip)] (tbip = 5-*tert*-butylisophthalate) has been synthesized under hydrothermal conditions in an attempt to generate hydrophobic channels.[32] *tert*-Butyl groups of tbips are exposed to the channels to generate a hydrophobic surface (Figure 7.26a). Although the aperture size (4.0 Å) of the channels is large enough for adsorption of methanol and water but not dimethyl ether (DME), [Zn(tbip)] shows only adsorption of methanol and DME but not water (<~1 mg/g at $P/P_0 = 0.65$) (Figure 7.26b). The authors explained that this unprecedented observation is due to the hydrophobic nature of the channels. However, the surface property of the channels is only important at low pressure. Owing to the window size large enough for water, [Zn (tbip)] may be capable of adsorption of water at higher pressures $P/P_0 > 0.65$. The low uptake for MeOH in low pressure region implies weak interactions between MeOH and framework; however, a sudden increase is observed at 40 Torr, which is believed to

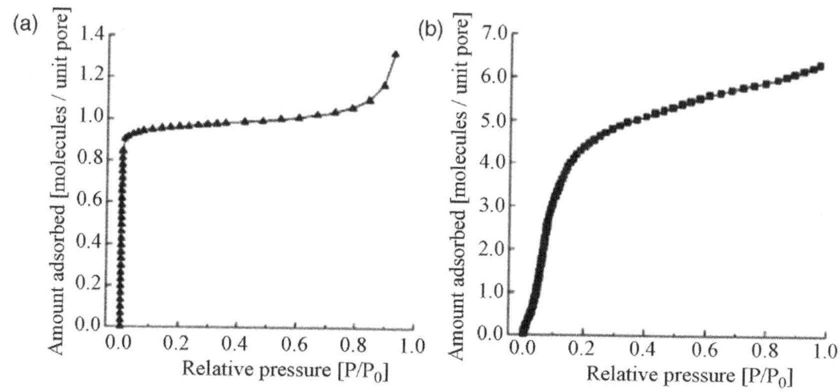


**FIGURE 7.25** (a) Adsorption isotherm for benzene of CPL-2 at 300 K in the relative pressure range from $10^{-4}$ to 0.95 ($P_0 = 14.6$ kPa). (b) Adsorption isotherm for water of CPL-2 at 300 K in the relative pressure range from $10^{-3}$ to 1.0 ($P_0 = 3.53$ kPa).

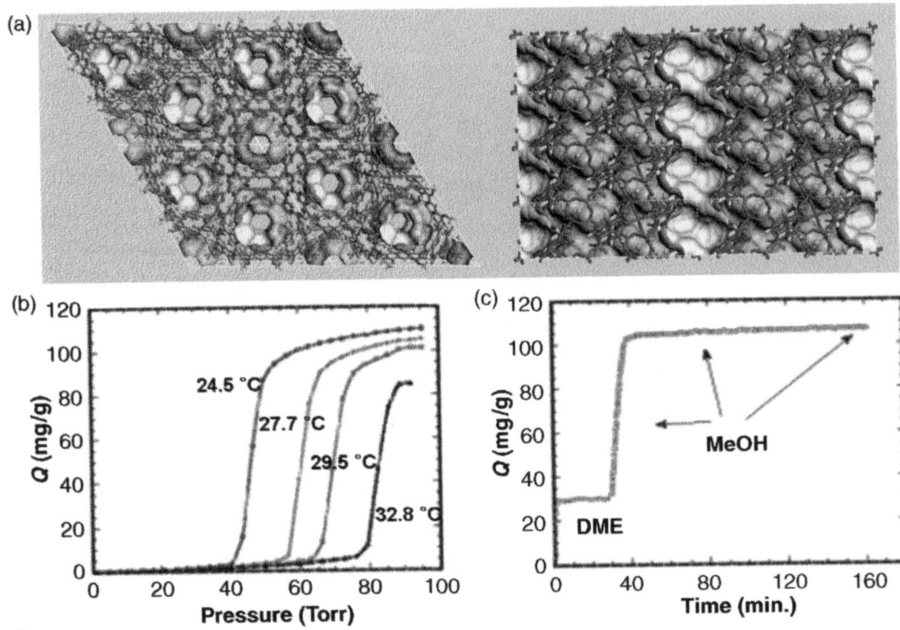


**FIGURE 7.26** (a) The Connolly surface of [Zn(tbip)] with the pore walls along the *c*-axis. (b) MeOH adsorption isotherms at different temperatures. (c) Adsorption of DME at 30 °C followed by MeOH on [Zn(tbip)].

be a capillary condensation. The sorption properties of Zn(tbip) may be compared with those of pure silica ZSM-5, which is hydrophobic but adsorbs a small amount of water at $P/P_0 = 0.65$ and shows moderate affinity for MeOH.

### 7.3.4 Selectivity for Specific Functional Groups

It is a well-known fact that the separation of molecular mixtures using microporous materials depends on specific interactions between the framework and the guest molecules. The framework of $[V(O)(bdc)](H_2bdc)_{0.71}$ (Figure 7.27a) includes benzenedicarboxylic acid ($H_2bdc$) as the guest in the channel and the guest-free framework shows specific interactions with sulfur-containing molecules such as thiophene.[33] $[V(O)(bdc)](H_2bdc)_{0.71}$ is activated on a thermobalance in flowing air to 350 °C to remove the $H_2bdc$ guest molecules. After cooling to room temperature, the gas stream was switched to a 5% $CH_4$/He stream. The data in Figure 7.27b show that methane is not absorbed under these conditions. When the gas stream is switched to a 5% $CH_4$/He stream saturated with thiophene at ambient temperature and 1 atm, a rapid absorption occurs corresponding to the uptake of 0.88 molecules of thiophene per [V(O)(bdc)]. Similar results have been obtained for the uptake of dimethyl sulfide and thiophene. These results imply that the removal of sulfur components in natural gas can be effectively carried out by the porous MOF.

(a)

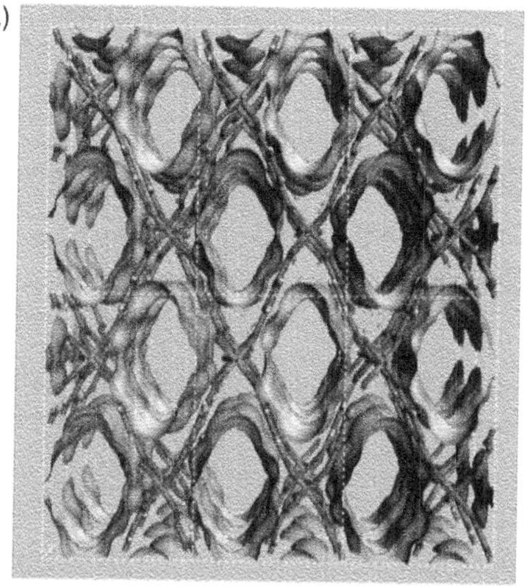

(b)

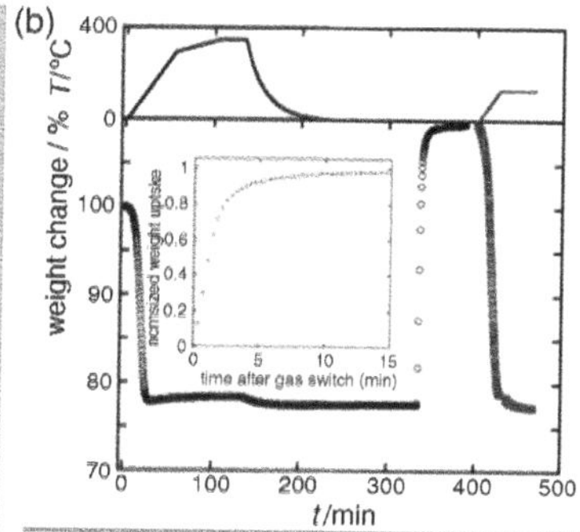


| Formula | wt % | Molecules [f.u.][a] | P [kPa][b] |
|---|---|---|---|
| $(CH_3)_2S$ | +34.0 | 1.12 | 66.9 |
| $C_4H_4S$ | +31.9 | 0.88 | 10.0 |
| $C_4H_4S$ | +23.4 | 0.65 | 1.0 |

**FIGURE 7.27** (a) The Connolly surface of [V(O)(bdc)] with pore walls along the *a*-axis. (b) Thiophene uptake by [V(O)(bdc)] from methane saturated with thiophene at ambient temperature. The temperature profile is shown in the top panel and the corresponding weight change in the bottom panel (left). The inset shows an expanded view of the thiophene uptake kinetics. The data in blue show a second experiment in which a sample saturated with thiophene at ambient temperature is heated to 150 °C (right).

In industry, the *ortho*-, *meta*-, and *para*-isomers of xylene that have similar boiling points (144, 139, and 138 °C for *o*-, *m*-, and *p*-xylene, respectively, and 136 °C for ethylbenzene) are separated using Zeolite X and Y. $[V^{III}(OH)(bdc)](H_2bdc)_{0.75}$ (MIL-47) (Figure 7.28a) first synthesized by Férey and coworkers is found to be capable of selective adsorption for these isomers.[34] Chromatographic experiments shown in Figure 7.28b clearly indicate the strong adsorption of *p*-xylene versus ethylbenzene

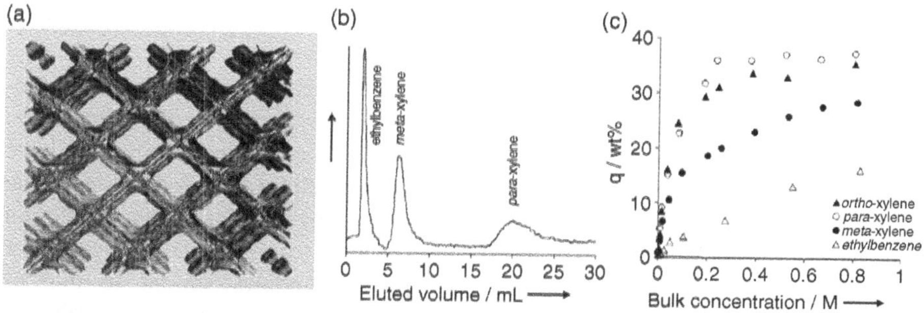


**FIGURE 7.28** (a) The Connolly surface of $[V^{III}(OH)(bdc)]$ with pore walls along the *a*-axis. (b) Chromatographic separation of a mixture of ethylbenzene, *meta*- and *para*-xylene on a column packed with MIL-47 in the liquid phase, with hexane as the desorbent at 298 K. The signal intensity of the refractive index detector is shown versus the eluted volume. (c) Adsorption isotherms of guest molecules on MIL-47 at 298 K.

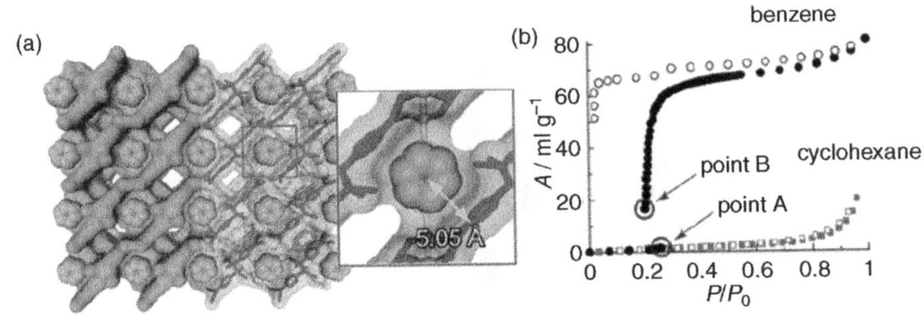


**FIGURE 7.29** (a) Benzene molecules arranged in the channels of [Zn($\mu_4$-TCNQ-TCNQ)(bpy)]. (b) Sorption isotherms for each adsorbate at 298 K. The gate-opening behavior is noted on going from point A to point B.

and *m*-xylene. The selectivity ratios are determined to be 3.1:1 for *p*-xylene versus *m*-xylene, and 9.7:1 for *p*-xylene versus ethylbenzene. These selectivities result from the molecular packing effects of these isomers inside MIL-47 cavities that have been characterized by Rietveld refinements of powder XRD profiles. The adsorption isotherms of these isomers also indicate that *o*- and *p*-xylene are packed more efficiently inside the MIL-47 cavities than *m*-xylene or ethylbenzene (Figure 7.28c).

Benzene and cyclohexane are a pair of organic compounds that have similar boiling points, molecular geometry and Lennard–Jones collision diameters. Therefore, specific binding interactions such as cation–π, H–π, or hydrogen bonding would be necessary to separate these two species. [Zn($\mu_4$-TCNQ-TCNQ)(bpy)]·1.5benzene (TCNQ = 7,7,8,8-tetracyano-*p*-quinodimethane) reported by Kitagawa and coworkers possesses such properties since the material has a large π electron surface and suitable cavity size for π–H interactions with benzene molecules.[35] Adsorption isotherms of the MOF are illustrated in Figure 7.29.

The $^1$H-NMR measured after digesting the vapor-loaded materials in DMSO-$d_6$ confirms that more than 95% benzene was separated from the mixture of benzene/cyclohexane. Moreover, the guest-free coordination polymer selectively adsorbs 77% of 1,4-cyclohexadiene from a mixture of 1,3-cyclohexadiene and 1,4-cyclohexadiene (50%:50%) based on H–π interactions.

Similar selectivities are observed in triply interpenetrated metal-organic frameworks [$Co_3(ndc)_3(bipyen)_{1.5}$]·$H_2O$ and [$Co_2(ndc)_2$(bipyen)]·$C_6H_6$·$H_2O$ (bipyen = *trans*-1,2-bis(4-pyridyl)ethylene, ndc = 2,6-naphthalenedicarboxylate) that have been prepared by hydrothermal reactions.[36] The structural transformation of the framework upon guest removal has been monitored by powder X-ray diffraction studies. When guest-free [$Co_2(ndc)_2$(bipyen)] is exposed to a vapor of cyclohexene for 10 days at room temperature, the XRD pattern becomes similar to that of [$Co_2(ndc)_2$(bipyen)]·$C_6H_6$·$H_2O$. On the contrary, similar experiments with toluene, xylenes isomers, cycloheptatriene, and cyclohexane do not change the XRD pattern of guest-free [$Co_2(ndc)_2$(bipyen)]. Clearly, these experiments demonstrate that [$Co_2(ndc)_2$(bipyen)] has selective adsorption properties for cyclohexene over toluene, xylenes, cycloheptatriene, or cyclohexane.

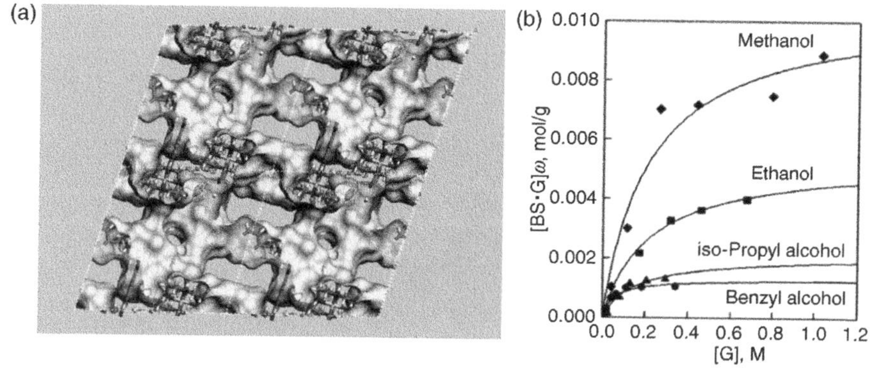


**FIGURE 7.30** (a) The Connolly surface of **BOF-1** with the pore walls. (b) A plot of the amount of sorption by dried solid **BOF-1** (powder form) with different alcohols in toluene with respect to their concentration.

### 7.3.5 Selective Inclusion of Organic Guest Molecules

A closely related phenomenon is the selective inclusion of organic guest molecules in MOFs. A unique synthetic approach using a metal complex as the building unit has been adopted for the synthesis of porous MOFs.[37,38] The bilayer open framework $[Ni_2(C_{26}H_{52}N_{10})]_3[BTC]_4{\cdot}6C_5H_5N{\cdot}36H_2O$ (**BOF-1**) is synthesized using the two parallel vacant coordination sites of square-planar Ni(II) azamacrocyclic complex $[Ni_2(C_{26}H_{52}N_{10})(Cl)_4]{\cdot}H_2O$. When guest-free **BOF-1** is exposed to different alcohols in toluene solution, the amount of sorption decreases in order of methanol $\gg$ ethanol $\gg$ isopropanol > benzyl alcohol (Figure 7.30). The higher sorption for methanol and ethanol is attributed to the size and strong hydrogen bonding interactions.

In an approach similar to the one mentioned above, $[Ni(cyclam)(bpydc)]{\cdot}5H_2O$ (cyclam = 1,4,8,11-tetraazacyclotetradecane; bpydc = 2,2′-bipyridyl-5,5′-dicarboxylate) has been assembled, in which a three-fold braid of slanting rod results in the robust framework with a permanent porosity.[39] The structural integrity is maintained during the reversible dehydration–rehydration with the change of the color from pale yellow to pink and back to pale yellow, respectively. A Langmuir surface area of 817 $m^2$ was determined from the $N_2$ sorption isotherms at 77 K. Interestingly, the guest-free porous framework adsorbs ethanol, phenol, pyridine, and benzene, but not toluene, which is illustrated in Figure 7.31. The free carbonyl groups exposed to the channels are responsible for the observed selectivity for different guest molecules.

A microporous framework has also been prepared from metalloporphyrin subunits. $[CoT(p\text{-}CO_2)PPCo_{1.5}]$ (PIZA-1) is based on carboxylic acid-substituted tetraphenylporphyrins and Co(II) ions, and possesses hydrophilic pores (Figure 7.32a).[40] The accessible volume of evacuated PIZA-1 is estimated to be nearly 48% of the unit cell volume based on 1.40 Åprobe radius. PIZA-1 exhibits the extraordinary high capacity for the sorption of water due to the hydrophilic environments of pores. Interestingly, PIZA-1 also demonstrate that the amount of adsorbed molecules decreases in the order of methanol > ethanol > propanol $\gg$ butanol > hexanol (Figure 7.32b). The order is

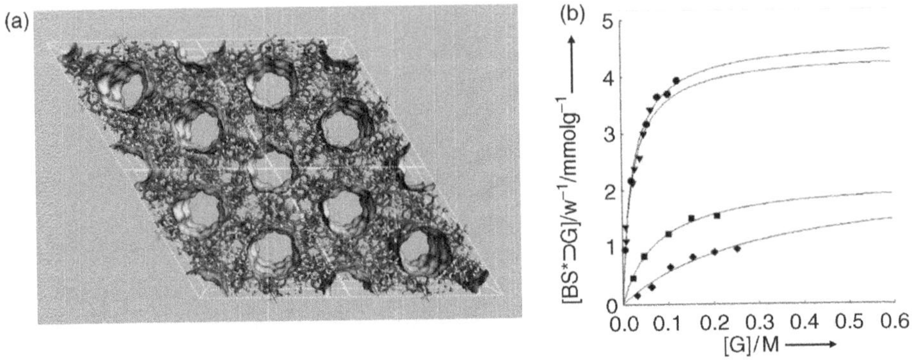


**FIGURE 7.31** (a) The Connolly surface of [Ni(cyclam)(bpydc)] with the pore walls. (b) Binding of host solid [Ni(cyclam)(bpydc)] with the organic guests EtOH (●), PhOH (▼), pyridine (■), and benzene (◆). $K_f$, [G], and [BS ⊃ G]/$\omega$ are defined as binding constant, concentration of guest (G), and binding sites of the host to the guest (G), respectively.

in line with the increasing hydrophobicity and steric hindrance. Adsorption properties of amines are also classified in terms of the size of the molecules.

A nickel(II) hexaazamacrocyclic complex with pendant pyridine groups has been pillared with *cis,cis*-1,3,5-cyclohexanetricarboxylic acid ($H_3CTC$) to form either a 1D chain structure (**A**; $[Ni(C_{20}H_{32}N_8)][C_6H_9(COOH)_2(COO)]_2 \cdot 4H_2O$) through hydrogen bonding of a discrete complex or a 2D layer structure

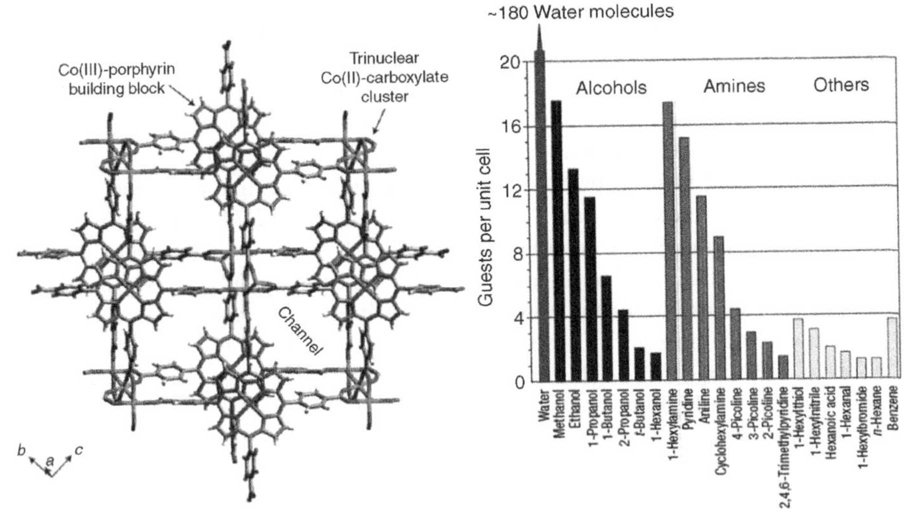


**FIGURE 7.32** (a) The structure of the PIZA-1 network. (b) Size-, shape-, and functional-group selectivity as probed by thermal desorption of guest molecules.

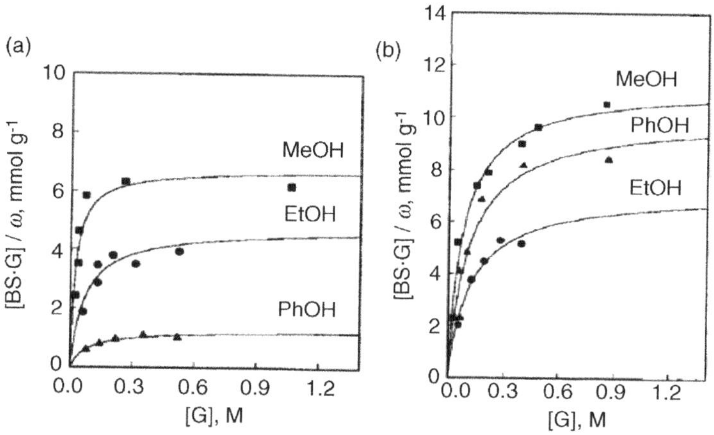


**FIGURE 7.33** The sorption uptakes of MeOH, EtOH, and PhOH by $[Ni(C_{20}H_{32}N_8)][C_6H_9(COOH)_2(COO)]_2$ (a) and $[Ni(C_{20}H_{32}N_8)]_3[C_6H_9(COO)_3]_2$ (b).

(**B**; $[Ni(C_{20}H_{32}N_8)]_3[C_6H_9(COO)_3]_2 \cdot 16H_2O$) depending on the coordination mode of CTC.[41] The X-ray powder diffraction patterns indicate that both frameworks are deformed upon the removal of water guests and restored upon rebinding of water. When evacuated, both materials do not interact with benzene and toluene, but they accommodate methanol, ethanol, and phenol in toluene solvent with the binding constants of 42, 14, and 12 $M^{-1}$, respectively, for **A**, and 13, 8.2, and 8.9 $M^{-1}$, respectively, for **B** (Figure 7.33).

## 7.4 POTENTIAL APPLICATIONS IN PRACTICAL SEPARATION PROCESSES

A seminal report on the industrial perspectives of using MOFs for practical applications has been published recently.[42] According to the report, gas separation and purification are among the areas for which MOFs can grow to be competitors to conventional porous materials based on aluminosilicates or carbons. For example, $[Cu_3(BTC)_2(H_2O)_3]$ is found not only active for the removal of sulfur odorant from natural gas, but the volume-specific uptake of the odorant is considerably higher than those for commercially available carbon materials (Figure 7.34). Similarly, electron-rich compounds such as ammonia, amines, water traces, and alcohols are removed by $[Cu_3(BTC)_2(H_2O)_3]$. Accompanying the uptake of the polar components from a mixture with methane is the color change of $[Cu_3(BTC)_2(H_2O)_3]$ from deep blue to light green, which is reversed during the regeneration of the saturated adsorbent. The presence of open metal sites in the paddlewheel SBUs of $[Cu_3(BTC)_2(H_2O)_3]$ explains the observed color change.

$[Cu_3(BTC)_2(H_2O)_3]$ has also been tested for the separation of rare gas mixtures. When a mixture of Kr and Xe (94:6 mol%) is fed to a tubular reactor filled with the

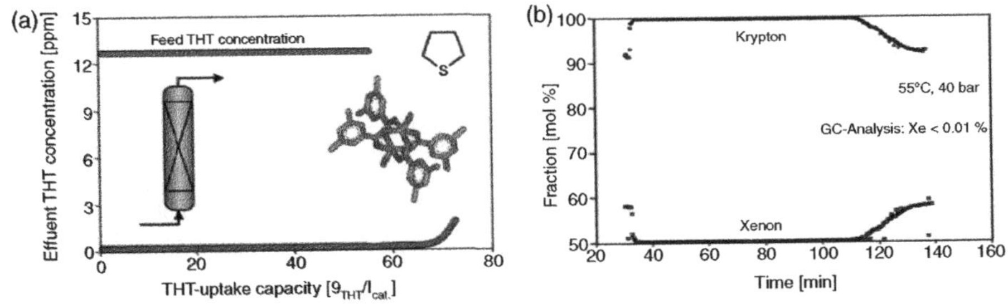


**FIGURE 7.34** (a) Breakthrough curve of continuous tetrahydrothiophene removal from nature gas using $[Cu_3(BTC)_2(H_2O)_3]$ out of electrochemical synthesis. (b) Separation of Kr-Xe gas mixture by continuous adsorption on electrochemically produced $[Cu_3(BTC)_2(H_2O)_3]$.

porous MOF, the Xe content in the gas stream leaving the adsorbent is found to be less than 0.01% implying a highly selective sorption of Xe over Kr. The uptake capacity of $[Cu_3(BTC)_2(H_2O)_3]$ for Xe (~60 wt%) is nearly twice that of an active carbon having the surface area of 2000 $m^2/g$. These results suggest that a pressure-swing adsorption process using porous MOFs is possible and may offer a much simpler and cost-saving alternative to the currently used cryogenic distillation methods. The high mobility of gas molecules in MOFs in general would be further advantages because a faster swing operation period between adsorption and desorption cycles means the reduction of both time and energy consumption.

If porous MOFs are to be utilized for large-scale separation processes, it would be desirable to fabricate the materials into the form of membranes or thin films. This is because membrane- or thin film-based processes can effectively avoid a tradeoff between selectivity and throughput.[43] An interesting work in this regard has been published recently in which the oriented crystallization of microporous $Mn(HCOO)_2$ has been attempted on porous $\alpha$-$Al_2O_3$ or porous graphite discs.[44] Proper orientations of the $Mn(HCOO)_2$ crystals having 1D channels on supports are necessary in order to ensure the mass transport through the MOF-membrane. After careful studies that include the modification of synthesis route, the authors were able to obtain a composite membrane with reasonable crystal densities and the tilt angle of the 1D channels (34°). Although the work has not led to membrane-quality coatings, important factors have been identified. That is, the substrates have to be carefully selected and properly treated ahead of in situ crystallizations – hydrolysis with NaOH for alumina and oxidation for graphite. Also, in case of $Mn(HCOO)_2$, a new synthesis route using sodium formate instead of formic acid has been found to be more effective with the given substrates. Considering the vast library of porous MOFs having higher symmetry and multi-dimensional pore systems, this area may develop into a topical issue quickly.

The purification of synthetic gas obtained from the reforming of natural gas presents another intriguing challenge. The reason is because this process accounts for most of the hydrogen used for fuel cells in industry, and because $CH_4$ and $CO_2$ have to be removed from $H_2$ from the gas mixture. The separation of $CO_2$ from $CH_4$ is also important in order to prevent the corrosion of natural gas pipelines. Some insights in

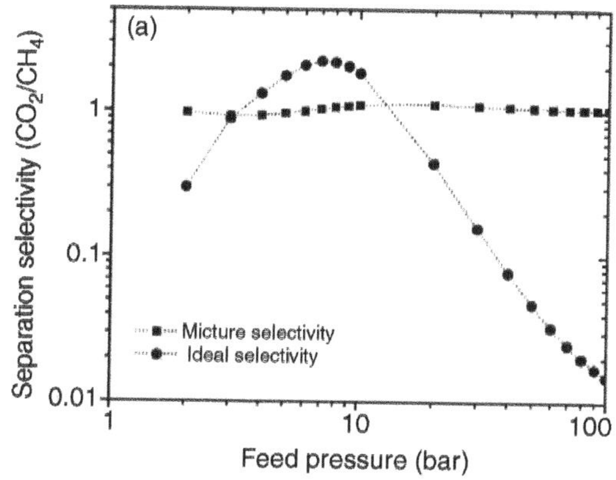


**FIGURE 7.35** Predicted ideal (circles) and mixture (squares) permeation selectivities for a 10-$\mu$m-thick MOF-5 membrane with a transmembrane pressure drop of 1.38 bar. Mixture selectivities are based on an equimolar feed.

these regards have been provided by two reports of theoretical studies using two representative MOFs, $[Cu_3(BTC)_2(H_2O)_3]$ and MOF-5. First, it is suggested from the simulations using single-component flux that the selectivity for $CO_2/CH_4$ by MOF-5 varies depending on the feed pressure, and may even be reversed at 50 bar or above (Figure 7.35).[45] However, the author also found that the simulations return very poor selectivity when a stream of equimolar mixture is used for the feeds. This is probably because MOF-5 is a porous material with large, straight channels with uniform sizes.

Second, it is expected that $[Cu_3(BTC)_2(H_2O)_3]$ would be more selective for $CH_4/H_2$, $CO_2/H_2$, or $CO_2/CH_4$ than MOF-5 (Figure 7.36).[46] Again, this result is justified in terms of the geometry and size of the pores in the frameworks. Similar to early findings on zeolites, gas molecules first occupy corner regions, and the organic linkers and channels start to accumulate adsorbate molecules under much higher pressures. Therefore, $[Cu_3(BTC)_2(H_2O)_3]$, which has both cavity-like pores and open channels, is merited over MOF-5 under the same simulation conditions.

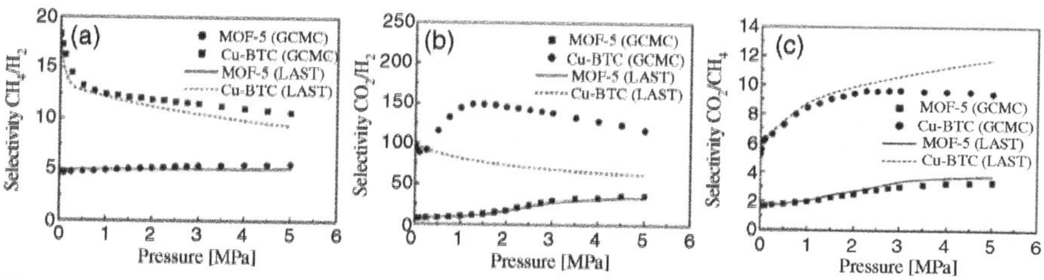


**FIGURE 7.36** (a) Selectivity for methane from the equimolar mixture of hydrogen and methane in MOF-5 and $[Cu_3(BTC)_2(H_2O)_3]$ at 298 K, (b) selectivity for $CO_2$ from the equimolar mixture of $H_2$ and $CO_2$ in MOF-5 and $[Cu_3(BTC)_2(H_2O)_3]$ at 298 K, and (c) selectivity for $CO_2$ from equimolar mixture of $CH_4$ and $CO_2$ in MOF-5 and $[Cu_3(BTC)_2(H_2O)_3]$ at 298 K.

## 7.5 CONCLUSIONS

It has been found that most cases of the selective gas sorption by MOFs reported in literature are based on size-exclusion. There are only a handful of reports, though without a conclusive evidence, in which MOFs with large, simple channel-type pores discriminate light gases. Unlike purely inorganic frameworks in which the size of pores and windows is determined by topologies, both the shape and the connectivity of organic spacers are responsible for delimiting exogenous guests in porous MOFs. Kinetic diameters of most gases of interest are in the range 3–4 Å, and therefore, rigid frameworks with the openings in comparable sizes may exhibit selective sorption properties. The catenation or interpenetration of frameworks is a convenient route to such an end. However, mechanically interlocked systems tend to show dynamic behavior depending on temperature and pressure or on the presence of guest molecules. If this is the case, care has to be taken whenever sorption measurements are interpreted in terms of the crystal structures of frameworks. The same is true for non-interpenetrated MOFs that are flexible and dynamic in nature. Preferably, various in situ measurements of bulk diffraction patterns should be in hand to fully understand the observed gas sorption selectivity.

Separation of organic guest molecules was also widely investigated in the industrial aspects. For selective sorption of organic guest molecules, the pore environments of MOFs have been finely controlled by the choice of organic building units. The hydrophobic and hydrophilic pores of MOFs mainly aim to target separations of alcohols, hydrophobic guests, stereoisomers, etc. In some cases where the window size of frameworks is limited for the inclusion of organic guest molecules, interactions with specific functional groups of organic guests may open the small apertures thanks to the flexibility of MOFs.

Assessments of the selectivity of gas sorption in a realistic sense, that is, near room temperature and moderately high pressures, suggest that either temperature- or pressure-swing adsorption in a cyclic process should be possible to separate or at least enrich a specific component in a gaseous mixture. The most convenient way of utilizing porous MOFs for such purposes would be to use them as column-packing agents for which a bulk synthesis is a prerequisite. More challenging, however, better suited for industrial applications, tasks would be to fabricate porous MOFs as gas-permeation membranes. The growth of MOF crystals on a solid support has already been attempted, and the results of the work may be pointing a direction for future research in this area. High-symmetry networks with three-dimensionally interconnected channels may be advantageous in this regard because the orientation of crystals with respect to the gas flow does not become problematic for efficient mass transport.

## ACKNOWLEDGMENTS

Our work described here has been supported by the Creative Research Initiatives, Brain Korea 21 and World Class University (Project No. R31-2008-000-10059-0) programs of the Korean Ministry of Education, Science and Technology. We also acknowledge Steel Science Program of POSCO for support of this work (KK). HC thanks the support from the Basic Science Research Program of the National Research Foundation of Korea funded by the Ministry of Education, Science and Technology (2009-0087194).

## LIST OF ABBREVIATIONS

| | |
|---|---|
| anthdc | 9,10-anthracene-dicarboxylate |
| aptz | 4-aminophenyl-1H-tetrazole |
| azdc | 4,4′-azobenzenedicarboxylate |
| azpy | 4,4′-azobispyridine |
| bdc | benzene dicarboxylate |
| bipyen | *trans*-1,2-bis(4-pyridyl)ethylene |
| bpe | *trans*-bis-(4-pyridyl)ethylene |
| bpee | *trans*-1,2 bis(4-pyridyl)ethylene |
| bpp | 1,3-bis(4-pyridyl)-propane |
| bpy | 4,4′-bipyridyl |
| bpydc | 2,2′-bipyridyl-5,5′-dicarboxylate |
| 4-btapa | 1,3,5-benzene tricarboxylic acid tris[N-(4-pyridyl)amide |
| btb | benzenetribenzoate |
| BTC | 1,3,5-benzenetricarboxylate |
| CTC | *cis,cis*-1,3,5-cyclohexanetricarboxylate |
| cyclam | 1,4,8,11-tetraazacyclotetradecane |
| def | diethylformamide |
| dhbz | 2,5-dihydroxybenzoate |
| DME | dimethyl ether |
| DMF | dimethylformamide |
| dpyg | 1,2-dipyridylglycol |
| fum | fumarate |
| ip | isophthalte |
| MEM | Maximum Entropy Method |
| ndc | 2,6-naphthalenedicarboxylate |
| pda | 1,4-phenylenediacetate |
| 2,4-pdc | pyridine-2,4-dicarboxylate |
| pur | purinate |
| py | pyridine |
| 2-pymo | 2-pyrimidinolate |
| pyrdc | pyridine-2,3-dicarboxylate |
| pyz | pyrazine |

| | |
|---|---|
| pzdc | pyrazine-2,3-dicarboxylate |
| tbip | 5-*tert*-butylisophthalate |
| TCNQ | 7,7,8,8-tetracyano-*p*-quinodimethane |
| t(*p*-$CO_2$)PP | tetra(4-carboxyphenyl) porphyrin |

## Note Added in Proof

Since this chapter was prepared, several important papers on the subject have been published. Britt et al. showed the selective adsorption of six metal-organic frameworks (MOF-5, IRMOF-3, MOF-74, MOF-177, MOF-199 and IRMOF-62) for harmful gases such as $SO_2$, $NH_3$, $Cl_2$, tetrahydrothiophene, $C_6H_6$, $CH_2Cl_2$ and ethylene oxide. The reported selectivities are attributed to the strong interactions between coordinatively unsaturated metal sites (MOF-74 and MOF-199) or amino functionality (IRMOF-3) with the toxic gases or vapors.[47] Kim and coworkers demonstrated an unusual temperature-dependent gas adsorption behavior of MOFs, which may be attributed to the dynamic opening of the pore aperture and/or sufficient kinetic energy of adsorbates to overcome a diffusion barrier above a critical temperature.[48] Babarao et al. reported a molecular simulation study for the separation of industrially important gas mixtures ($CO_2/H_2$, $CO_2/CH_4$, and $CO_2/N_2$) in *rho*-ZMOF.[49] The coadsorption studies using a $CO_2$–$CH_4$ mixture with MIL-53(Cr)[50] and amino-MIL-53(Al)[51] showed a stronger affinity for $CO_2$, demonstrating their potential in pressure-swing adsorption (PSA) processes. A similar study was reported by Yaghi and coworkers in which a $Mg^{2+}$ analogue of MOF-74 was used as a sorbent. The interaction between $CO_2$ and open metal sites of MOF-74(Mg) was proposed to be the reason for the high selectivity.[52] Yan and coworkers showed the GC separation of xylene isomers and ethylbenzene with a MIL-101 coated capillary column at 160 °C.[53] Deng, et al. reported that MOFs with a large number of different functionalities in one structure exhibits unusual increases in the selective uptake of $CO_2$ over CO.[54] Chun et al. reported a flexible 3D net [Zn(pydc)(dma)] discriminating small gas molecules not by the sizes, but by the interactions between the gas molecules and the framework.[55] Banerjee, et al. reported that the $CO^2$ uptake capacity of ZIF-78 is 4 times higher than that of $CH_4$ uptake and 8 and 10 times higher than that of $O_2$ and $N_2$ uptake, respectively, at 800 Torr and 298 K.[56]

## REFERENCES


1. Connolly, M. L. Analytical molecular surface calculation. *J. Appl. Cryst.* **1983**, *16*, 548–558.
2. Oak Ridge National Laboratory, BCS. Materials for Separation Technologies: Energy and Emission Reduction Opportunities. **2005**. http://www1.eere.energy.gov/industry/imf/pdfs/separations_ report.pdf.
3. Pan, L.; Adams, K. M.; Hernandez, H. E.; Wang, X.; Zheng, C.; Hattori, Y.; Kaneko, K. Porous lanthanide–organic frameworks: synthesis, characterization, and unprecedented gas adsorption properties. *J. Am. Chem. Soc.* **2003**, *125*, 3062–3067.

4. Dybtsev, D. N.; Chun, H.; Yoon, S. H.; Kim, D.; Kim, K. Microporous manganese formate: a simple metal-organic porous material with high framework stability and highly selective gas sorption Properties. *J. Am. Chem. Soc.* **2004**, *126*, 32–33.
5. Samsonenko, D. G.; Kim, H.; Sun, Y.; Kim, G. -H.; Lee, H. -S.; Kim, K. Microporous magnesium and manganese formates for acetylene storage and separation. *Chem. Asian J.* **2007**, *2*, 484–488.
6. Kortunov, P. V.; Heinke, L.; Arnold, M.; Nedellec, Y.; Jones, D. J.; Caro, J.; Kärger, J. Intracrystalline diffusivities and surface permeabilities deduced from transient concentration profiles: methanol in MOF manganese formate. *J. Am. Chem. Soc.* **2007**, *129*, 8041–8047.
7. Rood, J. A.; Noll, B. C.; Henderson, K. W. Synthesis, structural characterization, gas sorption and guest-exchange studies of the lightweight, porous metal-organic framework $\alpha$-[$Mg_3(O_2CH)_6$]. *Inorg. Chem.* **2006**, *45*, 5521–5528.
8. Wang, Z.; Zhang, B.; Kurmoo, M.; Green, M. A.; Fujiwara, H.; Otsuka, T.; Kobayashi, H. Synthesis and characterization of a porous magnetic diamond framework, $Co_3(HCOO)_6$, and its $N_2$ sorption characteristic. *Inorg. Chem.* **2005**, *44*, 1230–1237.
9. Chen, B.; Ma, S.; Zapata, F.; Fronczek, F. R.; Lobkovsky, E. B.; Zhou, H. C. Rationally designed micropores within a metal-organic framework for selective sorption of gas molecules. *Inorg. Chem.* **2007**, *46*, 1233–1236.
10. Chen, B.; Ma, S.; Hurtado, E. J.; Lobkovsky, E. B.; Zhou, H. C. A triply interpenetrated microporous metal-organic framework for selective sorption of gas molecules. *Inorg. Chem.* **2007**, *46*, 8490–8492.
11. Chen, B.; Ma, S.; Hurtado, E. J.; Lobkovsky, E. B.; Liang, C.; Zhu, H.; Dai, S. Selective gas sorption within a dynamic metal-organic framework. *Inorg. Chem.* **2007**, *46*, 8705–8709.
12. Ma, S.; Wang, X. S.; Collier, C. D.; Manis, E. S.; Zhou, H. C. Ultramicroporous metal-organic framework based on 9,10-anthracenedicarboxylate for selective gas adsorption. *Inorg. Chem.* **2007**, *46*, 8499–8501.
13. Navarro, J. A. R.; Barea, E.; Salas, J. M.; Masciocchi, N.; Galli, S.; Sironi, A.; Ania, C. O.; Parra, J. B. $H_2$, $N_2$, CO, and $CO_2$ sorption properties of a series of robust sodalite-type microporous coordination polymers. *Inorg. Chem.* **2006**, *45*, 2397–2399.
14. Dincă, M.; Long, J. R. Strong $H_2$ binding and selective gas adsorption within the microporous coordination solid $Mg_3(O_2C\text{-}C_{10}H_6\text{-}CO_2)_3$. *J. Am. Chem. Soc.* **2005**, *127*, 9376–9377.
15. Ma, S.; Sun, D.; Wang, X. S.; Zhou, H. C. A mesh-adjustable molecular sieve for general use in gas separation. *Angew. Chem. Int. Ed.* **2007**, *46*, 2458–2462.
16. Yoon, J. W.; Jhung, S. H.; Hwang, Y. K.; Humphrey, S. M.; Wood, P. T.; Chang, J. S. Gas-sorption selectivity of CUK-1: a porous coordination solid made of cobalt(II) and pyridine-2,4-dicarboxylic acid. *Adv. Mater.* **2007**, *19*, 1830–1834.
17. Humphrey, S. M.; Chang, J. S.; Jhung, S. H.; Yoon, J. W.; Wood, P. T. Porous cobalt (II)–organic frameworks with corrugated walls: structurally robust gas-sorption materials. *Angew. Chem. Int. Ed.* **2007**, *46*, 272–275.
18. Zou, Y.; Hong, S.; Park, M.; Chun, H.; Lah, M. S. Selective gas sorption property of an interdigitated 3-D metal-organic framework with 1-D channels. *Chem. Commun.* **2007**, *43*, 5182–5184.

19. Matsuda, R.; Kitaura, R.; Kitagawa, S.; Kubota, Y.; Belosludov, R. V.; Kobayashi, T. C.; Sakamoto, H.; Chiba, T.; Takata, M.; Kawazoe, Y.; Mita, Y. Highly controlled acetylene accommodation in a metal-organic microporous material. *Nature* **2005**, *436*, 238–241.

20. Hayashi, H.; Côté, A. P.; Furukawa, H.; O'Keeffe, M.; Yaghi, O. M. Zeolite A imidazolate frameworks. *Nature Mater.* **2007**, *6*, 501–506.

21. Ma, S.; Wang, X. S.; Manis, E. S.; Collier, C. D.; Zhou, H. C. Metal–organic framework based on a trinickel secondary building unit exhibiting gas-sorption hysteresis. *Inorg. Chem.* **2007**, *46*, 3432–3434.

22. Moon, H. R.; Kobayashi, N.; Suh, M. P. Porous metal-organic framework with coordinatively unsaturated $Mn^{II}$ sites: sorption properties for various gases. *Inorg. Chem.* **2006**, *45*, 8672–8676.

23. Li, Y.; Yang, R. T. Gas adsorption and storage in metal-organic framework MOF-177. *Langmuir* **2007**, *23*, 12937–12944.

24. Kitaura, R.; Seki, K.; Akiyama, G.; Kitagawa, S. Porous coordination-polymer crystals with gated channels specific for supercritical gases. *Angew. Chem. Int. Ed.* **2003**, *42*, 428–431.

25. Hasegawa, S.; Horike, S.; Matsuda, R.; Furukawa, S.; Mochizuki, K.; Kinoshita, Y.; Kitagawa, S. Three-dimensional porous coordination polymer functionalized with amide groups based on tridentate ligand: selective sorption and catalysis. *J. Am. Chem. Soc.* **2007**, *129*, 2607–2614.

26. Maji, T. K.; Uemura, K.; Chang, H. C.; Matsuda, R.; Kitagawa, S. Expanding and shrinking porous modulation based on pillared-layer coordination polymers showing selective guest adsorption. *Angew. Chem. Int. Ed.* **2004**, *43*, 3269–3272.

27. Kitaura, R.; Fujimoto, K.; Noro, S. I.; Kondo, M.; Kitagawa, S. A pillared-layer coordination polymer network displaying hysteretic sorption: $[Cu_2(pzdc)_2(dpyg)]_n$ (pzdc= pyrazine-2,3-dicarboxylate; dpyg=1,2-di(4-pyridyl)-glycol). *Angew. Chem. Int. Ed.* **2002**, *41*, 133–135.

28. Maji, T. K.; Mostafa, G.; Matsuda, R.; Kitagawa, S. Guest-induced asymmetry in a metal-organic porous solid with reversible single-crystal-to-single-crystal structural transformation. *J. Am. Chem. Soc.* **2005**, *127*, 17152–17153.

29. Uchida, S.; Mizuno, N. Zeotype ionic crystal of $Cs_5[Cr_3O(OOCH)_6(H_2O)_3]$ $[\alpha\text{-}CoW_{12}O_{40}]\cdot 7.\ 5H_2O$ with shape-selective adsorption of water. *J. Am. Chem. Soc.* **2004**, *126*, 1602–1603.

30. Horike, S.; Tanaka, D.; Nakagawa, K.; Kitagawa, S. Selective guest sorption in an interdigitated porous framework with hydrophobic pore surfaces. *Chem. Commun.* **2007**, *43*, 3395–3397.

31. Matsuda, R.; Kitaura, R.; Kitagawa, S.; Kubota, Y.; Kobayashi, T. C.; Horike, S.; Takata, M. Guest shape-responsive fitting of porous coordination polymer with shrinkable framework. *J. Am. Chem. Soc.* **2004**, *126*, 14063–14070.

32. Pan, L.; Parker, B.; Huang, X.; Olson, D. H.; Lee, J.; Li, J. Zn(tbip) ($H_2$tbip= 5-tert-butyl isophthalic acid): a highly stable guest-free microporous metal-organic framework with unique gas separation capability. *J. Am. Chem. Soc.* **2006**, *128*, 4180–4181.

33. Wang, X.; Liu, L.; Jacobson, A. J. Intercalation of organic molecules into vanadium(IV) benzenedicarboxylate: adsorbate structure and selective absorption of organosulfur compounds. *Angew. Chem. Int. Ed.* **2006**, *45*, 6499–6503.

34. Alaerts, L.; Kirschhock, C. E. A.; Maes, M.; van der Veen, M. A.; Finsy, V.; Depla, A.; Martens, J. A.; Baron, G. V.; Jacobs, P. A.; Denayer, J. F. M.; De Vos, D. E. Selective adsorption and separation of xylene isomers and ethylbenzene with the microporous vanadium(IV) terephthalate MIL-47. *Angew. Chem. Int. Ed.* **2007**, *46*, 4293–4297.

35. Shimomura, S.; Horike, S.; Matsuda, R.; Kitagawa, S. Guest-specific function of a flexible undulating channel in a 7,7,8,8-tetracyano-p-quinodimethane dimer-based porous coordination polymer. *J. Am. Chem. Soc.* **2007**, *129*, 10990–10991.

36. Choi, E. Y.; Park, K.; Yang, C. M.; Kim, H.; Son, J. H.; Lee, S. W.; Lee, Y. H.; Min, D.; Kwon, Y. U. Benzene-templated hydrothermal synthesis of metal-organic frameworks with selective sorption properties. *Chem. Eur. J.* **2004**, *10*, 5535–5540.

37. Choi, H. J.; Suh, M. P. Dynamic and redox active pillared bilayer open framework: single-crystal-to-single-crystal transformations upon guest removal, guest exchange, and framework oxidation. *J. Am. Chem. Soc.* **2004**, *126*, 15844–15851.

38. Suh, M. P.; Ko, J. W.; Choi, H. J. A metal-organic bilayer open framework with a dynamic component: single-crystal-to-single-crystal transformations. *J. Am. Chem. Soc.* **2002**, *124*, 10976–10977.

39. Lee, E. Y.; Suh, M. P. A robust porous material constructed of linear coordination polymer chains: reversible single-crystal to single-crystal transformations upon dehydration and rehydration. *Angew. Chem. Int. Ed.* **2004**, *43*, 2798–2801.

40. Kosal, M. E.; Chou, J. H.; Wilson, S. R.; Suslick, K. S. A functional zeolite analogue assembled from metalloporphyrins. *Nature Mater.* **2002**, *1*, 118–121.

41. Min, K. S.; Suh, M. P. Self-assembly and selective guest binding of three-dimensional open-framework Solids from a macrocyclic complex as a trifunctional metal building block. *Chem. Eur. J.* **2001**, *7*, 303–313.

42. Mueller, U.; Schubert, M.; Teich, F.; Puetter, H.; Schierle-Arndt, K.; Pastré, J. Metal–organic frameworks–prospective industrial applications. *J. Mater. Chem.* **2006**, *16*, 626–636.

43. Lai, Z.; Bonilla, G.; Diaz, I.; Nery, J. G.; Sujaoti, K.; Amat, M. A.; Kokkoli, E.; Terasaki, O.; Thompson, R. W.; Tsapatsis, M.; Vlachos, D. G. Microstructural optimization of a zeolite membrane for organic vapor separation. *Science* **2003**, *300*, 456–460.

44. Arnold, M.; Kortunov, P.; Jones, D. J.; Nedellec, Y.; Kärger, J.; Caro, J. Oriented crystallisation on supports and anisotropic mass transport of the metal-organic framework manganese formate. *Eur. J. Inorg. Chem.* **2007**, 60–64.

45. Keskin, S.; Sholl, D. S. Screening metal-organic framework materials for membrane-based methane/carbon dioxide separations. *J. Phys. Chem. C* **2007**, *111*, 14055–14059.

46. Yang, Q.; Zhong, C. Molecular simulation of carbon dioxide/methane/hydrogen mixture adsorption in metal-organic frameworks. *J. Phys. Chem. B* **2006**, *110*, 17776–17783.

47. Britt, D.; Tranchemontagne, D.; Yaghi, O. M. Metal-organic frameworks with high capacity and selectivity for harmful gases.*Proc. Natl. Acad. Sci. USA* **2008**, *105*, 11623–11627.

48. Kim, H.; Samsonenko, D. G.; Yoon, M.; Yoon, J. W.; Hwang, Y. K.; Chang, J.-S.; Kim, K. Temperature-triggered gate opening for gas adsorption in microporous manganese formate. *Chem. Commun.* **2008**, *44*, 4697–4699.

49. Babarao, R.; Jiang, J. Unprecedentedly high selective adsorption of gas mixtures in *rho* zeolite-like metal-organic framework: a molecular simulation study. *J. Am. Chem. Soc.* **2009**, *131*, 11417–11425.

50. Hamon, L.; Llewellyn, P. L.; Devic, T.; Ghoufi, A.; Clet, G.; Guillerm, V.; Pirngruber, G. D.; Maurin, G.; Serre, C.; Driver, G.; van Beek, W.; Jolimaître, E.; Vimont, A.; Daturi, M.; Férey G. Co-adsorption and separation of $CO_2$-$CH_4$ mixtures in highly flexible MIL-53 (Cr) MOF. *J. Am. Chem. Soc.* **2009**, *131*, 17490–17499.

51. Couck, S.; Denayer, J. F. M.; Baron, G. V.; Rèmy, T; Gascon J.; Kapteijn, F. An amine-functionalized MIL-53 metal-organic framework with large separation power for $CO_2$ and $CH_4$.*J. Am. Chem. Soc.* **2009**, *131*, 6326–6327.

52. Britt, D.; Furukawa, H.; Wang, B.Glover, T. G.; Yaghi, O. M. Highly efficient separation of carbon dioxide by a metal-organic framework replete with open metal sites. *Proc. Natl. Acad. Sci. USA* **2009**, *106*, 20637–20640.

53. Gu, Z.-Y.; Yan, X.-P. Metal-organic framework MIL-101 for high-resolution gas-chromatographic separation of xylene isomers and ethylbenzene. *Angew. Chem. Int. Ed.* **2010**, *49*, 1477–1480.

54. Deng, H.; Doonan, C. J.; Furukawa, H.; Ferreira, R. B.; Towne, J.; Knobler, C. B.; Wang, B.; Yaghi O. M. Multiple functional groups of varying ratios in metal-organic frameworks. *Science*, **2010**, *327*, 846–850.

55. Chun, H.;Seo, Discrimination of small gas molecules through adsorption: reverse selectivity for hydrogen in a flexible metal-organic framework.*J. Inorg. Chem.* **2009**, *48*, 9980–9982.

56. Banerjee, R.; Furukawa, H.; Britt, D.; Knobler, C.; O'Keeffe, M.; Yaghi O. M. Control of pore size and functionality in isoreticular zeolitic imidazolate frameworks and their carbon dioxide selective capture properties. *J. Am. Chem. Soc.* **2009**, *131*, 3875–3877.

# 8

# HYDROGEN AND METHANE STORAGE IN METAL-ORGANIC FRAMEWORKS

DAVID J. COLLINS, SHENGQIAN MA, AND HONG-CAI ZHOU
*Department of Chemistry, Texas A&M University, College Station, TX 77843, USA*

## 8.1 INTRODUCTION[1,2]

Since 1990, rapid advancement has been made in the field of metal-organic frameworks (MOFs), a new type of functional materials. MOFs, also known as coordination polymers or coordination networks, are highly crystalline hybrid organic–inorganic materials constructed from the assembly of metal ions or clusters (called secondary building units or SBUs) and multi-topic organic ligands via coordination bonds.[3–6] Most have three-dimensional structures incorporating uniform pores and a network of channels, often containing guest species introduced during syntheses. If these guests can be removed under vacuum or at elevated temperature, and permanent porosity is retained, then guest molecules of interest, such as hydrogen, nitrogen, or methane, can be re-introduced into the network.

By varying the choice of metal ion and the shape and terminal functionality of the organic linker, a variety of MOF topologies and structures can be produced. Systematic variation in pore size and pore-wall functionalization can be achieved; this versatility has led to proposed application of MOFs in a number of practical applications, including catalysis,[7,8] gas storage,[9–11] and gas separations.[12–14] This chapter will discuss the development of porous MOFs for storage of hydrogen and methane for fuel applications, including relevant design variables and considerations, specific examples, and evaluation of emerging understanding in the field.

---

*Metal-Organic Frameworks: Design and Application,* Edited by Leonard R. MacGillivray

## 8.2 HYDROGEN STORAGE

### 8.2.1 Hydrogen Storage Goals

Decreasing stockpiles of fossil fuels and the increasing threat of global warming have prompted the global community to search for alternative energy carriers to supplement those currently used, namely to replace the use of petroleum-based gasoline and diesel fuel in road vehicles such as automobiles, buses, and trucks. Among various alternatives, hydrogen stands at the forefront: hydrogen is ubiquitous (one dihydrogen molecule can be obtained from each molecule of water), oxidation of hydrogen in an engine or fuel cell releases only water as a byproduct (and thus generates no greenhouse gases), and hydrogen has a relatively high-energy capacity.

In 2003, the U.S. government launched the Hydrogen Fuel Initiative for developing clean, hydrogen-powered automobiles to replace those currently powered by fossil fuels. The successful commercialization of hydrogen fuel-cell powered vehicles, however, largely relies on the development of a safe, efficient, and economical on-board hydrogen storage system. To maintain a typical driving range of 400–500 km, it is estimated that about 5 kg of hydrogen would be needed. Based on the idea that vehicles similar to those in use today will be powered by future high-efficiency hydrogen fuel-cell power sources, the U.S. Department of Energy (DOE) has set a number of targets for a potential hydrogen storage system (including the container and all necessary components): 4.5 wt.% (or 0.045 kg/kg) or 0.028 kg/L by the year 2010 and 5.5 wt.% (or 0.055 kg/kg) or 0.040 kg/L by the year 2015 at near-ambient temperatures (−50–80°C) and reasonable pressures (less than 100 bar). Additionally, the kinetics of hydrogen release and recharging must meet the requirements for practical applications; in other words, the hydrogen fueling and release should be totally reversible, and refueling or recharging of hydrogen should be completed within minutes.[15]

Several methods are being explored for on-board hydrogen storage. Although high-pressure hydrogen gas and cryogenically stored liquid hydrogen are currently utilized in demonstration fuel-cell powered vehicles, each of these technologies would be difficult to implement in a typical small personal vehicle. Compression of 5 kg of hydrogen to the volume of a standard 45-L automotive fuel tank requires extremely high pressure (>700 bar), and the tank itself would be quite heavy in order to contain the pressure (barring development of high-strength, lightweight new tank materials); liquefaction requires extreme cooling (to 21 K) and highly efficient insulation. Additionally, the large amount of energy input for either the cooling of liquid hydrogen or compression of gaseous hydrogen to high pressures negates most gains in energy content from the hydrogen itself.

Metal hydrides and chemical hydrides have been actively studied as hydrogen carriers in the past decades; however, the irreversibility of hydrogen uptake and poor kinetics of hydrogen release necessitate continued investigation to improve uptake/release kinetics and retention of cycling capacity.

### 8.2.2 Physisorption of Hydrogen

Compared with chemical means for hydrogen storage, physisorption of hydrogen using porous materials has the advantage of fast charge–recharge processes as well as an appreciable amount of hydrogen molecules held in the pores. In the past decade, activated carbons,[16] carbon nanotubes,[17] and inorganic zeolites[18] have been widely investigated as potential candidates for hydrogen storage. These microporous materials exhibit Type I gas-adsorption isotherms. For materials with this pore size, the potential fields of attraction between pore walls and adsorbate molecules overlap, increasing the attractive force acting on the adsorbate and in turn increasing adsorption. In some cases the adsorbate molecules may pack nearly as closely as in the bulk liquid. However, for some adsorbate gases, the pores and passages may be small enough to render some portions of the interior volume inaccessible; obviously, this is more a problem for larger adsorbate molecules than for small molecules such as hydrogen. Microporous materials typically have an internal surface area on the order of tens to thousands of square meters per gram. The internal surface area of a microporous material is relatively easy to measure using the BET method; however, surface area does not necessarily correlate well with hydrogen uptake.

The weak interactions (via van der Waals forces) between hydrogen molecules and these porous materials yield limited hydrogen uptake even at low temperatures and high pressures despite high-surface areas. A crucial parameter for measuring the interaction of hydrogen with the pore walls is the heat of adsorption. Physisorbents interact with adsorbed hydrogen weakly, with $\Delta H_{ads}$ typically considerably less than 10 kJ/mol; significant adsorption of hydrogen can only occur at cryogenic temperatures. For the binding of hydrogen on a homogenous surface, Bhatia and Meyers[19] have calculated the optimum $\Delta H_{ads}$ at room temperature and 30 atm pressure to be ~15 kJ/mol.

### 8.2.3 Hydrogen Storage in MOFs

Porous MOFs have been recently deemed one of the most promising candidates to approach the DOE targets for on-board hydrogen storage, due to their high-specific surface areas, tunable pore sizes, functionalizable pore walls, and well-defined framework–hydrogen interaction sites. In May 2003, Yaghi and coworkers reported what is believed to be the first measurements of hydrogen adsorption on an MOF: a remarkable 4.5 wt% at 77 K and pressures less than 1 atm, and 1.0 wt% at room temperature at 20 bar on the material $Zn_4O(bdc)$ (bdc = 1,4-benzenedicarboxylate) (also referred to as MOF-5 and IRMOF-1).[20] These values were later adjusted downward based on follow-up studies,[21] but the idea remained: for hydrogen storage, these porous MOFs were a competitive alternative to other physisorption-based materials such as zeolites or activated carbon. Since 2003, at least 70 unique MOFs have been evaluated for their ability to store hydrogen.[2] Coupled with measurements of porosity and surface area based on nitrogen adsorption, some understanding of the many factors that determine the hydrogen uptake by a porous MOF has been developed. Recent computational studies, involving both electronic-structure methods (*ab initio* and density functional theory) and molecular mechanics (Grand Canonical Monte Carlo methods), in addition to increasingly detailed structural characterization of the hydrogen-adsorbed species (including neutron scattering and synchrotron X-ray diffraction) have added insight to these remarkable materials and the mechanisms of hydrogen adsorption.

***8.2.3.1 Low Pressure Cryo-Temperature Hydrogen Adsorption Studies*** Although the DOE targets for hydrogen storage are for systems operating at near-ambient temperatures and high pressures, the uptake values of hydrogen at 77 K and 1 atm have been widely investigated and generally used as benchmarks to compare the hydrogen adsorption capacities of different MOF materials. These values can be very useful and instructive at this early stage of exploration for hydrogen storage materials. Several factors influencing the hydrogen uptake of porous MOFs at 77 K/1 atm, such as specific surface area/pore volume, pore sizes, catenation, etc., have been extensively studied.

*Specific Surface Area/Pore Volume* Generally speaking, pore volume is proportional to specific surface area. Most porous MOFs exhibit exceptionally high-specific surface areas compared to carbon materials and inorganic zeolites. The record for highest published surface area in MOF has been broken repeatedly over the past several years. For example, a porous MOF with specific surface area of 4500 $m^2/g$ and pore volume of 1.61 $cm^3/g$ was reported for MOF-177 in 2004;[22] these values were eclipsed less than 1 year later by MIL-101, with surface area and pore volume of 5500 $m^2/g$ and 1.9 $cm^3/g$, respectively.[23]

In studies of zeolites and activated carbons, it was proposed that a higher gas adsorption could be achieved within materials possessing a large volume of micropores with an "appropriate diameter."[18] The influence of surface area and pore volume on hydrogen uptake at 77 K/1 atm have been extensively studied in MOFs; however, it has been found that for MOFs with high-specific surface areas (above 1000 $m^2/g$) and large pore volumes (over 1.0 $cm^3/g$), there is no direct correlation between specific surface area/pore volume and hydrogen adsorption.[2] For example, MOF-177 can only adsorb 1.25 wt% hydrogen at 77 K/1 atm, despite its high-surface area of 4500 $m^2/g$ and pore volume of 1.61 $cm^3/g$; however, IRMOF-8, whose surface area (1466 $m^2/g$) and pore volume (0.52 $cm^3/g$) are less than one-third of those of MOF-177, can uptake 1.5 wt% hydrogen under similar conditions.[21] The lack of a linear correlation between hydrogen adsorption capacity and surface area/pore volume strongly indicates that low-pressure hydrogen adsorption is controlled by other factors, which will be discussed below.

*Pore Size* Low hydrogen-adsorption capacities in porous MOFs with high-surface areas and large pore volumes are presumably due to weak interactions between hydrogen molecules and the frameworks. Perhaps counterintuitively, smaller pores actually take up hydrogen more effectively than very large ones. Reducing the pore size allows the dihydrogen molecule to interact with multiple portions of the framework; in a smaller pore, the attractive potential fields of opposite walls overlap.[24,25] This has been extensively explored as a strategy to increase hydrogen-framework interactions, thereby improving hydrogen uptake.[26,27] Systematic investigation of pore sizes on hydrogen uptake was recently exemplified in a series of NbO-type MOFs based on tetracarboxylate organic ligands and dicopper paddlewheel SBUs. Extension of biphenyl-3,3′,5,5′-tetracarboxylate to terphenyl-3,3′,5,5′-tetracarboxylate and quaterphenyl-3,3′,5,5′-tetracarboxylate leads to a proportional increase in pore size but decrease in hydrogen uptake at 77 K/1 atm.[28]

The ideal pore size seems to be 4.5–5 Å, or approximately 2.8–3.3 Å when the van der Waal radii of the atoms composing the pore walls are excluded; this is comparable to the ~2.8 Å kinetic diameter of $H_2$. This leads to optimal interaction between the dihydrogen molecule and the framework, thus maximizing the total van der Waals forces acting on dihydrogen and increasing $\Delta H_{ads}$.[29]

One problem encountered when attempting to quantify the exact relationship between pore size and hydrogen uptake is the variety of methods used to calculate and report the size of pores. One set of methods, the application of Dubnin–Astakhov analysis or the Horvath–Kawazoe model to gas sorption data, provides an estimation of pore size, but is limited by the quality of the adsorption data and the gas used. Given that MOFs typically have a highly ordered crystalline structure, it is generally not difficult to determine a high-quality single-crystal X-ray structure of these materials. Many researchers apply a variety of software tools to estimate accessible pore volume and pore sizes based on these structures; however, this method relies both on estimations of the van der Waals radii of the atoms along the pore or channel walls, and the particular algorithms of the software packages used. We have found perhaps the most useful information to be simple atom-to-atom distance measurements across the pores or channels, as measured directly from the crystal structure; this allows the reader or reviewer to make "apples-to-apples" comparisons from one paper to the next, and to apply any additional assumptions or estimations as necessary.

*Catenation* Catenation is the physical entanglement of two or more identical frameworks, and is a common structural motif in MOFs.[30] The use of longer ligands favors the generation of catenation, and is often used as an alternate strategy for reducing pore sizes in porous MOFs.[31] The typical effect of interpenetration on porosity is to subdivide large single pores, each bounded by the entire organic linker, into several smaller ones, each bounded by smaller portions of the organic linker. The effects of catenation on hydrogen adsorption was illustrated by Yaghi and Rowsell[31] demonstrating that the catenated MOFs IRMOF-9, IRMOF-11, and IRMOF-13 showed higher hydrogen adsorption capacities than non-catenated IRMOF-1, with the effect directly related to the reduction of pore diameter due to catenane formation.

Conceptually, a catenated MOF and its non-catenated counterpart can be viewed as a supramolecular pair of stereoisomers. Recently, we developed a templating strategy to predictably synthesize catenation isomer-pairs by using copper paddlewheel SBUs and two trigonal-planar ligands (TATB and HTB) (Figure 8.1). This allowed us to evaluate catenation as an independent criterion in the hydrogen uptake of an MOF. The catenation isomerism is controlled by the presence or absence of oxalic acid. Although the non-cantenated form (PCN-6′) has a higher overall porosity, based on

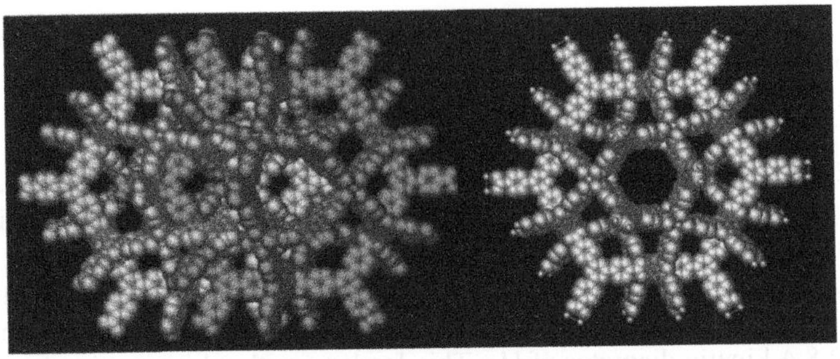

**FIGURE 8.1** Catenated (left) and non-catenated (right) frameworks. [Reproduced by permission of the American Chemical Society from Reference. 32]

the solvent-accessible volume calculated from the single-crystal X-ray structure, its catenated counterpart (PCN-6) exhibits a 41% increase in surface area, 133% increase in volumetric hydrogen uptake, and 29% increase in gravimetric hydrogen uptake.[32] Similar findings have been reported by Long and coworkers[33] in a study of ligand-controlled catenation in sodalite-type MOFs; simulations also suggest that new adsorption sites and small pores formed as a result of catenation may strengthen the overall interaction between gas molecules and the pore walls, thereby increasing the apparent surface area and hydrogen uptake.[34–36]

*Unsaturated Metal Centers (UMCs)* In addition to solvent molecules trapped within the pores of the material, as-synthesized MOFs may also have solvent molecules attached as ligands to the metal centers or incorporated as part of the SBU. Removal of these solvent ligands is referred to as thermal activation, and is often a necessary step to access the full gas-adsorption potential of a material. In some cases, these coordinated solvent molecules may merely protrude into windows or channels, blocking access of dihydrogen molecules into the larger spaces within the framework. In other cases, removal of these ligands (often aqua ligands) leaves the metal cation coordinatively unsaturated and open to direct approach by the dihydrogen molecule. This metal cation is referred to as an unsaturated metal center (UMC).

Chen et al.[37] have shown that the removal of axial aqua ligands from dicopper paddlewheel SBUs via thermal activation exposes the copper cations in MOF-505; Long and coworkers have demonstrated the same phenomenon in an Mn-containing MOF with tetrazolate ligands,[38] as have Bordiga and coworkers in HKUST-1.[39] These UMCs can be seen as analogous to entatic metal centers in bioinorganic chemistry, in which metal ions (such as the iron in hemoglobin) are forced into an unusual coordination geometry (see Figure 8.2)—such a concept has been advanced in our lab by the study of PCN-9, an MOF containing a coordinatively unsaturated $Co_4(\mu_4\text{-}O)(CO_2)_8$ SBU.[40] This study also demonstrated via IR spectroscopy that the cobalt site can bind probe molecules $CN^-$ and CO; similar results were found in CO binding to the copper UMC of HKUST-1.[39]

One of the advantages in porous MOFs when compared to carbon materials is that metal ions incorporated in porous MOFs have much higher hydrogen bonding energies

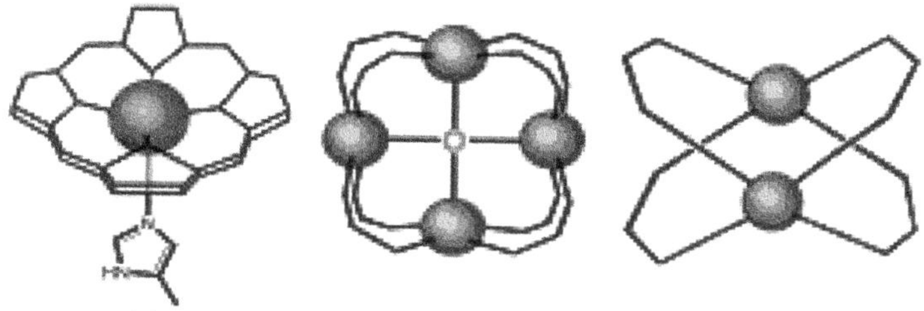

(a) (b) (c)

**FIGURE 8.2** (a) Schematic drawing of the active center of hemoglobin; the gold sphere represents an iron atom. (b) Schematic drawing of the tetranuclear $M_4(\mu_4\text{-}O)(CO_2)_8$ SBU containing coordinatively unsaturated metal atoms. (c) The $M_2(CO_2)_4$ paddlewheel SBU.

than carbon materials do.[26,27] Recent neutron-diffraction and -scattering studies on $H_2$ or $D_2$-loaded MOFs have revealed that hydrogen adsorption is highly dependent on the nature of the metal cation or oxide of the SBUs in porous MOFs. Single-crystal neutron diffraction of MOF-5 revealed two hydrogen-binding sites, one higher energy site over the center of the $Zn_4(\mu_4\text{-}O)(CO_2)_8$ SBU, and a second site over the face of a $ZnO_4$ tetrahedron.[41] Neutron powder diffraction reveals two additional sites in MOF-5 at increased loading: one associated with the zinc-carboxylate moiety $Zn(CO_2)$, and one over the phenyl ring of the ligand.[42] Generally these agree with an inelastic neutron scattering experiment performed on the same material, differing only in the preferred order of site occupation. MP2 *ab initio* calculations of simplified MOF models predict the metal–carboxylate linker to be a high-energy binding site, with additional lower-energy sites located around the phenyl ring of the ligand.[43]

Additional studies probing $H_2$ binding sites have been performed on a number of other MOFs as well. Long *et al.*[44] have employed neutron powder diffraction to confirm that $H_2$ is closely associated with the UMCs found in a Mn-tetrazole MOF. A combination of temperature-programmed desorption and inelastic neutron scattering of $H_2$-loaded $NaNi(sip)_2$ (sip = 5-sulfoisophthalate) has revealed a number of discrete $H_2$ binding sites, the strongest of which can be associated with an unsaturated Ni site.[45] A combined DFT and GCMC dynamical study of the dicopper-paddlewheel-containing MOF-505 shows that, as expected based on the earlier thermal-activation study (*vide supra*), the binding energy of $H_2$ on MOF-505 is highest at the copper UMC sites exposed by thermal activation.[37]

Most recently, neutron powder diffraction of $D_2$-loaded HKUST-1 identified six distinct $D_2$ sites, shown in Figure 8.3. The first, highest energy site is associated with the copper UMCs (on the axes of the paddlewheel SBUs), near enough to indicate significant interaction with the $d^9$ Cu(II) center. The remaining sites fill competitively from the smallest to largest pores, with these sites located near the benzene ring and carboxylate moieties of the ligand.[46]

*Ligand Functionalization* Although the metal sites and/or the SBU are the preferential adsorption sites for hydrogen, the organic linker can play an important secondary role in increasing adsorption further. Organic linkers with aromatic fragments, such as phenylene, naphthylene, and biphenylene, are widely used in

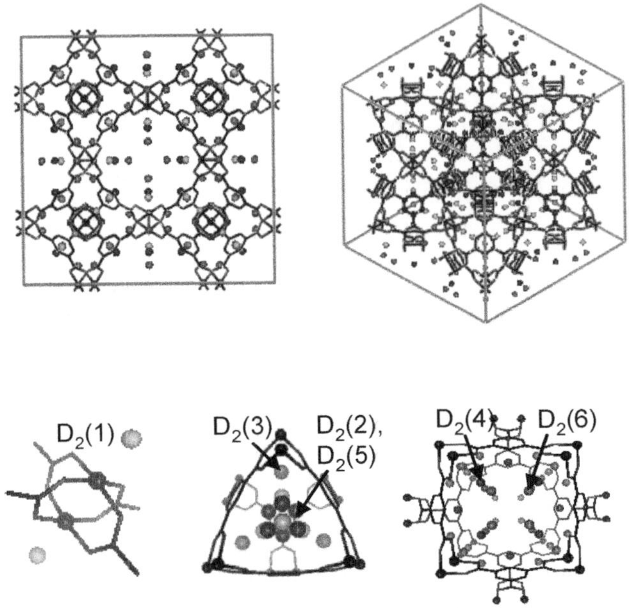


**FIGURE 8.3** $D_2$ sites in HKUST-1, identified via neutron powder diffraction, numbered in order of occupation with increased loading. Top: shown along [001] (left) and [111] (right). Bottom: axial Cu(II) paddlewheel UMC site (left), along [111] in the 5 Å small pore with 3.5 Å side windows (middle), and along [100] showing the 9 Å pore. Reprinted with permission from Reference 46. Copyright 2006, American Chemical Society.

the synthesis of MOFs to form a rigid three-dimensional porous framework.[21,31] Increasing the aromaticity of these organic ligands has been theoretically predicted and experimentally proven as an effective way to improve hydrogen adsorption capacity.[21,47,48] In the IRMOF series developed by Yaghi and coworkers,[21,31] the basic structural motif of $Zn_4(\mu_4\text{-}O)(CO_2)_8$ SBUs connected by aromatic phenyl-containing linkers is repeated to generate a series of isostructural materials, which differ only in the central portion of the ligand. Increasing the aromaticity of this central portion, from a simple phenyl ring (MOF-5/IRMOF-1) to cyclobutylbenzene (IRMOF-6) to naphthalene (IRMOF-8) increases the hydrogen uptake dramatically, from 0.5 to 1.0 and 1.5 wt%, respectively.[21] In a second example, maximum adsorption increases from 4.2 molecules of $H_2$ per formula unit in IRMOF-18 (2,3,5,6-tetra-methylpheylene-1,4-dicarboxylate) to 9.8 in IRMOF-13 (pyrene-2.7-dicarboxylate); however, the gravimetric hydrogen capacity of IRMOF-13 (1.73 wt%) is almost double that of the IRMOF-18 (0.89 wt%), which contains multiple aromatic rings.[21,31] Recently, our research group designed a porous MOF-containing nanoscopic cages, PCN-14, built from an anthracene-containing ligand, 5,5′-(9,10-anthracenediyl)-diisophthalate. PCN-14 exhibits significant enhancement of both hydrogen adsorption capacity and hydrogen affinity when compared to those of other NbO-type MOFs also based on tetracarboxylate organic ligands and dicopper paddlewheel SBUs.[49] The improved hydrogen uptake in PCN-14 is most likely ascribed to the central anthracene aromatic rings.

In addition to increasing the aromaticity of the organic ligands, chemical modification of the organic linkers by introducing an electron-donating group (or groups) has been suggested, based on *ab initio* calculations, as another way to further enhance framework affinity for the dihydrogen molecule. This is illustrated in the hydrogen adsorption studies of the IRMOF series. Adding one –Br, one $-NH_2$, or four methyl groups to the central benzene ring of the linker in IRMOF-1 affords IRMOF-2, -3, and -18, respectively,[21] while replacing the phenyl ring of bdc with a thieno-[3,2*b*] thiophene moiety affords IRMOF-20.[31] The increased polarizability of the hetero-polycyclic ligand improves hydrogen sorption on a molar basis in IRMOF-20 due to a stronger interaction of hydrogen with the organic linker, despite a reduction in gravimetric capacity due to the heavy sulfur atom. Little enhancement, however, was found in IRMOF-2, IRMOF-3, or IRMOF-18, although MP2 computational studies suggest that functionalizing the phenylene ring with electron-donor groups, such as $NH_2$ or Me, can improve hydrogen affinity by $\sim$15%.[43] A similar lack of hydrogen adsorption enhancement was found in pillared MOFs constructed by ligands with all phenyl H atoms replaced with either –F or $-CH_3$.[24] This may be attributed to restricted pores or blocking of some high-affinity binding sites by the larger ligand, thus canceling out the benefit derived from electronic enhancement of the ligand. It has also been proposed that *N*-heterocyclic ligands may have a higher hydrogen affinity than purely graphitic ligands, based on studies of carbon, carbon nitride, and boron nitride nanotubes.[50,51] This idea has been illustrated, with moderate success, by porous MOFs constructed from triazine ligands developed in our lab,[32,52] and by tetrazole-containing ligands by Long and coworkers.[38]

The versatility of organic ligands has provided infinite space for the construction of MOFs with various topologies. Instead of just modifying the organic ligands to build MOFs with similar topology, utilizing flexible organic ligands having different steroisomerism under external stimuli can result in porous MOFs with quite different topologies. This phenomenon is referred to as supramolecular isomerism. As structure determines property, supramolecular isomers are expected to exhibit different hydrogen adsorption capacities. Recently, we designed a tetracarboxylate ligand, *N*,*N*,*N*′,*N*′-tetrakis(4-carboxyphenyl)-1,4-phenylenediamine (tcppda), which has three stereoisomers with a pair of enantiomers and a diastereomer.[53] The diastereomer has $C_{2h}$ symmetry with three phenyl rings oriented as left- and right-handed propellers around the two nitrogen atoms, and a plane of symmetry through the central phenyl ring, reflecting one *N*-centered propeller to the other. The pair of enantiomers possesses $D_2$ point group symmetry with the two *N*-centered propellers being either right handed ($\Delta D_2$) or inverted ($\Delta D_2$). Under solvothermal conditions, the reaction between $Cu(NO_3)_2 \cdot 2.5H_2O$ and $H_4$tccpda in DMSO at 115°C gives rise to a porous MOFs with NbO topology, in which only the $C_{2h}$ isomer of the tcppda ligand exists. Increasing the reaction temperature to 120°C generates a second MOF with PtS topology, in which only the $D_2$ tcppda isomer is found in a racemic combination of $\Delta D_2$ and $\Delta D_2$, as shown in Figure 8.4. The temperature-dependent supramolecular isomerism of the two MOFs can be attributed to the thermally activated

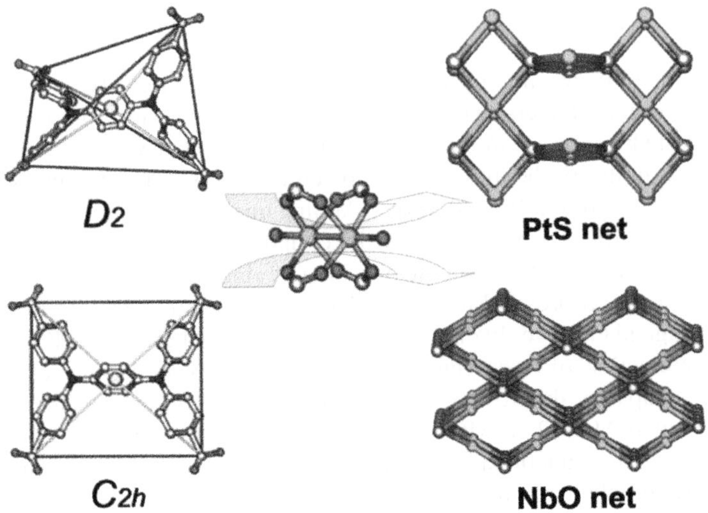


**FIGURE 8.4** Different conformations of the tcppda ligand lead to two porous MOFs with different topologies. [Reproduced by permission of the Royal Society of Chemistry from Reference 53.]

interconversion of the $D_2$ and $C_{2h}$ isomers of tcppda. Nitrogen and hydrogen adsorption studies at 77 K revealed that both surface area and hydrogen uptake of the porous MOFs with PtS topology are ~20% higher those of the NbO typed porous MOF. These studies suggest that designing flexible organic linkers is a promising way to construct porous MOFs with high-hydrogen uptake.

***8.2.3.2 Hydrogen Saturation at Cryogenic Temperature*** Although extensive studies have been focused on hydrogen uptake at low temperature/low pressure (77 K/1 atm), increasing attention is being drawn to high-pressure hydrogen studies because of their direct application in practical on-board hydrogen storage, as required by the DOE goals. In addition to evaluating high-pressure gravimetric adsorption capacity in porous MOFs, volumetric adsorption capacity has also been widely assessed as another important criterion for vehicular on-board hydrogen storage.

*Excess Adsorption and Absolute Adsorption* In high-pressure studies, two quantities, excess adsorption and absolute adsorption, are frequently used to describe hydrogen adsorption in porous MOFs. In brief, excess adsorption is the amount of adsorbate gas interacting with the framework, while absolute adsorption is the amount of gas both interacting with the framework and confined within pores in the absence of gas–solid intermolecular forces. The majority of reported experimental adsorption data in the literature are excess adsorption isotherms. The absolute adsorption amount can be estimated for systems with known crystal structure (i.e., known absolute porosity volume fraction). From the viewpoint of hydrogen storage, the total amount that a material can store or its absolute adsorption is another informative value,

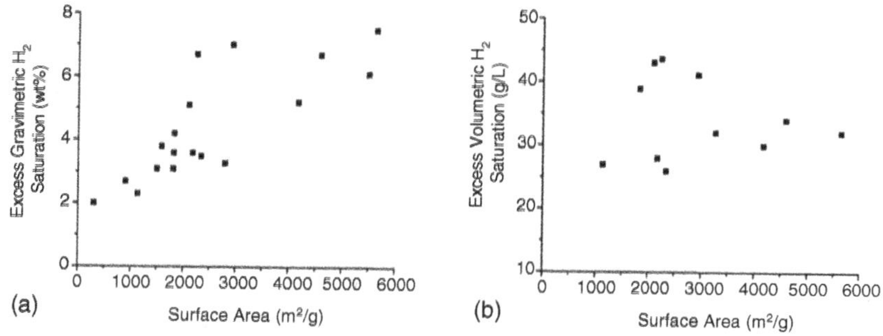


**FIGURE 8.5** Correlation between surface area and (a) excess gravimetric saturation at 77 K, (b) excess volumetric hydrogen saturation at 77 K.

although the mass and volume of a storage tank must be taken into account for real applications.[54]

*Hydrogen Saturation at 77 K* Hydrogen sorption behavior at saturation is a critical parameter for judging the practicality of porous MOF materials.[55] At room temperature, hydrogen saturation is very difficult to achieve due to the rapid thermal motion of dihydrogen molecules; for this reason, current research focuses on investigating hydrogen saturation uptake at 77 K. Existing studies indicate that hydrogen saturation uptake at cryogenic temperatures is roughly correlated with surface area.[55] As shown in Figure 8.5, some MOFs can reach or exceed the 2010 DOE gravimetric system goal of 6 wt%, albeit at 77 K. The excess gravimetric uptake of MOF-177 is as high as 7.5 wt %, while the absolute gravimetric uptake is considerably higher: 11.3 wt% at 77 K, 70 bar.[54]

Most porous MOFs are extremely lightweight, many with densities less than 1 g/cm$^3$. Generally speaking, the higher the surface area, the lower the crystal density. In most cases, the low density serves to reduce the volumetric hydrogen uptake of the MOF material despite its high-gravimetric uptake. As indicated in Figure 8.5, few porous MOFs can reach the 2010 DOE volumetric hydrogen uptake system target, even at 77 K. A typical example is MOF-177, which has an excess gravimetric uptake of 7.5 wt%; its low crystallographic density of 0.427 g/cm$^3$, however, leads to an excess volumetric uptake of 32 g/L, which is far from the 2010 DOE volumetric system goal of 45 g/L.[55] A compromise between the surface area and crystal density should be met in search of porous MOF materials with both high-gravimetric and volumetric hydrogen uptake. Nevertheless, in terms of absolute adsorption, there exist some MOF materials with both gravimetric and volumetric hydrogen uptake at 77 K which surpass the 2010 DOE targets for hydrogen storage. $Mn_3[(Mn_4Cl)_3(btt)_3(CH_3OH)_{10}]_2$ (btt = 1,3,5-benzenetristetrazolate) reported by Long and coworkers[44] exhibits high-hydrogen adsorption capacity at 77 K under high pressure, with an absolute gravimetric uptake of 6.9 wt% and volumetric uptake of 60 g/L. The absolute gravimetric uptake and volumetric uptake in MOF-177 can reach 11 wt% and 48 g/L respectively at 77 K, both of which also pass the DOE goals.[54]

It would seem very promising to store hydrogen in porous MOFs at 77 K and high pressure; however, the cost and weight of the cryogenic pressure vessel precludes a practical on-board application, and porous MOFs with high-hydrogen uptake near ambient temperature are badly needed.

*8.2.3.3* ***High-Pressure Hydrogen Adsorption at Room Temperature*** Room-temperature hydrogen adsorption studies under high pressure have been carried out for porous MOF materials. Unfortunately, these materials have very low hydrogen uptake at room temperature, less than 1.5 wt%. At non-cryogenic temperatures, high-surface area is insufficient to achieve high-capacity ambient-temperature storage; the low interaction energy between the framework and the dihydrogen molecule is overwhelmed by the thermal energy of the hydrogen gas. Instead, many of the same techniques described above, which have the effect of increasing $\Delta H_{ads}$, allow increased interaction between the pore walls and the hydrogen molecule, thereby increasing room-temperature adsorptive capacity.

Small pore size is more favorable for ambient-temperature hydrogen adsorption due to enhanced interaction energy. This is well-illustrated by [Cu(hfipbb)-($H_2$hfipbb)$_{0.5}$], a microporous MOF which contains small pores of two types: small (~3.5 × 3.5 Å) and large (5.1 × 5.1 Å). At room temperature and 48 atm, it can adsorb ~1 wt% of hydrogen, which is more than three times that of MOF-5 (0.28 wt%, 60 atm) which contains pore size of ~7.7 × 7.7 Å and has a high-surface area of 2300 $m^2$/g.[56] As described above, an effective way to reduce pore size is by utilizing interpenetration (or catenation), which has been proposed as a strategy to improve room-temperature hydrogen uptake. One extreme example of a highly interpenetrated structure with small pores is $Zn_4O(L^1)_3$ ($L^1$ = 6,6′-dichloro-2,2′-diethoxy-1,1′-bi-naphthyl-4,4′-dibenzoate), with a fourfold interpenetrating structure, open channels of less than 5 Å, and BET surface area of only 502 $m^2$/g; this material adsorbs 1.12 wt% of hydrogen at room temperature and 48 bar, among the highest of reported MOF materials.[25]

The exposition of UMCs has been shown to be a promising way to improve hydrogen affinity, and has been widely explored for room temperature hydrogen adsorption studies. The ability of UMCs to adsorb significant amounts of hydrogen is well-demonstrated by Long and coworkers in the porous MOF $Mn_3[(Mn_4Cl)_3(btt)_3(CH_3OH)_{10}]_2$. Upon thermal activation, this MOF can adsorb hydrogen 1.4 wt% at 298 K and 90 bar. This high-uptake capacity can be ascribed to exposed $Mn^{2+}$ sites within the framework which interact strongly with $H_2$ molecules.[44]

Despite the increasing number of investigations of hydrogen adsorption in porous MOFs, the reported near-ambient-temperature hydrogen uptakes by MOFs fall far short of the 2010 DOE system targets. Future research must focus on increasing the interaction energy ($\Delta H_{ads}$) between the framework and the dihydrogen molecule in order to improve room-temperature adsorption amounts. Exploration of porous MOFs with high-hydrogen uptake for on-board storage applications still has a long way to go.

## 8.3 METHANE STORAGE

### 8.3.1 Methane Storage Goals

As with hydrogen, methane is considered an ideal energy gas for future applications. Methane is the primary component of natural gas; as such, an extensive system of collection, purification, and distribution infrastructure already exists, capable of delivering methane to the majority of homes and businesses in the United States and many other countries worldwide. Deposits of methane-containing natural gas are more widespread globally than those of petroleum, and its refinement (purification) to an energy fuel is much simpler than that of crude petroleum oil to gasoline or diesel fuels. Methane is also produced by decomposition of organic waste and by bacteria in the guts of ruminants and termites. In fact, methane and natural gas are often considered waste products in crude oil collection and refining and other industrial processes, and are often burned off in giant flares with no secondary energy capture. Finally, compressed natural gas (CNG) vehicles already exist, and make up a small fraction of commercial and personal vehicle fleets in Argentina, Brazil, Pakistan, Italy, Iran, and the United States. However, as in the case of hydrogen, current vehicles store the methane CNG in high-pressure (greater than 200 atm) tanks which are heavy and potentially explosive. To address the needs for better methane-storage technology, the U.S. DOE has set targets for methane storage systems at 180 v(STP)/v (STP equivalent of methane per volume of adsorbent material storage system) under 35 bar and near ambient temperature, with the energy density of adsorbed natural gas comparable to that of current CNG technology.[57]

As in the case of hydrogen storage, a variety of porous materials have been extensively evaluated as methane storage materials, including activated carbon,[58] carbon nanotubes,[59] and zeolites.[60] However, with the exception of activated carbon,[57] none of these materials have been able to store quantities equal to the DOE targets, and a high-capacity adsorbent remains elusive; like the hydrogen molecule, the methane molecule is small and interacts only weakly with the pore walls of the adsorbent.

### 8.3.2 Methane Storage in MOFs

MOFs are a relative newcomer to the field of methane adsorbants, while carbon materials have been extensively studied for methane storage since the early 1990s, the first reported measurement of methane uptake by an MOF was in 1999 by Kitagawa and coworkers.[61] The methane uptake of this pyrazine-based MOF was low, comparable to several zeolites. A second MOF reported by Kitagawa and coworkers[62] the next year represented a serious effort at rationally synthesizing a material which would have a higher uptake of methane; this material exceeded by nearly 100% the most absorptive zeolite, zeolite 5A. However, the field of methane storage on MOFs has not expanded as quickly as the hydrogen-storage field, perhaps due to the existence of a successful, albeit unattractive, storage system for methane, namely the CNG cylinder.

As in the case of hydrogen storage, a variety of factors influence the ability of the framework to adsorb methane, namely surface area and pore volume, pore size, and heat of adsorption (with contributions from both framework topology and chemical functionality).[63] For example, the contribution of catenation/interpenetration was demonstrated by Kitagawa and coworkers by a series of azopyridine-based MOFs, with the highest of the series adsorbing ~60 v(STP)/v.[64]

The ability of IRMOF-6 to adsorb a higher amount of methane than the other members of the IRMOF series was attributed to both the accessible surface area and the functionality of the ligand: in IRMOF-6, the phenyl ring of the typical bdc ligand was modified to generate 1,2-cyclobutane-3,6-benzenedicarboxylate. The resulting MOF was found to adsorb 155 v(STP)/v methane at 298 K and 36 atm, considerably higher than any zeolite material or any other MOF at the time.[11] Molecular simulations indicated that further functionalization of the ligand by inclusion of an anthracene ring would increase methane uptake further, perhaps within reach of the DOE goal.[63] Synthesis of this proposed MOF, however, resulted in a material with only limited methane uptake, due to the ultramicroporous nature of the MOF, with pores too small to accommodate methane.[65]

Recently, our research group endeavored to continue to pursue this proposed anthracene-based MOF with high-methane adsorption; in an effort to surmount the problems associated with extremely small pores, the ligand was extended by additional phenyl rings to form 5,5′-(9,10-anthracenediyl)-diisophthalate (adip). The resulting MOF, dubbed PCN-14, was found to contain nanoscopic cages of a size suitable for methane storage, with an adsorption capacity of 230 v(STP)/v, 28% higher than the DOE target, at 290 K and 35 bar.[66] Additionally, the heat of adsorption of methane on the framework is ~30 kJ/mol, higher than any other reported MOF—this indicates the validity of using $\Delta H_{ads}$ as a benchmark for evaluating potential saturation uptake at room temperature.[2]

## 8.4 OUTLOOK

As a relatively new class of materials, MOFs continue to attract interest and inquiry by both academia and industry. They exhibit considerable potential for the adsorptive storage of both hydrogen and methane in energy applications, in addition to the myriad of other proposed applications. The emerging ability to tune pore size and pore wall functionality allows researchers to focus on those factors which hold to the most promise, increasing both the volume available for storage and the affinity of the network for stored gas. As alternative fuels such as hydrogen and methane continue to be developed in automotive and other applications, needs for effective storage technologies will continue to increase, and MOFs are well-positioned to continue to be at the forefront of this research.

## REFERENCES

1. (a) The material in this chapter includes significant contribution from the following sources and the references therein: Ma, S.; Collier, C. D.; Zhou, H. -C. Design and

construction of metal-organic frameworks for hydrogen storage and selective gas-adsorption. In *Design and Construction of Coordination Polymers*, Hong, M., Ed.; PN Wiley: New York, 2008. (b) Collins, D. J.; Zhou, H.-C. Nano/microporous materials: hydrogen storage materials. In *Nanomaterials: Inorganic and Bioinorganic Perspectives*, Lukehart, C. M.; Scott, R. A., Eds.; Wiley: New York, 2008.

2. Collins, D. J.; Zhou, H.-C. Hydrogen storage in metal-organic frameworks. *J. Mater. Chem.* **2007**, *17*, 3154.
3. Yaghi, O. M.; Li, H.; Davis, C.; Richardson, D.; Groy, T. L. Synthetic strategies, structure patterns, and emerging properties in the chemistry of modular porous solids. *Acc. Chem. Res.* **1998**, *31*, 474.
4. Eddaoudi, M.; Moler, D. B.; Li, H.; Chen, B.; Reineke, T. M.; O'Keeffe, M.; Yaghi, O. M. Modular chemistry: secondary building units as a basis for the design of highly porous and robust metal-organic carboxylate frameworks. *Acc. Chem. Res.* **2001**, *34*, 319.
5. James, S. L. Metal-organic frameworks. *Chem. Soc. Rev.* **2003**, *32*, 276.
6. Yaghi, O. M.; O'Keeffe, M.; Ockwig, N. W.; Chae, H. K.; Eddaoudi, M.; Kim, J. Reticular synthesis and the design of new materials. *Nature* **2003**, *423*, 705.
7. Seo, J. S.; Whang, D.; Lee, H.; Jun, S. I.; Oh, J.; Jeon, Y. J.; Kim, K. A homochiral metal-organic porous material for enantioselective separation and catalysis. *Nature* **2000**, *404*, 982.
8. Hu, A.; Ngo, H. L.; Lin, W. Chiral porous hybrid solids for practical heterogeneous asymmetric hydrogenation of aromatic ketones. *J. Am. Chem. Soc.* **2003**, *125*, 11490.
9. Millward, A. R.; Yaghi, O. M. Metal-organic frameworks with exceptionally high capacity for storage of carbon dioxide at room temperature. *J. Am. Chem. Soc.* **2005**, *127*, 17998.
10. Bourrelly, S.; Llewellyn, P. L.; Serre, C.; Millange, F.; Loiseau, T.; Férey, G. Different adsorption behaviors of methane and carbon dioxide in the isotypic nanoporous metal terephthalates MIL-53 and MIL-47. *J. Am. Chem. Soc.* **2005**, *127*, 13519.
11. Eddaoudi, M.; Kim, J.; Rosi, N.; Vodak, D.; Wachter, J.; O'Keeffe, M.; Yaghi, O. M. Systematic design of pore size and functionality in isoreticular MOFs and their application in methane storage. *Science* **2002**, *295*, 469.
12. Matsuda, R.; Kitaura, R.; Kitagawa, S.; Kubota, Y.; Belosludov, R. V.; Kobayashi, T. C.; Sakamoto, H.; Chiba, T.; Takata, M.; Kawazoe, Y.; Mita, Y. Highly controlled acetylene accommodation in a metal-organic microporous material. *Nature* **2005**, *436*, 238.
13. Pan, L.; Adams, K. M.; Hernandez, H. E.; Wang, X.; Zheng, C.; Hattori, Y.; Kaneko, K. Porous lanthanide-organic frameworks: synthesis, characterization, and unprecedented gas adsorption properties. *J. Am. Chem. Soc.* **2003**, *125*, 3062.
14. Ma, S.; Sun, D.; Wang, X. -S.; Zhou, H. -C. A mesh-adjustable molecular sieve for general use in gas separation. *Angew. Chem. Int. Ed.* **2007**, *46*, 2458.
15. U.S. Department of Energy, Targets for on-board hydrogen storage systems: Current R&D focus is on 2015 targets with potential to meet ultimate targets http://www1.eere.energy.gov/hydrogenandfuelcells/storage/current_technology.html).
16. Hynek, S.; Fuller, W.; Bentley, J. Hydrogen storage by carbon sorption. *Int. J. Hydrogen Energy* **1997**, *22*, 601.
17. Züttel, A.; Wenger, P.; Sudan, P.; Mauron, P.; Orimo, S.-i. Hydrogen density in nanostructured carbon, metals and complex materials. *Mater. Sci. Eng. B* **2004**, *108*, 9.

18. Nijkamp, M. G.; Raaymakers, J. E. M. J.; van Dillen, A. J.; de Jong, K. P. Hydrogen storage using physisorption—materials demands. *Appl. Phys. A* **2001**, *72*, 619.
19. Bhatia, S. K.; Myers, A. L. Optimum conditions for adsorptive storage. *Langmuir* **2006**, *22*, 1688.
20. Rosi, N. L.; Eckert, J.; Eddaoudi, M.; Vodak, D. T.; Kim, J.; O'Keefe, M.; Yaghi, O. M. Hydrogen storage in microporous metal-organic frameworks. *Science* **2003**, *300*, 1127.
21. Rowsell, J. L. C.; Millward, A. R.; Park, K. S.; Yaghi, O. M. Hydrogen sorption in functionalized metal-organic frameworks. *J. Am. Chem. Soc.* **2004**, *126*, 5666.
22. Chae, H. K.; Siberio-Perez, D. Y.; Kim, J.; Go, Y. B.; Eddaoudi, M.; Matzger, A. J.; O'Keeffe, M.; Yaghi, O. M. A route to high surface area, porosity and inclusion of large molecules in crystals. *Nature* **2004**, *427*, 523.
23. Férey, G.; Mellot-Draznieks, C.; Serre, C.; Millange, F.; Dutour, J.; Surblé, S.; Margiolaki, I. A chromium terephthalate-based solid with unusually large pore volumes and surface area. *Science* **2005**, *309*, 2040.
24. Chun, H.; Dybtsev, D. N.; Kim, H.; Kim, K. Synthesis, X-ray crystal structures, and gas sorption properties of pillared square grid nets based on paddle-wheel motifs: implications for hydrogen storage in porous materials. *Chem. Eur. J.* **2005**, *11*, 3521.
25. Kesanli, B.; Cui, Y.; Smith, M. R.; Bittner, E. W.; Bockrath, B. C.; Lin, W. Highly interpenetrated metal-organic frameworks for hydrogen storage. *Angew. Chem. Int. Ed.* **2005**, *44*, 72.
26. Isaeva, V. I.; Kustov, L. M. Metal-organic frameworks—new materials for hydrogen storage. *Russ. J. Gen. Chem.* **2007**, *77*, 721.
27. Rowsell, J. L. C.; Yaghi, O. M. Strategies for hydrogen storage in metal-organic frameworks. *Angew. Chem. Int. Ed.* **2005**, *44*, 4670.
28. Lin, X.; Jia, J.; Zhao, X.; Thomas, K. M.; Blake, A. J.; Walker, G. S.; Champness, N. R.; Hubberstey, P.; Schröder, M. High $H_2$ adsorption by coordination-framework materials. *Angew. Chem. Int. Ed.* **2006**, *45*, 7358.
29. Chapman, K. W.; Chupas, P. J.; Maxey, E. R.; Richardson, J. W. Direct observation of adsorbed $H_2$-framework interactions in the Prussian Blue analogue $Mn^{II}{}_3[Co^{III}(CN)_6]_2$: the relative importance of accessible coordination sites and van der Waals interactions. *Chem. Commun.* **2006**, 4013.
30. Batten, S. T.; Robson, R. Interpenetrating nets: ordered, periodic entanglement. *Angew. Chem. Int. Ed. Engl.* **1998**, *37*, 1460.
31. Rowsell, J. L. C.; Yaghi, O. M. Effects of functionalization, catenation, and variation of the metal oxide and organic linking units on the low-pressure hydrogen adsorption properties of metal-organic frameworks. *J. Am. Chem. Soc.* **2006**, *128*, 1304.
32. Ma, S.; Sun, D.; Ambrogio, M. W.; Fillinger, J. A.; Parkin, S.; Zhou, H.-C. Framework-catenation isomerism in MOFs and its impact on hydrogen uptake. *J. Am. Chem. Soc.* **2007**, *129*, 1858.
33. Dincă, M.; Dailly, A.; Tsay, C.; Long, J. R. Expanded sodalite-type metal-organic frameworks: increased stability and H2 adsorption through ligand-directed catenation. *Inorg. Chem.* **2008**, *47*, 11.
34. Jung, D. H.; Kim, D.; Lee, T. B.; Choi, S. B.; Yoon, J. H.; Kim, J.; Choi, K.; Choi, S. H. Grand canonical Monte Carlo Simulation study on the catenation effect on hydrogen adsorption onto the interpenetrating metal-organic frameworks. *J. Phys. Chem. B* **2006**, *110*, 22987.

35. Yang, Q.; Zhong, C. Understanding hydrogen adsorption in metal-organic frameworks with open metal sites: a computational study. *J. Phys. Chem. B* **2006**, *110*, 655.
36. Frost, H.; Duren, T.; Snurr, R. Q. Effects of surface area, free volume, and heat of adsorption on hydrogen uptake in metal-organic frameworks. *J. Phys. Chem. B* **2006**, *110*, 9565.
37. Chen, B.; Ockwig, N. W.; Millward, A. R.; Contreras, D. S.; Yaghi, O. M. High $H_2$ adsorption in a microporous metal-organic framework with open metal sites. *Angew. Chem. Int. Ed.* **2005**, *44*, 4745.
38. Dincă, M.; Yu, A. F.; Long, J. R. Microporous metal-organic frameworks incorporating 1,4-benzeneditetrazolate: syntheses, structures, and hydrogen storage properties. *J. Am. Chem. Soc.* **2006**, *128*, 8904.
39. Prestipino, C.; Regli, L.; Vitillo, J. G.; Bonino, F.; Damin, A.; Lamberti, C.; Zecchina, A.; Solari, P. L.; Kongshaug, K. O.; Bordiga, S. Local structure of framework Cu(II) in HKUST-1 metallorganic framework: spectroscopic characterization upon activation and interaction with adsorbates. *Chem. Mater.* **2006**, *18*, 1337.
40. Ma, S.; Zhou, H. -C. A metal-organic framework with entatic metal centers exhibiting high gas adsorption affinity. *J. Am. Chem. Soc.* **2006**, *128*, 11734.
41. Spencer, E. C.; Howard, J. A. K.; McIntyre, G. J.; Rowsell, J. L. C.; Yaghi, O. M. Determination of the hydrogen absorption sites in $Zn_4O$(1,4-benzenedicarboxylate) by single crystal neutron diffraction. *Chem. Commun.* **2006**, 278.
42. Yildirim, T.; Hartman, M. R. Direct observation of hydrogen adsorption sites and nanocage formation in metal-organic frameworks. *Phys. Rev. Lett.* **2005**, *95*, 215504.
43. Sagara, T.; Klassen, J.; Ortony, J.; Ganz, E. Binding energies of hydrogen molecules to isoreticular metal-organic framework materials. *J. Chem. Phys.* **2005**, *123*, 014701.
44. Dincă, M.; Dailly, A.; Liu, Y.; Brown, C. M.; Neumann, D. A.; Long, J. R. Hydrogen storage in a microporous metal-organic framework with exposed $Mn^{2+}$ coordination sites. *J. Am. Chem. Soc.* **2006**, *128*, 16876.
45. Forster, P. M.; Eckert, J.; Heiken, B. D.; Parise, J. B.; Yoon, J. W.; Jhung, S. H.; Chang, J. S.; Cheetham, A. K. Adsorption of molecular hydrogen on coordinatively unsaturated Ni(II) sites in a nanoporous hybrid material. *J. Am. Chem. Soc.* **2006**, *128*, 16846.
46. Peterson, V. K.; Liu, Y.; Brown, C. M.; Kepert, C. J. Neutron powder diffraction study of D2 sorption in $Cu_3$(1,3,5-benzenetricarboxylate)$_2$. *J. Am. Chem. Soc.* **2006**, *128*, 15578.
47. Han, S. S.; Deng, W. -Q.; Goddard, W. A. Improved designs of metal-organic frameworks for hydrogen storage. *Angew. Chem. Int. Ed.* **2007**, *46*, 6289.
48. Han, S. S.; Goddard, W. A. Lithium-doped metal-organic frameworks for reversible $H_2$ storage at ambient temperature. *J. Am. Chem. Soc.* **2007**, *129*, 8422.
49. Ma, S.; Sun, D.; Simmons, J. M.; Yuan, D.; Zhou, H. -C. *Inorg. Chem.* **2009**, *48*, 5263
50. Bai, X. D.; Zhong, D.; Zhang, G. Y.; Ma, X. C.; Liu, S.; Wang, E. G. Hydrogen storage in carbon nitride nanobells. *Appl. Phys. Lett.* **2001**, *79*, 1552.
51. Jhi, S. H.; Kwon, Y. K. Hydrogen adsorption on boron nitride nanotubes: a path to room-temperature hydrogen storage. *Phys. Rev. B* **2004**, *69*, 245407.
52. Sun, D.; Ma, S.; Ke, Y.; Collins, D. J.; Zhou, H.-C. An interweaving MOF with high hydrogen uptake. *J. Am. Chem. Soc.* **2006**, *128*, 3896.
53. Sun, D.; Ke, Y.; Mattox, T. M.; Ooro, B. A.; Zhou, H.-C. Temperature-dependent supramolecular stereoisomerism in porous copper coordination networks based on a designed carboxylate ligand. *Chem. Commun.* **2005**, 5447.

54. Furukawa, H.; Miller, M. A.; Yaghi, O. M. Independent verification of the saturation hydrogen uptake in MOF-177 and establishment of a benchmark for hydrogen adsorption in metal-organic frameworks. *J. Mater. Chem.* **2007**, *17*, 3197.

55. Wong-Foy, A. G.; Matzger, A. J.; Yaghi, O. M. Exceptional $H_2$ saturation uptake in microporous metal-organic frameworks. *J. Am. Chem. Soc.* **2006**, *128*, 3494.

56. Pan, L.; Sander, M. B.; Huang, X.; Li, J.; Smith, M.; Bittner, E.; Bockrath, B.; Johnson, J.K. Microporous metal-organic materials: promising candidates as sorbents for hydrogen storage. *J. Am. Chem. Soc.* **2004**, *126*, 1308.

57. Burchell, T.; Judkins, R. R.; Rogers, M. *PCT Int. Appl.* **2000**.

58. Quinn, D. F.; MacDonald, J. A. Natural gas storage. *Carbon* **1992**, *30*, 1097.

59. Muris, M.; Dupont-Pavlovsky, N.; Bienfait, M.; Zeppenfeld, P. Where are the molecules adsorbed on single-walled nanotubes?. *Surf. Sci.* **2001**, *492*, 67.

60. Dunne, J. A.; Rao, M.; Sircar, S.; Gorte, R. J.; Myers, A. L. Calorimetric heats of adsorption and adsorption isotherms. $O_2$, $N_2$, Ar, $CO_2$, $CH_4$, $C_2H_6$, and $SF_6$ on NaX, H-ZSM-5, and Na-ZSM-5 Zeolites. *Langmuir* **1996**, *12*, 5896.

61. Kondo, M.; Okubo, T.; Asami, A.; Noro, S.-i.; Yoshitomi, T.; Kitagawa, S.; Ishii, T.; Matsuzaka, H.; Seki, K. Rational synthesis of stable channel-like cavities with methane gas adsorption properties: $[\{Cu_2pzdc_2(L)\}_n]$; (pzdc=pyrazine-2, 3-dicarboxylate; L=a Pillar Ligand). *Angew. Chem. Int. Ed.* **1999**, *38*, 140.

62. Noro, S.-i.; Kitagawa, S.; Kondo, M.; Seki, K. A new, methane adsorbent, porous coordination polymer $[\{CuSiF_6(4,4'\text{-bipyridine})_2\}_n]$. *Angew. Chem. Int. Ed.* **2000**, *39*, 2082.

63. Düren, T.; Sarkisov, L.; Yaghi, O. M.; Snurr, R. Q. Design of new materials for methane storage. *Langmuir* **2004**, *20*, 2683.

64. Kondo, M.; Shimamura, M.; Noro, S. -i.; Minakoshi, S.; Asami, A.; Seki, K.; Kitagawa, S. Microporous materials constructed from the interpenetrated coordination networks. Structures and methane adsorption properties. *Chem. Mater.* **2000**, *12*, 1288.

65. Ma, S.; Wang, X. S.; Collier, C. D.; Manis, E. S.; Zhou, H. C. Ultramicroporous metal-organic framework based on 9,10-anthracenedicarboxylate for selective gas adsorption. *Inorg. Chem.* **2007**, *46*, 8499.

66. Ma, S.; Sun, D.; Simmons, J. M.; Collier, C. D.; Yuan, D.; Zhou, H. C. Metal-organic framework from an anthracene derivative containing nanoscopic cages exhibiting high methane uptake. *J. Am. Chem. Soc.* **2008**, *130*, 1012.

# 9

# TOWARD MECHANOCHEMICAL SYNTHESIS OF METAL-ORGANIC FRAMEWORKS: FROM COORDINATION POLYMERS AND LATTICE INCLUSION COMPOUNDS TO POROUS MATERIALS

Tomislav Friščić

*Department of Chemistry, University of Cambridge, Lensfield Road, Cambridge CB2 1EW, UK*

## 9.1 INTRODUCTION

The research and development of functional materials can be roughly divided into three broad aspects: (1) the discovery of potential applications; (2) the development of reliable and general synthetic designs and (3) the development of efficient synthetic strategies. In the case of metal-organic frameworks (MOFs),[1,2] the first aspect is beyond doubt their permanent porosity that enables numerous applications in gas storage or catalysis.[3,4] The second aspect is illustrated by well-established MOF design strategies, such as the isoreticular[5,6] or the inverted (IMOF) designs.[7] Whilst the applications and design of MOFs have often been discussed (e.g.[1,2]), the third, synthetic aspect of MOF chemistry is seldom the subject of review. While MOF synthesis is traditionally dominated by solvothermal solution-based methods requiring elevated temperatures and pressures,[8,9] recent years have witnessed a rapid increase in interest in alternative synthetic methods that would be faster, cleaner and less expensive. Such methods include mechanochemical methods[10,11]

*Metal-Organic Frameworks: Design and Application,* Edited by Leonard R. MacGillivray

of solid-state grinding and liquid-assisted grinding (LAG),[12,13] sonochemical methods[14] and microwave-assisted synthesis.[15,16] This chapter will attempt to partially fill the literature gap concerning MOF synthesis and describe the development of green mehanochemical methodologies, that is, environmentally friendly approaches that achieve reactivity through mechanochemical force, rather than thermal energy.[10,11]

Although scattered reports on mechanochemical transformations and construction of metal-organic compounds can be found in the literature since 1970s,[17–19] intensive development of this area begun less than a decade ago with simultaneous reports of Orita et al. on mechanosynthesis of a coordination bowl, and of Belcher et al. who first reported the formation of a coordination polymer by grinding.[20,21] This development recently culminated in two approaches for the mechanosynthesis of porous MOFs.[22,23] The rapid development of metal-organic mechanosynthesis not only has presented synthetic chemists and materials scientists with faster and more efficient routes to metal-organic materials, but also has provided significant advances to mechanochemistry and environmentally friendly synthesis.[24,25] As a result, metal-organic mechanosynthesis now encompasses a respectable number of chemical reactions, methodologies and materials. Consequently, this chapter will take the opportunity to provide, for the first time, a systematic analysis of metal-organic mechanosynthesis. The analysis follows two criteria: the type of mechanochemical methodology and the type of chemical transformation that was implemented.

### 9.1.1 Scope

Mechanochemical construction of MOFs is enabled by the ability to construct metal–ligand bonds by grinding.[10,11] Consequently, this chapter focuses on methods that provide this ability, and their diversity in terms of procedure and types of reactants, products and chemical reactions. We will pay special attention to the formation of metal–ligand bonds in the context of extended architectures (i.e., coordination polymers).[26] These encompass diverse one-dimensional (1D), two-dimensional (2D) and three-dimensional (3D) topologies. The latter can further be divided into porous and non-porous (e.g., interpenetrated) frameworks. These are schematically presented in Scheme 9.1.

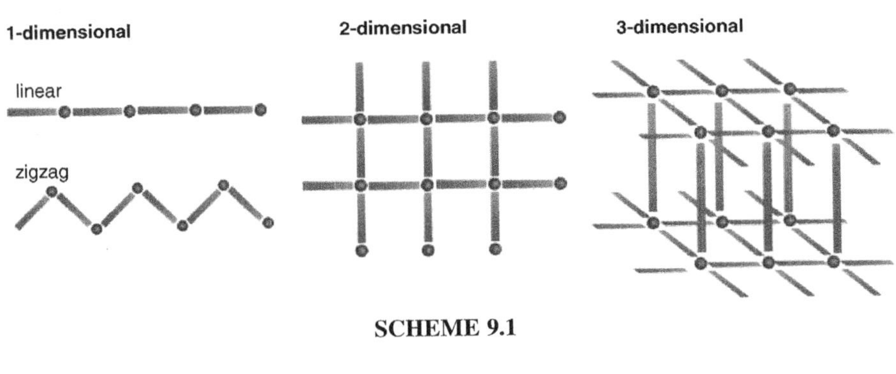


**SCHEME 9.1**

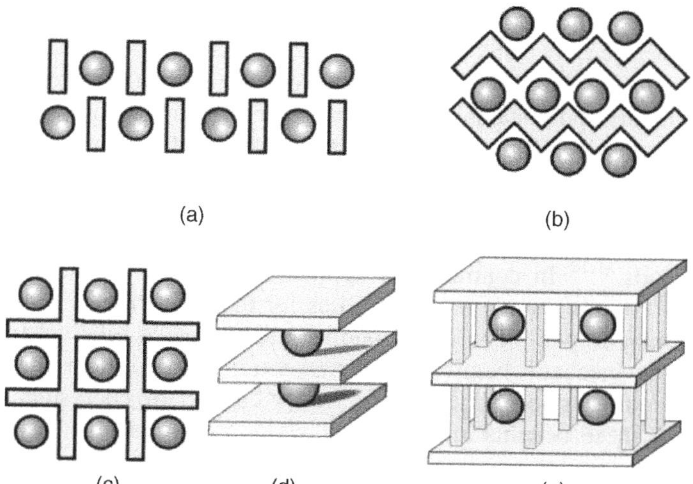


**FIGURE 9.1** Types of inclusion compounds involving discrete (A), one-dimensional (B), two-dimensional (C and D), and three-dimensional (E) metal–organic hosts. The host structure is shown in black, while the gray circles represent guest molecules.

Having in mind that one of the principal driving forces behind MOF research is molecular inclusion, particular attention will be given to mechanosynthesis of coordination compounds resulting in the formation of inclusion compounds.[27] In doing so, we will not restrict ourselves to inclusion within 3D porous frameworks, but will also investigate the formation of host–guest inclusion compounds of metal-organic lattice hosts assembled via hydrogen bonding or van der Waals interactions. There are several types of such molecular inclusion, depicted in Figure 9.1. Type A inclusion compounds involve the assembly of guest molecules with monomeric metal complexes. This type of materials, however, does not involve extended arrays of metal–ligand bonds, and is not relevant for this study. Thus, type A inclusion compounds will be only briefly mentioned. Inclusion compounds of type B are based on a host assembled from 1D coordination polymer chains. These chains can adopt different topologies, as illustrated in Figure 9.1 for linear and zigzag topologies. Inclusion compounds of types C and D correspond to two modes of molecular inclusion expected for 2D coordination polymers. Finally, inclusion in 3D porous MOFs is depicted as type E molecular inclusion.

In addition to mechanochemical approaches to coordination frameworks, this overview will also address recently reported MOF syntheses via sonication[14] and microwave irradiation.[15,16] In certain aspects these methods bear similarity to mechanosynthesis, and their further development may provide valuable insight into the mechanisms underlying synthesis by grinding.[28]

## 9.2 ADVANTAGES AND LIMITATIONS OF MECHANOSYNTHESIS

The interest in mechanochemistry results from the growing interest in environmentally friendly and sustainable chemical processes. In that context, the obvious benefit of mechanosynthesis is the ability to either completely avoid the use of solvent (as in neat grinding),[29] or reduce the amount of solvent to catalytic or near-stoichiometric amounts (e.g. in LAG or kneading).[11–13,30] As a result, mechanosynthesis has been extensively applied as a 'green' method for the construction of covalent bonds.[31,32] In contrast, the application of mechanochemistry for the synthesis of coordination bonds, as well as for the formation of cocrystals [33,34] built up from hydrogen- or halogen-bonding interactions, is relatively recent. The advantages of mechanochemical over conventional solution-based approaches have been demonstrated largely in the latter context of cocrystals. The most important of these benefits are increased yields and product purity,[35] reduced reaction times,[36] and the availability of reactants and products that are difficult to encounter in conventional solution-based synthesis.[37–40] Although the application of mechanochemistry for the construction of coordination compounds is not yet as extensive as it is for cocrystal synthesis, it is likely that the same advantages apply in both areas and, consequently, analogies can be drawn.[41] For example, as mechanochemistry has enabled cocrystal formation with molecules that are considered insoluble in most organic solvents,[38] mechanochemical grinding has allowed the use of insoluble metal oxides as precursors in the one-step construction of coordination polymers and MOFs.[23] The application of mechanochemistry is additionally attractive for MOF synthesis as it provides an alternative to high temperatures and pressures of solvothermal syntheses.[1,2,23,42]

Probably the greatest disadvantage that is encountered in the implementation of mechanosynthesis is the difficulty of structural characterization of products. Indeed, the product obtained by grinding is inherently not suitable for characterization via conventional methods of single crystal X-ray diffraction. Nevertheless, as described in Section 9.3.4, this problem is rapidly alleviated by the continuous development of methodologies for crystal structure solution from powder diffraction data,[38] solid-state NMR spectroscopy[43] and molecular simulations.[44,45]

## 9.3 METHODS FOR MECHANOSYNTHESIS OF COORDINATION BONDS

Three distinct mechanochemical methodologies have so far been applied for the construction of coordination polymers and frameworks: neat grinding,[29] LAG[12,13] or kneading[11] (other names[46] sometimes encountered are solvent-drop grinding[36] or solvent-assisted[47] grinding), and neat grinding followed by annealing.[48–51] These methods and their development are briefly described in this section, with selected examples, while detailed case studies are provided in Section 9.5.

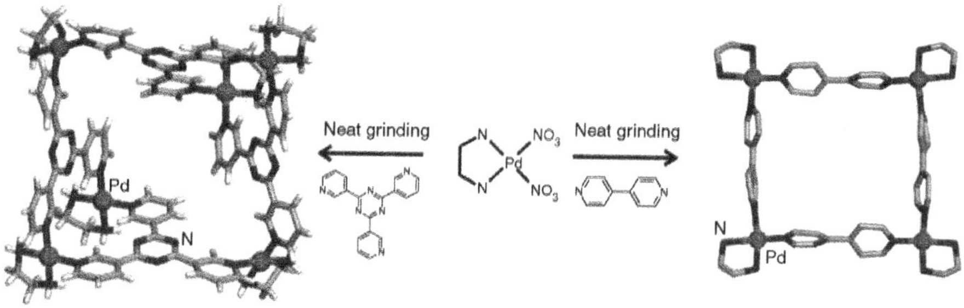


**FIGURE 9.2** Mechanochemical construction of a molecular bowl (left) and a molecular (square) via neat grinding[20].

### 9.3.1 Neat Grinding

Neat grinding is the simplest form of mechanosynthesis, as it involves grinding a physical mixture of two (or more) reactants. Grinding can be performed manually, using a mortar and a pestle. Alternatively, a mechanical ball mill can be employed. Historically, neat grinding was the first method applied for the mechanosynthesis of metal–ligand bonds. In particular, Orita et al. applied neat grinding for the construction of a tetranuclear metal-organic square and a hexanuclear metal-organic bowl by grinding (ethylenediamine)platinum(II) nitrate with suitable di- and tri-pyridine ligands, respectively (Figure 9.2).[20]

Neat grinding has also been applied to the construction of 1D and 2D-coordination polymers, as well as of 3D porous MOFs.[21,22] Although the field of mechanosynthesis of coordination compounds by neat grinding is still too small to allow broad generalizations, a recent array-based study of neat grinding reactions between diverse organic ligands and metal salts suggested that reactivity is favored with solid reactants that have lower melting points and/or are solvated.[52]

Although neat grinding reactions are considered to occur in a dry environment, it is noteworthy that most neat grinding reactions involve the use of hydrated metal salts (e.g., acetates or nitrates) as precursors.[22,52] It has been suggested that water or acid, which can be liberated during a mechanochemical reaction when using hydrated acetates or formates as reagents, could serve as a liquid phase that facilitates the mechanochemical reaction (as in LAG, Section 9.3.3) or templates the formation of a porous structure.

In some cases, neat grinding of anhydrous reactants in air can lead to the formation of hydrated products, as reported by Braga et al. for the solid-state reaction of silver acetate and 1,4-diaminocyclohexane (**dace**).[53] Grinding of the two results in the formation of metal–ligand bonds and absorption of moisture from air to form a coordination polymer material of composition $Ag(CH_3COO)(\mathbf{dace})\cdot nH_2O$.

### 9.3.2 Grinding-Annealing

Heating of the product initially obtained by neat grinding can lead to a subsequent reaction and the formation of a new product. In particular, Kuroda and coworkers have demonstrated that grinding of 3-cyanoacetylacetone with acetates of divalent transition metals can sometimes lead to the formation of discrete monomeric (i.e., 0-dimensional) β-diketonate complexes with axially coordinated water molecules.[48,49] Annealing of this product results in the expulsion of water and the formation of 3D metal-organic polymer.

A different type of reactivity obtained by a neat grinding-annealing sequence was demonstrated by Orpen and coworkers,[50,51] who mechanochemically constructed hydrogen-bonded pyridinium and imidazolium salts of tetrachlorometallates(II), by grinding of corresponding hydrochlorides with metal chlorides. Annealing of the tetrachlorometallate(II) salts was found to result in the reversible loss of HCl gas[54] to provide 1D coordination polymers consisting of metal chloride units connected to pyridine or imidazole ligands.

### 9.3.3 Liquid-Assisted Grinding and Kneading

LAG and kneading were introduced to the toolbox of mechanosynthesis in the context of cocrystal synthesis. In contrast to neat grinding, LAG utilizes the addition of small quantities of a liquid phase to assist or enable the mechanochemical reaction. Overall, the use of LAG for cocrystal synthesis has resulted in quantitative yields, short reaction times, and a greater scope of reactants and products. For cocrystals, the enhancement of reaction rate using LAG in comparison to neat grinding was recently monitored using terahertz (THz) spectroscopy.[55] The same method also demonstrated the high crystallinity of products obtained via LAG, suggesting a means to avoid amorphous impurities that typically result from neat grinding. The mechanism through which a liquid accelerates mechanochemical reactivity and produces a highly crystalline product are not yet clear, but are likely related to enhanced molecular diffusion and the plasticizing effect of the liquid phase.

Similar observations have recently been made in the context of coordination chemistry, where LAG was found to enhance the rate of mechanochemical synthesis of ethylenethiourea adducts of silver halides.[47] In particular, the solution reaction of AgCl and ethylenethiourea yields a single product, AgCl(ethylenethiourea)$_{1.5}$, composed of 1D polymeric cation tapes [ClAg$_4$(ethylenethiourea)$_6$]$^{3+}$ surrounded by Cl$^-$ anions (Figure 9.3). Neat grinding of this cationic polymer with further 1.5

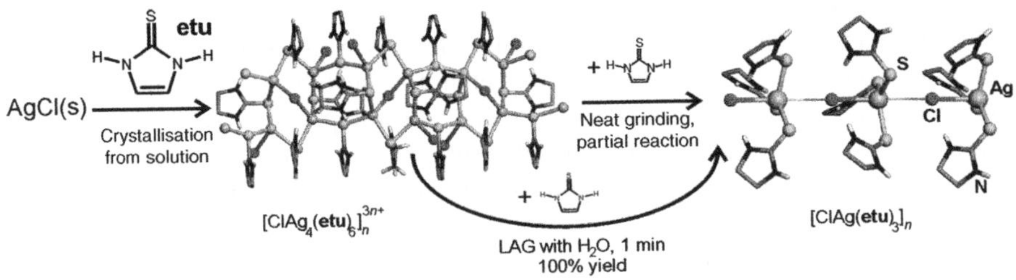


**FIGURE 9.3** Mechanochemical transformation of the [ClAg$_4$(ethylenethiourea)$_6$]$^{3+}$ coordination polymer into AgCl(ethylenethiourea)$_3$, accelerated by small quantities of water.[47]

equivalents of ethylenethiourea results in slow partial formation of a different product with composition AgCl(ethylenethiourea)$_3$ (Figure 9.3b). However, LAG in the presence of a small amount of water resulted in the quantitative formation of the AgCl(ethylenethiourea)$_3$ product within 1 min. Enhancement of reactivity upon the addition of a liquid phase was also observed in the reaction of silver iodide with ethylenethiourea. In this case, neat grinding did not lead to a reaction, whereas grinding in the presence of a small amount of water resulted in the quantitative formation of AgI(ethylenethiourea)$_2$ product, isostructural to AgBr(ethylenethiourea)$_2$.

LAG is particularly interesting for mechanochemical synthesis of porous materials, as the liquid can sometimes become incorporated in the final product as a guest. Such mechanochemical formation of host–guest inclusion has been described by Braga et al. who observed the formation of inclusion compounds upon kneading of copper(II) chloride and **dace** in the presence of *S*,*S*-dimethylsulfoxide (DMSO) or water.[56]

The ability of LAG to enhance mechanochemical reactivity as well as lead to molecular inclusion has been recently exploited to screen for inclusion compounds of coordination polymers, and porous MOFs, from a mixture of reactants that do not react under neat grinding.[23]

### 9.3.4 Characterization

Synthesis of metal-organic materials by grinding introduces additional challenges in terms of product characterization. In particular, the possibilities of solvate formation, polymorphism or dissociation/association processes upon dissolution make it difficult to characterize with confidence the product of mechanosynthesis in any other way except in the solid form obtained immediately from the reaction. In that way, mechanosynthesis of coordination compounds and cocrystals contrasts organic mechanosynthesis,[31,32,57,58] where the product can typically be analyzed in solution after extraction from the solid reaction mixture. Consequently, most products of metal-organic mechanosynthesis are characterized primarily by X-ray powder diffraction (PXRD), as well as reflectance FT-IR spectroscopy and solid-state NMR spectroscopy.[21,23] Although the latter methods immediately provide information on the chemical composition of the product, full structural characterization usually depends on the ability to obtain an identical product by single crystal growth from solution.[59] In such cases, structural characterization is readily achieved through single crystal X-ray diffraction. In cases where reaction in solution provides a different product than mechanosynthesis, it is, in principle, possible to obtain single crystals from solution by seeding it with fine powder of the grinding product.[60] This provides single crystals of the grinding product through heterogeneous nucleation. However, in cases where crystallization from solution consistently fails to provide single crystals of the desired solid phase, a possible alternative is given by methods for crystal structure solution from PXRD data.[23] Indeed, due to recent advances in laboratory technology and software, PXRD structure solution, aided by information gathered through spectroscopic or thermal analysis, has become a viable method for structural characterization of molecular materials on an almost routine basis.[38]

## 9.4 MECHANOCHEMICAL REACTIVITY LEADING TO COORDINATION POLYMERS

The mechanochemical construction of coordination polymers and frameworks is achieved by applying the synthetic methodologies, presented in Section 9.3, to different types of reactions that lead to the formation of coordination bonds. This section will provide a brief overview of such bond-forming reactions that have been applied to date for the construction of coordination polymers.

### 9.4.1 Direct Formation of Metal–Ligand Bonds

The simplest approach for the construction of coordination compounds is adduct formation, that is, reactions in which the reactants mutually bind to form the final product, without any byproducts. As adduct-forming reactions display the highest level of atom efficiency, they are particularly interesting for environmentally friendly synthesis. In the context of mechanosynthesis of coordination compounds, a recent example is the construction of a transition metal derivative of a pharmaceutical compound gabapentin, by grinding with anhydrous zinc chloride (Figure 9.4a). Upon grinding, the molecules of gabapentin act as neutral monodentate ligands and bind, through carboxylate oxygen atoms, to mononuclear $ZnCl_2$ units to yield a tetrahedral mononuclear complex Zn(gabapentin)$_2$Cl$_2$.[61] As anhydrous $ZnCl_2$ is not composed of discrete units, it is obvious that adduct formation reactions, albeit conceptually simple, require significant rearrangement of coordination bonds.

Accordingly, the construction of coordination polymers by mechanochemical adduct formation is expected to occur by using divergent polydentate ligands that can bridge two or more metal centers. For example, neat grinding of $ZnCl_2$ with a bridging ligand 4,4′-dipyridyl (**bipy**) results in the formation of the orthorhombic form of the coordination polymer [ZnCl$_2$(**bipy**)]$_n$, composed of 1D zigzag chains (Figure 9.4b). The formation of an analogous polymer from $CoCl_2$ was not possible

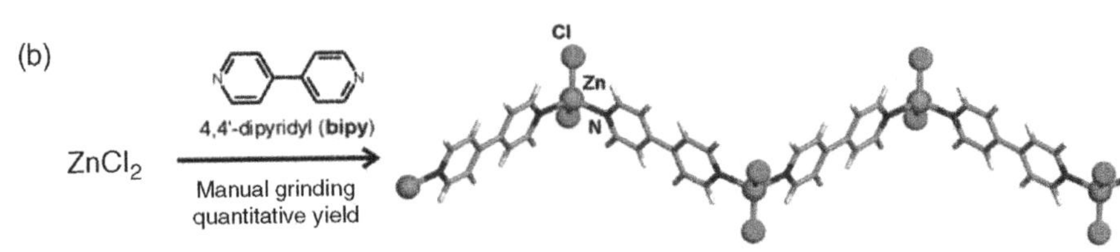


**FIGURE 9.4** Manual grinding synthesis of: (a) a metal–organic derivative of the drug gabapentin[61] and (b) the orthorhombic polymorph of the [ZnCl$_2$(**bipy**)]$_n$ polymer.[50]

by neat grinding. However, LAG of anhydrous $CoCl_2$ and **bipy** with the addition of a small quantity of ethanol, or the use of $CoCl_2{\cdot}6H_2O$ as a reactant in neat grinding, readily provided the 2D sheet polymer $[CoCl_2(bipy)]_n$.[50,51] The formation of ethylenethiourea complexes with AgCl and AgI, described in Section 9.3.3., also belongs to mechanochemical reactions involving the direct formation (and rearrangement) of coordination bonds.[47]

### 9.4.2 Mechanochemical Ligand Exchange

Probably the most frequently encountered reaction type in the mechanosynthesis of coordination polymers is ligand substitution. In particular, as most commercially available metal precursors are in the form of hydrates, the mechanochemical formation of new metal–ligand bonds often involves breaking metal–water linkages and water removal. The basic example of this type of reactivity is the neat grinding reaction of green $Ni(NO_3)_2{\cdot}6H_2O$ with 1,10-phenantroline, to provide the red solid Ni(1,10-phenantroline)$_3$(NO$_3$)$_2$, consisting of discrete Ni(1,10-phenantroline)$_3^{2+}$ cations (Figure 9.5).[62]

All known mechanochemical reactions of the popular "paddlewheel" complex copper(II) acetate monohydrate involve an exchange of ligands, either in the form of water, leading to the appearance of the paddlewheel building block in the resulting product, or in the form of water and acetic acid. The latter type of reactivity is observed in the mechanochemical synthesis of copper isonicotinate MOF by grinding $Cu(AcO)_2{\cdot}2H_2O$ with isonicotinic acid.[22] However, the latter reaction also involves proton transfer to a basic acetate ion and, consequently, can also be classified as a neutralization reaction described in Section 9.4.4.

### 9.4.3 Dehydrohalogenation

The construction of coordination bonds in a dehydrohalogenation process involves the removal of hydrogen chloride gas from crystalline salts based on pyridinium or imidazolium cations and tetrachlorometallate(II) cations.[54] This reaction type has

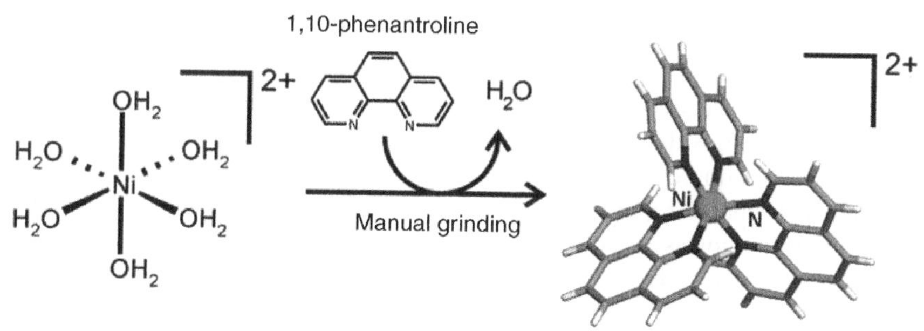


**FIGURE 9.5** Mechanochemical replacement of water ligands on an octahedral $Ni^{2+}$ ion with 1,10-phenantroline.[62] The counter ion is nitrate.

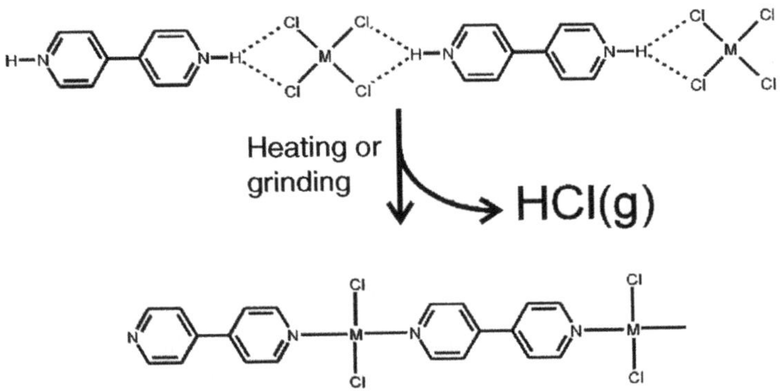


**FIGURE 9.6** A schematic representation of a dehydrohalogenation reaction leading to the formation of a coordination polymer.

been exploited by Orpen and coworkers for the construction of 1D polymers based on **bipy** (Figure 9.6).[50,51]

Since dehydrohalogenation is the thermal step in the grinding-annealing methodology, it can be argued that it should not be considered as a mechanochemical reaction. Nevertheless, the report by Brammer and coworkers that the dehydrochlorination of 3-Cl-pyridinium and 3-Br-pyridinium tetrachlorocuprates(II) is initiated by manual grinding suggests that mechanochemistry could bear relevance to such processes.[54]

Dehydrohalogenation reactions are reversible. Consequently, prolonged storage of reactant ammonium metallate salts does not lead to significant decomposition, as the equilibrium is readily achieved. However, the removal of the gaseous product from the solid–gas equilibrium by conducting the reaction in open air provides a suitable means to achieve quantitative conversion, in accordance with Le Chatelier's principle. As dehydrohalogenation reactions involve proton transfer, along with the construction of coordination bonds, they might also be classified in Section 9.4.4 among neutralization reactions.

### 9.4.4 Neutralization

The most general description of a neutralization reaction in the context of this review would be a process that, in addition to the construction of metal–ligand bonds, also involves a proton transfer. However, some reactions of this type, such as dehydrohalogenation by removal of the gaseous phase or the replacement of an entire ligand by protonation, for example removal of an acetate through acetic acid formation, are more conveniently discussed separately.[22,50] Accordingly, we will adopt a working definition of a neutralization as a reaction that involves proton transfer to a species generally considered as a strong base, for example hydroxide, carbonate or oxide. Such a definition also unmasks the thermodynamic role of neutralization as an additional driving force to achieve metal–ligand bond formation when using metal oxides or carbonates as reactants.

There are two approaches to utilize neutralization as means to drive the formation of coordination polymers. The first is through the addition of an external base, for example potassium hydroxide, to achieve the dehydrohalogenation of 4,4′-bipyridinium tetrachlorometallate salts (Figure 9.7a). Such reactions provide an identical coordination polymer to that obtained by thermal removal of HCl gas, with the obvious advantage of avoiding high temperatures. This advantage is somewhat offset by the additional workup needed to eliminate KCl and water byproducts.[51]

An alternative way to use neutralization in the mechanosynthesis of coordination polymers is by incorporating basic anions into reactants. This is readily accomplished by using metal carbonates or oxides as reactants. In such a scenario, the LAG reaction of a metal precursor with an acidic reagent, such as a bis(pyridinium) salt or a carboxylic acid, leads to the formation of the desired product (Figure 9.7b). The use of metal carbonates provides an additional driving force for the reaction, as the loss of the gaseous byproduct carbon dioxide from the reaction mixture steers the reaction to completion.

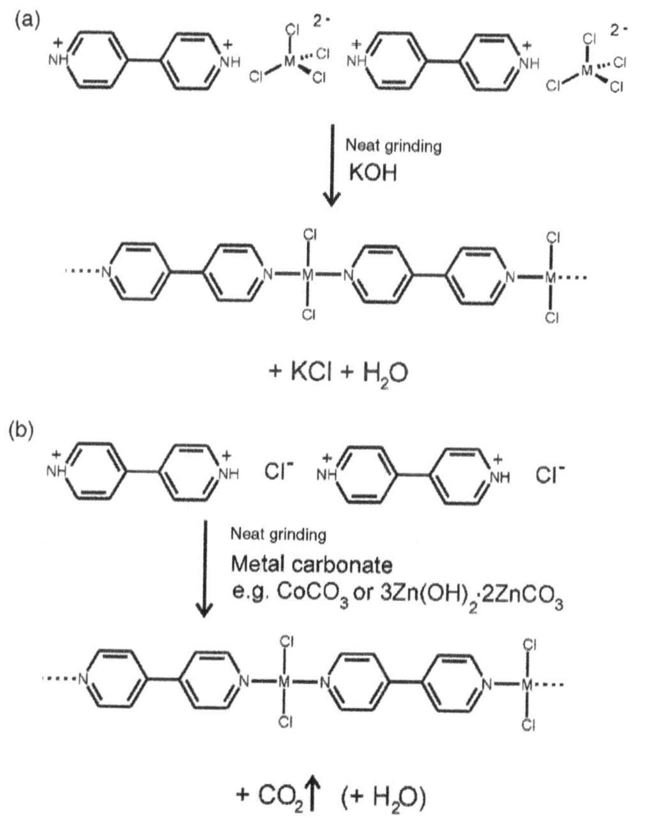


**FIGURE 9.7** Two applications of neutralization in assisting the mechanosynthesis of coordination polymers: (a) by the addition of an external reagent (KOH) and (b) by using a metal salt of a basic anion, such as carbonate or hydroxide.

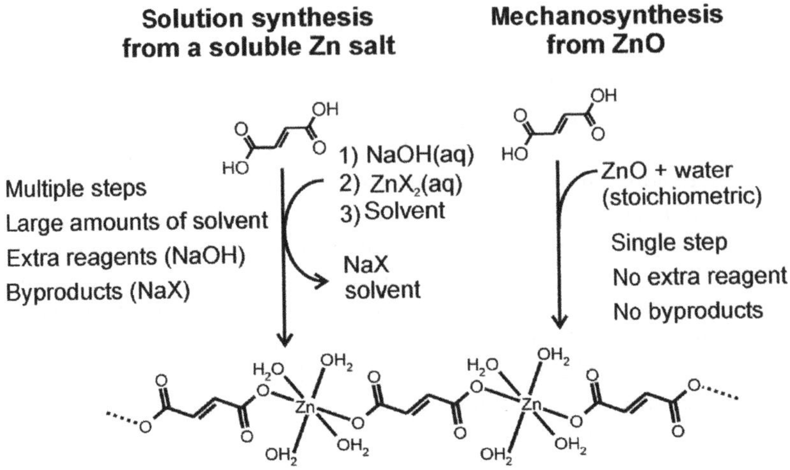


**FIGURE 9.8** Comparison of solution-based synthesis of a coordination polymer to the mechanosynthesis starting from the metal oxide.

Whereas LAG reactions of transition metal carbonates have been applied for the construction of 1D polymers involving **bipy** as the bridging ligand, mechanochemical reactivity of zinc oxide (ZnO) was utilized for the LAG synthesis of coordination polymers and MOFs.[23] The latter highlighted the environmental significance of mechanosynthesis from a metal oxide, as mechanosynthesis from slightly soluble ZnO was advantageous to solution-based synthesis by providing the product in a single step, without the need for additional solvent, neutralization reagents and related waste salts (Figure 9.8).

## 9.5 CONSTRUCTION OF COORDINATION POLYMERS BY GRINDING

The following section provides detailed case-by-case descriptions of reported mechanochemical syntheses of coordination polymers, and also attempts to categorize each one of them according to the methodologies and reactions delineated in Sections 9.3 and 9.4. For clarity, mechanochemical reactions leading to the formation of 1D and 2D coordination polymers and their inclusion compounds are treated separately from the ones resulting in the formation of 3D porous and non-porous MOFs.

### 9.5.1 Coordination Polymers by Neat Grinding

The first report of a coordination polymer constructed by grinding was given by Steed and coworkers, who described the formation of an inclusion compound of a zigzag 1D polymer by 15 min manual grinding of copper(II) acetate monohydrate with 1,3-bis(4-pyridyl)propane (**pn**) (Figure 9.9).[21] Grinding resulted in the replacement of water molecules bonded to the $Cu_2(AcO)_4$ paddlewheel fragment by bridging bis(pyridine)

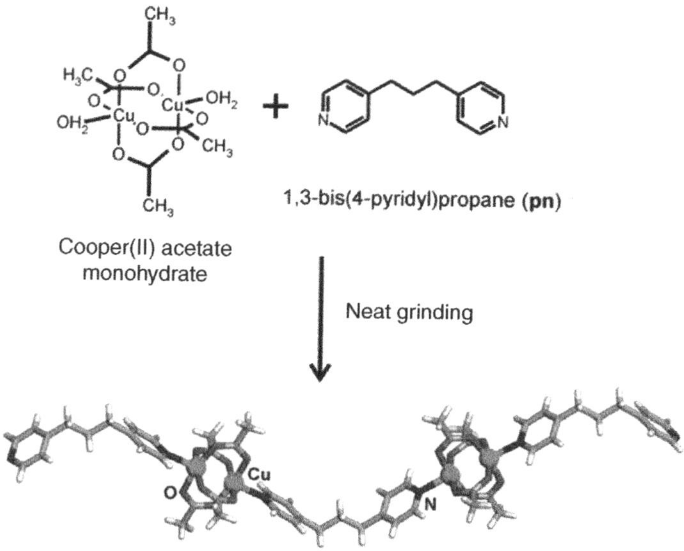


**FIGURE 9.9** Mechanochemical construction of a 1D coordination polymer.[21]

ligands. Ligand exchange was evident by the change of the color of the reaction mixture from blue to blue-green. The analogous reaction in refluxing methanol resulted in a type B inclusion compound with methanol molecules enclosed as guests between the self-assembled chains of the zigzag $[Cu_2(AcO)_4(\mathbf{pn})]_n$ polymer. As evidenced by PXRD, the mechanochemical product was isostructural to the inclusion compound obtained from methanol solution, indicating that the water produced by mechanochemical ligand exchange became incorporated in the product as a guest. This was further supported by $^{13}C$ MAS-NMR spectroscopy which revealed that the spectrum of the mechanochemically obtained inclusion compounds was identical to the one of the solution-grown methanol solvate, excluding the resonances corresponding to included methanol molecules.

The use of bis(pyridines) as bridging ligands in the mechanosynthesis of coordination polymers was also reported by Pichon et al., who utilized **bipy** as a reagent with copper(II) acetylacetonate, Cu(**acac**)$_2$, and hexafluoroacetylacetonate, Cu(**hfac**)$_2$.[52] In both cases, the **bipy** ligand bridged the planar copper β-diketonate moieties by forming axial Cu–N bonds, so as to provide linear 1D polymers. The products were identified through PXRD patterns that corresponded to those calculated for the known crystal structures of $[Cu(\mathbf{acac})_2(\mathbf{bipy})]_n$ and $[Cu(\mathbf{hfac})_2(\mathbf{bipy})]_n$ (Figure 9.10).

Neat grinding of anhydrous $ZnCl_2$ with an aliphatic nitrogen-based ligand, [2.2.2]-diazabicyclooctane (**dabco**) provides an interesting example of a stepwise mechanosynthesis[63] of a coordination polymer.[64] The first step in the reaction is the formation of a new crystalline material, characterized by PXRD, which has not yet been identified. Upon heating or upon further grinding, the initial product transforms to the 1D zigzag coordination polymer $[Zn(\mathbf{dabco})Cl_2]_n$ that was previously obtained

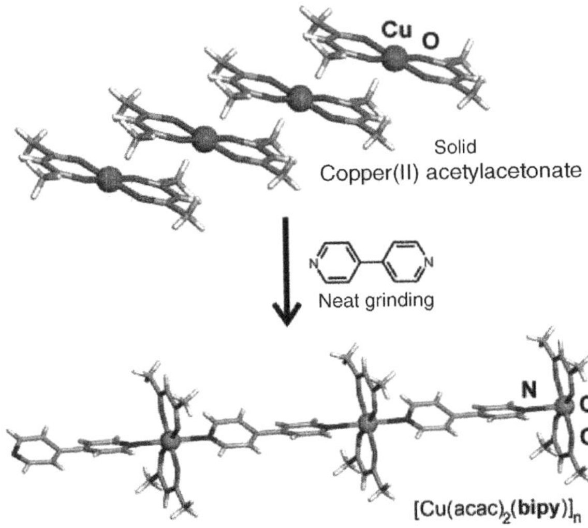


**FIGURE 9.10** Mechanochemical construction of a 1D coordination polymer from copper(II) acetylacetonate and **bipy**, reported by Pichon and James.[52]

from solution and was characterized via single crystal X-ray diffraction (Figure 9.11). The formation of an intermediate was ascribed to the hygroscopic nature of **dabco** under grinding conditions. Indeed, thermogravimetric analysis indicated that the intermediate is most likely a tetrahydrate of the final product (i.e., Zn(**dabco**)$Cl_2$·$4H_2O$), that can be thermally dehydrated to yield [Zn(**dabco**)$Cl_2]_n$. The active role of moisture in the formation of the intermediate was further confirmed by conducting the grinding reaction in a dry atmosphere, with carefully dried reactants. Under such dry conditions, grinding immediately provided [Zn(**dabco**)$Cl_2]_n$, without observable intermediates.[64]

In contrast, no intermediate was reported in the formation of an analogous zigzag 1D coordination polymer by manual neat grinding of anhydrous $ZnCl_2$ with **bipy**,

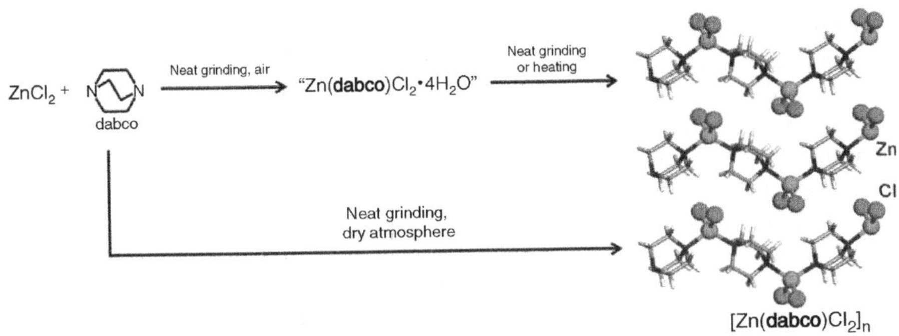


**FIGURE 9.11** The formation of the [Zn(**dabco**)$Cl_2]_n$ coordination polymer by manual grinding in air and grinding in a dry atmosphere.[64]

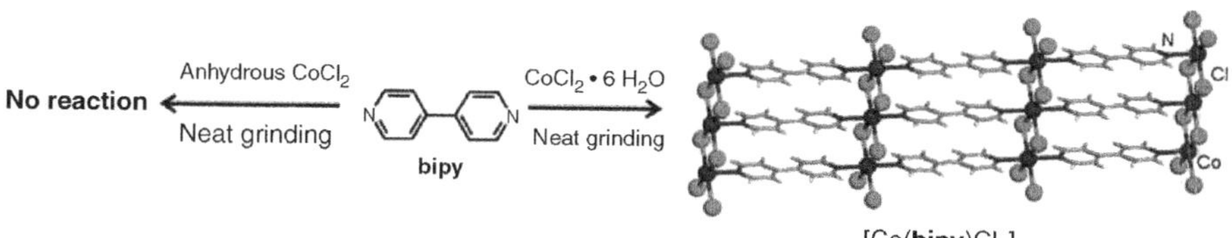


**FIGURE 9.12** Mechanochemical reactivity of **bipy** towards anhydrous $CoCl_2$ and $CoCl_2{\cdot}6H_2O$.[50]

reported by Orpen and coworkers.[50] Interestingly, the mechanochemical method provided only one of the three known polymorphic forms of the [Zn(**bipy**)$Cl_2]_n$ polymer. The construction of a 2D sheet polymer from anhydrous $CoCl_2$ and **bipy** was not possible by grinding. The sheet polymer [Co(**bipy**)$Cl_2]_n$ was, nevertheless, readily obtained by neat grinding of $CoCl_2{\cdot}6H_2O$ and **bipy**, suggesting that the water produced by desolvation of the reagents can play an important role in achieving mechanochemical reactivity via neat grinding (Figure 9.12).

Grinding of silver acetate, AgOAc, with **dabco** provided a coordination polymer with a metal:**dabco** ratio of 1:2.[64] The crystalline product obtained by manual grinding was formulated as a type B inclusion compound Ag(**dabco**)$_2$(OAc)$\cdot$5$H_2O$, through isostructurality with single crystals obtained from solution. The structure consists of 1D zigzag cationic polymer chains, [Ag(**dabco**)$_2(H_2O)]_n^{n+}$, involving tetrahedrally coordinated $Ag^I$ cations bridged by **dabco** ligands. In addition to bridging **dabco** ligands, each $Ag^+$ cation is coordinated by a water molecule and a **dabco** molecule that participates in an extended hydrogen-bonded network with guest water molecules and acetate ions (Figure 9.13). Similar to the mechanochemical reaction of **dabco** and $ZnCl_2$, the formation of a highly hydrated product upon neat grinding in air represents an excellent example of how the surrounding atmosphere can affect the course of

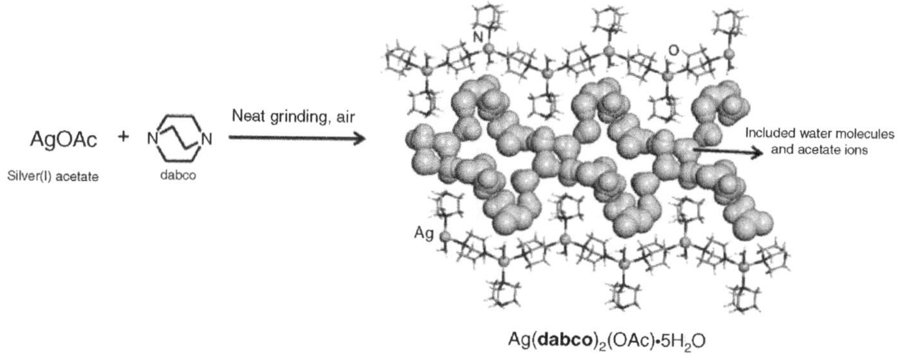


**FIGURE 9.13** Formation of a hydrated coordination polymer by neat manual grinding of silver(I) acetate and **dabco** in air. The water molecules and acetate ions that act as guests in the lattice host self-assembled from cationic chains [Ag(**dabco**)$_2(H_2O)]_n^{n+}$ are shown using the space-filling model.[64]

mechanosynthesis. Indeed, that included water molecules form coordination bonds to $Ag^+$, as well as hydrogen bonds to some of the **dabco** molecules in Ag(**dabco**)$_2$(OAc)·5$H_2O$, suggests that the metal-to-ligand ratio in the final product is determined by the absorption of moisture upon grinding.

Absorption of moisture into a product of mechanosynthesis was also observed in the mechanochemical reaction of AgOAc with a monocyclic ligand resembling **dabco**, 1,4-diaminocyclohexane (**dace**).[53] However, in contrast to **dabco** and **bipy** that are rigid bridging ligands, **dace** exhibits a higher degree of conformational flexibility, suitable to explore the structural diversity of coordination polymers obtained by grinding and from solution. Neat manual grinding of silver(I) acetate and **dace** in air provided a coordination polymer that was tentatively characterized as Ag(**dace**)(OAc)·$n$$H_2O$. The crystal structure of Ag(**dace**)(OAc)·$n$$H_2O$ is not yet known. However, re-crystallization from anhydrous methanol either by slow evaporation or in a stream of argon gas resulted in the formation of two structurally similar crystalline products: Ag(**dace**)(OAc)·3$H_2O$ and Ag(**dace**)(OAc)·$\frac{1}{2}$$H_2O$·$CH_3OH$, respectively.[53] Single crystal X-ray structure analysis revealed that both consist of 1D coordination polymer chains wherein each silver atom is coordinated by two equatorial amino groups of two bridging **dace** ligands. In addition, a solvent molecule that is either water (in Ag(**dace**)(OAc)·3$H_2O$) or methanol (in Ag(**dace**)(OAc)·$\frac{1}{2}$$H_2O$·$CH_3OH$) is coordinated to the Ag(I) ion. In each chain, the **dace** ligands around an $Ag^+$ center adopt a *cisoid* transformation, that is, are positioned exactly on top of each other and approximately perpendicular to the direction of the chain. The polymer chains in Ag(**dace**)(OAc)·$\frac{1}{2}$$H_2O$·$CH_3OH$ come together via short Ag···Ag bonds of 3.32 Å to produce 2D sheets (Figure 9.14). The additional solvent water molecules in the structure that are located within each sheet suggest this compound

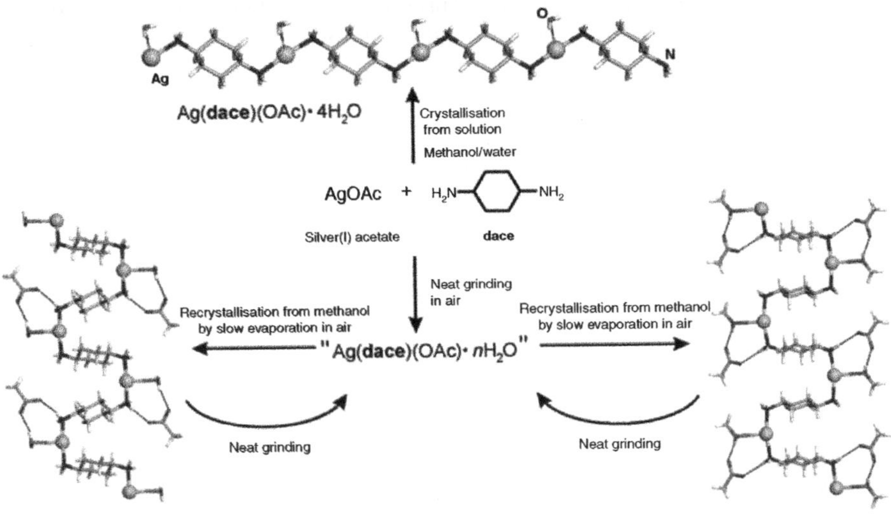


**FIGURE 9.14** Construction of solvated coordination polymers from silver(I) acetate and **dace** by mechanochemical and solution routes.[53]

should be classified as a type C inclusion compound (Figure 9.1). In Ag(**dace**)(OAc)·$3H_2O$, solvent water molecules are placed between chains of Ag(**dace**)$^+$, resulting in a type B inclusion compound. The similarity of Ag(**dace**)$^+$ chains in Ag(**dace**)(OAc)·$3H_2O$ and Ag(**dace**)(OAc)·$\frac{1}{2}H_2O$·$CH_3OH$ suggests that the original product mechanosynthesis, Ag(**dace**)(OAc)·$nH_2O$, is also composed of identical *cisoid* chains.

In contrast to grinding synthesis, cocrystallization of AgOAc and **dace** from a mixture of water and methanol provides a coordination polymer Ag(**dace**)(OAc)·$4H_2O$. The polymer is composed of Ag(**dace**)$^+$ chains with **dace** ligands arranged in a *transoid* conformation (Figure 9.14). Thus, mechanochemical grinding and reaction from solution provide two different isomeric structures of the Ag(**dace**)(OAc) polymer, illustrating the potential of mechanosynthesis in exploring modes of coordination-driven self-assembly that are not encountered via synthesis in solution.

The formation of type B inclusion compounds via ligand exchange was also reported by Pichon et al. in an extensive study of mechanochemical metal–ligand reactions.[52] Specifically, grinding of either copper(II) acetate, formate or trifluoroacetate with acetylenedicarboxylic acid (**H$_2$adc**) resulted in the formation of a solvated form of the linear copper(II) acetylenedicarboxylate polymer, Cu(**adc**)$(H_2O)_3$·$H_2O$. The same material was previously obtained from aqueous solution and characterized via single crystal X-ray diffraction[65] (Figure 9.15), facilitating the characterization through PXRD methods.

### 9.5.2 Mechanosynthesis of 3D Polymers and Porous MOFs by Neat Grinding

The formation of a porous 3D coordination polymer MOF by neat grinding was described by James, who conducted the reaction between copper(II) acetate monohydrate and isonicotinic acid (**Hina**).[22] Grinding of the reactants in a ball mill over a period of 10 min resulted in the change of the color of the reaction mixture from green to dark blue. The PXRD pattern of the product almost completely coincided with the one simulated for the microporous 3D MOF of copper(II) isonicotinate with included water molecules, Cu(**ina**)$_2$·$2H_2O$. The minor differences between the two patterns were interpreted as a consequence of partial inclusion of the acetic acid byproduct into

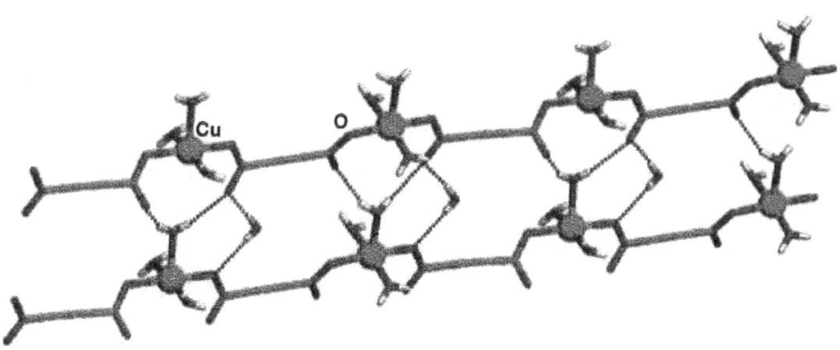


**FIGURE 9.15** Fragment of the crystal structure of the hydrated Cu(**adc**)$(H_2O)_3$·$H_2O$ polymer, obtained by grinding copper(II) acetate, trifluoroacetate or formate with **H$_2$adc**.[52]

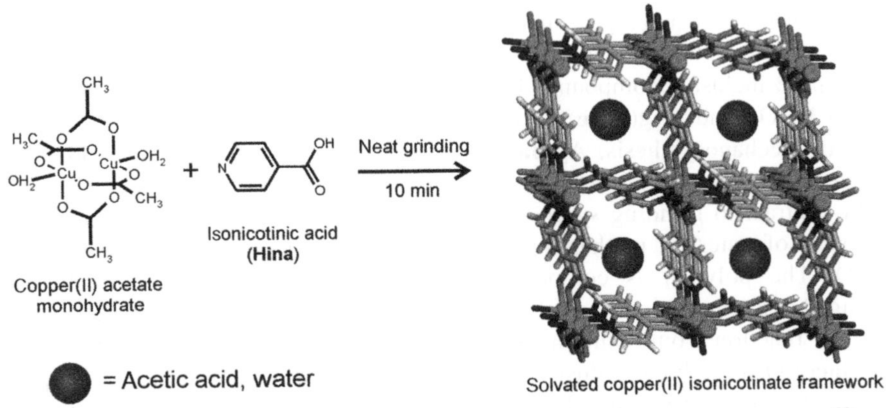


**FIGURE 9.16** Mechanochemical synthesis of a porous MOF by neat grinding.[22]

the pores of the Cu(**ina**)$_2$ MOF. Indeed, desolvation of the grinding product resulted in a material with a PXRD pattern identical to the one expected for the porous, desolvated Cu(**ina**)$_2$ framework (Figure 9.16).

In addition to continuous grinding, the quantitative formation of the microporous Cu(**ina**)$_2$ could also be achieved by mechanochemical activation and ageing. Specifically, grinding the reaction mixture for 1 min resulted in a partial reaction to form the MOF, which continued up to 6 h after grinding to provide Cu(**ina**)$_2$ in quantitative yield. Such approach of mechanochemical activation and ageing provided a product composed of larger crystals, with more developed crystal morphology than in case of continuous grinding synthesis.

That the carboxylic acid formed by the ligand exchange might play an active role in the reaction course is suggested by the mechanochemical reactivity of copper(II) formate and copper(II) trifluoroacetate toward **Hina**. While the reaction with copper (II) formate also resulted in the formation of the Cu(**ina**)$_2$ MOF, the trifluoroacetate salt yielded a previously unknown material. A tentative reason for such a difference is the ability of trifluoroacetic acid to template a different type of a porous structure.[52] Further evidence that the produced acid acts as a template in the formation of porous products was given by comparison of the mechanochemical reactions of benzene-1,3,5-tricarboxylic acid (**H$_3$bta**) toward hydrated forms of copper(II) acetate, formate and trifluoroacetate. In particular, $Cu(OAc)_2{\cdot}H_2O$ leads to the formation of the microporous MOF Cu$_3$(**bta**)$_2$, while copper formate leads to the quantitative formation of a new and yet unidentified product. Cu $(CF_3COO)_2{\cdot}H_2O$ resulted in only partial conversion, again to a yet unidentified product. The inability to form the Cu$_3$(**bta**)$_2$ from any other salt but the acetate was explained by specific templating of the Cu$_3$(**bta**)$_2$ microporous structure by acetic acid.[52]

An example of a solid-state grinding reaction that provides a non-porous 3D framework has been provided by Yoshida et al., who conducted neat manual grinding

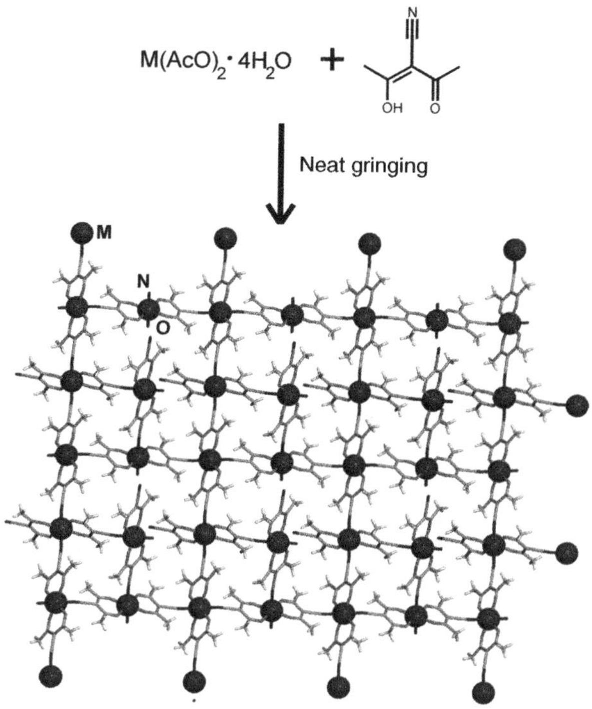


**FIGURE 9.17** Neat grinding synthesis of the non-porous 3D MOFs by grinding.[48]

of hydrated acetates of iron(II), cobalt(II) and nickel(II) with the tridentate ligand 3-cyanoacetylacetone (**HCNacac**) (Figure 9.17).[48]

Similar to the reaction of copper(II) acetate with **Hina**, grinding results in proton transfer and ligand exchange to form corresponding bis(3-cyanoacetylacetonates) and the byproduct acetic acid. Planar metal 3-cyanoacetylacetonate complexes self-assemble into a non-porous 3D framework through coordination bonds involving the 3-cyano substituents.

The ability to mechanochemically construct the [Ni(**CNacac**)$_2$]$_n$ framework in pure form is noteworthy, as solution methods typically provide a product contaminated with the sodium salt of the [Ni(**CNacac**)$_3$]$^-$ anion.[48,49,66,67] Analogous frameworks of Mn(II) and Zn(II) cannot be obtained by simple grinding of hydrated acetates with the β-diketone ligand, but can be obtained in a grinding-annealing procedure, described in Section 9.5.3.

The neat grinding construction of a non-porous MOF that, however, contains solvent in the form of solvated building blocks has been demonstrated by Pichon et al. in the mechanochemical reaction of **Hadc** with hydrated nickel(II) acetate ($Ni(OAc)_2{\cdot}4H_2O$) or nitrate ($Ni(NO_3)_2{\cdot}6H_2O$).[52] In both cases the proton and

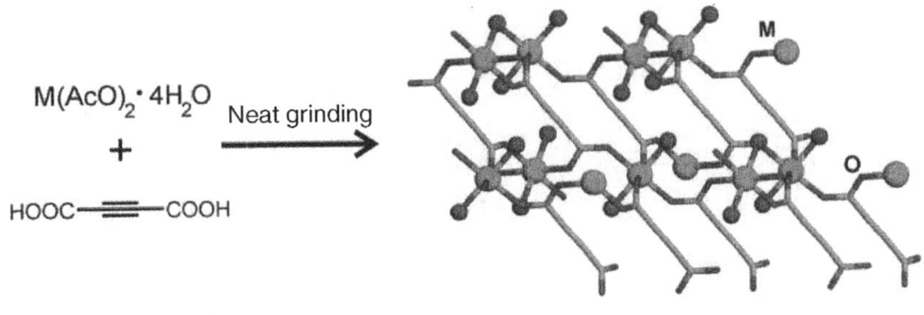


**FIGURE 9.18** Mechanochemical synthesis of a 3D non-porous polymer by neat grinding of zinc(II) or nickel(II) acetate tetrahydrate with **Hadc**. The metal(II) ions and oxygen atoms of water molecules are shown as black and dark gray spheres, respectively, whereas the oxygen (light gray) and carbon (gray) atoms of **adc**$^-$ ligands are displayed using the wireframe model.[52]

ligand exchange reactions result in the formation of the 3D coordination polymer Ni(**adc**)$(H_2O)_2$ (Figure 9.18).[68] The polymer is composed of octahedrally coordinated nickel(II) ions, with four equatorial positions occupied by **adc**$^-$ ligands, and the two axial ones with attached water molecules. The grinding reaction of **Hadc** with zinc acetate leads to the formation of a previously unknown 3D polymer Zn(**adc**)$(H_2O)_2$, identified by isostructurality to its Ni(II) analogue, demonstrating the use of mechanochemical synthesis in the discovery of new materials.[52]

### 9.5.3 Coordination Polymers by Grinding-Annealing

As mentioned in Section 9.3.2, the grinding-annealing technique is very recent in the mechanosynthesis of coordination compounds, and there are very few examples of its application to the construction of metal-organic polymers. Typically, the grinding-annealing technique is utilized when the product formed by grinding can be further transformed in a second, thermal step.

To date, only two types of reactions have been utilized as the second step in a grinding-annealing procedure: ligand exchange involving dehydration, as reported by Kuroda's group, and dehydrochlorination reaction, as reported by Orpen's group. Both reactions are reversible.

In contrast to hydrated acetates of Co(II), Ni(II) and Fe(II) that provided 3D coordination polymers by grinding with 3-cyanoacetylacetone (**HCNacac**), hydrated acetates $Mn(OAc)_2{\cdot}4H_2O$, $Cu_2(OAc)_4{\cdot}2H_2O$ and $Zn(OAc)_2{\cdot}2H_2O$ provided previously unknown mononuclear hydrated complexes Mn(**CNacac**)$_2{\cdot}2H_2O$, Cu(**CNacac**)$_2{\cdot}H_2O$ and Zn(**CNacac**)$_2{\cdot}H_2O$, respectively (Figure 9.19a–c). Upon heating to 100 °C, these hydrated 3-cyanoacetylacetonates readily lose water to provide anhydrous [M(**CNacac**)$_2]_n$ complexes (where M=Mn, Cu, Zn). The complexes [Mn(**CNacac**)$_2]_n$ and [Zn(**CNacac**)$_2]_n$ are non-porous 3D MOFs, isostructural

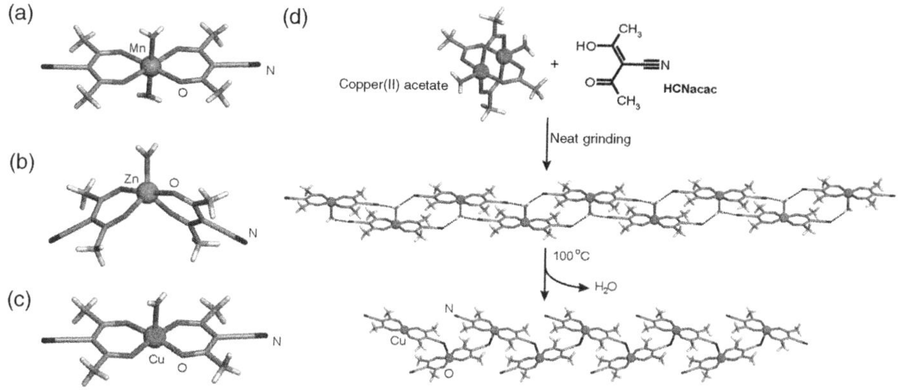


**FIGURE 9.19** Molecular structures of: (a) Mn(**CNacac**)$_2$(H$_2$O)$_2$; (b) Zn(**CNacac**)$_2$(H$_2$O); (c) Cu(**CNacac**)$_2$(H$_2$O) and (d) construction of the Cu(**CNacac**)$_2$ coordination polymer by a neat grinding-annealing sequence, starting from copper(II) acetate and **HCNacac**.[48]

to the analogous compounds of Co, Ni and Fe (Figure 9.17, Section 5.2). In contrast, Cu(**CNacac**)$_2$ (Figure 9.19d) is a zigzag 1D polymer.[66]

The use of dehydrochlorination as the annealing step in the construction of coordination polymers is illustrated by heating isomorphous 4,4′-bipyridinium salts of $FeCl_4^{2-}$, $CoCl_4^{2-}$ and $ZnCl_4^{2-}$ anions that are prepared by neat grinding of 4,4′-bipyridinium chloride with $FeCl_2{\cdot}4H_2O$, anhydrous $CoCl_2$ (or $CoCl_2{\cdot}6H_2O$) and $ZnCl_2$, respectively. The thermal step results in the elimination of HCl gas and the formation of a 1D zigzag (in case of Zn) or 2D sheet (in case of Fe and Co) polymers (Figure 9.20).[50,51]

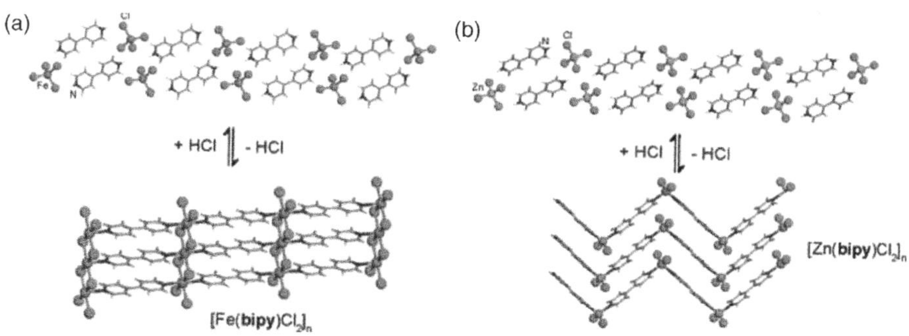


**FIGURE 9.20** The construction of 2D and 1D coordination polymers by thermal dehydrochlorination of (a) 4,4′-bipyridinium tetrachloroferrate(II) and (b) 4,4′-bipyridinium tetrachlorozincate.[50]

### 9.5.4 Screening for and Construction of Coordination Polymers by LAG

The ability to use LAG for rapid screening[12,13] for molecular inclusion within *in situ* assembled hydrogen-bonded networks inspired a similar attempt in the context of coordination-driven self-assembly.[23] Different liquid phases were found to steer the LAG neutralization reaction of zinc oxide and fumaric acid (**Hfum**) toward the formation of different products. Grinding in the presence of methanol or ethanol leads to the formation of previously[69] not characterized anhydrous form of zinc fumarate, Zn(**fum**). Crystal structure solution revealed Zn(**fum**) is a 3D non-porous coordination polymer based on tetrahedrally coordinated $Zn^{II}$ ions and bridging fumarate ligands. In contrast, grinding of ZnO and **Hfum** in the presence of a 1:1 mixture of ethanol and water resulted in the formation of a previously unknown dihydrate form Zn(**fum**)$(H_2O)_2$. The structure of the dihydrate was also determined by means of X-ray powder diffraction that revealed a 2D sheet polymer based on octahedrally coordinated zinc(II) ions. Furthermore, grinding of equimolar amounts of ZnO and **Hfum** in the presence of three and four equivalents of water selectively produced the previously known hydrated forms of zinc fumarate: the tetrahydrate and the pentahydrate, respectively (Figure 9.21).[70,71] Of particular interest was the formation of Zn(**fum**) pentahydrate, a type B inclusion compound of a self-assembled host composed of zigzag chains of [Zn(**fum**)$(H_2O)_4]_n$ with additional water as inclusion guest.

Consequently, the use of LAG enabled the construction of four different coordination polymer topologies from the same set of reactants: a non-porous 3D polymer, a 2D sheet polymer, a linear 1D polymer (in the form of

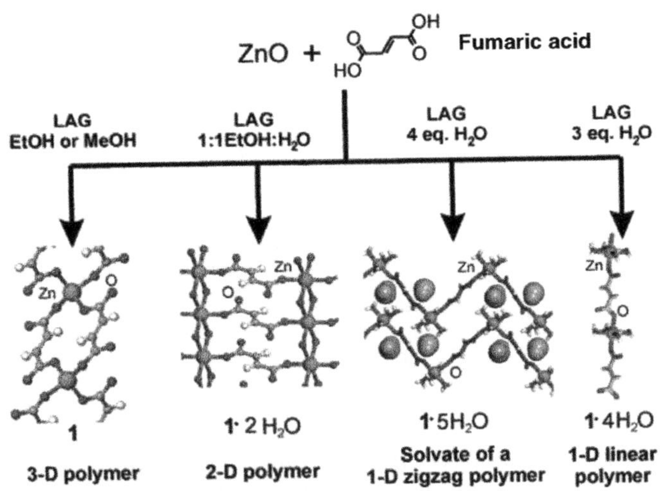


**FIGURE 9.21** Screening for coordination polymers from ZnO, using LAG.[23]

Zn(**fum**)($H_2O$)$_4$)[70] and a type B inclusion compound of a zigzag 1D polymer (in the form of Zn(**fum**)($H_2O$)$_4$·$H_2O$).[71]

LAG construction of coordination polymers from basic carbonate reactants was conducted by Adams et al. who obtained the 2D sheet polymer of [Co(**bipy**)$Cl_2$]$_n$ by grinding cobalt(II) carbonate with bipyridinium chloride in the presence of a small amount of water.[51] LAG resulted in the formation of the polymer, along with water and $CO_2$ gas as the byproducts (Figure 9.22). The same polymer could also be constructed from anhydrous $CoCl_2$ and **bipy** by LAG with a small quantity of ethanol.

Grinding of the basic zinc carbonate with bipyridinium chloride in the presence of small amount of water resulted in the formation of a mixture of two polymorphs (Figure 9.22) of the 1D zigzag polymer [Zn(**bipy**)$Cl_2$]$_n$. This contrasts the neat grinding reaction involving $ZnCl_2$ and **bipy**, which results in the formation of a single (orthorhombic) polymorph (Section 9.5.1), suggesting a surface templating effect in the reaction.

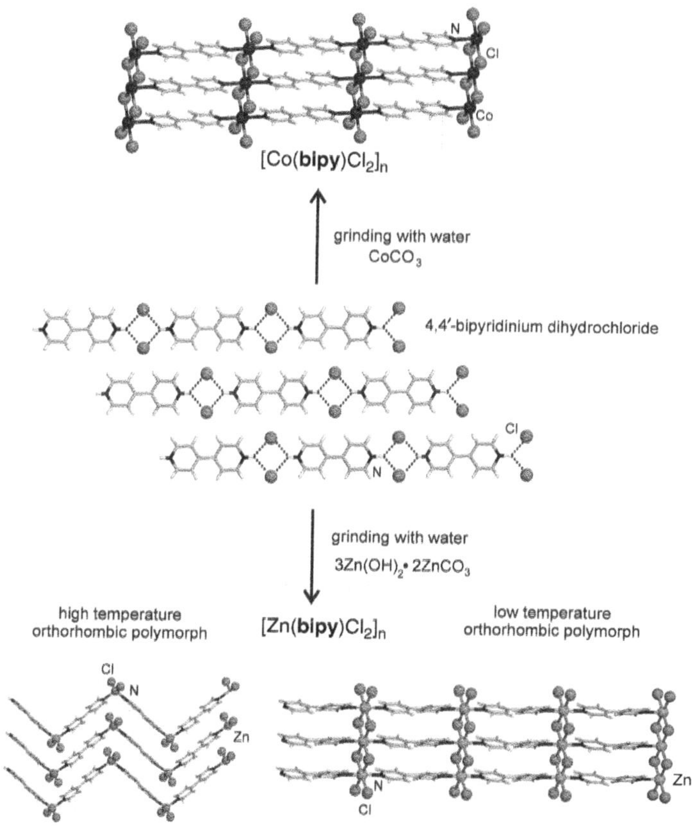


**FIGURE 9.22** Mechanosynthesis of coordination polymers using a transition metal carbonate or a basic carbonate as starting materials.[51]

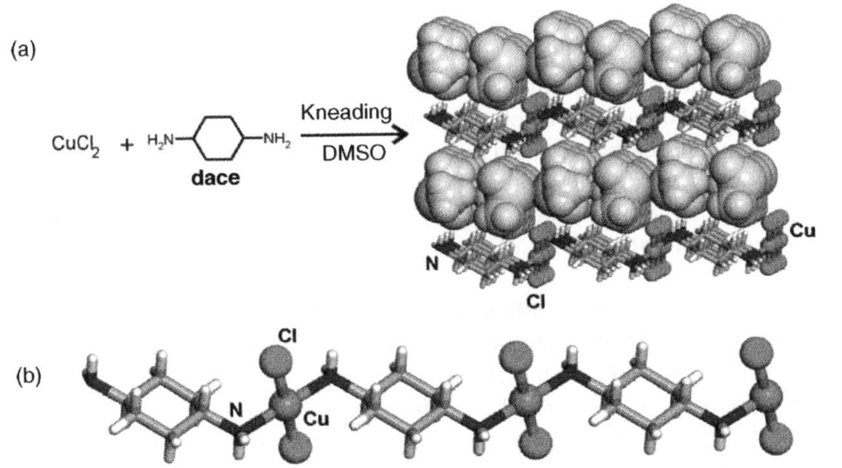


**FIGURE 9.23** (a) The construction of a metal–organic clay mimic material by kneading copper(II) chloride and dace in the presence of DMSO (or water) and (b) a single chain of $[Cu(\mathbf{dace})Cl_2]_n$. Solvent molecules included between the layers of $[Cu(\mathbf{dace})Cl_2]_n$ are shown using the space-filling model.[56]

### 9.5.5 Construction of Metal-Organic Inclusion Hosts by Kneading

The addition of a liquid phase to the grinding mixture presents an excellent opportunity to achieve molecular inclusion within a coordination polymer host. This was elegantly demonstrated by Braga et al.[56] who described a versatile 1D coordination polymer host with composition $[Cu(\mathbf{dace})Cl_2]$. Although the polymer could not be obtained by neat grinding of $CuCl_2$ and **dace**, kneading of the two components in a small amount of DMSO resulted in the formation of an inclusion compound $[Cu(\mathbf{dace})Cl_2]\cdot n$DMSO. Similarly, kneading in water produced the inclusion compound $[Cu(\mathbf{dace})Cl_2]\cdot nH_2O$. Both inclusion compounds could be obtained in the form of single crystals upon crystallization from corresponding liquid guests. In both cases, the structural analysis revealed a layered structure composed of layers of juxtaposed 1D chains of $[Cu(\mathbf{dace})Cl_2]$, separated by layers of included solvent guest (Figure 9.23).

Thermal desolvation of $[Cu(\mathbf{dace})Cl_2]\cdot n$DMSO and $[Cu(\mathbf{dace})Cl_2]\cdot nH_2O$ provided the same non-solvated polymer, $[Cu(\mathbf{dace})Cl_2]_n$. Kneading of thus formed $[Cu(\mathbf{dace})Cl_2]_n$ in a variety of solvents, followed by suspension overnight, resulted in the reversible formation of corresponding inclusion compounds. Interestingly, the same inclusion compounds could be prepared directly by LAG of $CuCl_2$ and **dace** in only a few cases, suggesting that the preformation of self-assembled layers in $[Cu(\mathbf{dace})Cl_2]_n$ was a prerequisite for successful inclusion. In that respect, $[Cu(\mathbf{dace})Cl_2]_n$ represents a very simple analogue of natural layered minerals with inclusion properties, such as clays.[56]

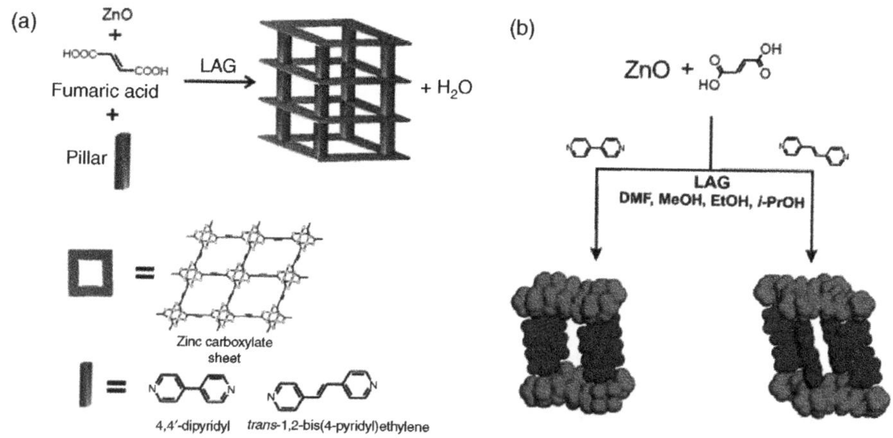


**FIGURE 9.24** (a and b) Synthesis of pillared open MOFs from zinc oxide and fumaric acid by using LAG.[23]

### 9.5.6 MOF Synthesis by LAG

The ability of different liquid phases to steer the formation of diverse coordination polymer topologies using LAG, demonstrated for zinc fumarate,[23] was interpreted as a result of binding of water molecules in the liquid phase to the *in situ* formed coordination polymer. This conclusion was subsequently utilized for the LAG construction of pillared MOF materials,[72,73] that is, porous 3D coordination polymers based on two different types of ligands, by providing additional ligands to the grinding mixture (Figure 9.24a). The addition of **bipy** and *trans*-1,2-(4-pyridyl)ethylene (**bpe**) to the LAG reaction mixture of ZnO and **Hfum** resulted in the quantitative formation of expected pillared MOFs $Zn_2$(**fum**)$_2$(**bipy**) and $Zn_2$(**fum**)$_2$(**bpe**), respectively (Figure 9.24b). These MOFs, composed of 2D zinc fumarate layers pillared by the bis(pyridine) ligands, were obtained in the solvated form, with the solvent used for grinding (DMF, methanol, ethanol or isopropanol) incorporated in the pores of the framework. Similar to the observation made for LAG synthesis of hydrated zinc fumarate polymers, different grinding liquids lead to the formation of products with slightly different PXRD patterns, interpreted as different framework breathing modes.[74]

Heating the solvated materials to 150 °C results in the complete loss of included guest, and the formation of evacuated frameworks. The pillared MOF $Zn_2$(**fum**)$_2$(**bipy**) was readily recognized through its PXRD pattern that completely coincided to the one calculated for the known crystal structure. Although the MOF $Zn_2$(**fum**)$_2$(**bpe**) was not previously known, its quantitative formation in the LAG reaction was recognized through the PXRD pattern that indicated isostructurality to the previously characterized copper(II) analogue. Subsequently, the crystal structure of $Zn_2$(**fum**)$_2$(**bpe**) was determined by refining the coordinates of the Cu analogue to the PXRD pattern measured for the LAG product (Figure 9.25).

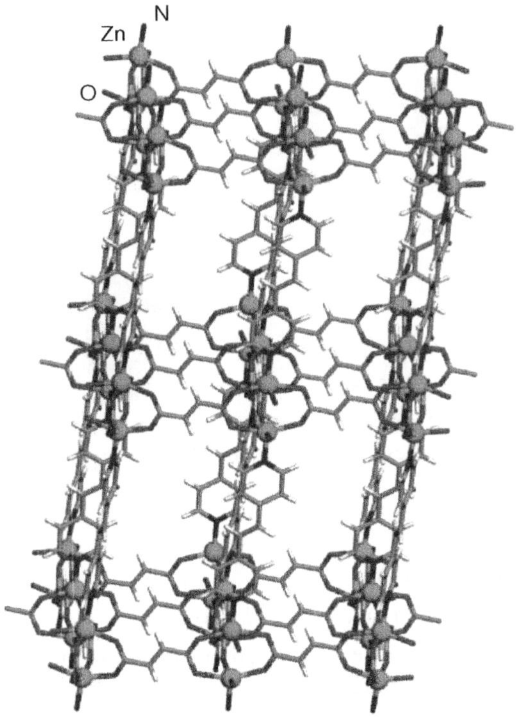


**FIGURE 9.25** Fragment of the crystal structure of the open MOF $Zn_2$(**fum**)$_2$(**bpe**), as determined by powder X-ray diffraction.[23] For clarity, only one of two interpenetrating networks is shown.

## 9.6 RELATED NONCONVENTIONAL TECHNIQUES

In addition to the mechanochemical methods of constructing coordination polymers, reviewed in previous sections, there have recently been advances in utilizing other alternative and environmentally friendly methodologies for the construction of MOFs. An overview of coordination polymer mechanosynthesis would not be complete without placing it in perspective along with such alternative (i.e., non-solvothermal) approaches.[28] Consequently, this section will briefly introduce the two most prominent alternative methods of MOF synthesis: through sonication and microwave irradiation.

### 9.6.1 MOF Synthesis via Sonication

The use of sonication as means to construct MOF materials was recently explored by Son et al. in the synthesis of the popular MOF-5 material from solutions of zinc nitrate and terephthalic acid.[14] Sonication allowed the use of a *N*-methylpyrrolidine (NMP) as an alternative solvent to *N*,*N*-diethylformamide (DEF), traditionally used in solution synthesis. By using NMP solvent, sonication resulted in significantly shorter reaction times of MOF-5 synthesis (typically in the range 8–30 min) than typically observed under solvothermal conditions (24 h). The reaction time and crystal quality during sonication synthesis in NMP could also be readily controlled by modifying the sonication power level. MOF-5 synthesis under sonication conditions in NMP occurred at a significantly higher temperature (155 °C) than during conventional solution synthesis (100 °C) and resulted in the formation of approximately 60 times smaller crystals. Attempts to conduct MOF-5 synthesis from DEF lead to significantly higher reaction times and poorer quality crystals. Consequently, NMP was recognized as the most suitable solvent for MOF synthesis, most likely due to a relatively lower boiling of DEF (177 °C).

### 9.6.2 Microwave Synthesis of Metal-Organic Frameworks

The application of microwave radiation to conduct MOF synthesis was investigated by Jhung et al. in the construction of the giant pore chromium terephthalate material MIL-101 (Figure 9.26).[15] Under microwave irradiation, the formation of MIL-101 occurred in significantly shorter time than in case of solvothermal synthesis. In particular, the optimum time for microwave synthesis of MIL-101 was found to be 60 min, compared to days using traditional solvothermal methods. However, microwave irradiation times longer than 60 min resulted in the disintegration of MIL-101 and the formation of unidentified products, as evidenced by PXRD.

The construction of zinc-based MOFs via microwave-assisted solvothermal synthesis was reported by Ni et al. who obtained microcrystals of zinc(II) frameworks with terephthalic (IRMOF-1), 2-bromoterephthalic (IRMOF-2) and

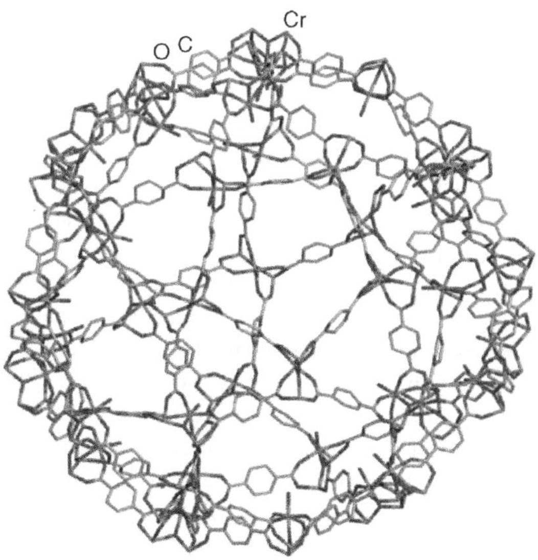


**FIGURE 9.26** Fragment of the giant pore MOF MIL-101, constructed by microwave-assisted synthesis.[15]

2-aminoterephthalic (IRMOF-3) acids.[16] The syntheses were conducted in DEF solvent and lasted minutes, compared to traditional solvothermal approaches that required days. In addition, the crystalline products, which were all characterized by PXRD methods, exhibited a highly regular cubic morphology and a very uniform distribution of particle sizes, demonstrated by SEM imaging. Crystallite size could also be controlled through varying the concentrations of starting materials. The uniform size and shape of MOF crystals was explained by the ability of microwave irradiation to initiate nucleation throughout the volume of solution, in contrast to solvothermal synthesis, where the nucleation process is dominated by heterogeneous nucleation on vessel walls.

The use of microwave irradiation to construct thin films of MOF crystallites was recently reported by Yoo and Jeong, who obtained oriented growth of the popular MOF-5 material on surfaces of pure, as well as graphite, amorphous carbon and gold-coated anodized aluminium oxide substrates. The formation of thin MOF layers on the substrates occurred within 30 s and was enhanced by the presence of a conductive (i.e., carbon, graphite, or gold) coating. Controlled deposition of a conductive carbon coating also allowed the construction of patterned MOF-covered surfaces.[75]

## 9.7 CONCLUSION

It is the author's hope that this short overview has successfully demonstrated the already established as well as the still growing potential of mechanochemical methods in the construction of metal-organic materials. Although the primary intent of this chapter was to provide an overview and propose a systematic organization of mechanochemical approaches to inclusion and porosity in metal-organic solids, a careful analysis of provided examples also suggests a strong relationship between metal-organic mechanosynthesis and molecular inclusion. Indeed, the formation of metal-organic inclusion compounds by mechanochemistry is more a rule than an exception: the use of common, typically hydrated metal salts as precursors often leads to the accidental formation of hydrated products, as well as the presence of moisture in air when grinding is performed manually with a mortar and a pestle. In case of kneading and LAG approaches, the molecules of the liquid phase, which is inherent to each method, readily result in, and even template, the formation of inclusion solvates and MOFs. Finally, the byproducts of mechanochemical reactions, such as water or acetic acid are also very likely to play an active role in the formation of microporosity. Consequently, it appears that mechanochemical reactions are not only suitable for MOF synthesis because of environmental or economic issues,[76] but they are particularly poised to achieve molecular inclusion necessary for permanent porosity.

Therefore, it is this author's firm belief that mechanochemical methods will be of growing significance in the design of new, efficient and versatile approaches to porous materials. The development of mechanochemical methods for the construction of metal-organic materials has now grown out of its infancy and is ready to enter a mature phase that will develop fundamental knowledge of underlying processes. Presumably, such knowledge will be in the form of mechanistic information obtained through extensive screening experiments and a plethora of modern solid-state analytical methods, such as NMR, FT-IR, PXRD and THz spectroscopy.

## ACKNOWLEDGMENTS

Professor William Jones, University of Cambridge, and Dr Robin Stein, Bruker, UK, are kindly acknowledged for helpful suggestions.

## LIST OF ABBREVIATIONS

$AcO^-$ = acetate ion
AcOH = acetic acid
**adc**$^{2-}$ = acetylenedicarboxylate anion
**bipy** = 4,4′-bipyridyl
**bpe** = *trans*-1,2-bis(4-pyridyl)ethylene
**bta**$^{3-}$ = 1,3,5-benzenetricarboxylate anion
**CNacac**$^-$ = 3-cyanoacetylacetonate anion
**dabco** = [2.2.2]diazabicyclooctane
**dace** = 1,4-diaminocyclohexane
DEF = *N,N*-diethylformamide
DMF = *N,N*-dimethylformamide
DMSO = *S,S*-dimethylsulfoxide
**fum**$^{2-}$ = fumarate anion
**Hadc** = acetylenedicarboxylic acid
**$H_3$bta** = 1,3,5-benzenetricarboxylic acid
**HCNacac** = 3-cyanoacetylacetone
**Hfum** = fumaric acid
**Hina** = isonicotinic acid
**ina**$^-$ = isonicotinate anion
NMP = *N*-methylpyrrolidine
**pn** = 1,3-bis(4-pyridyl)propane

## REFERENCES

1. Férey, G. Hybrid porous solids: past, present, future. *Chem. Soc. Rev.* **2008**, *37*, 191–214.
2. Li, H.; Eddaoudi, M.; O'Keeffe, M.; Yaghi, O. M. Design and synthesis of an exceptionally stable and highly porous metal-organic framework. *Nature* **1999**, *402*, 276–279.
3. Rowsell, J. L. C.; Yaghi, O. M. Strategies for hydrogen storage in metal-organic frameworks. *Angew. Chem. Int. Ed.* **2005**, *44*, 4670–4679.

4. Kitagawa, S.; Kitaura, R.; Noro, S. I. Functional porous coordination polymers. *Angew. Chem. Int. Ed.* **2004**, *43*, 2334–2375.

5. Eddaoudi, M.; Kim, J.; Rosi, N.; Vodak, D.; Wachter, J.; O'Keeffe, M.; Yaghi, O. M. Systematic design of pore size and functionality in isoreticular MOFs and their application in methane storage. *Science* **2002**, *295*, 469–472.

6. Yaghi, O. M.; O'Keeffe, M.; Ockwig, N. W.; Chae, H. K.; Eddaoudi, M.; Kim, J. Reticular synthesis and the design of new materials. *Nature* **2003**, *423*, 705–714.

7. Papaefstathiou, G. S.; MacGillivray, L. R. Inverted metal-organic frameworks: solid-state hosts with modular functionality. *Coord. Chem. Rev.* **2003**, *246*, 169–184.

8. Rosseinsky, M. J. Recent developments in metal-organic framework chemistry: design, discovery, permanent porosity and flexibility. *Micropor. Mesopor. Mater.* **2004**, *73*, 15–30.

9. Yaghi, O. M.; Li, H.; Davis, C.; Richardson, D.; Groy, T. L. Synthetic strategies, structure patterns, and emerging properties in the chemistry of modular porous solids. *Acc. Chem. Res.* **1998**, *31*, 474–484.

10. Lazuen-Garay, A.; Pichon, A.; James, S. L. Solvent-free synthesis of metal complexes. *Chem. Soc. Rev.* **2007**, *36*, 846–855.

11. Braga, D.; Giaffreda, S. L.; Grepioni, F.; Pettersen, A.; Maini, L.; Curzi, M.; Polito, M. Mechanochemical preparation of molecular and supramolecular organometallic materials and coordination networks. *Dalton Trans.* **2006**, 1249–1263.

12. Friščić, T.; Jones, W. Recent advances in understanding the mechanism of cocrystal formation via grinding. *Cryst. Growth Des.* **2009**, *9*, 1621–1637.

13. Friščić, T.; Trask, A. V.; Jones, W.; Motherwell, W. D. S. Screening for inclusion compounds and systematic construction of three-component solids by liquid-assisted grinding. *Angew. Chem. Int. Ed.* **2006**, *45*, 7546–7550.

14. Son, W. J.; Kim, J.; Kim, J.; Ahn, W. S. Sonochemical synthesis of MOF-5. *Chem. Commun.* **2008**, 6336–6338.

15. Jhung, S. H.; Lee, J. H.; Yoon, W.; Serre, C.; Férey, G.; Chang, J. S. Microwave synthesis of chromium terephthalate MIL-101 and its benzene sorption ability. *Adv. Mater.* **2007**, *19*, 121–124.

16. Ni, Z.; Masel, R. I. Rapid production of metal-organic frameworks via microwave-assisted solvothermal synthesis. *J. Am. Chem. Soc.* **2006**, *128*, 12394–12395.

17. Lemay, H. E.; Ash, L. A.; Jones, W. Concerning the extent of solid-state metal exchange between cobalt(II) chloride and tris(oxinato) iron. *Inorg. Nucl. Chem. Lett.* **1979**, *15*, 191–193.

18. Fernández-Betrán, J.; Castellanos-Serra, L.; Yee-Madeira, H.; Reguera, E. Proton transfer in solid state: mechanochemical reactions of imidazole with metallic oxides. *J. Solid State Chem.* **1999**, *147*, 561–564.

19. Fernández-Betrán, J. F. Mechanochemistry: an overview. *Pure Appl. Chem.* **1999**, *71*, 581–586.

20. Orita, A.; Jiang, L.; Nakano, T.; Ma, N.; Otera, J. Solventless reaction dramatically accelerates supramolecular self-assembly. *Chem. Commun.* **2002**, 1362–1363.

21. Belcher, W. J.; Longstaff, C. A.; Neckenig, M. R.; Steed, J. W. Channel-containing 1D coordination polymers based on a linear dimetallic spacer. *Chem. Commun.* **2002**, 1602–1603.

22. Pichon, A.; Lazuen-Garay, A.; James, S. L. Solvent-free synthesis of a microporous metal-organic framework. *CrystEngComm.* **2006**, *8*, 211–214.

23. Friščić, T.; Fábián, L. Mechanochemical conversion of metal oxide into coordination polymers and porous frameworks using liquid-assisted grinding (LAG). *CrystEngComm.* **2009**, *11*, 743–745. DOI: 10.1039/b822934c.
24. Braga, D.; Grepioni, F. Reactions between or within molecular crystals. *Angew. Chem. Int. Ed.* **2004**, *43*, 4002–4011.
25. Braga, D.; Grepioni, F. Making crystals from crystals: a green route to crystal engineering and polymorphism. *Chem. Commun.* **2005**, 3635–3645.
26. Moulton, B.; Zaworotko, M. J. From molecules to crystal engineering: supramolecular isomerism and polymorphism in network solids. *Chem. Rev.* **2001**, *101*, 1629–1658.
27. Kitagawa, S.; Matsuda, R. Chemistry of coordination space of porous coordination polymers. *Coord. Chem. Rev.* **2007**, *251*, 2490–2509.
28. Friščić, T.; Childs, S. L.; Rizvi, S. A.; Jones, W. The role of solvent in mechanochemical and sonochemical cocrystal formation: a solubility-based approach for predicting cocrystallization outcome. *CrystEngComm.* **2009**, *11*, 418–426.
29. Trask, A. V.; Jones, W. Crystal engineering of organic cocrystals by the solid-state grinding approach. *Topics Curr. Chem.* **2005**, *254*, 41–70.
30. Braga, D.; D'Addario, D.; Maini, L.; Polito, M.; Giaffreda, S.; Rubini, K.; Grepioni, F. Applications of crystal engineering strategies in solvent-free reactions: toward a supramolecular green chemistry. In *Frontiers in Crystal Engineering*, Tiekink, E. R. T.; Vittal J. J.,Eds.; John Wiley & Sons Ltd.: Chichester, England, 2006.
31. Tanaka, K.; Toda, F. Solvent-free organic synthesis. *Chem. Rev.* **2000**, *100*, 1025–1074.
32. Toda, F. Solid state organic chemistry: efficient reactions, remarkable yields, and stereoselectivity. *Acc. Chem. Res.* **1995**, *28*, 480–486.
33. Atkinson, M. B. J.; Bučar, D. K.; Sokolov, A. N.; Friščić, T.; Robinson, C. M.; Bilal, M. Y.; Sinada, N. G.; Chevannes, A.; MacGillivray, L. R. General application of mechanochemistry to templated solid-state reactivity: rapid and solvent-free access to crystalline supermolecules. *Chem. Commun.* **2008**, 5713–5715.
34. Weyna, D. R.; Shattock, T.; Vishweshwar, P.; Zaworotko, M. J. Synthesis and structural characterization of cocrystals and pharmaceutical cocrystals: mechanochemistry vs. slow evaporation from solution. *Cryst. Growth Des.* **2009**, *9*, 1106–1123.
35. Karki, S.; Friščić, T.; Jones, W. Control and interconversion of cocrystal stoichiometry in grinding: stepwise mechanism for the formation of a hydrogen-bonded cocrystal. *CrystEngComm* **2009**, *11*, 470–481.
36. Shan, N.; Toda, F.; Jones, W. Mechanochemistry and co-crystal formation: effect of solvent on reaction kinetics. *Chem. Commun.* **2002**, 2372–2373.
37. Patil, A. O.; Curtin, D. Y.; Paul, I. C. Formation of crystalline complexes between polymethylated quinones and hydroquinones. *J. Chem. Soc. Perkin Trans.* **1986**, *2*, 1687–1692.
38. Karki, S.; Fábián, L.; Friščić, T.; Jones, W. Powder X-ray diffraction as an emerging method to structurally characterize organic solids. *Org. Lett.* **2007**, *9*, 3133–3136.
39. Trask, A. V.; Motherwell, W. D. S.; Jones, W. Solvent-drop grinding: green polymorph control of cocrystallization. *Chem. Commun.* **2004**, 890–891.
40. Trask, A. V.; Haynes, D. A.; Motherwell, W. D. S.; Jones, W. Screening for crystalline salts via mechanochemistry. *Chem. Commun.* **2006**, 51–53.
41. Baisch, U.; Rubini, K.; Braga, D. Remarkable structural similarities between organic cocrystals and a metal-organic coordination network-insights into hydrogen bonded aliphatic ammonium chlorides. *CrystEngComm.* **2008**, *10*, 1939–1947.

42. Tranchemontagne, D. J.; Hunt, J. R.; Yaghi, O. M. Room temperature synthesis of metal-organic frameworks: MOF-5, MOF-74, MOF-177, MOF-199, and IRMOF-0. *Tetrahedron* **2008**, *64*, 8553–8557.

43. Salager, E.; Stein, R. S.; Pickard, C. J.; Elena, B.; Emsley, L. Powder NMR crystallography of thymol. *Phys. Chem. Chem. Phys.* **2009**, *11*, 2610–2621.

44. Férey, G.; Serre, C.; Mellot-Draznieks, C.; Millange, F.; Surblé, S.; Dutour, J.; Margiolaki, I. A hybrid solid with giant pores prepared by a combination of targeted chemistry, simulation, and powder diffraction. *Angew. Chem. Int. Ed.* **2004**, *43*, 6296–6301.

45. Férey, G.; Mellot-Draznieks, C.; Serre, C.; Millange, F. Crystallized frameworks with giant pores: are there limits to the possible? *Acc. Chem. Res.* **2005**, *38*, 217–225.

46. Friščić, T.; Jones, W. Cocrystal architecture and properties: design and building of chiral and racemic structures by solid-solid reactions. *Faraday Discuss.* **2007**, *136*, 167–178.

47. Bowmaker, G. A.; Chaichit, N.; Pakawatchai, C.; Skelton, B. W.; White, A. H. Solvent-assisted mechanochemical synthesis of metal complexes. *Dalton Trans.* **2008**, 2926–2928.

48. Yoshida, J.; Nishikiori, S. I.; Kuroda, R. Formation of 1D and 3D coordination polymers in the solid state induced by mechanochemical and annealing treatments: bis(3-cyano-pentane-2,4-dionato) metal complexes. *Chem. Eur. J.* **2008**, *14*, 10570–10578.

49. Kuroda, R.; Yoshida, J.; Nakamura, A.; Nishikiori, S. I. Annealing assisted mechano-chemical synthesis of transition-metal coordination compounds and co-crystal formation. *CrystEngComm.* **2009**, *11*, 427–432.

50. Adams, C. J.; Colquhoun, H. M.; Crawford, P. C.; Lusi, M.; Orpen, A. G. Solid-state interconversions of coordination networks and hydrogen-bonded salts. *Angew. Chem. Int. Ed.* **2007**, *46*, 1124–1128.

51. Adams, C. J.; Kurawa, M. A.; Lusi, M.; Orpen, A. G. Solid state synthesis of coordination compounds from basic metal salts. *CrystEngComm.* **2008**, *10*, 1790–1795.

52. Pichon, A.; James, S. L. An array-based study of reactivity under solvent-free mechano-chemical conditions-insights and trends. *CrystEngComm.* **2008**, *10*, 1839–1847.

53. Braga, D.; Curzi, M.; Grepioni, F.; Polito, M. Mechanochemical and solution reactions between $AgCH_3COO$ and $[H_2NC_6H_{10}NH_2]$ yield three isomers of the coordination network $\{Ag[H_2NC_6H_{10}NH_2]^+\}_\infty$. *Chem. Commun.* **2005**, 2915–2917.

54. Espallargas, G. M.; Brammer, L.; van de Streek, J.; Shankland, K.; Florence, A. J.; Adams, H. Reversible extrusion and uptake of HCl molecules by crystalline solids involving coordination bond cleavage and formation. *J. Am. Chem. Soc.* **2006**, *128*, 9584–9585.

55. Nguyen, K. L.; Friščić, T.; Day, G. M.; Gladden, L. F.; Jones, W. Terahertz time-domain spectroscopy and the quantitative monitoring of mechanochemical cocrystal formation. *Nature Mater.* **2007**, *6*, 206–209.

56. Braga, D.; Curzi, M.; Johansson, A.; Polito, M.; Rubini, K.; Grepioni, F. Simple and quantitative mechanochemical preparation of a porous crystalline material based on a 1D coordination network for uptake of small molecules. *Angew. Chem. Int. Ed.* **2006**, *45*, 142–146.

57. Kaupp, G. Waste-free large-scale syntheses without auxiliaries for sustainable production omitting purifying workup. *CrystEngComm.* **2006**, *9*, 794–804.

58. Kaupp, G.; Herrmann, A.; Schmeyers, J. Waste-free chemistry of diazonium salts and benign separation of coupling products in solid salt reactions. *Chem. Eur. J.* **2002**, *8*, 1395–1406.

59. Braga, D.; Maini, L.; de Sanctis, G.; Rubini, K.; Grepioni, F.; Chierotti, M. R.; Gobetto, R. Mechanochemical preparation of hydrogen-bonded adducts between the diamine 1,4-

diazabicyclo[2.2.2]octane and dicarboxylic acids of variable chain length: an X-ray diffraction and solid-state NMR study. *Chem. Eur. J.* **2003**, *9*, 5538–5548.

60. Trask, A. V.; van de Streek, J.; Motherwell, W. D. S.; Jones, W. Achieving polymorphic and stoichiometric diversity in cocrystal formation: importance of solid-state grinding, powder X-ray structure determination, and seeding. *Cryst. Growth Des.* **2005**, *5*, 2233–2241.
61. Braga, D.; Grepioni, F.; Maini, L.; Brescello, R.; Cotarca, L. Simple and quantitative mechanochemical preparation of the first zinc and copper complexes of the neuroleptic drug gabapentin. *CrystEngComm.* **2008**, *10*, 469–471.
62. Nichols, P. J.; Raston, C. L.; Steed, J. W. Engineering of porous π-stacked solids using mechanochemistry. *Chem. Commun.* **2001**, 1062–1063.
63. Cinčić, D.; Friščić, T.; Jones, W. A stepwise mechanism for the mechanochemical synthesis of halogen-bonded cocrystal architectures. *J. Am. Chem. Soc.* **2008**, *130*, 7524–7525.
64. Braga, D.; Giaffreda, S. L.; Grepioni, F.; Polito, M. Mechanochemical and solution preparation of the coordination polymers $Ag[N(CH_2CH_2)_3N]_2[CH_3COO]\cdot 5H_2O$ and $Zn[N(CH_2CH_2)_3N]Cl_2$. *CrystEngComm.* **2004**, *6*, 458–462.
65. Billetter, H.; Hohn, F.; Pantenburg, I.; Ruschewitz, U. $[Cu\{C_2(COO)_2\}(H_2O)_3]\cdot H_2O$, the first copper complex of acetylenedicarboxylic acid. *Acta Cryst. C.* **2003**, *59*, m130–m131.
66. Angelova, O.; Petrov, G.; Macicek, J. Structure of bis(3-cyano-2,4-pentanedionato)copper (II). *Acta Cryst. C* **1989**, *45*, 710–713.
67. Angelova, O.; Macicek, J.; Atanasov, M.; Petrov, G. Chelating modes of 3-substituted 2,4-pentanediones. Crystal and electronic structure of bis(3-cyano-2,2-pentanedionato)cobalt (II). *Inorg. Chem.* **1991**, *30*, 1943–1949.
68. Hohn, F.; Billetter, H.; Pantenburg, I.; Ruschewitz, U. $Ni(C_2(COO)_2)(H_2O)_4\cdot 2H_2O$ and $Ni(C_2(COO)_2)(H_2O)_2$: two coordination polymers of the acetylenedicarboxylate dianion. *Z. Naturforsch. B Chem. Sci.* **2002**, *57*, 1375–1381.
69. Narkis, M.; Ron, I.; Siegmann, R. A.; Kaçir, L. Reactions of zinc oxide with molten diacids and their anhydrides. *Eur. Polym. J.* **1974**, *10*, 267–272.
70. Xu, W.; Zheng, Y. Q. Refinement of the crystal structure of catena-tetraaquofumarato-O, O′-zinc(II), $Zn(H_2O)_4(C_4H_2O_4)$. *Z. Kristallogr. NCS* **2004**, *219*, 235–236.
71. Xie, H. Z.; Zheng, Y. Q.; Shou, K. Q. Synthesis and crystal structure of $[Zn(H_2O)_4(C_4H_2O_4)]\cdot H_2O$. *J. Coord. Chem.* **2003**, *56*, 1291–1297.
72. Dybtsev, D. N.; Chun, H.; Kim, K. Rigid and flexible: a highly porous metal-organic framework with unusual guest-dependent dynamic behaviour. *Angew. Chem. Int. Ed.* **2004**, *43*, 5033–5036.
73. Ma, B. Q.; Mulfort, K. L.; Hupp, J. T. Microporous pillared paddle-wheel frameworks based on mixed-ligand coordination of zinc ions. *Inorg. Chem.* **2005**, *44*, 4912–4914.
74. Llewellyn, P. L.; Maurin, G.; Devic, T.; Loera-Serna, S.; Rosenbach, N.; Serre, C.; Bourelly, S.; Horcajada, P.; Filinchuk, Y.; Férey, G. Prediction of the conditions for breathing of metal-organic framework materials using a combination of X-ray powder diffraction, microcalorimetry, and molecular simulation. *J. Am. Chem. Soc.* **2008**, *130*, 12808–12814.
75. Yoo, Y.; Jeong, H. K. Rapid fabrication of metal-organic framework thin films using microwave-induced thermal deposition. *Chem. Commun.* **2008**, 2441–2443.
76. Sheldon, R. A. E factors, green chemistry and catalysis: an odyssey. *Chem. Commun.* **2008**, 3352–3365.

# 10

# METAL-ORGANIC FRAMEWORKS WITH PHOTOCHEMICAL BUILDING UNITS

SAIKAT DUTTA, IVAN G. GEORGIEV, AND LEONARD R. MACGILLIVRAY
*Department of Chemistry, University of Iowa, Iowa City, IA 52242, USA*

## 10.1 INTRODUCTION

Microporous materials (e.g., zeolites, activated carbons) have long played a prominent role in large-scale chemical separations and catalysis.[1,2] The functional behaviors of such porous materials are strongly dependent on the dimension and chemical environment of the pores.[3] The development of microporous materials with desired and unique activities will depend on the rational design of pore structures that have been elusive in traditional microporous materials. Over the last decade, metal-organic frameworks (MOFs) have attracted much attention as promising complements to existing classes of microporous solids.[4] MOFs consist of metal ions and/or clusters that form vertices of a framework and organic linkers that form bridges. One advantage of MOFs is their easily modifiable synthesis to control pore connectivity, structure, and dimension by varying the ligands, metals, and/or the counteranions. A wide range of functionality can be incorporated into the pores of MOFs by altering the coordination geometries of the metals and the topicities (e.g., ditopic, tritopic) of the ligands.

Developments in the field of coordination-driven supramolecular chemistry and crystal engineering in recent years have led to reports of MOFs with components that undergo reaction in the solid state.[5] The well-organized environment of the solid state, as well as the geometries adopted by the organic components of MOFs, can provide useful platforms to assemble molecules into suitable positions to react. In this chapter,

*Metal-Organic Frameworks: Design and Application*, Edited by Leonard R. MacGillivray

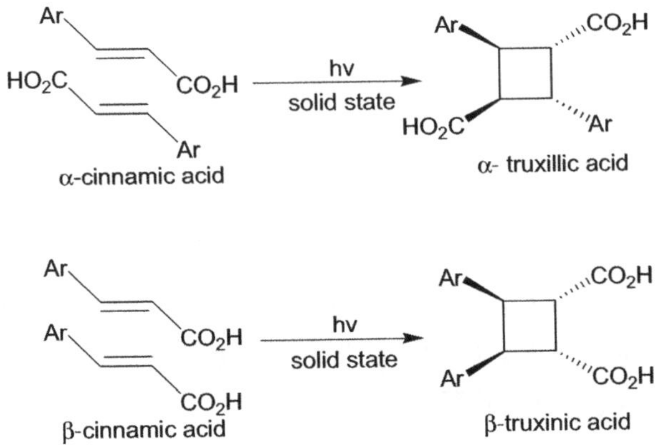


**SCHEME 10.1** Schematic view of solid-state reactions of two polymorphic forms of cinnamic acid.

we will demonstrate how chemical reactivity involving the [2 + 2] photodimerization can be integrated into MOFs. We will show how the metal–ligand interactions play a role in guiding organic molecules in close proximity to satisfy geometry criteria for the photoreaction. A main motivation to construct MOFs with reactive building units is that the reactions lead to changes in structure and, therefore, can affect properties of such solids (e.g., pore size). Applications in areas such as molecular sensing, controlled guest release, and imaging can be envisioned. In related work, we will show how molecular products of [2 + 2] photodimerizations conducted in the solid state can be used as ligands to build MOFs. The photoreaction can be used to synthesize ligands and afford MOFs that are, otherwise, markedly less available from other synthetic routes. Thus, the formation of MOFs with unique properties is possible using ligands rationally designed and synthesized in the solid state.

## 10.2 [2 + 2] PHOTODIMERIZATION IN THE SOLID STATE

Approximately four decades ago, Schmidt and coworkers determined general geometry criteria for a [2 + 2] photodimerization to proceed in the solid state.[6] A [2 + 2] photodimerization will typically occur if the reactive centers (i.e., olefins) are aligned parallel and separated by less than 4.2 Å. The criteria, being part of the topochemical postulate, were derived from studies involving α-, β-, γ- polymorphs of cinnamic acid (Scheme 10.1).

Since the work of Schmidt, there have been numerous studies that aim to control the [2 + 2] photodimerization in the solid state, which have been achieved with different levels of success. The sensitivity of solid-state structure to molecular structure has made it extremely difficult to control the reaction owing to structure demands of close packing.[7] In early work, suitable orientations of olefins for the photoreaction were achieved using substituents intended to guide the solid-state packing for reaction.[8] For example, interactions between chlorine substituents were shown to promote the

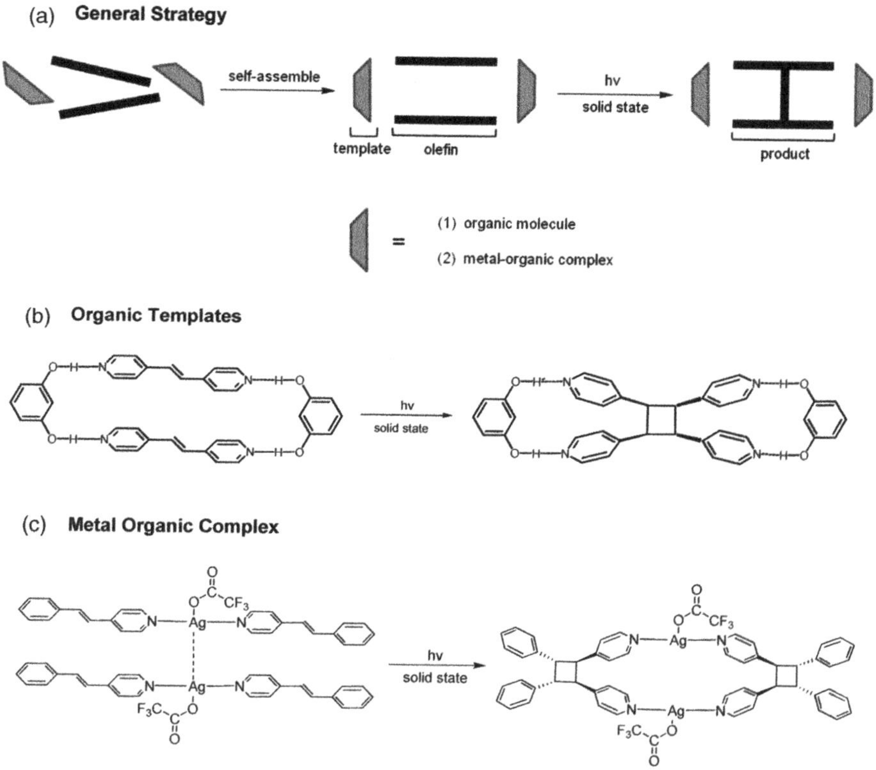


**SCHEME 10.2** Schematic of (a) general strategy of template-controlled solid-state synthesis, (b) organic templates, and (c) metal-organic complexes.

reaction in a series of cinnamic acids.[9] In more recent years, control of the photodimerization has been achieved using auxilliaries; specifically, organic templates (e.g., resorcinol) and metal-organic complexes (e.g., Ag · · · Ag interactions). The organic and metal-organic auxilliaries assemble the olefins for reaction into discrete complexes via non-covalent bonds (e.g., hydrogen bonds, coordination bonds) for the reaction (Scheme 10.2).[10] By assembling the olefins within discrete complexes, the reactivity can be largely decoupled from effects of long-range packing. The success of the auxilliary approach has enabled the synthesis of organic molecules by design and control of physical properties of solids (e.g., optical). The level of control that has been achieved using auxilliaries based on metal-organic complexes, in particular, has led to a recent movement to determine whether the [2 + 2] photodimerization can be integrated and controlled within the extended frameworks of MOFs.

## 10.3 [2 + 2] PHOTODIMERIZATIONS INTEGRATED INTO MOFs

The sizes and shapes of the cavities and pores of MOFs are defined by the metal atoms and organic bridges.[3] It follows that any changes to the structures of the components

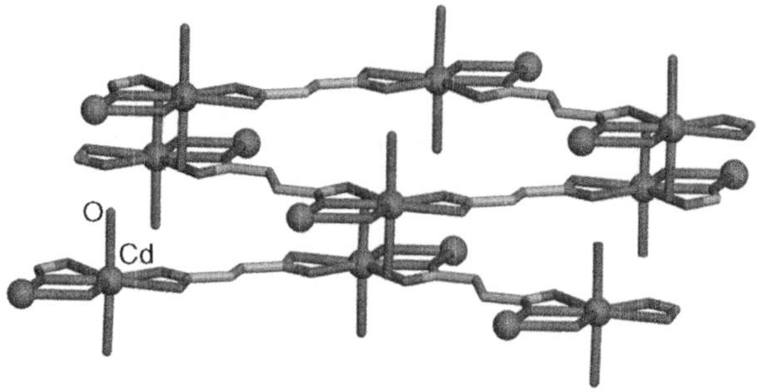


**FIGURE 10.1** Representation of three adjacent layers of $[Cd_2(O_2CCH{=}CHCO_2]\cdot 2H_2O$.

can be expected to lead to changes in properties and functions of pores. For example, it has been shown that the functionalization of mesoporous silica (i.e., MCM-41) pore outlets with photoactive groups in form of coumarins leads to controlled access to the pores.[11] Prior to a [2 + 2] photocycloaddition of the coumarins, the pores are open to appropriate guests (e.g., cholestane). Upon UV-irradiation, the cyclobutane products close the pores, thus, effectively storing guests and limiting access to the pore structure. Upon photocleavage, the pores reopen and release the guests. When considering a MOF, the sizes and shapes, as well as the number of pores, can be tuned using an organic group. A question remains, however, as to whether chemical reactivity can be incorporated into the extended framework of MOFs.

Michaelides et al. were the first to describe a photoreaction integrated into a MOF.[12] The structure $[Cd_2(O_2CCH{=}CHCO_2]\cdot 2H_2O$ contained Cd(II) cations connected to fumarate ligands that afford a rectangular-grid network (Figure 10.1). The coordination sphere at the metal centers contained axially disposed water molecules above and below the grid plane, respectively. The close proximity of olefin groups (3.37 Å) in adjacent layers allowed the ligands to undergo a [2 + 2] cycloaddition reaction forming *trans,trans*-1,2,3,4-cyclobutanetetracarboxylic acid in quantitative yield. Although the network did not contain cavities, this example established that photoreactive organic components can be incorporated into a MOF.

In related work, Vital et al. have reported the photoreactive coordination polymer $[\{CF_3CO_2)(\mu\text{-}O_2CCH_3)Zn\}_2(\mu\text{-bpe})_2]_n$ (where: 4,4′-bpe = *trans*-1,2-bis(4-pyridyl)ethylene).[13] A single crystal X-ray diffraction analysis revealed the formation of a molecular ladder-like polymer. Each Zn(II) center adopted a distorted octahedral geometry, with two acetate ligands bridging a pair of Zn(II) ions in $[Zn_2(\mu\text{-bpe})_2]_\infty$ with Zn $\cdots$ Zn distances of 3.85 Å. Each metal center was chelated by a trifluoroaceate anion to satisfy the coordination geometry (Figure 10.2). The ethylenic carbons of the bpe ligands of parallel chains were separated by 3.75 Å, with the closest distance between the alkene groups of neighboring ladders being 7.10 Å. UV-irradiation of the solid led to the stereocontrolled formation of *rctt*-tetrakis(4-pyridyl)cyclobutane (4,4′-tpcb) within the ladder framework in quantitative yield. The reaction also underwent a rare single-crystal-to-single-crystal (SCSC) reaction (i.e., a reaction where crystallinity of the reactant material is maintained during the reaction).

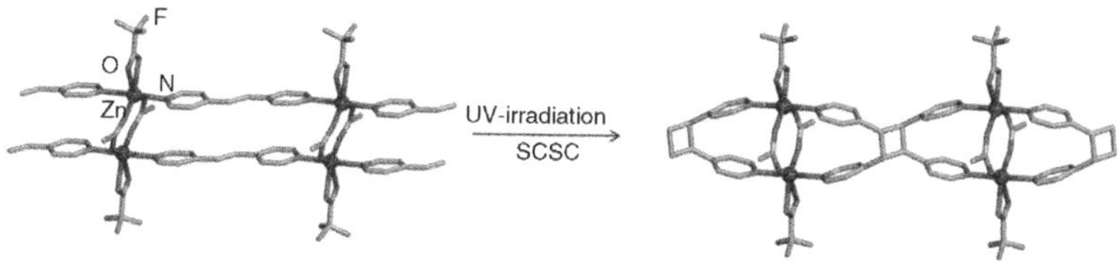


**FIGURE 10.2** Representation of SCSC [2 + 2] cycloaddition reaction of $[\{(CF_3CO_2)(\mu\text{-}O_2CCH_3)Zn\}_2(\mu\text{-bpe})_2]_n$. Hydrogen atoms omitted for clarity.

Following the work of Vittal, our group described a dinuclear metal complex that assembled and pre-organized two olefins within a 1D MOF to react in the solid state. Specifically, the linear photoreactive coordination polymer $\{[Zn_2L(OH)(4,4'\text{-bpe})_2](ClO_4)_2\}_\infty$ (Figure 10.3) {where: L = 2,6-bis[*N*-(2-pyridylethyl)formimidoyl]-4-methylphenol} was obtained when reacted with the Schiff-base complex $[Zn_2L(OH)](ClO_4)_2$ with 4,4′-bpe.[14] An X-ray diffraction analysis revealed that each metal ion adopted an octahedral coordination environment. The metals were tetra-coordinated by L in the basal plane while the apical sites were coordinated by pyridyl groups one from two different bpe molecules. The olefins were coordinated to two neighboring dinuclear complexes while being stacked parallel and separated by 3.71 Å. The formation of 4,4′-tpcb within the assemblies occurred in up to 95% yield upon UV-irradiation (broadband medium pressure Hg-lamp). In contrast to Vittal, however, the reaction did not proceed via a SCSC transformation, with included $H_2O$ molecules being partially liberated from the solid during the photoreaction.

Whereas argentophilic forces had been reported[15] to direct reactivity within a discrete binuclear complex, Vittal et al. reported a [2 + 2] photodimerization within the solvated 1D coordination polymer $[Ag(\mu\text{-bpe})(H_2O)](CF_3CO_2){\cdot}CH_3CN$.[16] The Ag(I) center was coordinated to a pyridyl N-atom of two 4,4′-bpe molecules and the oxygen atom of a water molecule. The hydrogen atoms of the coordinated water molecules from two neighboring polymeric strands bridged the O-atoms of two non-coordinated trifluoroacetate ions through hydrogen bonds. The bonding led to a 2D brickwall structure. The olefins of two consecutive layers were misaligned and separated at a distance of 5.15 Å. UV-irradiation, however, resulted in an unusual solid-state reorganization that led to the formation of cyclobutane-based ladderlike polymer, as supported by powder X-ray diffraction. The formation of the ladderlike

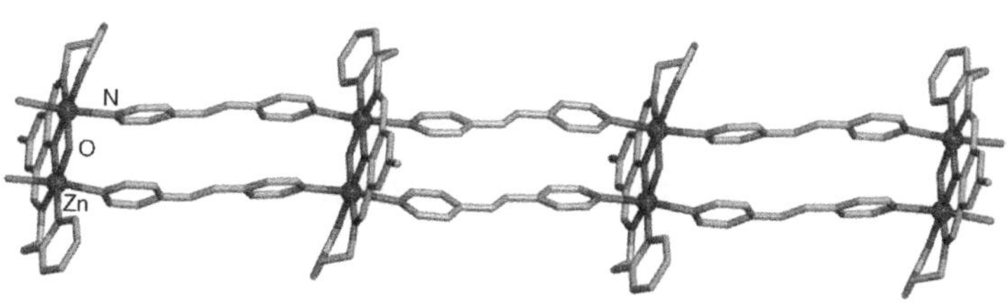


**FIGURE 10.3** Crystal structure of $[Zn_2L(OH)(4,4'\text{-bpe})_2](ClO_4)_2 \cdot 4H_2O$. Hydrogen atoms and perchlorate ions omitted for clarity.

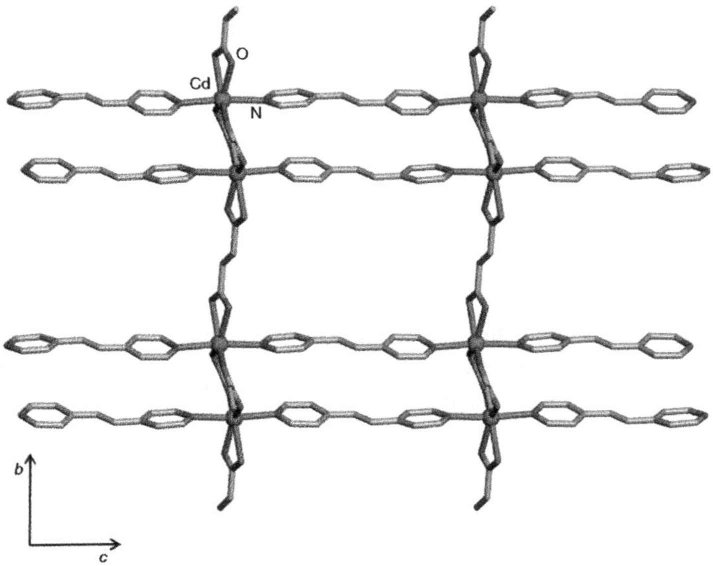


**FIGURE 10.4** 3D network $[Cd_2(O_2CCH{=}CHCO_2)_2(bpe)_2]$ along the *a*-axis. The olefins are directed along the *c*-axis and act as pillars of rectangular grids. Hydrogens are omitted for clarity.

polymer was ascribed to desolvation that occurred upon either application of light or standing under ambient conditions.

Michaelides et al. have recently shown that the previous reported 2D layered structure composed of Cd(II) dimers linked by fumarates can be effectively pillared to afford a 3D photoactive MOF. Specifically, the axial water molecules on the Cd(II) metal centers in $[Cd_2(O_2CCH{=}CHCO_2]{\cdot}2H_2O$ were substituted by bpe ligands to form a 3D non-cubic octahedral-like net (Figure 10.4).[17] The fumarate ions played the role of bridging ligands in the 2D grid while the double columns of bpe ligands acted as linkers between the grids. The close proximity between related pairs of Cd(II) ions brought the pillared 4,4′-bpe ligands approximately parallel and within a distance of 3.95 Å. UV-irradiation afforded 4,4′-tpcb stereospecifically and in quantitative yield.

The above examples are promising in terms of demonstrating how chemical reactivity can be integrated within the structures of MOFs. In each case a monodentate bridging olefin has been assembled within the framework for a crosslinking photodimerization. Given the somewhat limited number of examples reported to date, we expect additional frameworks to be designed so as to sustain reactivity within MOFs of increasing structural and functional (i.e., pore structure) complexity.

## 10.4 CYCLOBUTANES AS ORGANIC BRIDGES OF MOFs

In this section, we will show how the cyclobutane-based products of [2 + 2] photodimerizations obtained from the solid state can be used as organic building units of MOFs. Apart from the ease of synthesis of the cyclobutanes using auxiliaries in the

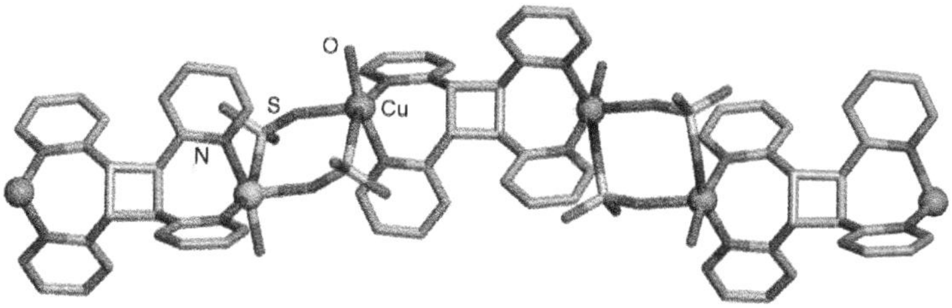


**FIGURE 10.5** View of a single strand of 1D polymer $[Cu_2(\mu\text{-}2\text{-}SO_4)_2(\mu\text{-}2,2'\text{-tpcb})(H_2O)_2]_\infty$.

form of templates in the solid state (e.g., quantitative yield, gram amounts, and the absence of byproducts), we expected that the products, being decorated with pyridine groups, could function as bi- and/or polydentate ligands and, thus, provide access to novel MOFs. A cyclobutane hub was expected to provide both acute and obtuse angles of a MOF that propagate the coordination geometry of a metal to yield discrete and/or infinite architectures. The positioning of the pyridyl nitrogen (e.g., 2-pyridyl, 4-pyridyl) on the cyclobutane product could modify the binding of the ligand to the metals and, thus, judicious choice of ligand could lead to the formation of MOFs of varying dimensionalities (e.g., 1D, 2D). The resulting MOFs would be expected to exhibit properties akin to zeolites and mesoporous materials with related applications.

### 10.4.1 1D MOF

Our first attempt to construct a MOF involving a cyclobutane decorated with pyridyl groups was 1,2,3,4-tetrakis-(2-pyridyl)cyclobutane (2,2′-tpcb). The molecule was generated using an organic template.[18] We showed that coordination driven self-assembly in the solid state afforded the 1D MOF $[Cu_2(\mu\text{-}2\text{-}SO_4)_2(\mu\text{-}2,2'\text{-tpcb})\text{-}(H_2O)_2]_\infty$ wherein 2,2′-tpcb served as a ditopic ligand (Figure 10.5). Square-pyramidal coordination geometry of each Cu(II) ion incorporated two N-atoms of two 2-pyridyl groups, one O-atom of a sulfate ion, and a O-atom of a water molecule in the basal plane, while the fifth coordination site was occupied by an O-atom of a second sulfate ion. The interstices of the parallel strands were occupied by solvent water molecules.

### 10.4.2 2D MOFs

Whereas a 1D framework was obtained using 2,2′-tpcb as a bridge, we obtained a 2D MOF using 4,4′-tpcb. We showed that the four 4-pyridyl groups of 4,4′-tpcb coordinated to four different Cu atoms, thus, enabling the cyclobutane to act as a 4-connected node in a porous 2D MOF. Specifically, reaction of 4,4′-tpcb with the copper paddle-wheel complex $[Cu_2(O_2CCH_3)_4(H_2O)_2]$ produced the 2D grid $[Cu_4(O_2CCH_3)_8(4,4'\text{-tpcb})]_\infty$ (Figure 10.6).[19] In the framework, the water molecules of the dicopper complex were substituted by the pyridyl groups of two 4,4′-tpcb molecules. The metal complex acted as a linear bridge while 4,4′-tpcb served as a node, which is the opposite

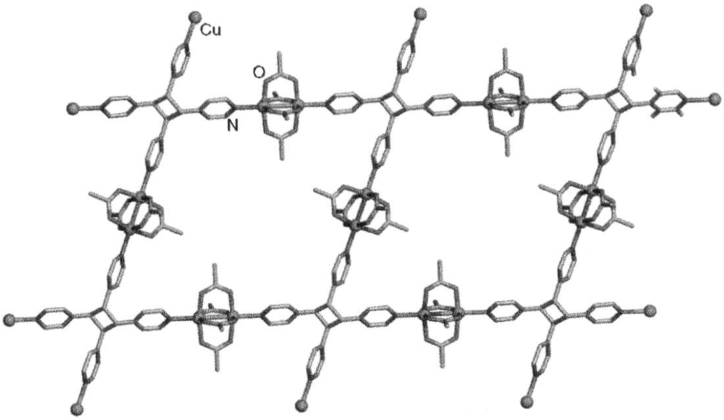


**FIGURE 10.6** Capped stick view of $[Cu_4(O_2CCH_3)_8(4,4'\text{-tpcb})]_\infty$ where metals are highlighted as spheres. Hydrogens are omitted for simplicity.

of a more conventional MOF. We defined the MOF as an inverted MOF (IMOF).[4] The cyclobutane and paddle-wheel complex produced a 2D grid with identical rhombic cavities (17.2 Å × 17.2 Å) with corner angles 75° and 104°. The 2D grids stacked to form a 3D framework with isolated 1D channels (10 Å × 12 Å) occupied by solvent benzene molecules. Furthermore, the cavities were functionalized and compartmentalized by the methyl groups of the acetate ions. This observation suggested that modification of the cavities could be achieved by changing the R-group of the paddle-wheel complex.

Later, we described a 2D MOF where 4,4′-tpcb and the metals both served as 4-connected nodes. Specifically, reaction of 4,4′-tpcb with $[Co(O_2CCH_3)_2(H_2O)_4]$ produced the 2D MOF $[Co(O_2CCH_3)_2(4,4'\text{-tpcb})]_\infty$.[20] Each Co atom conformed to an octahedral geometry being tetra-coordinated by pyridyl groups from four different 4.4′-tpcb molecules in the basal plane while the apical sites were occupied by monocoordinated acetate ions. The framework possessed rhombic cavities with edge lengths of 7.3 Å. In contrast to $[Cu_4(O_2CCH_3)_8(4,4'\text{-tpcb})]_\infty$, however, the framework was made up of two different rhombic cavities (e.g., A, B) (Figure 10.7). The cavities stacked in an ABAB manner producing small interstices occupied by solvent methanol molecules. The MOF retained the crystallinity upon removing the guests by heating.

Recently, we have demonstrated that tetrakis(4-pyridyl)-1,2,9,10-diethano[2.2]-paracyclophane (4,4′-tppcp), a [2.2]paracyclophane obtained from a template-directed solid-state synthesis, can act as a bridge of a MOF. We reacted the 4,4′-tppcp and $Co(O_2CCH_3)_2{\cdot}4H_2O$, which afforded the 2D MOF $[Co(O_2CCH_3)_2(4,4'\text{-tppcp})]_\infty$ (Figure 10.8).[21] The octahedral geometry of the Co(II) centers were satisfied by coordinating with four pyridyl N-atoms of four neighboring 4,4′-tppcp ligands in the basal plane whereas the apical sites were occupied by the monoligated acetates. In the 2D MOF, the Co(II) ion served as a 4-connected node while the cyclophane effectively served as two covalently fused 3-connected nodes. The resulting MOF possessed two

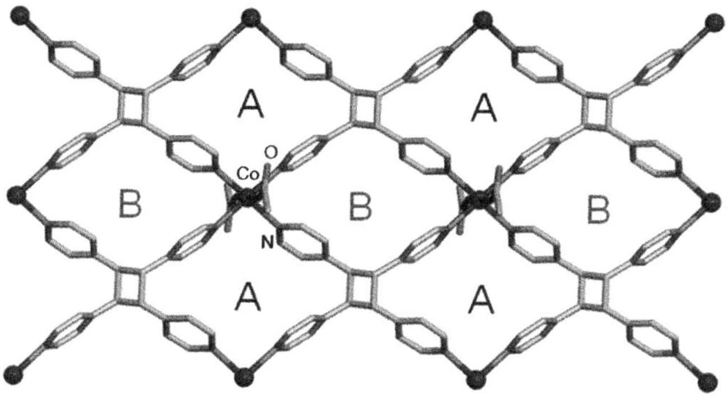


**FIGURE 10.7** Perspective view of $[Co(O_2CCH_3)_2(4,4'\text{-tpcb})]_\infty$ that illustrates two different cavities of the 2D MOF. Hydrogen atoms omitted for clarity.

different cavities (i.e., square and hexagonal) with opposite chemical environment. This was demonstrated by the guest methanol molecule filling hydrophilic square cavities, while toluene molecules filling hydrophobic hexagonal cavities. The use of the cyclophane as a bridge led to a structure that conformed to a less common non-regular net (i.e., two different polygonal cavities and two nodes of different connectivity). With the cyclophane acting as two fused 3-connected nodes, the formation of a possible higher-symmetry MOFs based on 3- and 4-connected nodes was avoided.

### 10.4.3 3D MOF

Prior to our work, Schröder et al. showed that a MOF can be formed by *in situ* generation of 4,4′-tpcb in solution in presence of a metal salt. Specifically, when a solution of bpe and $AgBF_4$ was irradiated by UV light, 4,4′-tpcb was produced *in situ* and crystallized with the salt to form a 3D cationic polymer (Figure 10.9).[21] Each Ag(I)

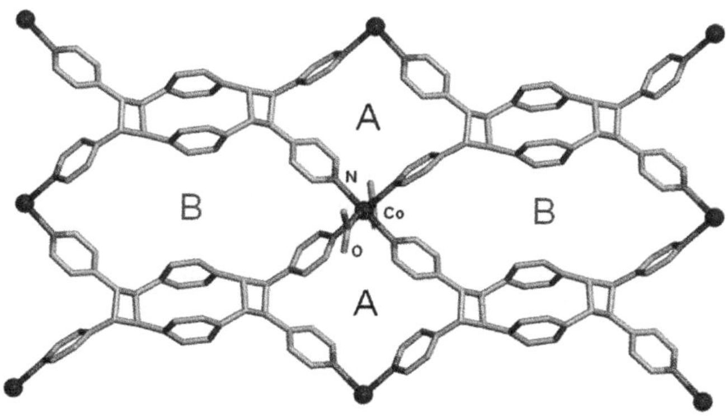


**FIGURE 10.8** Perspective of $[Co(O_2CCH_3)_2(4,4'\text{-tppcp})]_\infty$ showing two different cavities of the 2D framework.

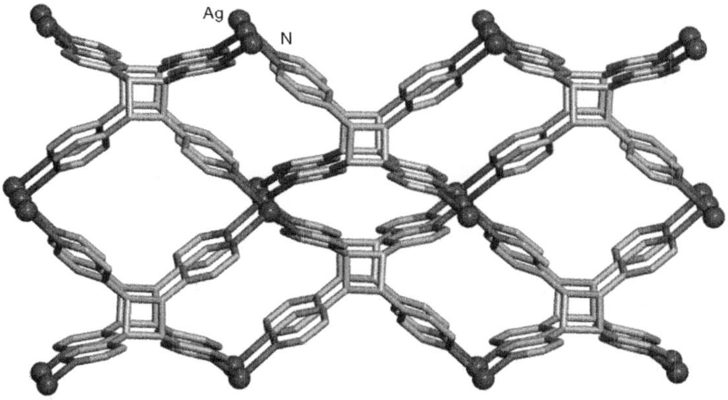


**FIGURE 10.9** View of the 3D cationic polymeric array [Ag(4,4′-tpcb)]$^{+}$ forming channels.

center was coordinated in a tetrahedral geometry to four pyridyl groups each from different cyclobutane ligands. Helical channels that were generated in the MOF contained counteranions (i.e., $BF_4$) and solvent molecules (i.e., $CH_3CN$).

## 10.5 CONCLUSION

In this chapter, we have focused on the integration of chemical reactivity into the structures of MOFs. We have demonstrated how the internal structures of MOFs can be used as platforms to assemble olefins that undergo [2 + 2] photodimerizations in the solid state. While considerable control of reactivity has been achieved in the context of using metal atoms to control reactivity, further studies to engineer reactive MOFs with designer pore size and function remains to be accomplished. Cyclobutane products obtained from reactive solids have also been demonstrated as general means to generate MOFs of varying topologies and dimensionalities.

## REFERENCES

1. Stein, A.; Melde, B. J.; Schoroden, R. C. Hybrid inorganic-organic mesoporous silicates-nanoscopic reactors coming of age. *Adv. Mater.* **2000**, *12*, 1403–1419.
2. (a) Feng, X.; Fryxell, G. E.; Wang, L. Q.; Kim, A. Y.; Liu, J.; Kemmer, K.M. From microporous to mesoporous molecular sieve materials and their use in catalysis. *Science* **1997**, *276*, 923–926; (b) Corma, A. Functionalized monolayers on ordered mesoporous supports. *Chem. Rev.* **1997**, *97*, 2373–2419.
3. (a) Rowsell, J. L. C.; Yaghi, O. M. Strategies for hydrogen storage in metal–organic frameworks. *Angew. Chem., Int. Ed.* **2005**, *44*, 4670–4679; (b) Millward, A. R.; Yaghi, O. M. Metal-organic frameworks with exceptionally high capacity for storage of carbon dioxide at room temperature. *J. Am. Chem. Soc.* **2005**, *127*, 17998–17999; (c) Schlichte,

K.; Kratzke, T.; Kaskel, S. Improved synthesis, thermal stability and catalytic properties of the metal-organic framework compound $Cu_3(BTC)_2$. *Microporous Mesoporous Mater.* **2004**, *73*, 81–88.

4. (a) Seidel, S. R.; Stang, P. J. High-symmetry coordination cages via self-assembly. *Acc. Chem. Res.* **2002**, *35*, 972–983; (b) Fujita, M.; Tominaga, M.; Hori, A.; Therrien, B. Coordination assemblies from a Pd(II)-cornered square complex. *Acc. Chem. Res.* **2005**, *38*, 371–380; (c) Dana, D. L.; Raymond, K. N. Supermolecules by design. *Acc. Chem. Res.* **1999**, *32*, 975–982; (d) Gianneschi, N. C.; Masar, M. S.; Mirkin, C. A. Development of a coordination chemistry-based approach for functional supramolecular structures. *Acc. Chem. Res.* **2005**, *38*, 825–837; (e) Kitagawa, S.; Noro, S.; Nakamura, T. Pore surface engineering of microporous coordination polymers. *Chem. Commun.* **2006**, 701–707; (f) Custelcean, R.; Gorbunova, M. G. A Metal-organic framework functionalized with free carboxylic acid sites and its selective binding of a $Cl(H_2O)_4$-cluster. *J. Am. Chem. Soc.* **2005**, *127*, 16362–16363; (g) Wu, C. D.; Hu, A.; Zhang, K.; Lin, W. A homochiral porous metal-organic framework for highly enantioselective heterogeneous asymmetric catalysis. *J. Am. Chem. Soc.* **2005**, *127*, 8940–8941; (h) Eddaoudi, M.; Kim, J.; Rosi, N.; Vodak, D.; Wachter, J.; O'Keeffe, M.; Yaghi, O. M. Systematic design of pore size and functionality in isoreticular MOFs and their application in methane storage. *Science* **2002**, *295*, 469–472.
5. (a) Vittal, J. J. Supramolecular structural transformations involving coordination polymers in the solid state. *Coord. Chem. Rev.* **2007**, *251*, 1781–1795; (b) MacGillivray, L. R.; Papaefstathiou, G. S.; Friscic, T.; Hamilton, T. D.; Bucar, D. -K.; Chu. Q.; Varshney, D. B.; Georgiev, I. G. Supramolecular control of reactivity in the solid state: from templates to ladderanes to metal-organic frameworks. *Acc. Chem. Res.* **2008**, *41*, 280–291; (c) Wang, X. Y.; Wang, Z. M.; Gao, S. A pillared layer MOF with anion-tunable magnetic properties and photochemical [2 + 2] cycloaddition. *Chem. Commun.* **2007**, 1127–1129.
6. (a) Cohen, M. D.; Schmidt, G. M. J. Photodimerization in the solid state. *Chem. Soc.* **1964**, 2000–2013; (b) Schmidt, G. M. J. The photochemistry of *trans*-cinnamic acids. *Pure Appl. Chem.* **1971**, *27*, 647–678.
7. Desiraju, G. R. Supramolecular synthons in crystal engineering—a new organic synthesis. *Angew. Chem. Int. Ed. Engl.* **1995**, *34*, 2311–2327.
8. Praetorius, P.; Korn, F. Exposure of unsaturated ketones in the presence of uranyl salts Ber. Dtsch. Chem. Ges. *Ber.* **1910**, *43*, 2744–2746.
9. Schmidt, G. M. J. The crystal chemistry of some *trans*-cinnamic acids. *J. Chem. Soc.* **1964**, 2014–2021.
10. (a) MacGillivray, L. R.; Reid, L. J.; Ripmeester, A. J. Supramolecular control of reactivity in the solid state using linear molecular templates. *J. Am. Chem. Soc.* **2000**, *122*, 7817–7818; (b) Vela, J. M.; Foxman, M. B. Bilayer formation in metal carboxylate structures: results from the Cambridge structural database. *Cryst. Eng.* **2000**, *3*, 11–31.
11. Mal, K. N.; Fujiwara, M.; Tanaka, Y. Photocontrolled reversible release of guest molecules from coumarin modified mesoporous silica. *Nature* **2003**, *421*, 350–353.
12. Michaelides, A.; Skoulika, S.; Siskos, M. Assembly of a photoreactive coordination polymer containing rectangular grids. *Chem. Commun.* **2004**, 2418–2419.
13. Toh, L. N.; Nagarathinam, M.; Vittal, J. J. Topochemical photodimerization in the coordination polymer $[\{(CF_3CO_2)(\mu\text{-}O_2CCH_3)Zn\}_2(\mu\text{-bpe})_2]_n$ through single-crystal to single-crystal transformation. *Angew. Chem., Int. Ed.* **2005**, *44*, 2237–2241.

14. Papaefstathiou, G. S.; Georgiev, I. G.; Friščić, T.; MacGillivray, L. R. Directed assembly and reactivity of olefins within a one-dimensional ladder-like coordination polymer based on a dinuclear Zn(II) platform. *Chem. Commun.* **2005**, 3974–3976.

15. Chu, Q.; Swenson, D. C.; MacGillivray, L. R. A single-crystal-to-single-crystal transformation mediated by argentophilic forces converts a finite metal complex into an infinite coordination network. *Angew. Chem. Int. Ed.* **2005**, *44*, 3569–3572.

16. Nagarathinam, M.; Vittal, J. J. Anisotropic movements of coordination polymers upon desolvation: solid-state transformation of a linear 1D coordination polymer to a ladderlike structure. *Angew. Chem. Int. Ed.* **2006**, *45*, 4443–4447.

17. Michaelides, A.; Skoulika, S.; Siskos, G. M. Designed self-assembly of a reactive metal–organic framework with quasi α-Po topology. *CrystEngComm* **2008**, *10*, 817–820.

18. Hamilton, T. D.; Papaefstathiou, G. S.; MacGillivray, L. R. Discrete and infinite coordination arrays derived from a template-directed solid-state organic synthesis. *CrystEngComm*. **2002**, *4*, 223–226.

19. Papaefstathiou, G. S.; MacGillivray, L. R. An inverted metal-organic framework with compartmentalized cavities constructed by using an organic bridging unit derived from the solid state. *Angew. Chem. Int. Ed.* **2002**, *41*, 2070–2073.

20. Papaefstathiou, G. S.; Milios, C.; MacGillivray, L. R. A 2D metal-organic framework with two different rhombus-shaped cavities: a rare example of a (4,4)-net with alternating metal and organic nodes. *Microporous Mesoporous Mater*. **2004**, *71*, 11–15.

21. Papaefstathiou, G. S.; Friščić, T.; MacGillivray, L. R. Design and construction of a 2D metal-organic framework with multiple cavities: a nonregular net with a paracyclophane that codes for multiply fused nodes. *J. Am. Chem. Soc.* **2005**, *127*, 14160–14161.

22. Blake, A. J.; Champness, N. R.; Chung, S. S. M.; Li, W. S.; Schroder, M. *In situ* ligand synthesis and construction of an unprecedented three-dimensional array with silver(i): a new approach to inorganic crystal engineering. *Chem. Commun.* **1997**, 1675–1676.

# 11

# MOLECULAR MODELING OF ADSORPTION AND DIFFUSION IN METAL-ORGANIC FRAMEWORKS

RANDALL Q. SNURR, A. ÖZGÜR YAZAYDIN, DAVID DUBBELDAM, AND HOUSTON FROST

*Department of Chemical and Biological Engineering, Northwestern University, Evanston, IL 60208, USA*

Molecular modeling is playing an important role in the young and rapidly expanding field of metal-organic frameworks (MOFs). As described in other chapters of this book and several excellent reviews,[1–7] MOFs are synthesized in a modular fashion from pre-formed metal or metal-oxide vertices and organic linker molecules in a self-assembly process. This building-block approach opens up the possibility to create an almost unlimited number of MOF structures. If modeling is sufficiently predictive, it can be used to screen existing MOFs for new applications or to predict their behavior under conditions that are difficult to attain in experiments. For example, if experimental results at 77 K and 1 atm suggest that a particular MOF is promising for hydrogen storage (a common set of conditions for preliminary experiments), molecular simulations can be used to test the MOF for hydrogen storage at 298 K and 100 atm (conditions more relevant to storing hydrogen on a vehicle, but where experiments require more specialized equipment). Similarly, adsorption for a wide variety of gases can be readily tested computationally on a particular MOF. Perhaps more intriguingly, molecular modeling can be performed on hypothetical MOFs. For any given MOF, one can easily imagine dozens of variations, obtained simply by changing the organic linker molecules. Many of these structures should be synthetically accessible due to

*Metal-Organic Frameworks: Design and Application,* Edited by Leonard R. MacGillivray

the modular nature of the synthesis. However, it still requires a significant effort to synthesize, characterize, and test a new MOF. Molecular simulations hold tremendous promise for screening hypothetical MOFs and focusing experimental efforts on the most promising candidates. In addition to this screening role, molecular-level simulations also provide a wealth of detailed information on the structure and dynamics of guest molecules within the MOF nanopores.

In this chapter, we focus on molecular modeling of adsorption and diffusion in MOFs. Much of the experimental work on MOFs to date has been devoted to applications in gas storage and chemical separations,[8,9] where adsorption thermodynamics and rates of diffusion of guest molecules play a key role. We have attempted to provide a summary of the current state-of-the-art as of early 2008, as well as future challenges. Molecular modeling studies of other properties, such as thermal properties,[10,11] mechanical properties,[12,13] or catalysis, are still in their infancy or nonexistent and are not discussed here. Modeling can also play an important role in solving crystal structures[14] but that is also beyond the scope of this chapter.

Before the advent of MOFs, molecular modeling was already a well-established tool to predict adsorption isotherms, heats of adsorption, and diffusion coefficients in other nanoporous materials. Particularly for zeolites, there is a large literature on these topics that is quite relevant for MOF researchers.[15–19] Molecular modeling has provided useful information on sorbate structure, dynamics, and diffusion mechanisms for many zeolite systems.

The remainder of this chapter is organized as follows. First, common molecular models and methods used for simulation of adsorption and diffusion in MOFs are briefly described. This is followed by an overview of molecular modeling studies of adsorption in MOFs. A brief review of diffusion simulations is next, followed by a discussion of modeling of hydrogen storage in MOFs. The chapter closes with a summary of future prospects and challenges.

## 11.1 MODELS AND METHODS

### 11.1.1 Developing a Molecular Model

Modeling adsorption and diffusion in nanoporous materials is generally based on classical mechanics. Because MOFs are crystalline, one can easily simulate several unit cells in full atomistic detail making use of well-developed simulation techniques.[20,21] Periodic boundary conditions allow one to predict properties valid for an extended crystal lattice. It is convenient to categorize energetic interactions as guest–guest and guest–host. There are also intramolecular and intraframework interactions if the guest molecules or the MOF structures are considered to be flexible. The MOF atoms are often fixed at their crystallographic positions, although models for treating MOF flexibility have been proposed recently.[10,13,22,23] Guest molecules can be modeled as rigid or flexible depending on their nature and the importance of high-frequency modes. A systematic approach to parameterize guest–guest interaction potentials is to fit them to experimental vapor–liquid equilibrium data as in the TraPPE

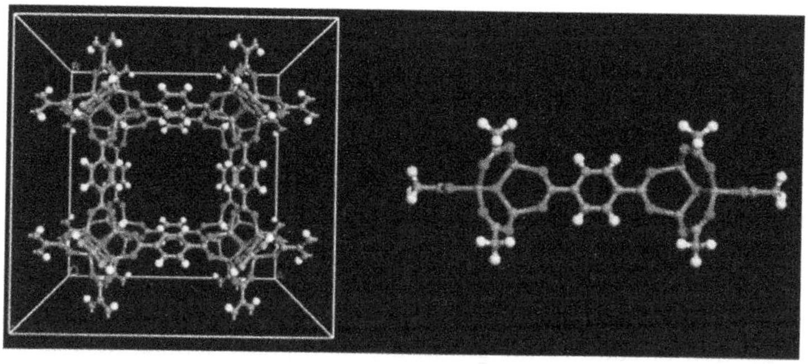

**FIGURE 11.1** IRMOF-1 unit cell (left) and the cluster extracted for calculating the atomic charges (right). Note the methyl terminated ends of the cluster to reduce termination effects.

force field.[24,25] These potentials typically include Lennard–Jones parameters and partial charges for all atoms, as well as any necessary expressions for describing bond stretching, bond bending, and torsional motions.

The guest–host interactions are usually modeled with a "Kiselev-type" potential, building on the work of Kiselev and others in zeolites.[16,26] In a Kiselev-type model, interactions between the sorbate molecules and the host are represented by placing Lennard–Jones sites and partial charges on all atoms of the framework and the sorbate molecules to capture the van der Waals and Coulombic forces. For MOFs, the Lennard–Jones parameters are often taken from generic force fields, particularly DREIDING,[27] UFF,[28] or OPLS-AA.[29] Dedicated force fields have been developed for zeolites and applied with great success, but the chemical diversity of MOFs has made these general force fields attractive. Partial charges for the MOF atoms are typically obtained from quantum chemical calculations on a cluster of atoms extracted from the periodic MOF structure. A single-point calculation is performed on a cluster such as in Figure 11.1 (right), followed by determination of the charges. Mulliken population analysis[30] or methods of fitting charges to match the electrostatic potential, such as ChelpG,[31] are the two most popular ways of deriving atomic partial charges. The latter is particularly suitable for use in molecular simulation because it tries to reproduce the quantum mechanical electrostatic potential using partial charges on the atomic centers.

The Kiselev-type models are widely used to predict adsorption isotherms, heats of adsorption, and other thermodynamic properties in zeolites and other nanoporous materials. A common strategy for simulating diffusion in zeolites is to start with a model that reproduces the thermodynamic properties. The Kiselev model is attractive because of its simplicity and computational efficiency.

### 11.1.2 Predicting Adsorption: GCMC Simulations

Grand canonical Monte Carlo (GCMC) simulations are widely used to obtain adsorption isotherms. In this method, the temperature, volume, and chemical potential are kept fixed while the number of molecules fluctuates.[20,21] For adsorption, the equilibrium conditions are that the temperature and chemical potential of the gas inside and outside the adsorbent must be equal. Thus, it is convenient to perform the simulations in the grand canonical ensemble where the chemical potential $\mu$ is fixed. The Monte Carlo simulation consists of millions of random moves that sample the chosen ensemble. For GCMC, these moves include molecular translations and rotations; molecule insertions, where it is attempted to add a new molecule to the system; and molecule deletions, where it is attempted to remove an existing molecule from the system. The attempted moves are accepted or rejected with appropriate criteria to sample the ensemble. The imposed chemical potential can be computed from the equation of state of the gas at the fluid-phase temperature and pressure. Alternatively, one can use the Gibbs ensemble formulation and explicitly simulate the surrounding fluid phase, too.[21] In either case, one calculates the average number of molecules in the adsorbed phase by averaging over the course of the simulation.

In simulations or theoretical treatments, absolute adsorption is usually considered, that is, the average number of particles in the pores. However, in experiment it is usually the "excess" adsorption that is measured, that is, the average number of molecules in the pores in excess above the number of molecules that would occupy the free pore volume at bulk-gas conditions.[32–35] These properties can be related by the equation $n_{ex} = n_{abs} - V^g \rho^g$, where $V^g$ is the pore volume and $\rho^g$ is the gas-phase density. The parameter $V^g$ is usually obtained experimentally through helium adsorption measurements, and this helium measurement can also be mimicked by simulation to obtain the pore volume in a consistent fashion.[36]

### 11.1.3 Predicting Diffusion: Molecular Dynamics Simulations

Diffusion in nanoporous materials is an activated process and is at least several orders of magnitude slower than in the gas phase. Dynamical properties such as the diffusion coefficients can be computed by using molecular dynamics (MD) simulations. In MD, successive configurations of the system are generated by integrating Newton's equations of motion, which then yield trajectories that describe the positions, velocities, and accelerations of the particles as they vary with time.[20] Many different diffusion coefficients can be defined for guest molecules in nanoporous materials, but it is useful to put them into two general classes: transport diffusivities and self-diffusivities.[37] The former describe the transport of mass and the decay of density fluctuations in the system; the latter describe the diffusive motion of individual particles at equilibrium. In MD simulations, one can measure the mean-squared displacement (MSD) of individual particles and calculate the self-diffusivity from the Einstein equation.

$$D_{s,i} = \lim_{t \to \infty} \frac{1}{6t} \left\langle \frac{1}{N_i} \sum_{l=1}^{N_i} [\mathbf{r}_{il}(t) - \mathbf{r}_{il}(0)]^2 \right\rangle \tag{11.1}$$

where $D_{s,i}$ is the self-diffusivity of species $i$, $\mathbf{r}_{il}(t)$ the position of molecule $l$ of species $i$ at time $t$, $N_i$ the number of molecules of species $i$, and the angular brackets denote that the quantity is an ensemble average property. In a similar fashion, the MSD of the center of mass of the system is related to the "corrected" diffusivity, which in turn is related to the Fickian transport diffusivity by a thermodynamic factor that can be computed from the adsorption isotherm.[19,37]

### 11.1.4 Calculating Surface Areas and Pore Volumes

The surface area and pore volume are important parameters for characterizing porous materials. The surface area can be obtained indirectly from experimental isotherms using the BET theory, and the pore volume can be estimated from the experimental saturation loading. These quantities can also be calculated in a geometric fashion from the crystal structure, either by sophisticated analytical treatments[38] or through simple Monte Carlo routines.[39–41] For example, the pore volume can be obtained by randomly shooting spherical probes into the MOF unit cell and keeping track of the fraction that do not overlap with the MOF atoms, taking into account appropriate sizes for the MOF atoms. This can be repeated for probes of arbitrary size, including point probes. The surface area is obtained by effectively rolling the probe atom over the surface of the framework. Comparison of the calculated and experimental pore volumes and surface areas can be a powerful tool for assessing the quality of synthesized samples, as discussed below.

The calculated surfaces are also quite useful for visualizing the pore structures, as shown in Figure 11.2 for the $[Cu_3(TMA)_2]_n$ structure of Chui et al.[42] (TMA = trimesic acid = benzene-1,3,5-tricarboxylic acid). This MOF is commonly known as Cu-BTC or HKUST-1. It has a cubic unit cell of 26.343 Å with an intersecting 3D system of large square-shaped pores and small side pockets. These side pockets are accessible only to

**FIGURE 11.2** The Cu-BTC metal–organic framework: (left) ball and stick, (middle) the adsorption surface, and (right) snapshot of $CO_2$ at 298 K close to maximum loading.

very small molecules. The left pictures in Figure 11.2 show Cu-BTC in a ball-and-stick representation, and the center pictures show the surface area of the structure. The surface area representation clearly shows the side pockets and the windows that connect them to the main channels.

## 11.2 MOLECULAR MODELING OF ADSORPTION IN MOFs

### 11.2.1 Single-Component Adsorption

The first molecular modeling study in MOFs was reported in 2001 by Kawakami et al.[43] They investigated the adsorption of $CO_2$, $N_2$, $H_2$, $O_2$, and Ar in a Zn(BDC) MOF synthesized by Yaghi and coworkers (BDC = 1,4-benzenedicarboxylate).[44] They performed GCMC simulations and tested the effects of different methods for obtaining partial charges. The predicted loadings were considerably larger than those reported from experiment. In another early molecular modeling study of MOFs, Vishnyakov et al.[45] investigated adsorption of argon in Cu-BTC[42] using GCMC simulations and high-resolution adsorption measurements at 87 K. Preferential adsorption sites were determined, and it was predicted that the side pockets fill first, followed by condensation in the main channels. Using a parameterized force field, the simulated isotherms agreed well with experiment over most loadings but overpredicted the saturation loading.

Since these early studies, there have been many simulations of adsorption in MOFs. A few examples are highlighted in this section. In 2004, Düren et al.[39] simulated the adsorption of $CH_4$ in a series of isoreticular metal-organic frameworks (IRMOFs) with the goal of developing new materials for methane storage. The structure of IRMOF-1 is shown in Figure 11.1. The predictions for $CH_4$ adsorption in IRMOF-1 and IRMOF-6 gave excellent agreement with experimental data at room temperature up to 40 bar, as shown in Figure 11.3. Düren et al. compared their results with other porous materials such as zeolites, MCM-41, and single-walled carbon nanotubes to help determine the

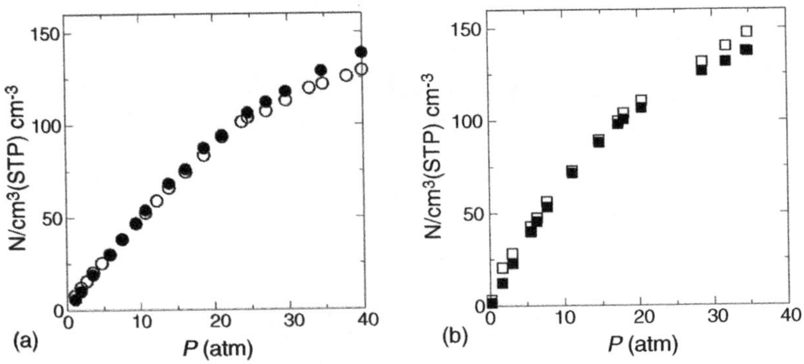


**FIGURE 11.3** Experimental and simulated methane adsorption isotherms at 298 K in (a) IRMOF-1 and (b) IRMOF-6 (open symbols, experimental results; closed symbols, simulation results). (Reprinted with permission from ref. 39. Copyright © 2004 American Chemical Society).

factors that promote high methane storage. They suggested that an ideal material for methane storage should have a large accessible surface area, high free volume, low framework density, and strong energetic interactions between the framework and the methane molecules. But they pointed out that these factors are related and changing the MOF to improve one factor might worsen others. Based on this understanding, they proposed three new hypothetical structures that were predicted by simulation to enhance $CH_4$ adsorption by up to 36% above the best results available at the time.

Jhon et al.[46] studied $CH_4$ adsorption in alkoxy-functionalized variations of IRMOF-1. They designed these materials on the computer by attaching methoxy, ethoxy, or propoxy groups to the linker of IRMOF-1. Note that the propoxy version of IRMOF-1 has been synthesized and is known as IRMOF-4.[47] The alkoxy-functionalized versions of IRMOF-1 were first structurally optimized. Then GCMC simulations of $CH_4$ adsorption were performed. Jhon et al. found that the constriction of the pores by the alkoxy-functionalized linkers promoted $CH_4$ adsorption at low to moderate pressures but the saturation capacities were decreased at room temperature. The propoxy-functionalized IRMOF-1, which has the longest functional group and smallest pore size, showed the largest volumetric adsorption at low to moderate pressures.

Adsorption of $CO_2$ in MOFs has attracted much attention because many important industrial separations involve this molecule, including natural gas purification, $H_2$ production through steam reforming, and carbon capture and sequestration. Yang et al.[48] studied adsorption of $CO_2$ in several different MOFs including IRMOFs-1, -8, -10, -11, -14, and -16,[47] Mn-MOF,[49] MOF-177,[50] and Cu-BTC. By investigating nine different MOFs they sought to reveal the effects of organic linker, pore size, pore topology, and the electrostatic field on the adsorption and diffusion of $CO_2$. They reported simulated adsorption isotherms for $CO_2$ up to 6.0 MPa at 298 K and concluded that the MOFs considered in this study show higher $CO_2$ capacity than most zeolites and carbon materials. A pore size between 1.0 and 2.0 nm was stated as the most suitable size. By artificially turning off the electrostatic interactions between the $CO_2$ molecules and the MOF atoms, they found that the electrostatic interactions enhance adsorption by up to 30% at low pressure. The effect is much smaller at high pressures.

Walton et al.[51] recently reported a comparison of GCMC simulations for $CO_2$ with extensive experimental results in IRMOF-1 at temperatures ranging from 195 to 298 K. As shown in Figure 11.4, the simulated isotherms are in excellent agreement with the experimental data. It should be noted that the simulation results were not fit in any way to these data, but the simulations are able to capture the complex shapes of these isotherms essentially quantitatively. The $CO_2$ isotherm in IRMOF-1 shows an unusual inflection at 298 K (Figure 11.5), which grows into a distinct step as temperature is decreased. Walton et al. found that the electrostatic interactions among $CO_2$ molecules were crucial to reproduce the inflection in the isotherm. To show this, they simulated $CO_2$ adsorption with and without the presence of electrostatic charges on the $CO_2$ molecules. As shown in Figure 11.5, the predicted isotherm using only Lennard–Jones interactions does not show an inflection, but when Coulombic interactions between $CO_2$ molecules are included, the isotherm exhibits an inflection and is

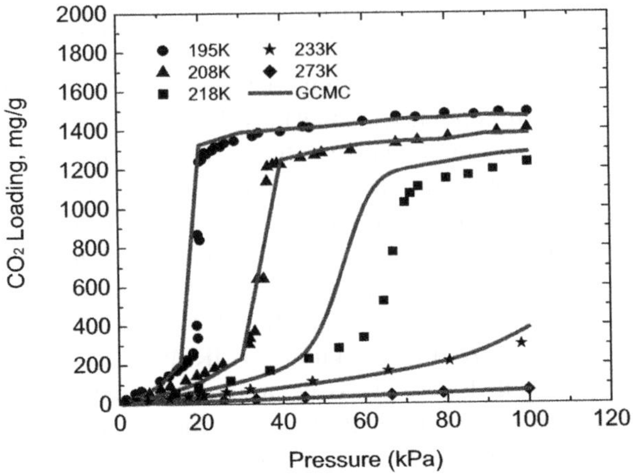


**FIGURE 11.4** Comparison of GCMC simulations and experimental adsorption isotherms for $CO_2$ in IRMOF-1. (Reprinted with permission from ref. 51. Copyright © 2008 American Chemical Society).

in good agreement with the experimental isotherm. The same model gave very good agreement for room temperature $CO_2$ isotherms in IRMOF-3 and MOF-177 as well. By simulating $CO_2$ adsorption in IRMOF-10 and IRMOF-16, which have two and three times, respectively, more pore volume per gram than IRMOF-1, they showed that

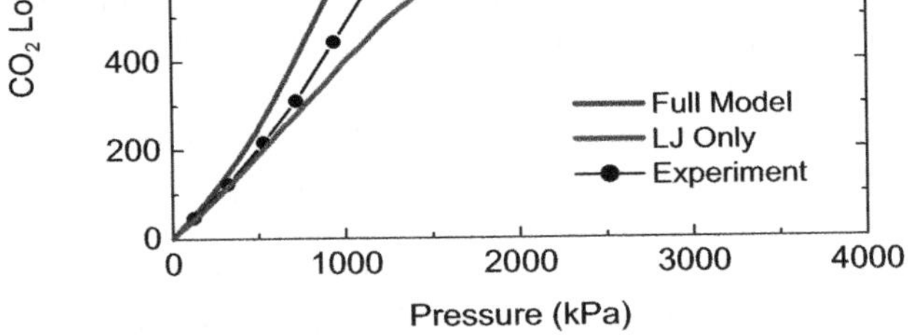


**FIGURE 11.5** Comparison of GCMC simulations and experimental adsorption isotherms for $CO_2$ in IRMOF-1 at 298 K. The top curve was calculated from a model that included electrostatic effects. The bottom curve was calculated considering only the Lennard–Jones interactions. (Reprinted with permission from ref. 51. Copyright © 2008 American Chemical Society).

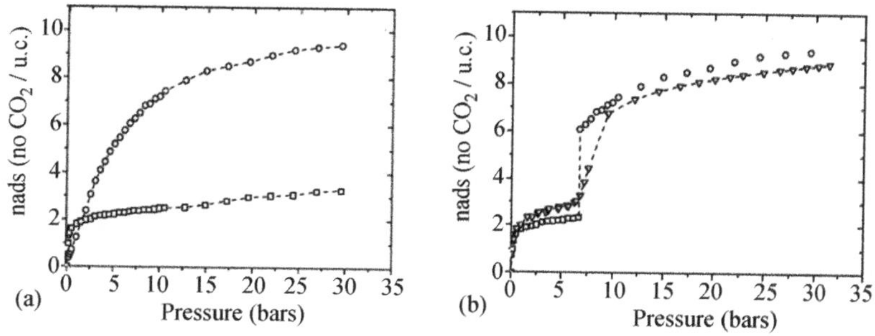


**FIGURE 11.6** (a) Simulated absolute adsorption isotherms for $CO_2$ in two different forms of MIL-53, MIL-53np (Al) (squares) and MIL-53lp (Al) (circles). (b) The simulations (circles) are compared with the experimental data (triangles).[56] (Reproduced by permission from the Royal Society of Chemistry).

the pore-filling pressure for $CO_2$ shifts toward the bulk condensation pressure with increasing pore size.

Another interesting study involving isotherms with steps was reported by Ramsahye et al.[52,53] In contrast to the work of Walton et al., in which the MOF did not undergo any structural change, Ramsahye investigated an interesting breathing mechanism in the MOF framework and its effect on the isotherms. MIL-53 (Al)[54] has been observed experimentally to have a structural interchange between a narrow-pore form, MIL-53np (Al), and a large-pore form, MIL-53lp (Al), upon hydration.[55] Similar behavior was experimentally observed during $CO_2$ adsorption.[56] Ramsahye and coworkers performed GCMC simulations to shed light on how this breathing affects $CO_2$ adsorption. Simulated adsorption isotherms of $CO_2$ were obtained for both forms of MIL-53(Al) up to 30 bars. (The crystal structures of both forms are known.) As expected, the isotherms showed quite different behavior, as shown in Figure 11.6a. The experimental isotherm (shown in Figure 11.6b) shows a sharp increase around 6 bar. When Ramsahye et al. combined the simulated adsorption data below 6 bar from MIL-53np (Al) and data above 6 bar from MIL-53lp (Al), a very good match with the experimental isotherm was obtained (Figure 11.6b). On the other hand, $CO_2$ adsorption in MIL-47 (V),[57] which is a vanadium analog of MIL-53lp (Al), did not show the breathing behavior. Besides the difference of the metals, MIL-47 (V) lacks the $\mu_2$-OH groups present in MIL-53 (Al). Snapshots from the simulations suggest that these $\mu_2$-OH sites are responsible for triggering the structural shift observed in MIL-53 (Al). $CO_2$ molecules were observed to bridge the $\mu_2$-OH groups across the pores of MIL-53np (Al), but the pores of MIL-53lp (Al) are too large to permit this (Figure 11.7). As pressure increases, interactions between the $CO_2$ molecules tend to break the interactions between the $CO_2$ molecules and the $\mu_2$-OH groups, leading to configurations with weaker adsorbate–adsorbent interactions. This is reported as the reason for the transition from the narrow-pore structure to the larger pore version.

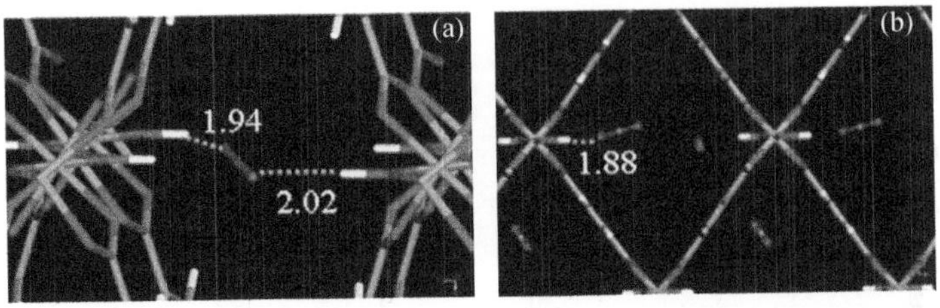


**FIGURE 11.7** (a) The double interaction of a $CO_2$ molecule with the MIL-53np (Al) structure and (b) the interaction of a $CO_2$ molecule with a single $\mu_2$-OH group in MIL-53lp (Al).[56] (Reproduced by permission from the Royal Society of Chemistry).

### 11.2.2 Comparisons of Predicted Isotherms with Experiment

An important question is how well GCMC simulations can predict adsorption in agreement with experiment. The isotherms shown in Figures 11.3–11.6 were all obtained without any fitting of the force field parameters to experimental data. For example, Düren et al.[39] used established methane parameters and took the MOF Lennard–Jones parameters from the DREIDING force field without adjusting them to obtain the results in Figure 11.3. Using the same approach Garberoglio et al.[58] simulated He, Ar, $CH_4$, and $H_2$ in a number of different MOFs. They mention the success of Düren and coworkers but point out that using standard force fields does not always result in good agreement with experimental isotherms. In particular, their simulated adsorption isotherms of Ar and $H_2$ in some MOFs showed large discrepancies with experimental data from the literature.

Liu et al.[59] investigated how the MOF activation process can have a large effect on the adsorption capacity, surface area, and pore volume. They developed a method for extracting the *N,N*-dimethylformamide-solvated Cu-BTC crystals with methanol. They reported experimentally measured isotherms and complemented them with simulations of $H_2$, $N_2$, and Ar adsorption. Their improved activation process resulted in the highest $H_2$ uptake reported for Cu-BTC in the literature. At 77 K they found good agreement between simulated and experimental $H_2$ adsorption capacities. They interpreted this as an indication that they had achieved a nearly pure and solvent-free Cu-BTC product after their activation process. Quantum diffraction effects were found to be important at cryogenic temperatures for $H_2$. Their simulations agreed fairly well with experimental adsorption isotherms also for $N_2$ and Ar.

Düren et al.[40] and Walton and Snurr[41] have put forth similar ideas. In particular, they suggest that the quality of MOF samples can be assessed by comparing the surface area calculated geometrically from the crystal structure with the BET surface area obtained from the 77 K experimental nitrogen isotherm. Large differences can indicate partial framework collapse, unexpected catenation, or pore blockage by solvent or unreacted molecules from the MOF synthesis. If the geometric and experimental surface areas do not agree, one should not expect agreement of simulated and experimental isotherms, as the simulations are usually based on the "perfect" X-ray crystal structure.

It should be kept in mind that the surface area is not a direct experimental observable. It is commonly obtained by applying the BET theory to nitrogen isotherms

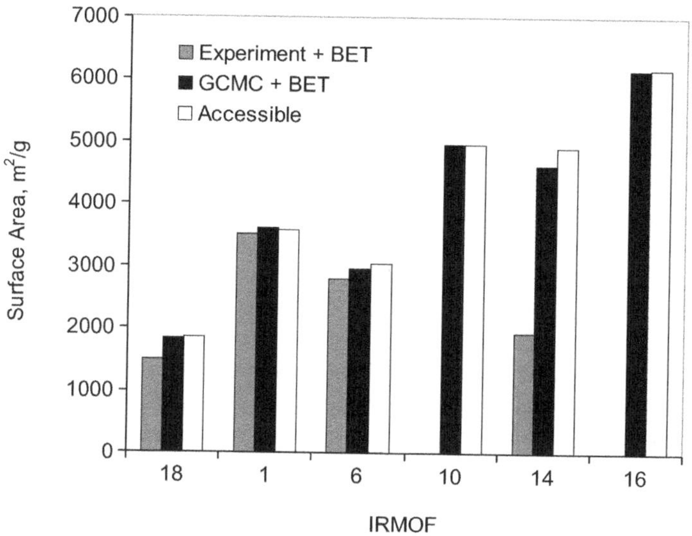


**FIGURE 11.8** Comparison of IRMOF surface areas obtained from the BET theory applied to experimental $N_2$ isotherms, surface areas from the BET theory applied to simulated $N_2$ isotherms, and accessible surface areas calculated from the crystal structures. (Reprinted with permission from ref. 41. Copyright © 2007 American Chemical Society).

measured at 77 K. This is a standard procedure that allows for comparisons among different materials. However, the BET analysis relies on several assumptions that may break down for microporous materials with ultrahigh surface areas such as MOFs. The BET analysis assumes that adsorption occurs by multilayer formation and that the number of adsorbed layers is infinite at the saturation pressure, that is, adsorption occurs as if on a free surface. Walton and Snurr[41] recently addressed the question of whether the surface area numbers obtained from BET analysis are meaningful in an absolute sense. To do this, they simulated nitrogen adsorption isotherms in a series of IRMOFs and then used the simulated isotherms in the BET analysis as if they were experimental data. The surface areas obtained agree surprisingly well with those calculated directly from the crystal structures in a geometric fashion, as shown in Figure 11.8. In addition, the surface areas agree well with experimental reports from the literature. These results provide a strong validation that the BET theory can be used to obtain reliable surface areas of MOFs.

Molecular simulations are often used to obtain molecular-level information, such as where molecules prefer to adsorb, with the implicit assumption that if the simulations predict macroscopic properties such as isotherms in agreement with experiments then the molecular-level predictions are also reliable. Dubbeldam et al.[60] used Monte Carlo methods to investigate the siting of small gas molecules in IRMOF-1. This system is unusual because there is also detailed experimental evidence for the siting of Ar and $N_2$ in this MOF from X-ray diffraction.[61] The simulations predicted that Ar and $N_2$ are localized at sites that are in very good agreement with the experimental positions at 30 K. At room temperature, the molecules are spread throughout the pore volume. In addition, they found that the preferred site for small molecules such as $H_2$, Ar, $N_2$, $CO_2$, methane, ethane, and propane is near the zinc–oxygen clusters in the cavities where the linkers point outward. The IRMOFs have alternating cavities where the linkers point inward or outward due to the structure of the $Zn_4O$ corners. The cavity where the linkers point outward is slightly bigger than the cage where the linkers point inward, which has implications for the siting of molecules. For example, up to high loadings benzene is found mostly in the larger cages, with almost no molecules in the smaller cages.[22]

To advance the field, additional systematic comparisons of simulated and experimental adsorption isotherms and heats of adsorption are needed. In particular, experimental samples should be carefully characterized, including comparison of the measured BET surface areas with those expected geometrically from the crystal structures. Measurements of a single probe molecule on a series of different MOFs from a single laboratory using an identical experimental protocol would provide a critical test of simulation, as would measurements on selected MOFs for a series of different probe molecules. At this point, one can tentatively say that if simulations are carefully performed for simple molecules like argon and methane, then very large discrepancies between simulation and experimental isotherms are likely due to poorly characterized samples or unexpected changes in the framework structure upon adsorption (framework flexibility), rather than deficiencies of the simulations.

### 11.2.3 Adsorption of Mixtures

Building on their work for single-component methane adsorption, Düren and Snurr[62] used molecular modeling to study $CH_4$/*n*-butane mixtures in IRMOFs-1, -8, -10, -14, and -16 at room temperature and up to 40 bar. They specifically focused on the effect of the linker molecule on the adsorption of pure $CH_4$, pure *n*-butane, and their mixtures. For the MOFs considered, the methane isotherms are simple type I isotherms and are still far from saturation even at 40 bar. The *n*-butane isotherms are more complex, showing sharp jumps where the pores completely fill. The pore-filling pressure shifts toward higher pressures with increasing cavity size, approaching the bulk condensation pressure in the largest pores. For a given pore size, the pore filling occurs at lower pressure for MOFs that have more carbon atoms in the linker molecule, as seen by comparing IRMOF-10 and IRMOF-14. Mixture simulations showed that selectivity for *n*-butane over $CH_4$ was favored by decreasing cavity size and increasing number of atoms in the linker. Based on these findings, they proposed a hypothetical structure with a 9,10-anthracenedicarboxylate linker and named it IRMOF-993. Predicted selectivities in IRMOF-993 for trace amounts of *n*-butane in methane gave selectivities as high as 2500. In adsorption, the selectivity is defined as $S_{A/B} = (x_A/x_B)/(y_A/y_B)$, where $x_i$ is the mole fraction of species $i$ in the adsorbed phase and $y_i$ is the mole fraction of species $i$ in the bulk-gas phase.

Jiang and Sandler[63] studied the adsorption of linear and branched alkane mixtures in IRMOF-1 using molecular modeling. Linear alkanes up to five carbons, isopentane, and neopentane were considered. The simulated adsorption isotherms of pure linear alkanes showed that the adsorption properties at infinite dilution varied linearly with carbon number. Mixture simulations containing all five linear alkanes were also reported. With increasing pressure, adsorption of the shorter alkanes gradually increased, whereas adsorption of $n$-$C_5$ first increased and then decreased, as it was replaced by the shorter alkanes due to size entropy effects. For the single-component $C_5$ isomers, the linear isomer adsorbed more than the two branched isomers, and this was attributed to configurational entropy effects. A three-component simulation of the $C_5$ isomers also showed selectivity for the linear molecule over isopentane and neopentane. While the adsorption capacity of IRMOF-1 for the alkanes was found to be much greater than silicalite or a bundle of single-wall carbon nanotubes, selectivities in the latter two are greater.

Yang, Zhong, and coworkers have published a series of molecular-level computational studies[64–66] to evaluate MOFs for applications such as purification of natural gas and removal of $CO_2$ from flue gas. In these papers, the authors refined the MOF Lennard–Jones parameters to match the single-component experimental isotherms and then focused on simulations of mixture adsorption. For natural gas purification, $CO_2$ and higher hydrocarbons such as ethane, propane, and butane often must be removed from methane. Yang and Zhong[65] simulated equimolar mixtures of $C_2H_6$/$CH_4$, $CO_2$/$CH_4$, and $CO_2$/$C_2H_6$ in Cu-BTC over a range of pressures. They found that the selectivity for ethane over methane decreases with increasing pressure, whereas the selectivity for $CO_2$ over methane increases with increasing pressure. To understand this difference, additional simulations were performed for $CO_2$/$CH_4$ mixtures, as shown in Figure 11.9. Case 3 is the full model, including all electrostatic interactions; this case shows increasing selectivity with increasing pressure. In the simulations for Case 1, all electrostatic interactions were turned off, and for Case 2 the electrostatic interactions between the $CO_2$ molecules and the MOF were turned off but electrostatic interactions among $CO_2$ molecules were included. As seen in Figure 11.9, Cases 1 and 2 show a decreasing selectivity, similar to that observed for $C_2H_6$/$CH_4$ (where electrostatic interactions do not play an important role). This demonstrates how electrostatic interactions, both $CO_2$/$CO_2$ and $CO_2$/MOF, can affect gas separation in

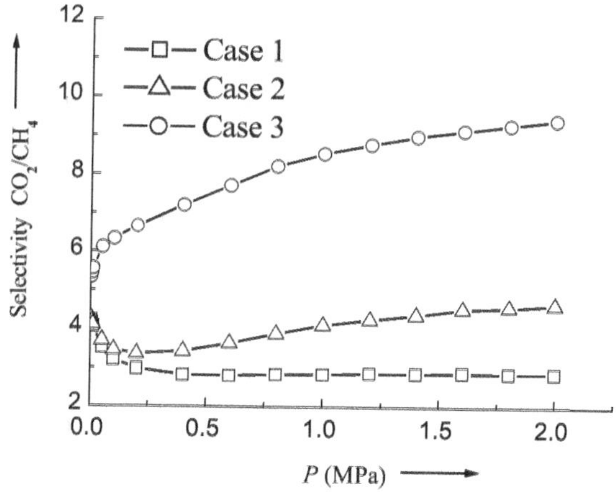


**FIGURE 11.9** Effect of electrostatic interactions on the selectivity of $CO_2$ from equimolar mixtures of methane and $CO_2$ in Cu-BTC at 298 K. (From ref. 65. Copyright © Wiley-VCH. Reproduced with permission).

MOFs. The authors found that all three molecules adsorb first in the small pockets of Cu-BTC and fill the larger pores at higher pressures. In methane/$CO_2$ mixtures, methane is pushed out of the pockets by $CO_2$ with increasing pressure.

For removal of $CO_2$ from flue gas, Yang et al.[64] also simulated $CO_2/N_2$ and $CO_2/O_2$ mixtures in Cu-BTC. Selectivities over 20 were predicted at 298 K over a wide range of pressures and gas-phase compositions. Again, the effects of electrostatics and molecular-level siting were investigated. In addition, Yang and Zhong[65,66] and Martín-Calvo et al.[67] compared adsorption of various mixtures in Cu-BTC and MOF-5 (IRMOF-1). They found that behavior in MOF-5 is simpler than that in Cu-BTC. For example, the selectivity for methane over hydrogen at 298 K is independent of pressure in MOF-5 but shows a decrease in Cu-BTC due to the more homogeneous environment in MOF-5. In all cases selectivities in Cu-BTC were higher than those in MOF-5.

For many years, researchers have worked to develop simple theoretical models that can predict mixture adsorption from single-component data. Such models can be used with either experimental or simulated single-component isotherms as inputs. Ideal adsorbed solution theory (IAST), developed by Myers and Prausnitz,[68] remains one of the most used methods for such predictions. It often works well for gas mixture adsorption in zeolites.[69] It is exact in the Henry's law regime, although deviations are frequently seen at higher loadings.[70] Several groups[64–66,71,72] have reported IAST calculations for binary mixtures in MOFs based on their simulated single-component isotherms. In these studies, binary mixture adsorption and selectivities from IAST calculations agreed well with data predicted from full GCMC simulations of the binary mixtures (Figure 11.10). Simulations are very useful for testing models like IAST due to the difficulty in performing multicomponent adsorption experiments.

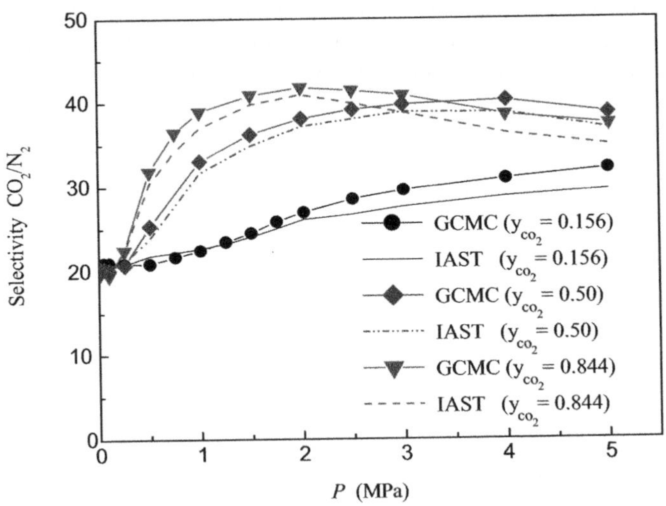


**FIGURE 11.10** Selectivities for $CO_2$ from binary mixtures of $CO_2$ and $N_2$ with three gas compositions in Cu-BTC at 298 K. Results from binary GCMC simulations are compared with results using simulated single-component isotherms and ideal adsorbed solution theory. (From ref. 64. Copyright © 2007 John Wiley & Sons, Inc. Reprinted with permission).

## 11.3 MOLECULAR MODELING OF DIFFUSION IN MOFs

Most of what we know about diffusion in MOFs comes from molecular simulation. In fact, Stallmach et al.[73] published the first experimentally measured diffusion coefficients in MOFs almost two years after the first simulation results.[74,75] Sarkisov et al.[74] reported the self-diffusion coefficients for $CH_4$ and $C_5$–$C_7$*n*-alkanes in IRMOF-1 at low loadings. The calculated self-diffusivities were on the order of $10^{-9}\,m^2\,s^{-1}$ for the $C_5$–$C_7$*n*-alkanes and $10^{-8}\,m^2\,s^{-1}$ for $CH_4$ at 300 K. Stallmach et al. reported a self-diffusivity of $1.8 \times 10^{-9}\,m^2\,s^{-1}$ for *n*-hexane in IRMOF-1 at 298 K, in very good agreement with the value $2.2 \times 10^{-9}\,m^2\,s^{-1}$ from Sarkisov et al. For methane, the experimental result is about an order of magnitude higher than the MD prediction. In their work on alkoxy-modified IRMOF-1, Jhon et al.[46] reported a decrease in the self-diffusivity of $CH_4$ in the presence of alkoxy chains.

Skoulidas[75] calculated the self and transport diffusivities of argon in Cu-BTC and compared the results with those for the zeolites silicalite, ITQ-3, and ITQ-7. Overall, he indicated many similarities between the diffusion of Ar in Cu-BTC and in all-silica zeolites, including magnitude, activation energies, concentration dependence, and temperature dependence. Yang and Zhong[76] reported similar conclusions regarding the similarity of diffusion in zeolites and MOFs from their work on the diffusion of $H_2$ in IRMOFs-1, -8, and -18.

Skoulidas and Sholl[77] studied the diffusivities of Ar, $CH_4$, $CO_2$, $N_2$, and $H_2$ in IRMOF-1 and the diffusivity of Ar in MOF-2, MOF-3, and Cu-BTC. Self, corrected, and transport diffusivities were reported as a function of pore loading at room temperature. Again, the general behavior is similar to that in zeolites. The results indicate that diffusion rates should not be the controlling factor for separation processes in MOF type materials.

## 11.4 MOLECULAR MODELING OF HYDROGEN STORAGE IN MOFs

In recent years, hydrogen has attracted much attention as a potential energy carrier. One of the most difficult challenges in implementing a "hydrogen economy" is the safe and efficient storage of hydrogen. A wide variety of storage methods have been considered, including high-pressure containers, liquid hydrogen, ammonia, metal hydrides, and physisorption in porous materials.[78,79] The U.S. Department of Energy has issued volumetric and gravimetric density targets for hydrogen storage for on-board vehicular applications along with fueling time, cost, safety, and other targets. It is a tremendous materials and engineering challenge to meet these targets. Molecular-level modeling is playing an important role in the evaluation and development of materials to meet these challenges.

### 11.4.1 Quantum Chemical Investigations of Hydrogen Storage in MOFs

Following initial experimental studies that showed MOFs were promising hydrogen storage materials, a number of groups have published investigations of hydrogen storage in MOFs using quantum chemical methods.[12,80–88] These studies aim to obtain a fundamental understanding of the nature of hydrogen adsorption in MOFs that will enable the design and optimization of frameworks for hydrogen storage. One of the most popular electronic structure tools today is density functional theory (DFT). Unfortunately, DFT does not handle weak interactions, such as dispersion interactions, well. This presents a problem for quantum chemical studies of hydrogen in MOFs, where weak interactions dominate. Many groups have turned to perturbation methods, such as MP2, although MP2 calculations are considerably more expensive than DFT, limiting the number of atoms that can be included. Many of these studies use molecular fragments from the MOF, such as in Figure 11.1, although a few fully periodic DFT studies have been published.

Several quantum chemical studies have addressed the question of where hydrogen adsorbs in MOFs, with a particular focus on the IRMOFs of Yaghi and coworkers. The IRMOFs have two general types of sites: near the metal corners and near the organic linkers. All of these studies point to stronger adsorption near the corners, in agreement with experiment.[89] In addition, quantum chemical studies have shown that larger linkers have larger binding energies. For example, Sagara et al.[88] evaluated hydrogen binding on a series of IRMOF linkers using high quality MP2 calculations with a QZVPP basis set. To obtain more accurate binding energies, they performed coupled cluster calculations for hydrogen/benzene interactions and then used this to estimate a correction factor for the MP2 calculations on the larger IRMOF linker molecules. They found that the linkers of IRMOFs-12, -993, and -14 can bind two to three, three, and four hydrogen molecules per side, respectively.

Several groups have examined how substituting different functional groups on aromatic linker molecules affects hydrogen adsorption, see for example Hübner et al.[90] and Sagara et al.[88] Hübner et al. concluded that larger aromatic linkers are preferred over single benzene rings to increase the interactions and that electron-donating groups give an increase in interaction energy, albeit that the energy enhancement is not expected to be substantial. Sagara et al.[88] found using MP2/QZVPP calculations that adding $NH_2$ or $CH_3$ groups can increase the hydrogen binding energy by up to 33%. Negri and Saendig[86] used MP2 and DFT calculations to explore the binding of $H_2$ to aromatic molecules representative of MOF linkers and to metal-oxide clusters representative of MOF corners. A moderate increase of the binding energy was computed for pyrrole, pyridine, ogligmers of pyridine, and *n*-oxide pyridine. For some hetero-aromatic molecules the edge sites, beside the top-sites, become competitive for hydrogen physisorption. These quantum chemical studies demonstrate that hydrogen adsorption in MOFs can be tuned by a variety of methods, especially choosing different linker molecules.

### 11.4.2 Monte Carlo Studies of Hydrogen Storage in MOFs

For MOFs where hydrogen adsorbs mainly by dispersion interactions, classical models can provide a useful complement to quantum chemical studies. Lennard–Jones interactions between MOF atoms and hydrogen account for dispersion and repulsion interactions, providing the essential physics of the system. At low

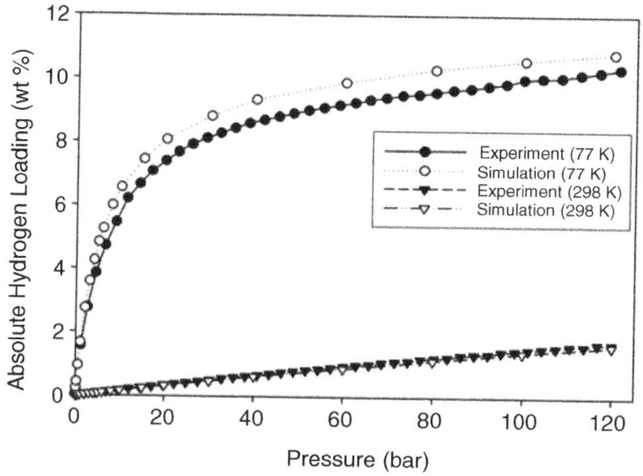


**FIGURE 11.11** Hydrogen adsorption isotherms (absolute quantities) in IRMOF-1 at 77 and 298 K. GCMC simulations[91] agree well with experimental measurements of Kaye et al.[92]

temperatures, quantum diffraction effects are important for very light molecules like hydrogen and can be included using a path integral formalism.[58] Figure 11.11 shows predicted hydrogen isotherms in IRMOF-1 at 77 and 298 K[91] along with experimental isotherms from Kaye et al.[92] Considering the simplicity of the model (which was not adjusted to match the experiments), the agreement between simulation and experiment is good (see also Section 11.2.2). The predicted heats of adsorption also agree well with those obtained from quantum chemical calculations and from experiment.

One particularly attractive feature of molecular modeling is that it is possible to screen a large number of systems (or conditions) more readily than with experiments. Frost et al.[91] took advantage of this and calculated hydrogen adsorption isotherms in a series of 10 IRMOFs up to 120 bar at 77 K. At this time, there were conflicting opinions in the literature about whether hydrogen adsorption in MOFs was correlated with the surface area or not. One goal of Frost and coworkers was to gain some insight into this question. The earlier work of Düren et al.[39] had highlighted the importance of free volume and the heat of adsorption, in addition to the surface area, for methane adsorption in MOFs. From the simulated hydrogen isotherms, Frost plotted the amount adsorbed in the 10 MOFs as functions of (a) the heat of adsorption, (b) the surface area, and (c) the free volume. This was done at three different pressures. As shown in Figure 11.12, they found that at low loadings (0.1 bar), the amount of adsorbed hydrogen correlates well with the heat of adsorption; at intermediate loadings (30 bar), the amount of adsorbed hydrogen correlates well with the surface area; and at high loadings (120 bar), the amount of adsorbed hydrogen correlates well with the free volume of the MOF. Thus, there are different regimes where different materials properties are most important. For example, at low loadings there are not enough molecules to utilize all of the surface area, so the amount adsorbed does not correlate with the surface area. Visualizations of molecular siting in IRMOF-10 are shown in Figure 11.13 for the three different pressure/loading regimes. Figure 11.13a

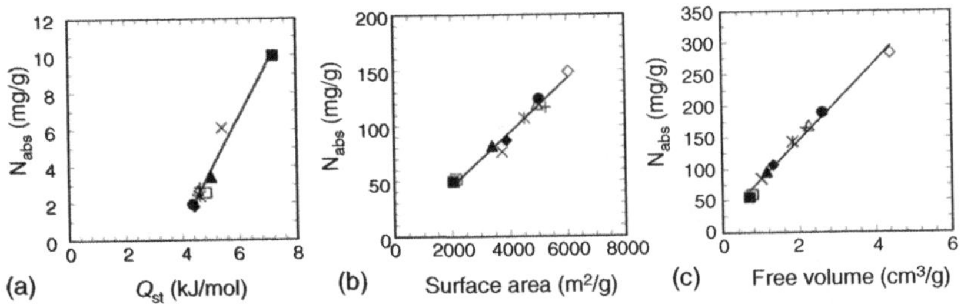


**FIGURE 11.12** Hydrogen adsorption in a series of 10 IRMOFs as predicted by GCMC simulation (a) Amount adsorbed at 0.1 bar and 77 K versus isosteric heat of adsorption. (b) Amount adsorbed at 30 bar and 77 K versus accessible surface area. (c) Amount adsorbed at 120 bar at 77 K versus free volume. ◆, IRMOF-1; ■, IRMOF-4; ▲, IRMOF-6; ×, IRMOF-7; *, IRMOF-8; ●, IRMOF-10; +, IRMOF-12; △, IRMOF-14; ◇, IRMOF-16; □, IRMOF-18. (Reprinted with permission from ref. 91. Copyright © 2006 American Chemical Society).

shows that at low loading, hydrogen molecules are mostly near the zinc corners, in agreement with quantum chemical calculations and experiment. Figure 11.13b shows that in the intermediate regime, molecules adsorb preferentially in the corners and along the linker molecules, with fewer molecules in the centers of the cavities. Figure 11.13c displays the limiting adsorption behavior where hydrogen fills the majority of the void regions of the material.

The analysis in Figure 11.12 was based on absolute adsorbed amounts at 77 K. Frost and Snurr[93] extended this analysis to room temperature and analyzed both absolute and excess adsorption. The correlations using absolute adsorption are somewhat easier to understand from a physical viewpoint, but the correlations using the excess adsorption provide a link to experiments. The difference between absolute and excess adsorption is of great importance in understanding adsorption data. For example, IRMOF-1 adsorbs more hydrogen in an absolute sense than Cu-BTC does, but Cu-BTC shows a larger excess adsorption than IRMOF-1. This is because IRMOF-1 has a substantially larger free volume per unit mass (lower framework density). The difference between the absolute and the excess amount adsorbed can be significant

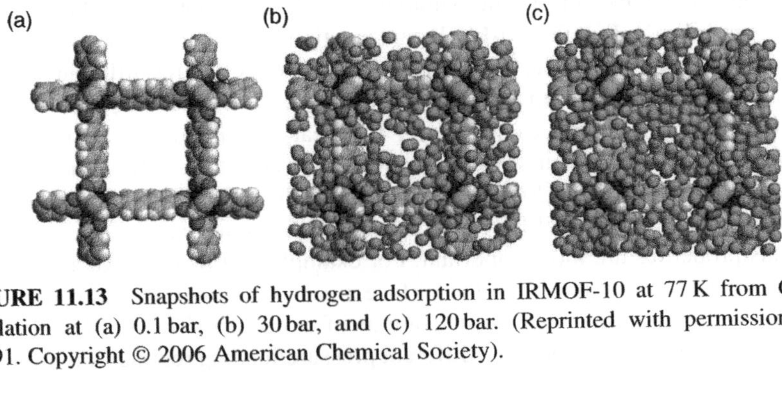


**FIGURE 11.13** Snapshots of hydrogen adsorption in IRMOF-10 at 77 K from GCMC simulation at (a) 0.1 bar, (b) 30 bar, and (c) 120 bar. (Reprinted with permission from ref. 91. Copyright © 2006 American Chemical Society).

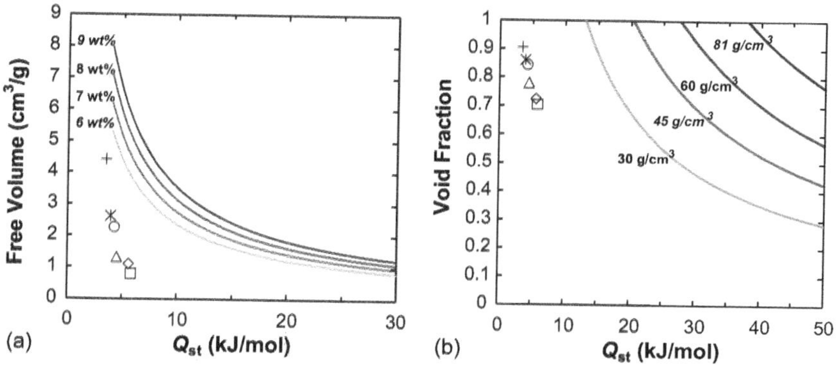


**FIGURE 11.14** (a) Requirements for target gravimetric loadings at 120 bar and 298 K. (b) Requirements for target volumetric loadings at 120 bar and 298 K. △, IRMOF-1; ◇, IRMOF-9; *, IRMOF-10; ○, IRMOF-14; +, IRMOF-16; □, Cu-BTC. (Reprinted with permission from ref. 93. Copyright © 2007 American Chemical Society).

for some systems. Examining the density of hydrogen within the pore void volume (rather than per total volume of material) can be a useful way to think about and compare different sorbents.

Frost and Snurr[93] also performed additional simulations at 298 K in which the Lennard–Jones interactions between $H_2$ and MOFs were artificially increased (leaving the $H_2/H_2$ interactions constant). This artificially increased the heat of adsorption. The goal was to provide guidance on how much the heat of adsorption must be increased in MOFs to meet current targets for hydrogen storage. From these studies, they found a correlation between the density of hydrogen in the pore void volume and the heat of adsorption. (Note this also points out that the heat of adsorption plays a role not only at low loading – something that may not be clear from the correlations in Figure 11.12.) Using this correlation, they prepared a graph showing combinations of heat of adsorption and free volume that meet target gravimetric and volumetric storage amounts at 298 K and 120 bar (see Figure 11.14). The graph suggests, for example, that if new materials can achieve an isosteric heat of 10–15 kJ/mol with a free volume between 1.6 and 2.4 $cm^3/g$, gravimetric $H_2$ uptake of 6% could be achieved.

### 11.4.3 Routes to Achieving Improved Hydrogen Storage

MOFs initially attracted interest for hydrogen storage because of their enormous internal surface areas (as high as 6000 $m^2/g$ as discussed above), which is an obvious advantage for gas storage by physisorption. There is now a consensus that higher heats of adsorption are needed if MOFs are to meet gravimetric and volumetric targets for hydrogen storage. Bhatia and Myers[94] performed a thermodynamic analysis from a systems viewpoint and calculated an optimum adsorption enthalpy of 15 kJ/mol for room temperature storage. This value is strong enough to store a large amount of hydrogen at the charging pressure (30 bar in their analysis) but weak enough to release most of that hydrogen at the discharge pressure (1.5 bar).

As discussed in Section 11.4.1, quantum chemical studies have shown that changing the organic linkers can increase the adsorption enthalpy, but the increases are not very large. To be able to reach the DOE targets it appears that a more substantial leap is needed than simple alterations of the organic building blocks. Several routes have been proposed to increase hydrogen-framework interactions and actively researched.[95] These include (1) increased surface area by increasing the exposed edges, (2) catenation, (3) impregnation of large pores to produce new internal sorption sites and higher surface areas, (4) open metal sites, (5) MOFs built from lighter elements, (6) chemically reduced frameworks, (7) incorporation of strongly binding metal ions like $Mg^{2+}$ and $Al^{3+}$, and (8) exploitation of framework flexibility, that is, kinetic trapping. The work of Frost indicates that any attempts to increase the adsorption enthalpy must not reduce the free volume too much. Thus, options 2 and 3 may have limitations. Many groups are currently exploring open metal sites.

Several modeling studies have investigated the idea of introducing cations into MOFs.[96–100] For example, Han and Goddard[98] reported that Li-doped MOFs can significantly improve $H_2$ uptake at ambient conditions. They performed classical GCMC simulations of hydrogen adsorption for MOF structure with up to nine fused rings (MOF-$C_{30}$). The force field was calibrated from DFT and MP2 calculations. They predicted that at $-30°$C and 100 bar the Li-MOF $C_{30}$ has a gravimetric uptake of 6.0 wt%, just reaching the 2010 DOE target. An ab initio study of Blomqvist et al.[96] also explored Li-doping. Reducing the framework by introduction of lithium increases adsorption, and most of the increase stems from the interaction of hydrogen with the lithium atom rather than the reduced linkers. Two lithium atoms are strongly adsorbed on the surfaces of the $C_6$ rings, one on each side, carrying a charge of $+0.9e$ per Li atom. Each Li can cluster three $H_2$ molecules around itself with a binding energy of 12 kJ/mol. This is more than twice the binding energy compared to a pure BDC linker. These works suggest that doping of MOFs with electropositive metals provides a promising route for practical hydrogen storage. Recently Mulfort and Hupp[101] demonstrated experimentally that lithium can be introduced into a MOF in a post-synthesis modification. Their Li-doped mixed-ligand MOF adsorbed twice as much hydrogen as the non-doped version and displayed an enhanced heat of adsorption over the entire range of loading.

Lochan and Head-Gordon[100] discuss the role of dispersion forces, electrostatics, and orbital interactions for molecular hydrogen binding affinities. They conclude that while orbital interactions are crucial for sustaining interaction energies with the framework, $H_2$ is an "intrinsically reluctant participant in such interactions." Based on elementary physical considerations, they note that dispersion interactions for $H_2$ are on the order of a few kJ/mol. For comparison, the electrostatic interaction of $H_2$ with a point charge at a distance of 3 Å is on the order of 3.5 kJ/mol, and the charge-induced dipole interaction at this distance is approximately 6.8 kJ/mol. Bare $Li^+$ cations show a strong affinity for $H_2$ of about 24 kJ/mol, but this is reduced to about 12 kJ/mol upon complexation (for example with a MOF). More highly charged metal ions like $Mg^{2+}$ and $Al^{3+}$ bind hydrogen too strongly to fall within the ideal binding range. However, their binding energy can be controlled by complexation with ligands like CO. These ideas may lead to future strategies to improve hydrogen uptake in MOFs.

## 11.5 SUMMARY AND FUTURE DIRECTIONS

Molecular modeling is playing an important role in the development of MOFs toward practical applications and in increasing our understanding of adsorption and diffusion of guest molecules in MOFs. For MOFs that are fairly rigid and for molecules that adsorb through simple van der Waals interactions, GCMC simulations predict adsorption isotherms, heats of adsorption, and siting of molecules within the pores that are generally in good agreement with experiment. For diffusion, there is little experimental data available, but the few comparisons between MD predictions and experiment are encouraging. These capabilities already open up the possibility to screen hypothetical MOFs on the computer before attempting to synthesize them for applications such as $CO_2$ capture. Development of faster high-throughput screening is likely in the near future.

A number of interesting phenomena in MOFs have been attributed to framework flexibility and dynamic framework movement during adsorption. Often it is difficult to verify this experimentally. This creates an opportunity for modeling. Recently, several groups have developed models that move away from the assumption of a rigid framework.[10,13,22,23] These models have been used to investigate the stability of MOFs in the presence of water[23] and the effect of framework flexibility on guest molecule diffusion.[22] Modeling also revealed the interesting finding that IRMOFs display negative thermal expansion and that the effect is quite large.[10] Future efforts should focus on even more challenging goals such as prediction of the dramatic structural changes observed in the MIL series of MOFs.

Other future developments are likely to focus on challenges in modeling chemisorption and the chemical diversity of MOFs and related COFs, ZMOFs, and ZIFs. This is closely related to the challenges in modeling of catalysis. Modeling of chemisorption and catalysis will naturally require quantum chemical methods. So-called embedding methods should be particularly helpful, in which a central region is treated quantum mechanically while the surrounding atoms are treated with a force field. This allows the computationally intensive QM calculations to focus on the region of bond breaking and bond forming, while also accounting for the importance of the MOF cavity. Development of force fields for chemically diverse frameworks will allow researchers to predict properties of new materials with increasing confidence and speed.

## ACKNOWLEDGMENTS

Parts of this work were supported by the Department of Energy (DEFG02–1ER15244) and the National Science Foundation (CTS-0507013). The authors thank Patrick Ryan for Figure 11.11.

## REFERENCES


1. Bradshaw, D.; Claridge, J. B.; Cussen, E. J.; Prior, T. J.; Rosseinsky, M. J. Design, chirality, and flexibility in nanoporous molecule-based materials. *Acc. Chem. Res.* **2005**, *38*, 273–282.
2. Ferey, G. Hybrid porous solids: past, present, future. *Chem. Soc. Rev.* **2008**, *37*, 191–214.
3. Kitagawa, S.; Kitaura, R.; Noro, S. Functional porous coordination polymers. *Angew. Chem. Int. Ed.* **2004**, *43*, 2334–2375.
4. Rosseinsky, M. J. Recent developments in metal-organic framework chemistry: design, discovery, permanent porosity and flexibility. *Microporous Mesoporous Mater.* **2004**, *73*, 15–30.
5. Rowsell, J. L. C.; Yaghi, O. M. Metal–organic frameworks: a new class of porous materials. *Microporous Mesoporous Mater.* **2004**, *73*, 3–14.
6. Stein, A. Advances in microporous and mesoporous solids – highlights of recent progress. *Adv. Mater.* **2003**, *15*, 763–775.
7. Yaghi, O. M.; O'Keeffe, M.; Ockwig, N. W.; Chae, H. K.; Eddaoudi, M.; Kim, J. Reticular synthesis and the design of new materials. *Nature* **2003**, *423*, 705–714.
8. Snurr, R. Q.; Hupp, J. T.; Nguyen, S. T. Prospects for nanoporous metal–organic materials in advanced separations processes. *AIChE J.* **2004**, *50*, 1090–1095.
9. Mueller, U.; Schubert, M.; Teich, F.; Puetter, H.; Schierle-Arndt, K.; Pastre, J. Metal–organic frameworks: prospective industrial applications. *J. Mater. Chem.* **2006**, *16*, 626–636.
10. Dubbeldam, D.; Walton, K. S.; Ellis, D. E.; Snurr, R. Q. Exceptional negative thermal expansion in isoreticular metal-organic frameworks. *Angew. Chem. Int. Ed.* **2007**, *46*, 4496–4499.
11. Huang, B. L.; McGaughey, A. J. H.; Kaviany, M. Thermal conductivity of metal-organic framework 5 (MOF-5): Part I. Molecular dynamics simulations. *Int. J. Heat Mass Transfer* **2007**, *50*, 393–404.
12. Samanta, A.; Furuta, T.; Li, J. Theoretical assessment of the elastic constants and hydrogen storage capacity of some metal-organic framework materials. *J. Chem. Phys.* **2006**, *125*, 084714.
13. Han, S. S.; Goddard, W. A. Metal–organic frameworks provide large negative thermal expansion behavior. *J. Phys. Chem. C* **2007**, *111*, 15185–15191.
14. Ferey, G.; Mellot-Draznieks, C.; Serre, C.; Millange, F.; Dutour, J.; Surble, S.; Margiolaki, I. A chromium terephthalate-based solid with unusually large pore volumes and surface area. *Science* **2005**, *309*, 2040–2042.
15. Auerbach, S. M. Modeling jump diffusion in zeolites: II. Applications. *Fluid Transport in Nanoporous Materials*. Springer: Dordrecht, **2006**; pp 535–551.
16. Fuchs, A. H.; Cheetham, A. K. Adsorption of guest molecules in zeolitic materials: computational aspects. *J. Phys. Chem. B* **2001**, *105*, 7375–7383.
17. Kärger, J.; Ruthven, D. M. *Diffusion in Zeolites and Other Microporous Solids*. John Wiley & Sons, Inc.: New York, 1992.
18. Keil, F. J.; Krishna, R.; Coppens, M. O. Modeling of diffusion in zeolites. *Rev. Chem. Eng.* **2000**, *16*, 71–197.

19. Theodorou, D. N.; Snurr R.Q.; Bell, A. T. Molecular dynamics and diffusion in microporous materials. In *Comprehensive Supramolecular Chemistry*; Alberti, G.; Bein, T. Eds.; Pergamon: Oxford, 1996; Vol. *7*, pp 507–548.
20. Allen, M. P.; Tildesley, D. J. *Computer Simulation of Liquids*. Oxford University Press: Oxford, 1987.
21. Frenkel, D.; Smit, B. *Understanding Molecular Simulation: From Algorithms to Applications*. Academic Press, Inc.: San Diego, California, 2002.
22. Amirjalayer, S.; Tafipolsky, M.; Schmid, R. Molecular dynamics simulation of benzene diffusion in MOF-5: importance of lattice dynamics. *Angew. Chem. Int. Ed.* **2007**, *46*, 463–466.
23. Greathouse, J. A.; Allendorf, M. D. The interaction of water with MOF-5 simulated by molecular dynamics. *J. Am. Chem. Soc.* **2006**, *128*, 10678–10679.
24. Martin, M. G.; Siepmann, J. I. Transferable potentials for phase equilibria. 1. United-atom description of n-alkanes. *J. Phys. Chem. B* **1998**, *102*, 2569–2577.
25. Potoff, J. J.; Siepmann, J. I. Vapor–liquid equilibria of mixtures containing alkanes, carbon dioxide, and nitrogen. *AIChE J.* **2001**, *47*, 1676–1682.
26. Bezus, A. G.; Kiselev, A. V.; Lopatkin, A. A.; Du, P. Q. Molecular statistical calculation of thermodynamic adsorption characteristics of zeolites using atom-atom approximation. 1. Adsorption of methane by zeolite NaX. *J. Chem. Soc. Faraday Trans. II* **1978**, *74*, 367–379.
27. Mayo, S. L.; Olafson, B. D.; Goddard, W. A. Dreiding – a generic force-field for molecular simulations. *J. Phys. Chem.* **1990**, *94*, 8897–8909.
28. Rappe, A. K.; Casewit, C. J.; Colwell, K. S.; Goddard, W. A.; Skiff, W. M. UFF, a full periodic-table force-field for molecular mechanics and molecular-dynamics simulations. *J. Am. Chem. Soc.* **1992**, *114*, 10024–10035.
29. Jorgensen, W. L.; Maxwell, D. S.; TiradoRives, J. Development and testing of the OPLS all-atom force field on conformational energetics and properties of organic liquids. *J. Am. Chem. Soc.* **1996**, *118*, 11225–11236.
30. Mulliken, R. S. Electronic population analysis on LCAO-MO molecular wave functions. 1. *J. Chem. Phys.* **1955**, *23*, 1833–1840.
31. Breneman, C. M.; Wiberg, K. B. Determining atom-centered monopoles from molecular electrostatic potentials – the need for high sampling density in formamide conformational-analysis. *J. Comp. Chem.* **1990**, *11*, 361–373.
32. Macedonia, M. D.; Moore, D. D.; Maginn, E. J.; Olken, M. M. Adsorption studies of methane, ethane, and argon in the zeolite mordenite: molecular simulations and experiments. *Langmuir* **2000**, *16*, 3823–3834.
33. Myers, A. L. Thermodynamics of adsorption in porous materials. *AIChE J.* **2002**, *48*, 145–160.
34. Myers, A. L.; Monson, P. A. Adsorption in porous materials at high pressure: theory and experiment. *Langmuir* **2002**, *18*, 10261–10273.
35. Talu, O.; Myers, A. L. Molecular simulation of adsorption: Gibbs dividing surface and comparison with experiment. *AIChE J.* **2001**, *47*, 1160–1168.
36. Talu, O.; Myers, A. L. Reference potentials for adsorption of helium, argon, methane, and krypton in high-silica zeolites. *Colloids Surf. A: Physicochem. Eng. Aspects* **2001**, *187*, 83–93.

37. Dubbeldam, D.; Snurr, R. Q. Recent developments in the molecular modeling of diffusion in nanoporous materials. *Mol. Simulat.* **2007**, *33*, 305–325.
38. Dodd, L. R.; Theodorou, D. N. Analytical treatment of the volume and surface area of molecules formed by an arbitrary collection of unequal spheres intersected by planes. *Mol. Phys.* **1991**, *72*, 1313–1345.
39. Düren, T.; Sarkisov, L.; Yaghi, O. M.; Snurr, R. Q. Design of new materials for methane storage. *Langmuir* **2004**, *20*, 2683–2689.
40. Düren, T.; Millange, F.; Ferey, G.; Walton, K. S.; Snurr, R. Q. Calculating geometric surface areas as a characterization tool for metal-organic frameworks. *J. Phys. Chem. C* **2007**, *111*, 15350–15356.
41. Walton, K. S.; Snurr, R. Q. Applicability of the BET method for determining surface areas of microporous metal-organic frameworks. *J. Am. Chem. Soc.* **2007**, *129*, 8552–8556.
42. Chui, S. S. Y.; Lo, S. M. F.; Charmant, J. P. H.; Orpen, A. G.; Williams, I. D. A chemically functionalizable nanoporous material $[Cu_3(TMA)_2(H_2O)_3]_n$. *Science* **1999**, *283*, 1148–1150.
43. Kawakami, T.; Takamizawa, S.; Kitagawa, Y.; Maruta, T.; Mori, W.; Yamaguchi, K. Theoretical studies of spin arrangement of adsorbed organic radicals in metal-organic nanoporous cavity. *Polyhedron* **2001**, *20*, 1197–1206.
44. Li, H.; Eddaoudi, M.; Groy, T. L.; Yaghi, O. M. Establishing microporosity in open metal-organic frameworks: gas sorption isotherms for Zn(BDC) (BDC = 1,4-benzenedicarboxylate). *J. Am. Chem. Soc.* **1998**, *120*, 8571–8572.
45. Vishnyakov, A.; Ravikovitch, P. I.; Neimark, A. V.; Bulow, M.; Wang, Q. M. Nanopore structure and sorption properties of Cu-BTC metal-organic framework. *Nano Lett.* **2003**, *3*, 713–718.
46. Jhon, Y. H.; Cho, M.; Jeon, H. R.; Park, I.; Chang, R.; Rowsell, J. L. C., Kim, J. Simulations of methane adsorption and diffusion within alkoxy-functionalized IRMOFs exhibiting severely disordered crystal structures. *J. Phys. Chem. C* **2007**, *111*, 16618–16625.
47. Eddaoudi, M.; Kim, J.; Rosi, N.; Vodak, D.; Wachter, J.; O'Keeffe, M.; Yaghi, O. M. Systematic design of pore size and functionality in isoreticular MOFs and their application in methane storage. *Science* **2002**, *295*, 469–472.
48. Yang, Q.; Zhong, C.; Chen, J. F. Computational study of $CO_2$ storage in metal-organic frameworks. *J. Phys. Chem. C* **2008**, *112*, 1562–1569.
49. Dybtsev, D. N.; Chun, H.; Yoon, S. H.; Kim, D.; Kim, K. Microporous manganese formate: a simple metal-organic porous material with high framework stability and highly selective gas sorption properties. *J. Am. Chem. Soc.* **2004**, *126*, 32–33.
50. Chae, H. K.; Siberio-Perez, D. Y.; Kim, J.; Go, Y.; Eddaoudi, M.; Matzger, A. J.; O'Keeffe, M.; Yaghi, O. M. A route to high surface area, porosity and inclusion of large molecules in crystals. *Nature* **2004**, *427*, 523–527.
51. Walton, K. S.; Millward, A. R.; Dubbeldam, D.; Frost, H.; Low, J. J.; Yaghi, O. M.; Snurr, R. Q. Understanding inflections and steps in carbon dioxide adsorption isotherms in metal-organic frameworks. *J. Am. Chem. Soc.* **2008**, *130*, 406–407.
52. Ramsahye, N. A.; Maurin, G.; Bourrelly, S.; Llewellyn, P. L.; Devic, T.; Serre, C.; Loiseau, T.; Ferey, G. Adsorption of $CO_2$ in metal-organic frameworks of different metal centres: Grand Canonical Monte Carlo simulations compared to experiments. *Adsorption* **2007**, *13*, 461–467.

53. Ramsahye, N. A.; Maurin, G.; Bourrelly, S.; Llewellyn, P. L.; Loiseau, T.; Serre, C.; Ferey, G. On the breathing effect of a metal-organic framework upon $CO_2$ adsorption: Monte Carlo compared to microcalorimetry experiments. *Chem. Commun.* **2007**, 3261–3263.

54. Serre, C.; Millange, F.; Thouvenot, C.; Nogues, M.; Marsolier, G.; Louer, D.; Ferey, G. Very large breathing effect in the first nanoporous chromium(III)-based solids: MIL-53 or $Cr^{III}(OH)\cdot\{O_2C\text{-}C_6H_4\text{-}CO_2\}\cdot\{HO_2C\text{-}C_6H_4\text{-}CO_2H\}_x\cdot H_2O_y$. *J. Am. Chem. Soc.* **2002**, *124*, 13519–13526.

55. Loiseau, T.; Serre, C.; Huguenard, C.; Fink, G.; Taulelle, F.; Henry, M.; Bataille, T.; Ferey, G. A rationale for the large breathing of the porous aluminum terephthalate (MIL-53) upon hydration. *Chem. Eur. J.* **2004**, *10*, 1373–1382.

56. Bourrelly, S.; Llewellyn, P. L.; Serre, C.; Millange, F.; Loiseau, T.; Ferey, G. Different adsorption behaviors of methane and carbon dioxide in the isotypic nanoporous metal terephthalates MIL-53 and MIL-47. *J. Am. Chem. Soc.* **2005**, *127*, 13519–13521.

57. Barthelet, K.; Marrot, J.; Riou, D.; Ferey, G. A breathing hybrid organic-inorganic solid with very large pores and high magnetic characteristics. *Angew. Chem. Int. Ed.* **2002**, *41*, 281–284.

58. Garberoglio, G.; Skoulidas, A. I.; Johnson, J. K. Adsorption of gases in metal-organic materials: comparison of simulations and experiments. *J. Phys. Chem. B* **2005**, *109*, 13094–13103.

59. Liu, J. C.; Culp, J. T.; Natesakhawat, S.; Bockrath, B. C.; Zande, B.; Sankar, S. G.; Garberoglio, G.; Johnson, J. K. Experimental and theoretical studies of gas adsorption in $Cu_3(BTC)_2$: an effective activation procedure. *J. Phys. Chem. C* **2007**, *111*, 9305–9313.

60. Dubbeldam, D.; Frost, H.; Walton, K. S.; Snurr, R. Q. Molecular simulation of adsorption sites of light gases in the metal-organic framework IRMOF-1. *Fluid Phase Equilib.* **2007**, *261*, 152–161.

61. Rowsell, J. L. C.; Spencer, E. C.; Eckert, J.; Howard, J. A. K.; Yaghi, O. M. Gas adsorption sites in a large-pore metal-organic framework. *Science* **2005**, *309*, 1350–1354.

62. Düren, T.; Snurr, R. Q. Assessment of isoreticular metal-organic frameworks for adsorption separations: a molecular simulation study of methane/*n*-butane mixtures. *J. Phys. Chem. B* **2004**, *108*, 15703–15708.

63. Jiang, J. W.; Sandler, S. I. Monte Carlo simulation for the adsorption and separation of linear and branched alkanes in IRMOF-1. *Langmuir* **2006**, *22*, 5702–5707.

64. Yang, Q. Y.; Xue, C. Y.; Zhong, C. L.; Chen, J. F. Molecular simulation of separation of $CO_2$ from flue gases in Cu-BTC metal-organic framework. *AIChE J.* **2007**, *53*, 2832–2840.

65. Yang, Q. Y.; Zhong, C. L. Electrostatic-field-induced enhancement of gas mixture separation in metal-organic frameworks: a computational study. *ChemPhysChem.* **2006**, *7*, 1417–1421.

66. Yang, Q. Y.; Zhong, C. L. Molecular simulation of carbon dioxide/methane/hydrogen mixture adsorption in metal-organic frameworks. *J. Phys. Chem. B* **2006**, *110*, 17776–17783.

67. Martín-Calvo, A.; García-Pérez, E.; Castillo, J. M.; Calero, S. Molecular simulations for adsorption and separation of natural gas in IRMOF-1 and Cu-BTC metal-organic frameworks. *Phys. Chem. Chem. Phys.* **2008**, *10*, 7085–7091.

68. Myers, A. L.; Prausnitz, J. M. Thermodynamics of mixed-gas adsorption. *AIChE J.* **1965**, *11*, 121–127.

69. Goj, A.; Sholl, D. S.; Akten, E. D.; Kohen, D. Atomistic simulations of $CO_2$ and $N_2$ adsorption in silica zeolites: the impact of pore size and shape. *J. Phys. Chem. B* **2002**, *106*, 8367–8375.
70. Murthi, M.; Snurr, R. Q. Effects of molecular siting and adsorbent heterogeneity on the ideality of adsorption equilibria. *Langmuir* **2004**, *20*, 2489–2497.
71. Babarao, R.; Hu, Z. Q.; Jiang, J. W.; Chempath, S.; Sandler, S. I. Storage and separation of $CO_2$ and $CH_4$ in silicalite, C-168 schwarzite, and IRMOF-1: a comparative study from monte carlo simulation. *Langmuir* **2007**, *23*, 659–666.
72. Bae, Y. S.; Mulfort, K. L.; Frost, H.; Ryan, P.; Punnathanam, S.; Broadbelt, L. J.; Hupp, J. T.; Snurr, R. Q. Separation of $CO_2$ from $CH_4$ using mixed-ligand metal-organic frameworks. *Langmuir* **2008**, *24*, 8592–8598.
73. Stallmach, F.; Gröger, S.; Künzel, V.; Kärger, J.; Yaghi, O. M.; Hesse, M.; Müller, U. NMR studies on the diffusion of hydrocarbons on the metal-organic framework material MOF-5. *Angew. Chem. Int. Ed.* **2006**, *45*, 2123–2126.
74. Sarkisov, L.; Düren, T.; Snurr, R. Q. Molecular modelling of adsorption in novel nanoporous metal-organic materials. *Mol. Phys.* **2004**, *102*, 211–221.
75. Skoulidas, A. I. Molecular dynamics simulations of gas diffusion in metal-organic frameworks: argon in CuBTC. *J. Am. Chem. Soc.* **2004**, *126*, 1356–1357.
76. Yang, Q. Y.; Zhong, C. L. Molecular simulation of adsorption and diffusion of hydrogen in metal-organic frameworks. *J. Phys. Chem. B* **2005**, *109*, 11862–11864.
77. Skoulidas, A. I.; Sholl, D. S. Self-diffusion and transport diffusion of light gases in metal-organic framework materials assessed using molecular dynamics simulations. *J. Phys. Chem. B* **2005**, *109*, 15760–15768.
78. Seayad, A. M.; Antonelli, D. M. Recent advances in hydrogen storage in metal-containing inorganic nanostructures and related materials. *Adv. Mater.* **2004**, *16*, 765–777.
79. van den Berg, A. W. C.; Arean, C. O. Materials for hydrogen storage: current research trends and perspectives. *Chem. Commun.* **2008**, 668–681.
80. Buda, C.; Dunietz, B. D. Hydrogen physisorption on the organic linker in metal-organic frameworks: ab initio computational study. *J. Phys. Chem. B* **2006**, *110*, 10479–10484.
81. Gao, Y.; Zeng, X. C. Ab initio study of hydrogen adsorption on benzenoid linkers in metal-organic framework materials. *J. Phys. Condens. Matter* **2007**, *19*, 386220.
82. Kim, D.; Kim, J.; Jung, D. H.; Lee, T. B.; Choi, S. B.; Yoon, J. H.; Kim, J.; Choi, K.; Choi, S. H. Quantitative structure-uptake relationship of metal-organic frameworks as hydrogen storage material. *Catal. Today* **2007**, *120*, 317–323.
83. Kim, D.; Lee, T. B.; Choi, S. B.; Yoon, J. H.; Kim, J. Choi, S. H. A density functional theory study of a series of functionalized metal-organic frameworks. *Chem. Phys. Lett.* **2006**, *420*, 256–260.
84. Mueller, T.; Ceder, G. A density functional theory study of hydrogen adsorption in MOF-5. *J. Phys. Chem. B* **2005**, *109*, 17974–17983.
85. Mulder, F. M.; Dingemans, T. J.; Wagemaker, M.; Kearley, G. J. Modelling of hydrogen adsorption in the metal-organic framework MOF5. *Chem. Phys.* **2005**, *317*, 113–118.
86. Negri, F.; Saendig, N. Tuning the physisorption of molecular hydrogen: binding to aromatic, hetero-aromatic and metal-organic framework materials. *Theor. Chem. Acc.* **2007**, *118*, 149–163.

87. Sagara, T.; Klassen, J.; Ganz, E. Computational study of hydrogen binding by metal–organic framework-5. *J. Chem. Phys.* **2004**, *121*, 12543–12547.

88. Sagara, T.; Klassen, J.; Ortony, J., Ganz, E. Binding energies of hydrogen molecules to isoreticular metal-organic framework materials. *J. Chem. Phys.* **2005**, *123*, 014701.

89. Rowsell, J. L. C.; Eckert, J.; Yaghi, O. M. Characterization of $H_2$ binding sites in prototypical metal-organic frameworks by inelastic neutron scattering. *J. Am. Chem. Soc.* **2005**, *127*, 14904–14910.

90. Hubner, O.; Gloss, A.; Fichtner, M.; Klopper, W. On the interaction of dihydrogen with aromatic systems. *J. Phys. Chem. A* **2004**, *108*, 3019–3023.

91. Frost, H.; Düren, T.; Snurr, R. Q. Effects of surface area, free volume, and heat of adsorption on hydrogen uptake in metal-organic frameworks. *J. Phys. Chem. B* **2006**, *110*, 9565–9570.

92. Kaye, S. S.; Dailly, A.; Yaghi, O. M.; Long, J. R. Impact of preparation and handling on the hydrogen storage properties of $Zn_4O$ (1,4-benzenedicarboxylate)$_3$ (MOF-5). *J. Am. Chem. Soc.* **2007**, *129*, 14176–14177.

93. Frost, H.; Snurr, R. Q. Design requirements for metal-organic frameworks as hydrogen storage materials. *J. Phys. Chem. C* **2007**, *111*, 18794–18803.

94. Bhatia, S. K.; Myers, A. L. Optimum conditions for adsorptive storage. *Langmuir* **2006**, *22*, 1688–1700.

95. Rowsell, J. L. C.; Yaghi, O. M. Strategies for hydrogen storage in metal-organic frameworks. *Angew. Chem. Int. Ed.* **2005**, *44*, 4670–4679.

96. Blomqvist, A.; Araujo, C. M.; Srepusharawoot, P.; Ahuja, R. Li-decorated metal-organic framework 5: a route to achieving a suitable hydrogen storage medium. *Proc. Natl. Acad. Sci.* **2007**, *104*, 20173–20176.

97. Dalach, P.; Frost, H.; Snurr, R. Q.; Ellis, D. E. Enhanced hydrogen uptake and the electronic structure of lithium-doped metal-organic frameworks. *J. Phys. Chem. C* **2008**, *112*, 9278–9284.

98. Han, S. S.; Goddard, W. A. Lithium-doped metal-organic frameworks for reversible $H_2$ storage at ambient temperature. *J. Am. Chem. Soc.* **2007**, *129*, 8422–8423.

99. Klontzas, E.; Mavrandonakis, A.; Tylianakis, E.; Froudakis, G. E. Improving hydrogen storage capacity of MOF by functionalization of the organic linker with lithium atoms. *Nano Lett.* **2008**, *8*, 1572–1576.

100. Lochan, R. C.; Head-Gordon, M. Computational studies of molecular hydrogen binding affinities: the role of dispersion forces, electrostatics, and orbital interactions. *Phys. Chem. Chem. Phys.* **2006**, *8*, 1357–1370.

101. Mulfort, K. L.; Hupp, J. T. Chemical reduction of metal-organic framework materials as a method to enhance gas uptake and binding. *J. Am. Chem. Soc.* **2007**, *129*, 9604–9605.

# INDEX



*Metal-Organic Frameworks: Design and Application,* Edited by Leonard R. MacGillivray

Printed and bound by CPI Group (UK) Ltd, Croydon, CR0 4YY
16/09/2026
14963460-0004